Audio and Speech Processing

Audio and Speech Processing

Edited by **Marcus Hintz**

Willford Press

New York

Published by Willford Press,
118-35 Queens Blvd., Suite 400,
Forest Hills, NY 11375, USA
www.willfordpress.com

Audio and Speech Processing
Edited by Marcus Hintz

International Standard Book Number: 978-1-68285-179-1 (Hardback)

Printed in China.

Contents

Preface

Audio and speech processing technologies are rapidly advancing. The researches and developments in this field have had a tremendous impact on audio engineering. This book has been written and edited by internationally renowned experts who can provide the readers with a constructive insight into audio and speech processing. It will provide the best practical understanding to the students. Various aspects of audio and speech processing like speech analysis and coding, room acoustics, speech enhancement, etc. have been discussed in detail. This book provides an in-depth analysis of the theoretical and practical aspects of processing of various audio signals. It will be beneficial to the students of sound engineering and associated disciplines.

The information contained in this book is the result of intensive hard work done by researchers in this field. All due efforts have been made to make this book serve as a complete guiding source for students and researchers. The topics in this book have been comprehensively explained to help readers understand the growing trends in the field.

I would like to thank the entire group of writers who made sincere efforts in this book and my family who supported me in my efforts of working on this book. I take this opportunity to thank all those who have been a guiding force throughout my life.

Editor

Exploiting contextual information for prosodic event detection using auto-context

Junhong Zhao[1,2*], Wei-Qiang Zhang[3], Hua Yuan[3], Michael T Johnson[4], Jia Liu[3] and Shanhong Xia[1]

Abstract

Prosody and prosodic boundaries carry significant information regarding linguistics and paralinguistics and are important aspects of speech. In the field of prosodic event detection, many local acoustic features have been investigated; however, contextual information has not yet been thoroughly exploited. The most difficult aspect of this lies in learning the long-distance contextual dependencies effectively and efficiently. To address this problem, we introduce the use of an algorithm called auto-context. In this algorithm, a classifier is first trained based on a set of local acoustic features, after which the generated probabilities are used along with the local features as contextual information to train new classifiers. By iteratively using updated probabilities as the contextual information, the algorithm can accurately model contextual dependencies and improve classification ability. The advantages of this method include its flexible structure and the ability of capturing contextual relationships. When using the auto-context algorithm based on support vector machine, we can improve the detection accuracy by about 3% and F-score by more than 7% on both two-way and four-way pitch accent detections in combination with the acoustic context. For boundary detection, the accuracy improvement is about 1% and the F-score improvement reaches 12%. The new algorithm outperforms conditional random fields, especially on boundary detection in terms of F-score. It also outperforms an n-gram language model on the task of pitch accent detection.

Keywords: Prosodic event detection; Auto-context; Pitch accent; Boundary; Support vector machines

1 Introduction

Speech is often characterized across two levels of expression: the segmental level encompassing basic phonetic meaning and the prosodic level with additional suprasegmental information. The prosodic level expression plays a crucial role in speech communication, carrying much linguistic and paralinguistic information. Prosody enables listeners to recover word meanings, emphasis, and speaker intent and attitude. In addition, prosody carries information about the speaker's emotional state. Prosody primarily manifests itself as pitch accent, pause, variations in speaking rate, and intonation. These prosodic events are realized by modulating acoustical correlates such as duration, pitch, and intensity at a syllable, word, or whole utterance level.

*Correspondence: junhong.iecas@gmail.com
[1] State Key Laboratory of Transducer Technology, Institute of Electronics, Chinese Academy of Sciences, No.19 North west Rd., Beijing 100190, China
[2] University of Chinese Academy of Sciences, No. 19A Yuquan Road, Beijing 100190, China
Full list of author information is available at the end of the article

In spoken language processing, the detection of prosodic events is a primitive step needed for computers to access the critical high-level information regarding human speech interaction. As such, this task has wide application. It can provide assistance for automatic prosody annotation. Since the manual annotation of prosody for speech synthesis or speech understanding is time-consuming and laborious, the detection of prosodic events can provide substantial time savings. For second language learning, there is potential for computer-assisted language learning (CALL) systems to incorporate prosodic event detection to help detect learner mispronunciations at a prosodic level and provide feedback for improving pronunciation naturalness. Prosodic event detection can also be used as foundation for the downstream spoken language processing tasks such as speech summarization and topic segmentation.

Due to the suprasegmental nature of prosody, contextual information is very important for prosodic event recognition. Here, 'context' refers to the correlation between each prosodic unit and its surroundings. There are several types contextual information, including nearby acoustic appearances ('acoustic context'), nearby prosodic event distributions ('prosodic context'), and nearby lexical and syntactic appearances ('textual context'). In this paper, we focus only on acoustic context and prosodic context, but do not address linguistic effects. Perceptually, pitch accent is perceived when the related acoustic features of a syllable stand out from its surroundings. In addition to the influence of the adjacent syllables, the overall surface realization of accent is also affected by many other broader-scale phenomena such as the presence of phrasal boundary, phrase structure, and topic [1,2].

In this paper, we investigate the utilization of contextual information for pitch accent and boundary detection by using the *auto-context* algorithm, which was first proposed in [3] for high-level computer vision tasks like image segmentation. In this algorithm, the classification probabilities obtained from the preceding iteration are used to provide possible contextual clues, together with acoustic features to improve the next iteration. Each detection object is supported by combinations of contextual probabilities from any contextual range. Our experimental results show that this algorithm enhances detection performance for both pitch accent and boundary detection tasks.

2 Related work

Many approaches have been explored for prosodic event detection. The target unit used to indicate the prosodic events can be a phone, syllable or word, with acoustic features typically modeled by statistical machine learning methods. In [4], Conkie et al. used short-frame speaker-normalized pitch and energy as acoustic representations modeled by prosodic context-dependent HMMs. In [5], Ananthakrishnan and Narayanan modeled the frame-level acoustic features using coupled hidden Markov models (CHMMs) to detect pitch accent and boundary in binary modes.

This work assumed that the modulation of prosody-related acoustic parameters, such as increase of the local energy, extension of the duration, and exaggeration of pitch movements, were all asynchronous and could be modeled by CHMMs with multiple data streams. In [6], bagging and boosting ensemble machine learning methods were adopted based on a decision tree learning algorithm. Both methods improved the overall accuracy of four-category pitch accent classification.

The work in [7] utilized a decision tree to model acoustic features. Using the posterior probabilities provided by the decision tree, a bigram prosodic label sequence model was combined to detect pitch accent and boundary tones at the syllable level. In [8], Ananthakrishnan and Narayanan used a maximum *a posteriori* (MAP) framework with multiple classifiers including GMMs, linear decision discriminants, and neural networks (NNs) to detect pitch accent and boundary, with NNs giving the best performance. When combined with 4-gram delexicalized language model, this method achieved an accuracy of 80.1% on binary pitch accent classification and 89.6% on boundary detection. Jeon and Liu used NNs [9] and support vector machines (SVMs) [10] for acoustic modeling. With NNs, pitch accent and boundary detection accuracy reached 83.5% and 84.8%, respectively. The performance of the SVM classifier was better than that of NNs, achieving 85.7% accuracy for pitch accent detection. Recently, conditional random fields (CRFs) have become popular in prosodic event detection. In [11], acoustic features were modeled by a linear-chain CRF as well as a two-level factorial CRF. Linear-chain CRF has been used extensively in recent work [2,12,13]. These reference approaches to prosodic event detection have been summarized in Table 1.

To investigate the importance of contextual information in prosodic event detection, the work in [16] examined the detection performance of pitch accent at word, syllable, and vowel levels, respectively. When using a constant amount of context, the results showed that detection in the word domain achieved the best performance, showing that acoustic excursions exist beyond syllable range. In [1], the contextual influence of local coarticulatory constraints and broader range phrasal effects were investigated for the detection of prominence. The results showed that the incorporation of local acoustic context can significantly improve the detection performance, with phrasal effects less significant.

With respect to utilizing contextual information in acoustic modeling, there are two representative methods: an n-gram language model [7,8] and CRF [2,11-13] model. An n-gram language model assumes that the current prosodic state is dependent on its finite histories, with dependencies established in the form of conditional probability. CRFs are a class of graphical models that are undirected and conditionally trained. For the commonly used linear chain CRF, the dependencies between prosodic labels are modeled in a pairwise neighborhood structure. CRFs have the advantage of modeling the relationships between sequential labels and have been proven efficient in prosodic prediction [2,11-13]. Figure 1 shows the dependency diagrams of the two models. Unlike these models, auto-context simultaneously integrates the acoustic features together with the context information by learning a series of classifiers. As discussed in [3], auto-context can integrate any mode of neighborhood structure, including long range, to make good use of contextual information. It is up to the learning algorithm to select and fuse the informative context and acoustic features.

Table 1 Summary of different approaches for prosodic event detection

Reference	Corpus	Domain	Model	Class	Pitch accent accuracy (%)	Boundary accuracy (%)
[4]	TTS and BN	Word	Prosodic context-dependent HMM	Binary	82.8	-
[5]	BURSC	Word, syllable	CHMM	Binary	72.0 74.0	77.3 86.0
[14]	BURSC	Syllable	GMM	Binary	77.3	68.2
[7]	BURSC	Syllable	Decision tree and n-gram language model	Binary	84.0	71.0
[8]	BURSC	Syllable	NN and n-gram language model	Binary	80.1	89.6
[9]	BURSC	Syllable	NN	Binary	83.5	84.8
[10]	BURSC	Syllable	SVM	Binary	85.7	-
[6]	BURSC	Syllable	Adaboost CART	Four	84.7	-
[15]	BURSC	Word	MaxEnt	Binary	80.1	82.7
[11]	BURSC	Syllable	CRF	Binary, four	79.5 77.1	92.4

3 Corpus and tasks

The data corpus used in this work is the Boston University Radio Speech Corpus (BURSC) [17], a standard corpus for prosody event detection and prediction studies. This corpus is composed of news stories read by seven FM radio news announcers. Each paragraphed-size utterance typically consists of several sentences and is hand-annotated with the orthographic transcription, phonetic alignments, part-of-speech (POS) tags, and prosodic labels based on ToBI conventions. Utterances from two females (F1, F2) and two males (M1, M2) constitute the training and testing dataset used here. The distribution of these utterances is listed in Table 2.

There are four parallel tiers for ToBI annotation conventions to describe the prosodic events [18]. Among these, the tone tier annotates the presence of pitch accent (* suffix) and phrase boundaries. There are two basic types of accent, high (H) and low (L), which can be further divided into subclasses such as downstepped accent (! prefix). The phrase boundaries include intermediate phrase boundary (- suffix) and intonational phrase boundary (% suffix), which follow the different types of phrase accent and boundary tones. The break tier is used to describe the disjuncture between words. The degree of disjuncture from weak to strong is marked by break indices ranging from 0 to 4. The phrase boundary locations usually score 3 or above, where '3' indicates an intermediate phrase boundary and '4' indicates a full intonational phrase boundary. These two kinds of boundaries are different in their degree of salience. Figure 2 shows a ToBI annotation example for the phrase 'design improvement and schedule'. The top three layers are the orthographic tier, the tone tier, and the break tier, respectively.

In our work, we have implemented syllable partitions and take the syllable as the domain of pitch accent and boundary detection. Using the detailed representation of prosodic types from the ToBI annotation framework can

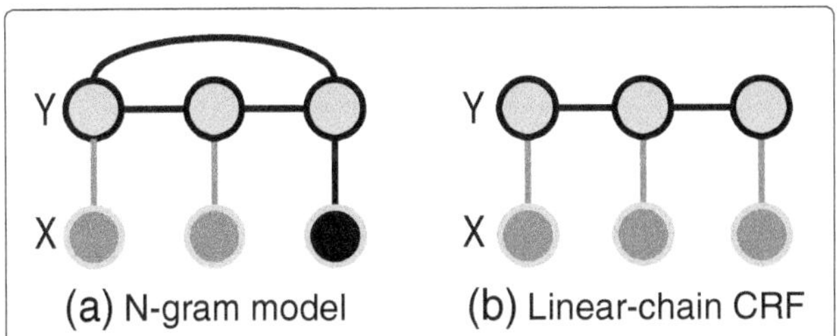

Figure 1 The relational structures adopted by n-gram and CRF models. Here, X represents the features, and Y represents the labels which are being predicted. **(a)** N-gram model. **(b)** Linear-chain CRF.

Table 2 The distribution of the dataset used in our experiment

	F1	F2	M1	M2
Number of utterances	74	166	72	51
Number of sentences	279	1,176	391	209
Number of words	3,993	12,060	5,059	3,608
Number of syllables	6,580	20,836	8,168	5,915
Number of accents	2,253	7,063	2,564	1,933
Number of boundaries	977	3,702	1,092	882
Accent occupancy (%)	34.3	33.9	31.4	32.7
Boundary occupancy (%)	14.9	17.8	13.4	14.9

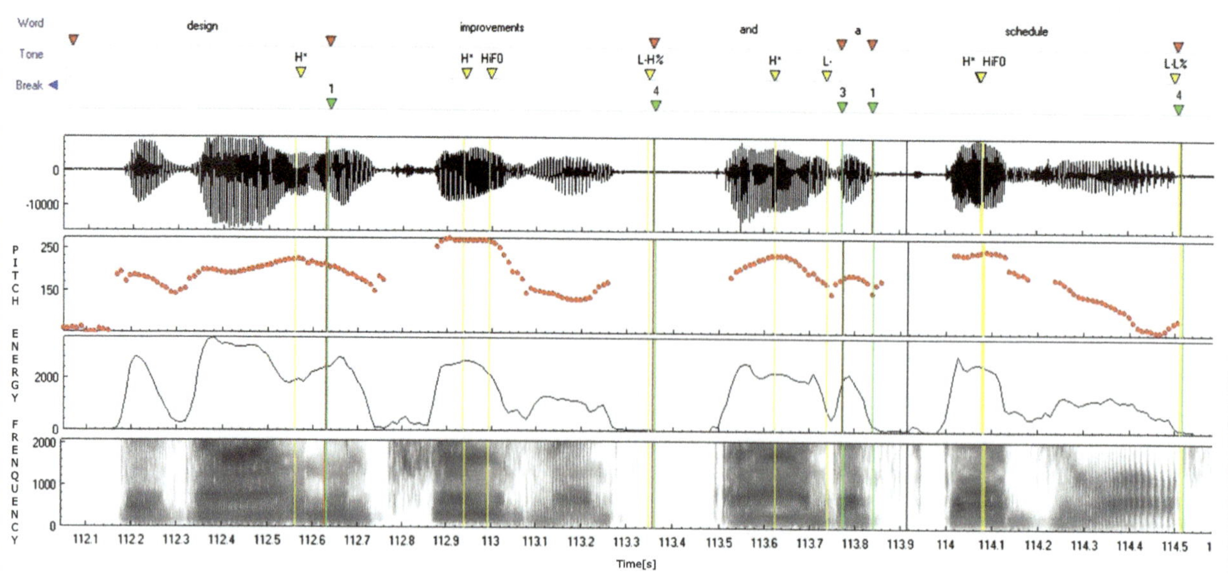

Figure 2 An example of ToBI annotation for the phrase 'design improvement and schedule'. The example includes the acoustic appearances of pitch, energy, and frequency spectrum.

cause serious data sparsity problems since there are only a few examples for some prosodic types. Considering a balance between the amount of the training data and detection fineness, in this work, we implement pitch accent detection tasks in two-way and four-way modes, following previous approaches [8,9,11,13]. For the two-way task, we divide the syllables into accented and unaccented based on the presence of an asterisk mark. As Table 2 shows, in this classification case, the percentage of the accented syllables is about 33%. For the four-way task, we decompose the pitch accent into three types: high, low, and downstepped, in addition to the unaccented type. Four-way classification increases the detection complexity and fineness, but causes some data sparsity. For boundary detection, we set our detection task as a binary (presence/absence) classification problem, which identifies whether the syllable is followed by a boundary or not. Although intonational phrase boundaries can be more reliably detected and have been widely applied in many downstream spoken language processing tasks such as speech summarization, intermediate phrase boundaries are also important elements for phrasal analysis and prosodic annotation. Here, we treat the intonational and intermediate phrase boundaries equally and group the break indices of '3' and '4' together to represent our 'boundary' category, as has been done in some previous works [8,13,19]. As shown in Table 2, the syllables with boundary presence represent about 15% of the dataset. The boundary detection results obtained in this way can be used directly or as preliminary information that can be followed by a further identification of the boundary salience at the presence positions. We summarize the clusters of ToBI labels and their mapping relationships with prosodic categories for detection in Table 3.

4 Prosodic event detection method: auto-context

In this paper, the auto-context algorithm is introduced for prosodic event detection. The basic objective of the auto-context algorithm is to maximize $p(y_i|\boldsymbol{X})$ for all samples, where $\boldsymbol{X} = (x_1, \cdots, x_n)$ is the input feature vector, and y_i is the class label for sample i. The auto-context algorithm provides an iterative way to asymptotically approach this objective as follows: Given a set of training samples with ground truth labels $S = \{(Y_i, \boldsymbol{X}(i)), i = 1, \ldots, m\}$, a classifier is first trained using local features so the probabilities of classes for each sample can be obtained. In the ensuing iterative process, the probabilities of both the current and surrounding samples obtained in the current iteration t are combined together as a probability vector $\boldsymbol{P}^t(i)$. This is then concatenated with the original acoustic features to construct a new feature vector. The new feature vectors of all the samples compose the training set $S_{t+1} = \{Y_i, [\boldsymbol{X}(i), \boldsymbol{P}^t(i)]\}$ for iteration $t+1$. Using this updated training set, the new model is trained.

During the iterative process, the auto-context algorithm selects the informative contexts automatically and fuses them with appearance cues. At first, samples with strong discriminant cues will be correctly classified by the initial model and obtain stable posterior probabilities. These probabilities can then influence their neighbors in subsequent iterations, especially when there are close correlations between them. Convergence has been proven,

Table 3 Mapping between clusters of ToBI labels and prosodic categories for detection

Annotation tier	ToBI labels	Four-way pitch accent	Two-way pitch accent	Boundary
Tone tier	H*, L+H*, *, *?, X*?	High	Accent	-
	L*, L*+!H, L*+H	Low		
	!H*, H+!H*, L+!H*	Downstepped		
	Others	Unaccent		
Boundary tier	3, 4	-		Boundary presence
	Others	-		Boundary absence

with a monotonically decreasing training error, as shown in [3].

We adopt this algorithm to explore contextual information for pitch accent and boundary detection. We choose syllables from one to five syllables away from the central syllable as contextual regions to conduct our analysis. In our tasks, the posterior probabilities $\boldsymbol{P}(i)$ are used as the prosodic context (as opposed to 'acoustic context') since they present the likelihoods of syllables belonging to different prosodic events. Acoustic features can include not only local features but also acoustic context. Correspondingly, $\boldsymbol{X}(i)$ can be decomposed as the local features vector $\boldsymbol{X}_{\text{local}}(i)$ and acoustic context vector $\boldsymbol{X}_{\text{context}}(i)$, as Eq. (1) denoted. Here, we will consider both of them. The local acoustic feature $\boldsymbol{X}_{local}(i)$ used in this work will be described in detail in Section 5.1. The first-order differential values are used as the acoustic context $\boldsymbol{X}_{\text{context}}(i)$. We use the following steps to implement the algorithm:

1. The acoustic features, including local features $\boldsymbol{X}_{\text{local}}(i)$ and acoustic context $\boldsymbol{X}_{\text{context}}(i)$, are used to train the initial acoustic model. After the first round of training and testing, we obtain the class probabilities.

$$\boldsymbol{X}(i) = [\boldsymbol{X}_{\text{local}}(i), \boldsymbol{X}_{\text{context}}(i)] \,, \tag{1}$$

2. The contextual information of each syllable is incorporated by combining the probabilities from its neighbors for the next iteration. The probability vector $\boldsymbol{P}(i)$ for the ith sample is constructed as follows:

$$\boldsymbol{P}(i) = [\boldsymbol{C}_*(i-n), \cdots, \boldsymbol{C}_*(i-1), \boldsymbol{C}_*(i), \boldsymbol{C}_*(i+1), \cdots, \boldsymbol{C}_*(i+n)] \,, \tag{2}$$

   where $\boldsymbol{C}_*(i)$ represents any collection of classification probabilities for the ith syllable, and n is the range parameter controlling the extent of the context. After extension, the new training set for the second stage becomes

$$S(i) = \{y_i, [\boldsymbol{X}(i), \boldsymbol{P}(i)], i = 1, 2, 3, \ldots, m\} \tag{3}$$

3. Using the $S(i)$ training set, we train the next acoustic model and update $\boldsymbol{P}(i)$.
4. Steps 2 and 3 are repeated until convergence to let the algorithm recursively learn the informative pitch accent context automatically.

The flowchart of the algorithm is depicted in Figure 3.

5 Acoustic representation

5.1 Features for pitch accent

Pitch accent typically correlates with a higher level of pitch and energy and an increased duration. We extracted acoustic features based on these acoustic measurements in syllable regions to detect pitch accent. These features can be categorized into two groups: frame-averaged features, including the mean values of pitch, energy and duration; and TILT features provided by parameterizing the pitch contour using the TILT model. In addition, the forward and backward difference values for both frame-averaged features and TILT features are extracted as acoustic contextual information. In the following section, we describe how to extract these features in detail.

5.1.1 Frame-averaged features

Loudness Pitch accent is closely related to human auditory characteristics. For perceptual accuracy, loudness can be used instead of intensity to detect pitch accent [20]. Here, we use a loudness model proposed by Zwicker and Fastl [21] to extract the loudness feature. This starts with using the short-time Fourier transform (STFT) to transfer the signal from the temporal to the frequency domain. The linear-scale frequency is grouped into a critical band rate scale to model the human hearing mechanism. This mapping relationship is given by

$$z(\text{Bark}) = T(f) = 13\tan^{-1}(0.00076f) + 3.5\tan^{-1}\left(\frac{f}{7500}\right)^2 , \tag{4}$$

where f denotes frequency in Hertz, z represents the critical band rate measured in Bark units, and $T(\cdot)$ is the transform function between them. Here, we divide the audible range into 24 critical bands. The intensity of each critical band is obtained by summing up all the frequency points that are distributed within the band range of

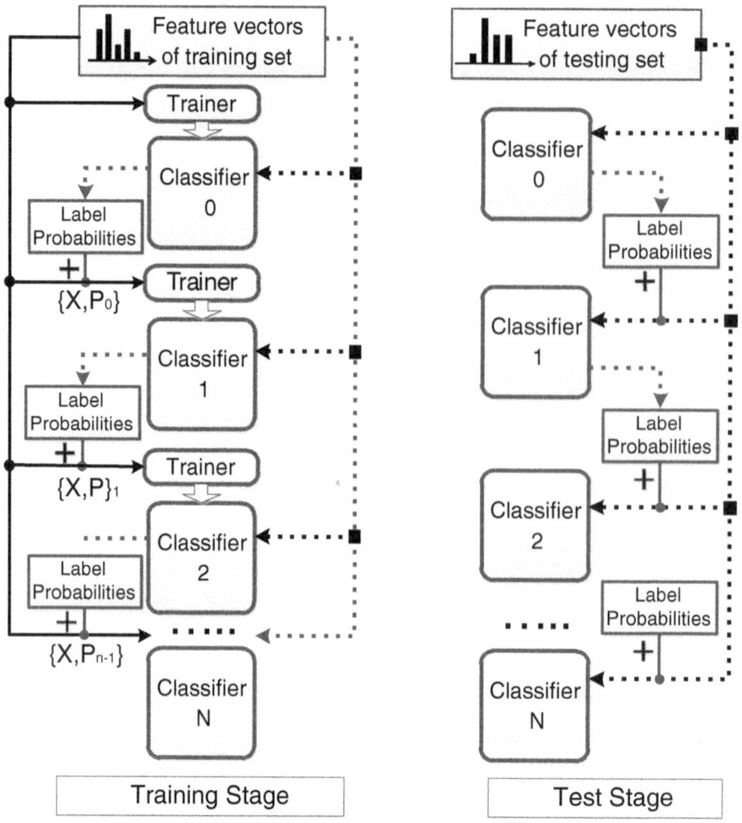

Figure 3 Flowchart of the auto-context algorithm.

$(z - 0.5, z + 0.5)$ and then by calculating the corresponding sound pressure level (SPL) according to

$$I(z) = 10 \log \left(\frac{\int_{L(z)}^{H(z)} I(f) df}{I_0} \right) \text{dB},$$
$$\text{where } H(z) = T^{-1}(z + 0.5),\ L(z) = T^{-1}(z - 0.5). \tag{5}$$

Here, $I_0 = 10^{-12}\ W/m^2$ is the standard threshold of hearing at 1 kHz.

Stevens pointed out that the relationship between intensity and perceptual loudness obeys the power law [22]. Following this law, we calculate the loudness of per Bark based on $I(z)$ by

$$L(z) = 0.08 \left(\frac{I_Q(z)}{I_0} \right)^{0.23} \left[\left(0.5 + 0.5 \frac{I(z)}{I_Q(z)} \right)^{0.23} - 1 \right], \tag{6}$$

where $L(z)$ denotes the specific loudness in the zth Bark, and $I_Q(z)$ is the correponding threshold of intensity in quiet enviroment. The total loudness L of the frame is given by the summation of every critical band loudness $L = \sum_{z=1}^{24} L(z)$.

Semitone Based on a similar perceptual consideration, we transform the pitch values in Hertz to the semitone scale to better match with human perception [23]. Raw pitch values are calculated first using Praat [24], then a log-scale transformation is conducted according to the following equation:

$$S = 69 + 12 \log_2 \left(\frac{f}{440} \right), \tag{7}$$

where f is the fundamental frequency in Hz, and S is in semitones.

Spectral emphasis Previous studies have shown that midfrequency energy is more effective in accent classification than full-range distributed energy [25]. Midfrequency refers to the frequencies between 500 and 2,000 Hz. In this work, we use a finite impulse response (FIR) filter with Kaiser window to extract the energy within this bandwidth as a spectral emphasis feature.

Duration We compute the syllable duration using the boundary information generated by forced alignment. The speaker-independent speech recognizer was trained using the data from BURSC and the corresponding manual transcriptions.

Among these four features, the loudness, spectral emphasis, and semitone values are all extracted at a frame-by-frame level in the first iteration. After this, the loudness and the spectral emphasis are averaged across a syllable scope, while the semitone is averaged across a syllable nucleus to obtain frame-averaged values. In order to reduce the negative impact caused by different speakers and speaking rates, these features are all normalized by the mean value across the whole sentence.

5.1.2 TILT features

We follow the rise/fall/connection (RFC) model proposed in [26] to extract the TILT parameters as the representation of the pitch variation [27]. As an intonation model, the RFC model first categorizes the F0 contour into rise (R), fall (F), and connection (C) cases according to pitch trend and then continuously parameterizes the contour based on this classification. Here, we still use the semitone-scaled F0 contour of each syllable nuclei to extract the TILT features.

Linear interpolation is implemented to smoothen the contour. Then, a labeling procedure is conducted to mark the contours with one of the three kinds of categories. Labels of successive frames are merged together to be a single interval. Within the range of these divided areas, the amplitude-related measurement (tilt_a) is calculated by

$$\text{tilt}_a = \frac{|A_{\text{rise}}| - |A_{\text{fall}}|}{|A_{\text{rise}}| + |A_{\text{fall}}|}, \tag{8}$$

the duration-related measurement (tilt_d) is calculated by

$$\text{tilt}_d = \frac{D_{\text{rise}} - D_{\text{fall}}}{D_{\text{rise}} + D_{\text{fall}}}, \tag{9}$$

and the overall measurement of tilt (tilt_t) is calculated by

$$\text{tilt}_t = \frac{|A_{\text{rise}}| - |A_{\text{fall}}|}{2\left(|A_{\text{rise}}| + |A_{\text{fall}}|\right)} + \frac{D_{\text{rise}} - D_{\text{fall}}}{2\left(D_{\text{rise}} + D_{\text{fall}}\right)}. \tag{10}$$

A_{rise} and A_{fall} are the sum of the rise and fall amplitudes, respectively, and D_{rise} and D_{fall} represent the sum of the rise and fall durations, respectively. We extracted these features using the Edinburgh Speech Tools Library (EST). In addition, we also included the maximum semitone that can be calculated directly by EST as one of the tilt features.

5.2 Features for boundary

As discussed in [28], the presence of a phrase boundary typically correlates with the presence of silence, the reset of pitch and energy, and the lengthening of pre-boundary duration. Each of these play a role in perception of increased disjuncture. We assume that acoustic variations caused by boundary phenomenon exist only within the region of syllables, and extract acoustic features for boundary detection using a syllable unit, as has been done in [7,8,19]. We assume that the region affected by the boundary covers the end of the pre-syllable, the silence interval (possibly absent), and the beginning of the post-syllable, while the candidate result of this region is assigned to the pre-syllable. The acoustic features include the acoustic measurement of the preceding and following syllables and their differential features across syllable boundaries for boundary detection. There are 25 feature dimensions in total, as follows:

1. The duration of the two syllables and their ratio value (3)
2. The duration of the two syllable nuclei and their ratio values (3)
3. The silence duration between the two syllables (1)
4. The means and maxima of pitch of the two syllables, and their differential values (6)
5. The loudness and spectral emphasis mean of the two syllables, and their differential values (6)
6. The amplitude, duration, and overall measurements of the TILT features of the two syllables (6).

Although one can describe a linguistic grammar that only word-final syllables can contain a boundary, we choose to take no linguistic constraints into the implementation of our algorithm and allow all possibilities for all syllables. We just let the algorithm itself to make decisions according to acoustic features and contextual information.

6 Experiments

We conduct a number of prosodic event detection experiments in this section. First, a classifier selection is conducted. The performances of the different classifiers are investigated, and the one with best performance is chosen as the baseline classifier for the auto-context experiments. The auto-context algorithm is then implemented for two-way and four-way pitch accent detections and for boundary detection. In these experiments, the effectiveness of the auto-context approach is verified from different aspects. Finally, comparisons are made between the auto-context, CRF, and n-gram methods.

The BURSC F1, F2, M1, and M2 data described in Section 3 is used for experimental evaluation. We use randomly selected fivefold cross-validation for each experiment. Accuracy and F-score are utilized to measure the performance of the pitch accent detection and boundary detection tasks. In addition, for boundary detection, we investigate the syllable-level detection performance, in which the measurement is presented as a fraction of all syllables, as well as the word-level detection performance, in which the measurement is presented as a fraction of word-final syllables. All test results presented here are obtained by averaging over the five cross-validation test

Table 4 Performance of prosodic event detection by NN and SVM classifiers using all combined features

	Two-way (%)		Four-way (%)		SL boundary (%)		WL boundary (%)	
	Accuracy	F-score	Accuracy	F-score	Accuracy	F-score	Accuracy	F-score
NN	80.5	66.8	75.0	58.9	87.7	43.7	80.3	44.1
SVM	81.2	70.8	76.3	59.9	88.1	45.3	81.2	46.2

Here 'Two-way' and 'Four-way' mean the pitch accent detection tasks in two-way and four-way modes, respectively, and 'SL boundary' and 'WL boundary' mean the boundary detection tasks measured at the syllable-level and word-level, respectively.

folds. Acoustic model parameters for both baseline and proposed methods are optimized using a development dataset. This development dataset is constructed from the F3 and M3 data of the BURSC corpus, including 57 utterances, 282 sentences, and about 8,000 syllables. All experiments are implemented by first optimizing the methods using the development data and then testing with fivefold cross-validation over the primary dataset.

6.1 Classifier selection

As mentioned in Section 4, the auto-context algorithm produces a sequence of classifiers through an iterative process. The performance of the chosen classifier will play a vital role in determining the final performance. The auto-context algorithm is not dependent on any specific classifier and can use either SVM or NN approaches. Since each of these have been used extensively in prosodic event

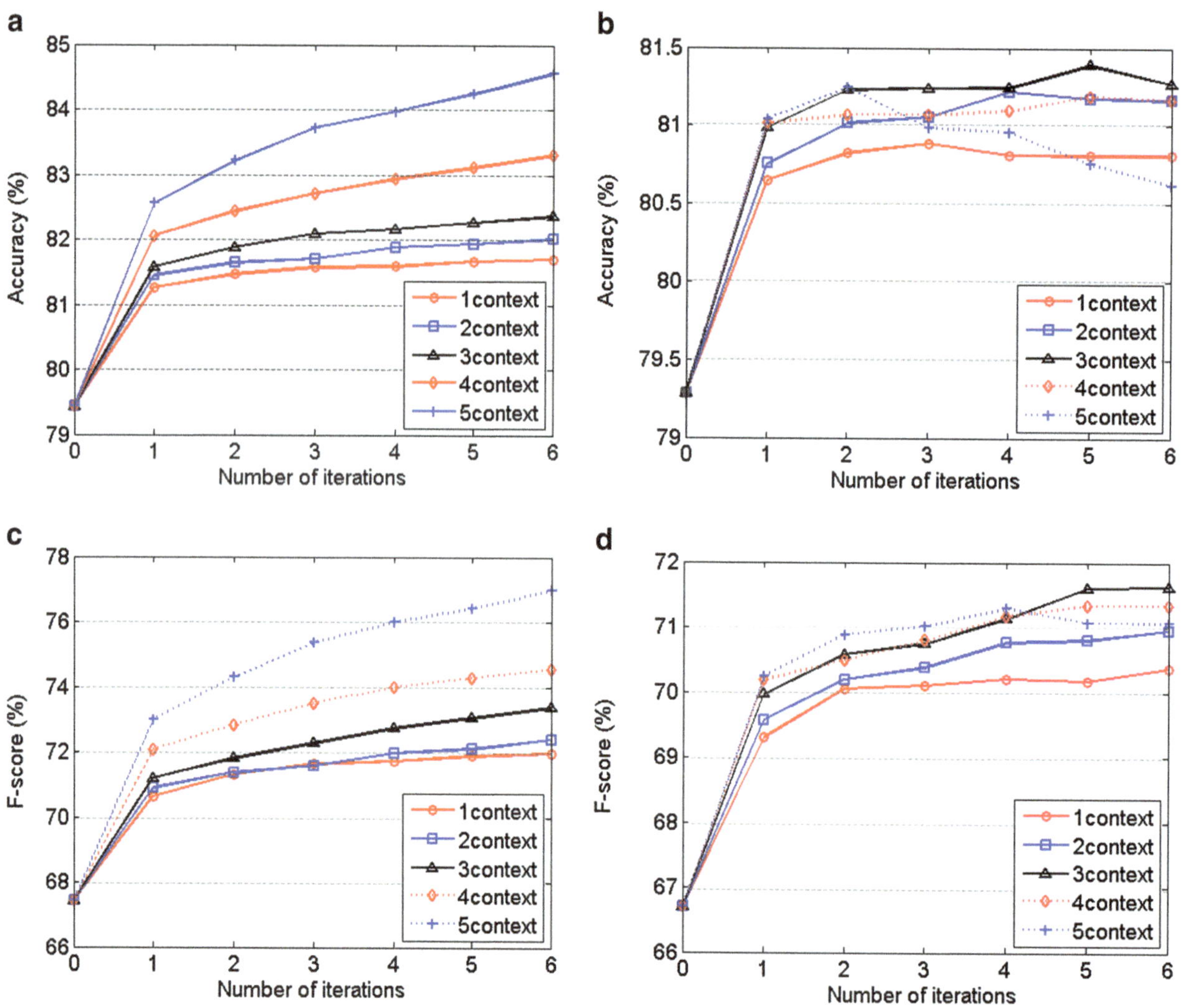

Figure 4 Performance of the auto-context algorithm for two-way pitch accent detection without acoustic features. (a) Accuracy performance on training data. **(b)** Accuracy performance on test data. **(c)** F-score performance on training data. **(d)** F-score performance on test data.

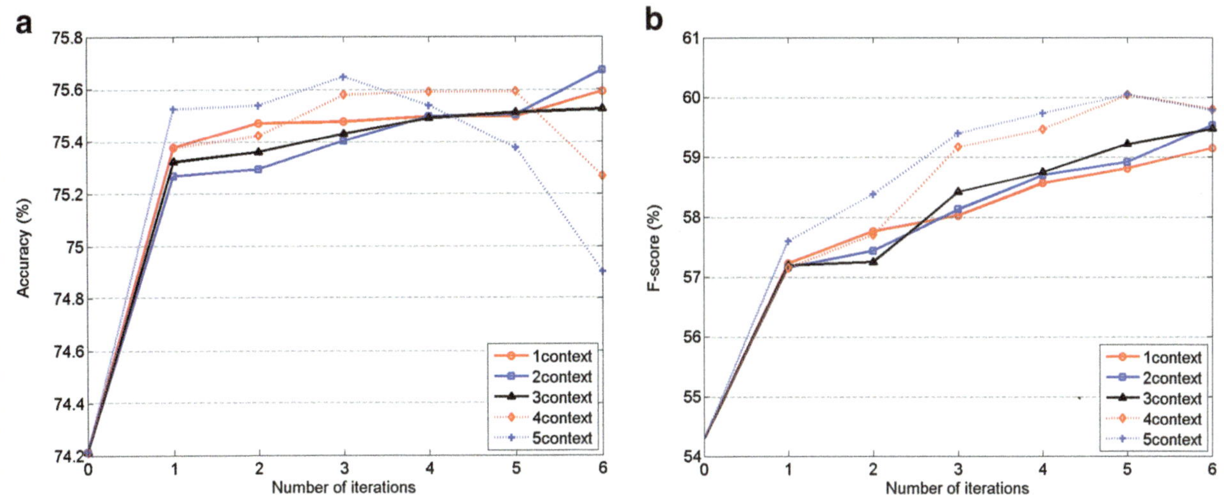

Figure 5 Performance of the auto-context algorithm on test dataset for four-way pitch accent detection without acoustic features. **(a)** Accuracy performance on test data. **(b)** F-score performance on test data.

detection, we will first conduct the investigation for their performance here.

We use LIBSVM with a radial basis function (RBF) kernel to implement SVM classification [29]. In the four-way pitch accent detection, the one-versus-the-rest mode is adopted to decompose multi-class classification into binary. For NN, we use a three-layer network with a fully connected structure. The sigmoid activation function is chosen for network nodes. The classical backpropagation is adopted to minimize the cross entropy error between outputs and targets. Many options related to the two classifiers, such as the number of hidden nodes, the learning rate, the momentum for NN and the cost, the gamma value of RBF function for SVM, are optimized with the development dataset. The results of the prosodic event detection by the two classifiers are given in Table 4.

We can see from Table 4 that on this task, SVM outperforms NN. Another advantage of SVM is that it has fewer parameters to tune and thus is not as sensitive to them as NN. Therefore, considering efficiency and accuracy, we adopt the SVM as our basic classifier in the following experiments. During the iterative process, the same configuration is used for all SVM classifiers. The posterior probabilities that the auto-context algorithm uses are obtained by mapping the distance between the sample and the classifying hyperplane with a sigmoid function.

6.2 Auto-context algorithm for pitch accent detection

We conduct the following experiments to investigate the ability of the auto-context algorithm to model contextual information for the task of pitch accent detection. The first experiment investigates the effect of prosodic context utilized by auto-context for pitch accent detection. Performance is verified only in independent-syllable conditions, where no acoustic context has been used in the baseline. Based on this, a further investigation is then conducted to identify the effect of different contextual locations, e.g., preceding and following contexts. In the final experiment, acoustic differential features are employed to investigate the combination effect of prosodic context and acoustic context.

6.2.1 Auto-context algorithm without acoustic context

To perform this experiment, we select the number of contextual syllables before and after the current syllable to range from $M = 1$ to $M = 5$, where M is the maximum contextual extent. For each selected syllable, the probabilities of all classes were included.

Figure 4a,b shows the average detection accuracy of the auto-context algorithm calculated across the training and test sets of the fivefold cross-validation. For each case, the training set accuracy increases with each iteration, and the wider the contextual range, the better the performance. Correspondingly, the test set accuracy also increases gradually across iterations, as illustrated in Figure 4b. This suggests that the auto-context algorithm is able to iteratively capture informative prosodic context. The larger the

Table 5 Accuracy performance of the auto-context algorithm using contextual information from different locations

Auto-context (%)	None	Preceding	Following	Both
Two-way ($M = 3$)	79.3	80.4	80.1	81.2 ($M = 3$ itr = 3)
Four-way ($M = 2$)	74.2	75.3	75.0	75.7 ($M = 2$ itr = 6)

Here, 'None' means not using auto-context, and 'Preceding', 'Following', and 'Both' refer to the directions of contextual prosodic information.

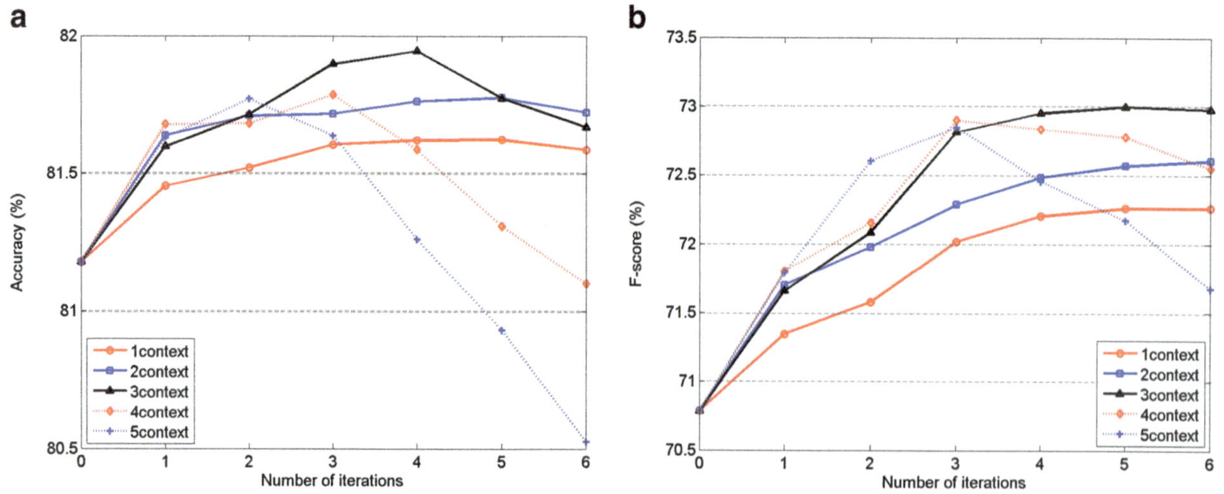

Figure 6 Performance of the auto-context algorithm on test dataset for two-way pitch accent detection with acoustic features. **(a)** Accuracy performance on test data. **(b)** F-score performance on test data.

range is, the greater ability this algorithm has to model these useful relationships.

Additionally, we can see that the greatest performance gain is often obtained in the first iteration. In our experiment, the first iteration produces no less than 1.5% net accuracy improvement (more than 70% of the total improvement) on test data. This is because it is at this stage that the context is first added to a baseline without any prior contextual information. It can also be observed that sometimes there is performance degradation on the test set when the classifiers are iterated too many times, which indicates the over-fitting of the model. To avoid this problem in practice, a development dataset needs to be used to tune and determine an appropriate configuration before each implementation. In our experiment under the fivefold cross-validation scenario, the optimal configuration is determined by the average performance of the development dataset. Final results are obtained under the constraint of the chosen configuration. In order to fully show the change trend, the curves shown in the figures are not preset with any stopping criterion. From Figure 4b, we can also see that $M = 3$ gives the best performance. This is consistent with statistics, showing that most pitch accent events occur once every two or three syllables. The final result is obtained at $M = 3$ and iteration number itr $= 3$, achieving accuracy of 81.2%.

The F-score performance on the training and test datasets are fully shown in Figure 4c,d, respectively. As

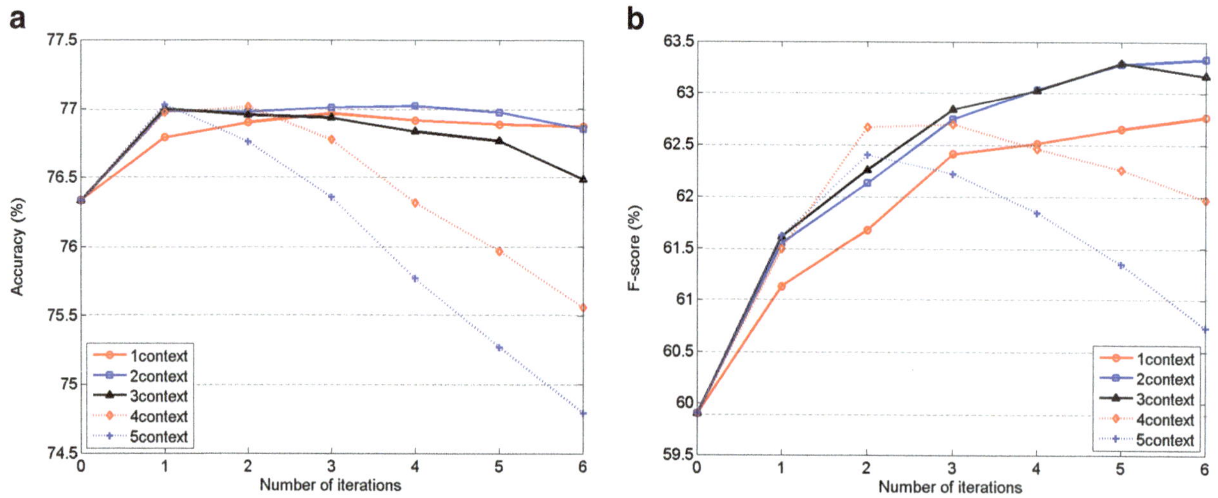

Figure 7 Performance of the auto-context algorithm on test dataset for four-way pitch accent detection with acoustic features. **(a)** Accuracy performance on test data. **(b)** F-score performance on test data.

Table 6 Performance comparison of auto-context, acoustic context, and their combinations

	No context	Acoustic context	Auto-context	Combination
Two-way (%)				
Accuracy	79.3	81.2	81.2 ($M = 3$ itr = 3)	82.0 ($M = 3$, itr = 4)
F-score	66.7	70.8	71.1 ($M = 3$ itr = 4)	73.0 ($M = 3$, itr = 4)
Four-way (%)				
Accuracy	74.2	76.3	75.7 ($M = 2$ itr = 6)	77.0 ($M = 2$, itr = 1)
F-score	54.3	59.9	59.5 ($M = 2$ itr = 6)	63.0 ($M = 24$, itr = 4)

with accuracy, we see a similar increasing trend. The first iteration again gives the most salient improvement, more than 3% net, representing more than 60% of the total improvement. The F-score improvement is more significant than the accuracy improvement. In our experiment, auto-context improves the F-score from 66.7% to 71.1% at $M = 3$ and itr = 4 on the test data.

The test accuracy and F-score performance on the four-way pitch accent detection are fully shown in Figure 5a,b. From this figure, we can see that the auto-context can substantially impact the four-way pitch accent detection result as well. In our experiments, it can improve the accuracy from 74.2% to 75.7% at $M = 2$ and itr = 6, and improve the F-score from 54.3% to 59.5% at $M = 2$ and itr = 6.

6.2.2 Effect of contextual location

Auto-context is sensitive to the choice of context. Section 6.2.1 has discussed the performance variation associated with contextual range influence. In this experiment, we conducted further analysis about how different contextual locations impact the auto-context effectiveness. The contextual location was split into preceding and following based on the position relative to the current syllable. We chose an upper contextual limit of $M = 3$ for the two-way task and $M = 2$ for the four-way task, based on the results of the previous experiments. In each experiment, only the preceding or following probabilities are used. The final results are given in Table 5. We see that both the preceding and following contexts are useful for the tasks of two-way and four-way extent detection. As expected, combining both of them gives the best performance.

6.2.3 Auto-context algorithm with acoustic context

The auto-context algorithm has the potential to explore relationships between prosodic events across a wider extent and combine these with acoustic information in a unified framework. In this experiment, under the same experiment setups to Section 6.2.1, we implement the auto-context algorithm based on not only local acoustic features but also acoustic context to investigate the combination effect between prosodic context and acoustic context. The testing accuracy and F-score results on the two-way task are shown in Figure 6a,b, and the testing performance of the four-way classification is shown in Figure 7a,b.

The final results are listed in Table 6. From these results, we can see that although the prosodic contexts used by auto-context achieves comparable performance to direct acoustic context when used separately, the combination of them yields further improvement. Based on the feature set that has included acoustic context, auto-context can give a further accuracy improvement on both of the

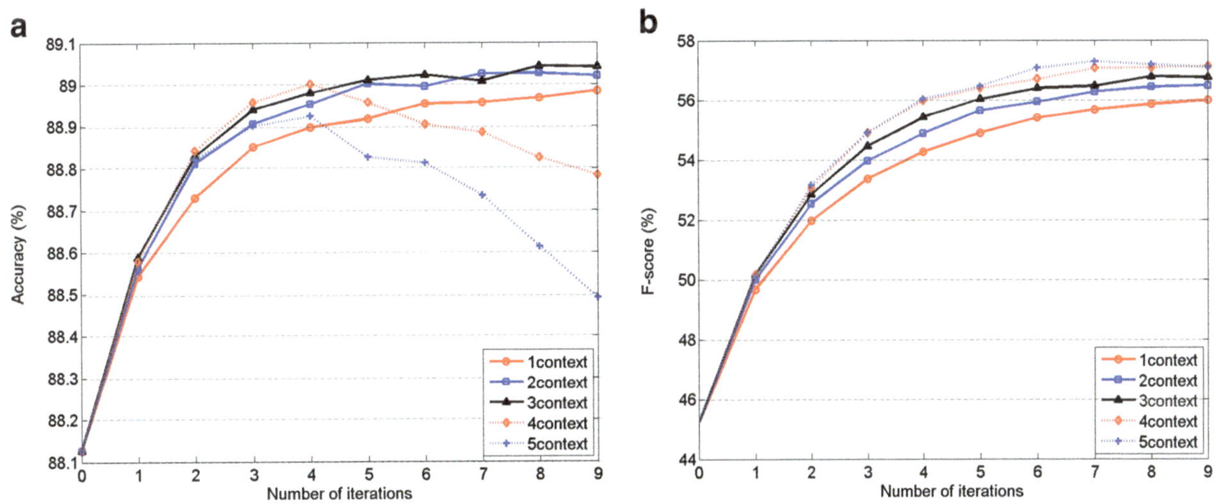

Figure 8 Performance of the auto-context algorithm on test dataset for overall boundary detection. (a) Accuracy performance on test data. **(b)** F-score performance on test data.

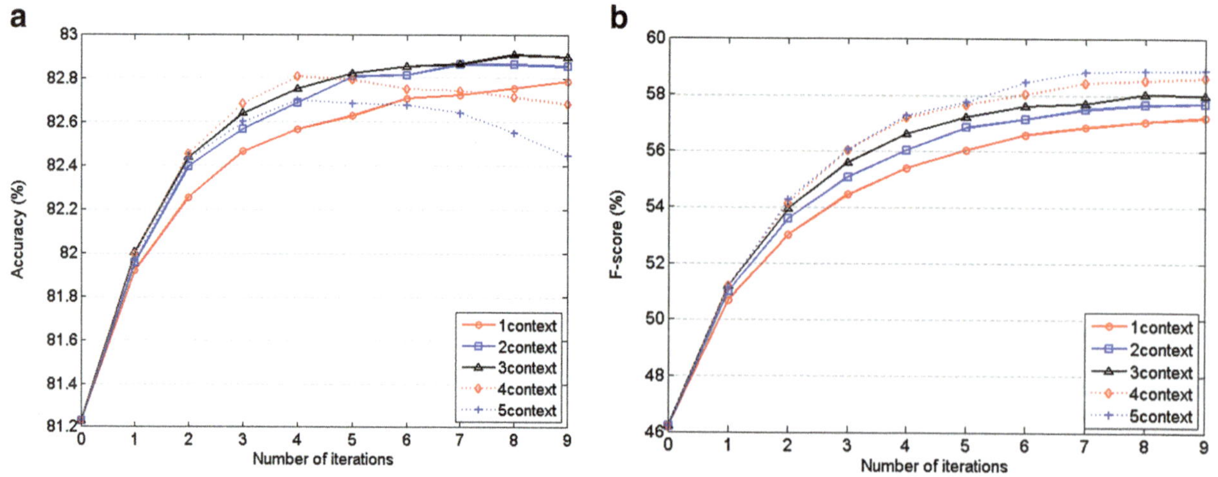

Figure 9 Performance of the auto-context algorithm on test dataset for word-final boundary detection. (a) Accuracy performance on test data. **(b)** F-score performance on test data.

pitch accent detection tasks. It improves the accuracy of two-way detection from 81.2% to 82.0% at $M = 3$ and itr $= 4$, and improves the accuracy of four-way detection from 76.3% to 77.0% at $M = 2$ and itr $= 1$. It also gives a F-score improvement of more than 2% (from 70.8% to 73.0% at $M = 3$ itr $= 4$) for the two-way task and 3% (from 59.9% to 63.0% at $M = 2$ itr $= 4$) for the four-way task. The addition of acoustic context improve the overall performance of auto-context as well, compared to the case of only utilizing prosodic context. For example, without acoustic context information, auto-context can achieve an accuracy of 75.7% on four-way pitch accent detection. When combined with acoustic context, however, the performance improves to 77.0%. These improvements also have been shown in other detection tasks and confirmed by F-score measurements. This demonstrates that prosodic context and acoustic context provide complementary contextual information and that the auto-context algorithm can effectively combine them together.

6.3 Auto-context on boundary detection

This experiment investigates the performance of auto-context algorithm on boundary detection. Using the boundary-related features introduced in Section 5.2, we obtain a syllable-level accuracy of 88.1% and F-score of 45.3% using overall statistics, and a word-level accuracy of 81.2% and F-score of 46.2% using word-final statistics. On this baseline, we implemented the auto-context algorithm across different contextual extents. The syllable-level accuracy and F-score performances are shown in Figure 8a,b, and the corresponding results of the word-level performance are shown in Figure 9a,b.

The final results are listed in Table 7. As we can observe, for boundary detection, the auto-context algorithm yields a nearly 1% improvement using overall statistics and 1.6% improvement using word-final statistics. For the former, it achieves an accuracy of 89.0% at $M = 4$ and itr $= 4$, while for the latter, it achieves accuracy of 82.9% at $M = 3$ and itr $= 7$. Similarly to pitch accent detection, the auto-context algorithm achieved a larger improvement on F-score than the accuracy for the boundary detection task, improving the F-score by about 12% for both statistics. It

Table 7 Performance of boundary detection when using the auto-context algorithm

Auto-context (%)	Accuracy (%)		F-score (%)	
	None	Auto-context	None	Auto-context
SL boundary	88.1	89.0 ($M = 4$, itr $= 4$)	45.3	57.3 ($M = 5$, itr $= 6$)
WL boundary	81.2	82.9 ($M = 3$, itr $= 7$)	46.2	58.9 ($M = 5$, itr $= 7$)

Here, 'None' means not using auto-context.

Table 8 Performance comparisons of n-gram, CRF, and auto-context methods

	N-gram [8]	CRF	Auto-context
Accuracy (%)			
Two-way	80.1	81.1	82.0
Four-way	-	76.9	77.0
SL boundary	89.6	88.2	89.0
WL boundary	84.0	82.2	82.9
F-score (%)			
Two-way	-	72.5	73.0
Four-way	-	62.9	63.0
SL boundary	-	55.1	57.3
WL boundary	-	56.3	58.9

achieves a syllable-level F-score of 57.3% and word-level F-score of 58.9%, respectively.

6.4 Methods comparison

In this section, we compare the performance of our SVM-based auto-context system with two other alternative methods, including an n-gram language model and the CRF approach. For the n-gram approach, we referred to the results of the representative work [8], in which the same two-way pitch accent detection and binary boundary detection are implemented on the BURSC dataset using the syllable-level acoustic features of F0, timing cues, and energy. This work used an NN as the classifier and applied the n-gram language model in the rescoring stage in order to utilize context information. The experiments were conducted without considering acoustic dependencies across syllables, and a five-fold cross-validation is used for the measurement of detection accuracy. By using a 4-gram language model, the accuracy of pitch accent detection was substantially improved from 74.1% to 80.1%. However, there was a slight performance degradation for the boundary detection accuracy from 90.0% to 89.6%. We conducted the CRF-related experiments using the CRF++ toolkit [30] with the same data and acoustic features (include first-order differential features), as were used in the auto-context experiments. The The CRF tool does not support continuous features, so we discretized them with a k-means approach. The linear chain CRF model with bigram mode was used, with the model trained using the limited memory BFGS algorithm. Options like cut-off threshold and number of quantized brackets for acoustic features are also optimally tuned using the development dataset. The final results are listed in Table 8. We can observe that the SVM-based auto-context system achieves better performance than CRF, especially in terms of F-score performance for boundary detection, which surpasses the CRF algorithm by 2%. There is also an advantage in binary pitch accent detection compared with the n-gram language model based on NN, although the performance on binary boundary detection is not improved.

7 Conclusions

In this paper, we introduce a flexible and effective algorithm called auto-context for prosodic event detection. This algorithm uses an iterative approach to incorporate the contextual information to improve prosodic event detection. The probabilities of neighboring syllables are integrated with acoustic features to recursively boost the classification performance of the acoustic models. The experiments on two-way and four-way pitch accent detection and binary boundary detection show that auto-context improves the performance both in terms of accuracy and F-score measurements. The combination of both prosodic and acoustic context together gives the best performance. For two-way pitch accent detection, accuracy is improved from 79.3% to 82.0% and F-score from 66.7% to 73.0%. For four-way pitch accent detection, accuracy is improved from 74.2% to 77.0% and F-score from 54.3% to 63.0%. Similar improvement is also shown for boundary detection. Using the overall statistical method, the detection accuracy is improved from 88.1% to 89.0% and F-score from 45.3% to 57.3%, while using the word-final statistical method, the detection is improved from 46.2% to 58.9%.

Competing interests

The authors declare that they have no competing interests.

Acknowledgements

This work was supported by the National Natural Science Foundation of China under grant nos. 61370034, 61273268, and 61005019.

Author details

[1]State Key Laboratory of Transducer Technology, Institute of Electronics, Chinese Academy of Sciences, No.19 North west Rd., Beijing 100190, China. [2]University of Chinese Academy of Sciences, No. 19A Yuquan Road, Beijing 100190, China. [3]National Laboratory for Information Science and Technology, Department of Electronic Engineering, Tsinghua University, Haidian, Beijing 100084, China. [4]Department of Electrical Engineering, Marquette University, 250 W Wisconsin Ave., Milwaukee, WI 53201, USA.

References

1. GA Levow, Context in multi-lingual tone and pitch accent recognition. Paper presented in the 9th International Speech Communication Association conference (Lisbon, Portugal, 4–8 Sept 2005), pp. 1809–1812
2. ML Gregory, Y Altun, Using conditional random fields to predict pitch accents in conversational speech. Paper presented in the 42nd meeting of the Association for Computational Linguistics (Barcelona, Spain, 21–26 July 2004), pp. 677–683
3. Z Tu, X Bai, Auto-context and its application to high-level vision tasks and 3d brain image segmentation. IEEE Trans. Pattern Anal. Mach. Intell. **32**(10), 1744–1757 (2009)
4. A Conkie, G Riccardi, RC Rose, Prosody recognition from speech utterances using acoustic and linguistic based models of prosodic events. Paper presented in sixth European conference on speech communication and technology (Budapest, Hungary, 5–9 Sept 1999), pp. 523–526
5. S Ananthakrishnan, SS Narayanan, An automatic prosody recognizer using a coupled multi-stream acoustic model and a syntactic-prosodic language model. ICASSP **1**, 269–272 (2005)
6. X Sun, Pitch accent prediction using ensemble machine learning. Paper presented at the 7th international conference on spoken language processing (Denver, Colorado, 16–20 Sept 2002), pp. 953–956
7. CW Wightman, M Ostendorf, Automatic labeling of prosodic patterns. IEEE Trans. Speech Audio Process. **2**, 469–481 (1994)
8. S Ananthakrishnan, S Narayanan, Automatic prosodic event detection using acoustic, lexical and syntactic evidence. IEEE Trans. Audio Speech Lang. Process. **16**, 216–228 (2008)
9. JH Jeon, Y Liu, Automatic prosodic events detection using syllable-based acoustic and syntactic features. Paper presented at the IEEE international conference on acoustics, speech and signal processing (Taipei, 19–24 April 2009), pp. 4565–4568
10. JH Jeon, Y Liu, Syllable-level prominence detection with acoustic evidence. Paper presented at the 11th annual conference of the International Speech Communication Association (Makuhari, Chiba Japan, 26–30 Sept 2010), pp. 1772–1775
11. GA Levow, Automatic prosodic labeling with conditional random fields and rich acoustic features. Paper presented in ACL-IJCNLP (Hyderabad, 7–12 Jan 2008), pp. 217–223

12. R Fernandez, B Ramabhadran, Discriminative training and unsupervised adaptation for labeling prosodic events with limited training data. Paper presented at the 11th annual conference of the International Speech Communication Association (Makuhari, Chiba, Japan, 26–30 Sept 2010), pp. 1429–1432
13. Y Qian, Z Wu, X Ma, F Soong, Automatic prosody prediction and detection with conditional random field models. Paper presented in the 7th international symposium on Chinese spoken language processing (Tainan, Taiwan, 29 Nov–3 Dec 2010), pp. 135–138
14. K Chen, M Hasegawa-Johnson, A Cohen, An automatic prosody labeling system using ANN-based syntactic-prosodic model and GMM-based acoustic prosodic model. ICASSP **1**, 509–512 (2004)
15. VKR Sridhar, S Bangalore, SS Narayanan, Exploiting acoustic and syntactic features for automatic prosody labeling in a maximum entropy framework. IEEE Trans. Audio Speech Lang. Process. **16**, 797–811 (2008)
16. A Rosenberg, J Hirschberg, Detecting pitch accents at the word, syllable and vowel level. Paper presented in the annual conference of the North American Chapter of the Association for Computational Linguistics: human language technologies (Boulder, Colorado, 31 May–5 June 2009), pp. 81–84
17. M Ostendorf, P Price, S Shattuck-Hufnagel, *The Boston University Radio News Corpus* (Linguistic Data Consortium, Philadelphia, 2013)
18. K Silverman, M Beckman, J Pitrelli, M Ostendorf, C Wightman, P Price, J Pierrehumbert, J Hirschberg, ToBI: A standard scheme for labeling prosody. Paper presented at the international conference on spoken language processing (Banff, Alberta, Canada, 13–16 Oct 1992), pp. 867–869
19. JH Jeon, Y Liu, Automatic prosodic event detection using a novel labeling and selection method in co-training. Speech Commun. **54**(3), 445–458 (2012)
20. K Li, S Zhang, M Li, WK Lo, H Meng, Prominence model for prosodic features in automatic lexical stress and pitch accent detection. Paper presented in the 12th annual conference of the International Speech Communication Association (Florence, Italy, 27–31 Aug 2011), pp. 2009–2012
21. E Zwicker, H Fastl, *Psychoacoustics—Facts and Models* (Springer, New York, 1999)
22. SS Stevens, On the psychophysical law. Psychological **64**(3), 153–181 (1957)
23. S Nooteboom, The prosody of speech: melody and rhythm, in *The Handbook of Phonetic Sciences*, ed. by WJ Hardcastle, J Laver (Wiley, New York, 1997), pp. 640–673
24. P Boersma, A Praat, system for doing phonetics by computer. Glot Int. **5**(9–10), 341–345 (2001)
25. AMC Sluijter, VJ van Heuven, Acoustic correlates of linguistic stress and accent in Dutch and American English. ICSLP **2**, 630–633 (1996)
26. P Taylor, The rise/fall/connection model of intonation. Speech Commun. **15**, 169–186 (1994)
27. P Taylor, The tilt intonation model. Paper presented in ICSLP (Sydney, Australia, 30 Nov–4 Dec 1998), pp. 1383–1386
28. A Rosenberg, Automatic detection and classification of prosodic events. PhD thesis (Columbia University, 2009)
29. RE Fan, PH Chen, CJ Lin, Working set selection using second order information for training support vector machines. J. Mach. Learn. Res. **6**, 1889–1918 (2005)
30. CRF++: Yet another CRF toolkit. http://crfpp.sourceforge.net/. Accessed 25 March 2012

2

Production and perception of velar stop (de)voicing in European Portuguese and Italian

Daniel Pape[1*] and Luis M T Jesus[1,2]

Abstract

Speech production and speech perception studies were conducted to compare (de)voicing in the Romance languages European Portuguese (EP) and Italian. For the speech production part, velar stops in two positions and four vowel contexts were recorded. The voicing status for 10 consecutive landmarks during stop closure was computed. Results showed that during the complete stop closure voicing was always maintained for Italian, and that for EP, there was strong devoicing for all vowel contexts and positions. Both *language* and *vowel context* had a significant effect on voicing during stop closure. The duration values and voicing patterns from the production study were then used as input factors to a follow-up perceptual experiment to test the effects of *vowel duration, stop duration* and *voicing maintenance* on voicing perception by EP and Italian listeners. Perceptual stimuli (VCV) were generated using biomechanical modelling so that they would include physically realistic transitions between phonemes. The consonants were velar stops, with no burst or noise included in the signal. A strong language dependency of the three factors on listeners' voicing distinction was found, with high sensitivity for all three cues for EP listeners and low sensitivity for Italian listeners. For EP stimuli with high voicing maintenance during stop closure, this cue was very strong and overrode the other two acoustic cues. However, for stimuli with low voicing maintenance (i.e. highly devoiced stimuli), the acoustic cues vowel duration and stop duration take over. Even in the absence of both voicing maintenance during stop closure and a *burst*, the acoustic cues vowel duration and stop duration guaranteed stable voicing distinction in EP. Italian listeners were insensitive to all three acoustic cues examined in this study, with stable voiced responses throughout all of the varying fully crossed factors. None of the examined acoustic cues appeared to be used by Italian listeners to obtain a robust voicing distinction, thus pointing to the use of other acoustic cues or combination of other cues to guarantee stable voicing distinction in this language.

Keywords: Voicing maintenance; Devoicing; Cue weighting; European Portuguese; Italian; Speech production; Speech perception

Introduction

Human perception does not rely only on a single perceptual cue, but also rather on a combination of different cues to guarantee a stable and robust perceptual outcome. One phonological contrast can be triggered by multiple acoustic cues or by a combination of these cues.

Several acoustic cues (separately or combined) form the perceptual construct of a voicing distinction between voiced and voiceless stops. For the perception of stop voicing in general, voice onset time (VOT) is seen as one of the major acoustic cues in a variety of languages, the other major cues being stop duration and adjacent vowel duration [1-5]. For stops in medial position, the classical definition of a voiced stop is *voicing during closure* [6]: intervocalic stops usually have voicing during closure in a number of languages, which influences perception [6,7]; however, other acoustic cues like *stop duration* [7] and preceding *vowel duration* [8] strongly contribute to voicing distinction. Adding to this complexity, Li et al. [9] also showed that a number of other acoustic cues can be extracted from natural speech which may also play a role as perceptual cues. Regarding

* Correspondence: danielpape@ua.pt
[1]Institute of Electronics and Informatics Engineering of Aveiro (IEETA), University of Aveiro, 3810-193 Aveiro, Portugal
Full list of author information is available at the end of the article

the interplay of several acoustic cues, Francis et al. [7] studied English simultaneous variation and cue weighting in the stop voicing distinction. They showed that if the stop closure duration in the word *rabid* is increased to more than 70 ms, then listeners hear the word *rapid*, but *only* when there is no voicing, i.e. in the absence of voicing maintenance during stop closure. Bailey and Summerfield [10] also showed in English that in fricative stop syllable initial clusters (e.g. *star*, *spar* or *scar*), all of the acoustic differences measured in natural production are able to provide correct stop perception results. No single cue was necessary, and many different cues in combination were sufficient for a correct perception. Redundant cues often enter into trading relationships that increase the magnitude of certain cues while decreasing the magnitude of others, i.e. the listener decision is the result of a certain cue trading of the perceptual system. The complexity increases when taking into account language differences; Oglesbee [11] showed that, in a multidimensional stop categorisation task, a comparison of listeners from different languages results in different preferences for the important perceptual cues, thus pointing to a language dependency for both selection and weighting of available stop contrast cues.

Even for various languages of the same family like Italian, Spanish and European Portuguese (EP), the role of acoustic cues and cue trading is rather difficult to predict. For example, Italian shows a stable *voicing maintenance* (the presence of voicing throughout the complete stop closure) for all word-medial phonologically voiced stops, as shown by Shih et al. [12] in a cross-linguistic database comparison. EP, in contrast, has completely different voicing maintenance for their phonologically voiced stops. Recent findings [13] showed that EP phonologically voiced stops often have no discernible burst, and more importantly, most of the EP phonologically voiced stops in speech production studies [13,14] are produced as devoiced. Thus, it would be reasonable to assume that the absence of voicing maintenance during stop closure in combination with the missing burst would lead EP listeners to identify the phonologically voiced consonants as voiceless. However, it is also possible that other cues take over to guarantee a robust perceptual outcome. Speech production results for EP point to the hypothesis that stop distinction would not be all that different from other languages, since both stop durations [5,13-16] and preceding vowel durations [14] are significantly different for the contrast of phonologically voiced and voiceless stops.

One aim of this study was to disentangle the described controversies of EP by examining the presence of several acoustic cues in the production of EP stops and compare the actual use of these acoustic cues in the perception of EP voicing distinction and to shed light on the interplay of the different cues for intervocalic medial velar stops, all in the absence of a facilitating burst. The second aim of this study was to contrast EP production and perception data from (matched) Italian production and perception data to find similarities and/or differences in the voicing behaviour of these two Romance languages. In other words, we were interested in comparing production and perception data from these two languages to find out how the perceptual system evaluates a voiced/voiceless distinction, which of the available acoustic cues are chosen and how a possible interplay of several cues is ordered.

This paper presents novel results on how particular languages select and weight available acoustic cues to achieve a robust perceptual outcome. This question is particularly relevant in the light of the results of Oglesbee [11], who showed that a comparison of listeners from different languages leads to different preferences for important perceptual cues, thus pointing to a language dependency for the selection and weighting of several stop contrast cues.

Method

Production study

Acoustic (MK 224 microphone and MV 181A preamplifier, Cirrus Research, Hunmanby, UK) and electroglottography (EGG) (EG2-PCX, Glottal Enterprises, Syracuse, NY, USA) signals of EP and Italian $C_1V_1C_2V_2$ clusters in the carrier sentence (EP, *Diga* $C_1V_1C_2V_2$ *outra vez*/'Say $C_1V_1C_2V_2$ again'; Italian, *Dite* $C_1V_1C_2V_2$ *ogni dì*/'Say $C_1V_1C_2V_2$ once more') were recorded (PMD 671 Solid State Recorder, Marantz, Eindhoven, The Netherlands; 16 Bits, 48 kHz) in a soundproof environment. As can be seen from the carrier sentences, both preceding and following vowels of the target $C_1V_1C_2V_2$ items were very similar for the two languages, thus excluding influences from differing vowel or consonant context. Both the velar stops /k g/ and the four vowel contexts /i e o a/ were pairwise identical for the target item (e.g. 'kaka' or 'gigi'), sentence stress was laid on the $C_1V_1C_2V_2$ pseudoword, and lexical stress was set to its first syllable. Thus, we obtained two different consonant positions (intervocalic *initial* and *medial* position). For both languages, each item was repeated nine times in randomised order by six native speakers with university education. The EP speakers were from Central Portugal (mean age 25 years, recorded at the Speech, Language and Hearing Laboratory (SLHlab), University of Aveiro, Portugal), and the Italian native speakers were from the Veneto region in Northern Italy (mean age 40 years, recorded at the Instituto di Scienze e Tecnologie della Cognizione (ISTC), Padova, Italy). Speech rate was held constant during all recordings; correct realisations of all items were supervised by a trained phonetician (both languages). One trained phonetician manually labelled the following landmarks:

- Onset and offset of the neutral lead-in vowel (preceding the target word)
- Onset of first stop (V_1) burst, if present
- Onset and offset of the first target vowel (V_1)
- Onset of the second stop (C_2) burst, if present
- Onset and offset of the second target vowel (V_2)

Similar to [12], we computed the time-dependent voicing status for all stops, sampled at 10 equidistant points throughout the complete stop duration. The first point was set to the beginning of the stop closure (i.e. the preceding vowel offset), whereas the 10th point was set to the following vowel onset. To determine the voicing status for each point a PRAAT 5.2 [17] automatic voicing detection algorithm (*AC pitch extraction* algorithm with the settings *voiceless decision* = 0.55 and *silence threshold* = 0.1) was used and then manually checked against the synchronised EGG signals.

Perception study

Thirty two native EP participants from Central Portugal and 10 native Italian participants from the Veneto region listened to stimuli in a soundproof room (EP, SLHlab Aveiro; Italian, ISTC Padova). They did not receive course credit or financial compensation for their participation. None of the listeners reported speech or hearing problems. Open headphones with a linear frequency response (Sennheiser HD 600, Wedemark Wennebostel, Germany) connected to the internal headphone output of a notebook computer (no other processes running and all networking interfaces disabled) were used. Listeners' responses were collected by means of mouse clicks placed on two screen buttons. The loudness of the stimuli presentation was held constant across all listeners to a comfortable level. The sampling frequency of the presented stimuli was 48,000 Hz at 16 bits.

The speech material generated for the perceptual experiments (extensively described in [18]) consisted of biomechanically modelled VCV stimuli [19] acoustically synthesised frame by frame with a parametric model of the vocal tract [20] driven by a three-mass vocal fold model [21]. The reason for using biomechanical modelling in contrast to, for example, a Klatt synthesiser, lies in the ability of the biomechanical models to generate physically realistic trajectories between consecutive phonemes. In other words, articulatory trajectories are not linearly interpolated, as is normally the case with other synthesis approaches. Research on trajectories has shown that the characteristics of curved paths are explained by anatomical factors and muscle mechanics, for arm movements [22-24] as well as for speech movements [19,25,26]. Biomechanical modelling, in contrast to other synthesis approaches, has the advantage that all obtained tongue movements, trajectories and phoneme targets are comparable to natural speech. This allows the manipulation of glottal source parameters while maintaining articulatory realism. Thus, the use of biomechanical modelling is the best compromise to guarantee highly realistic perceptual stimuli, without the risk of missing hidden perceptual cues (which cannot be controlled for) when using manipulated natural speech. In other words, the use of the biomechanical model has the main advantage that all obtained tongue movements, trajectories and phoneme targets are comparable to natural speech (see for example the modelling of articulatory loops), with the additional possibility to manipulate relevant temporal and glottal source parameters while at the same time maintaining articulatory realism. Thus, the use of biomechanical modelling is the best compromise to guarantee highly realistic perceptual stimuli and to independently control parameters such as duration, transition and targets. Figure 1 shows the comparison of synthesised waveforms and spectrograms of the generated /aga/ stimulus (top) with the /aga/ item as produced by an EP speaker (bottom).

Three different factors known to influence the perception of stop voicing were examined: stop duration, contextual vowel duration and voicing maintenance during stop closure. Each factor was laid out in a continuum with several levels and was combined with all levels of the other factors (i.e. fully crossed and non-adaptive design). The extremes of the continua (perception experiments) were determined according to the values of the speech production study described before:

1. Stop duration*:* mean durations (rounded to the closest decimal) of the voiceless and phonologically voiced velar stops /k g/ in the vowel contexts /a o/ were taken as the limits of the stop duration continuum, i.e. 100 ms (mean of the voiced stop) and 150 ms (mean of the voiceless stop). One intermediate value (125 ms) was introduced.
2. Vowel duration*:* mean durations (rounded to the closest decimal) of the preceding vowels /a o/ of the voiceless/voiced velar stops /k g/ were taken as the limits of the vowel duration continuum, i.e. 70 ms (mean of the preceding vowel of the voiceless velar stop) and 130 ms (mean of the preceding vowel of the voiced velar stop). One intermediate value (100 ms) was introduced.
3. Voicing maintenance*:* the voicing maintenance continuum was defined by the two endpoints fully voiced and fully devoiced/voiceless. For the intermediate values, five conditions were defined (12.5%, 25%, 37.5%, 50% and 75%) at which the stop

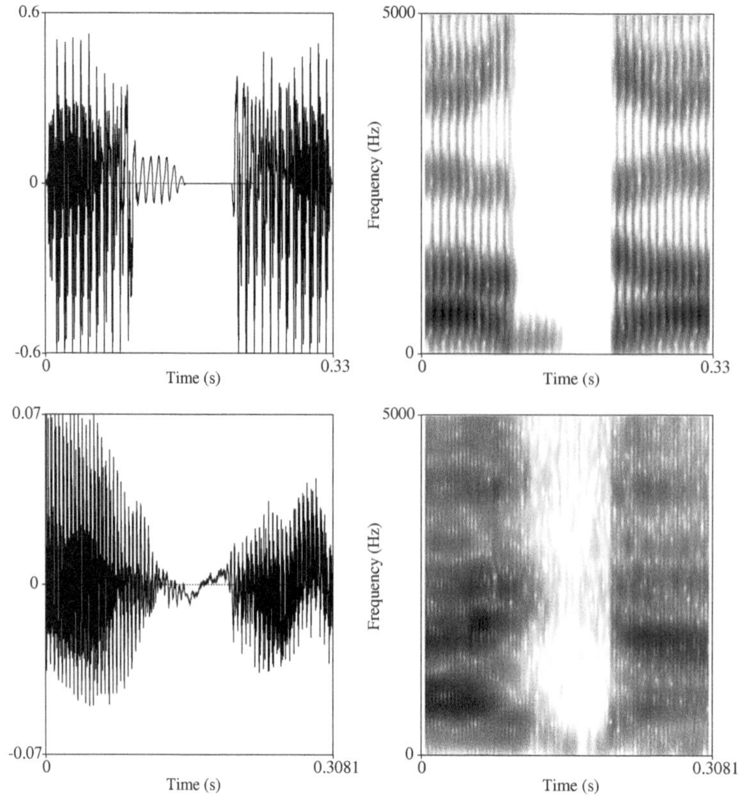

Figure 1 Waveforms and spectrograms of synthesised (top) and naturally produced (bottom) /aga/ EP tokens. The vowel length of the synthetic stimulus is 100 ms, and the stop duration is 125 ms; voicing maintenance is set to 50%.

voicing ceases and thus the stop devoicing begins (and remains until its offset). The unequal step sizes result from the hypothesis that the perceptual differences would be smaller towards higher voicing percentages of the stimuli, so smaller step sizes for higher voicing percentages were excluded to obtain a reduced total number of stimuli. The fully devoiced condition denotes different underlying control mechanisms than the voiceless condition [27,28], although the result, i.e. the voicing maintenance, is identical in both conditions.

A three-factor design with 3 × 3 × 7 levels of the corresponding continuum was thus used in the perception experiment (see Table 1). In this experiment, all possible combinations of contextual vowel duration (70, 100 and 130 ms) with all combinations of stop duration (100, 125 and 150 ms) and all voicing maintenance steps (0%, 12.5%, 25%, 37.5%, 50%, 75% and 100%) were tested. The experiment was performed in two different vowel conditions (/a/ and /o/). Five repetitions of the complete stimuli set for each of the listeners were played in randomised order. In summary, a total of 630 stimuli (three vowel durations × three *closure durations* × seven voicing maintenance conditions × two *vowel* identities × five *repetitions*) were generated. The average time to perform the task was 20 min. There was a practice session of 25 stimuli prior to the beginning of the main experiment. Listeners were informed that they would hear synthetic VCV items, and their task was to identify whether the consonant was /g/ or /k/ (forced choice). Speed of response was emphasised, asking listeners to respond as quickly and accurately as possible. Stimulus repetition was not possible. *Alvin* v 1.27 [29] open source software for stimulus and visual presentation was used. The computer screen for the *identification task* showed two buttons (labelled *g* and *k*) at identical distances around a *next* button at the screen centre. After selecting their response, listeners had to click on the next button to

Table 1 Three fully crossed factors for the perceptual study: vowel duration, stop duration and percentage of stop voicing maintenance

Factor	Continuum levels
Stop duration (ms)	100, 125, 150
Vowel duration (ms)	70, 100, 130
Voicing maintenance (%)	0, 12.5, 25, 37.5, 50, 75, 100

The 0% and 100% voicing maintenance correspond to a fully devoiced and a fully voiced stimulus, respectively.

proceed, thus placing the cursor at the exact centre of the screen before the next stimulus presentation (guaranteeing identical distances for the two answer possibilities). All button options and accompanying text were written in Portuguese for EP listeners and in Italian for Italian listeners in order not to confuse listeners' internal language presentation. The placement of all buttons was rotated 180° for one half of the participants, thus counterbalancing biases of horizontal movement and listener preference.

Statistical analysis

Production study

For the statistic validation of the (production) voicing patterns, we chose to examine the three central landmarks (i.e. point 5, point 6 and point 7) as the dependent variables for the statistical analyses. The obstruent onset (point 1) and offset (point 10) are voiced by definition (i.e. vowel formants offset and onset), but the more central points can be regarded as a valid representative to examine significant voicing differences during stop closure. To statistically analyse the devoicing behaviour at these landmarks, a series of logit models (function lmer [30]) with mixed effects were run in the R environment [31]. The logit models are based on binomial distributions (z-scores, generalized linear mixed model, GLMM). This allows modelling based on binary decisions [32,33], since a binary voicing decision (either *voiced* or *voiceless/devoiced*) is obtained for each of the 10 consecutive landmarks (point 1 to point 10). The devoicing occurrences of the three central points of the phonologically voiced velar stop /g/ (dependent variables) were analysed with a $p < 0.05$ significance threshold for effects of the factors *language* (EP, Italian), *consonant position* (initial, medial) and *vowel context* /i e o a/, and their interactions. All numerical fixed factors were centred (z-transformation). *Speaker* was chosen as random factor.

Perception study

A series of logit models with mixed effects was run to statistically analyse the listeners' response patterns. Again, logit models based on binomial distributions (z-scores, Generalized Linear Mixed Model) were used to model binary decisions [33] by our listener responses (i.e., the listeners' /g/ or /k/ response). The dependent variable was the listener response; fixed factors were language (EP, Italian), stop duration (100, 125 and 150 ms), contextual vowel duration (100, 125 and 150 ms) and voicing maintenance percentage during stop closure (0%, 12.5%, 25%, 37.5%, 50%, 75%, 100%) and their interactions. All numerical fixed factors were centred (z-transformation). *Listener* was chosen as random factor.

Results

Production study

Table 2 presents the mean durations for the medial position of the production database (initial position is not given due to the occurring word boundary). It can be seen that the preceding vowels are longer for Italian compared to EP, but velar stop durations are almost identical for the two languages. Furthermore, for both languages, the classical duration pattern can be observed: vowels are longer when preceding voiced stops in contrast to voiceless stops, and stop durations are longer for voiceless stops as compared to voiced stops.

The GLMM logit models computed on the devoicing behaviour of the phonologically voiced velar stop showed that the effect of language is significant for the three central landmarks of the stop closure ($z = 7.9$, $z = 7.3$, $z = 6.4$; all, $p < 0.001$) with higher voicing for Italian than EP. The effect of vowel context was also significant ($z = 2.3$, $z = 4.5$, $z = 3.6$, $p < 0.01$), but only when comparing the open vowel /a/ to the close/close-mid vowels /i e/, with lower voicing for the open vowel. There was no significant effect between initial and medial consonant position. No interaction between the significant factors was found. Figure 2 presents the complete voicing profiles over all EP and Italian speakers for both voiceless and voiced velar stop in medial position.

When comparing the two languages, it can be seen that, throughout the complete stop closure, both the EP and Italian voiceless consonants /k/ show nearly identical extinction of voicing. In contrast, the phonologically voiced velar stop /g/ shows for EP a substantial devoicing throughout the complete stop closure, whereas the Italian /g/ maintains voicing, i.e. no devoicing is present for Italian. The observations throughout the complete stop closure confirm the statistical results performed on the central parts of the closure. Figure 3 shows the interactions between the factors *language* and *contextual vowel*: While for Italian no effect of contextual vowel height can be observed (i.e. no devoicing), EP shows higher devoicing for /a/ as compared to /i e/. Further,

Table 2 Durations in medial stop position (all values in ms)

	EP	Italian
/g/		
Preceding vowel duration	133 (26)	192 (46)
Velar stop duration	97 (19)	103 (15)
/k/		
Preceding vowel duration	72 (22)	151 (36)
Velar stop duration	153 (22)	146 (28)

Mean durations (standard deviation in brackets) over all EP and Italian speakers (pooled over the three vowel contexts) in medial stop position (all values in ms).

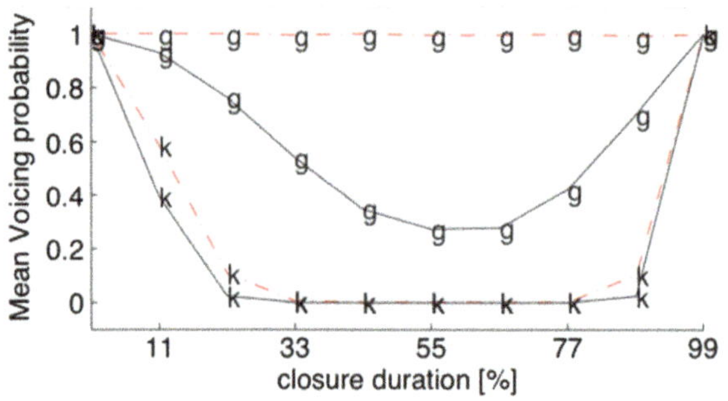

Figure 2 Mean voicing probability of the production study for EP (solid curves) and Italian (dashed curves). Over the 10 discrete stop closure landmarks (*x*-axis), shown for both voiceless velar stop /k/ and voiced velar stop /g/ in medial position.

the observed pattern of overall higher voicing for Italian compared to EP (see Figure 2) is maintained in all four vowel contexts /i e o a/.

Perception study

Figure 4 shows the results of the perception experiment for the languages EP (top panels) and Italian (bottom panels). In Figure 4, the effects of the three factors (vowel duration in different panels, stop duration as different lines in each panel, voicing maintenance on the

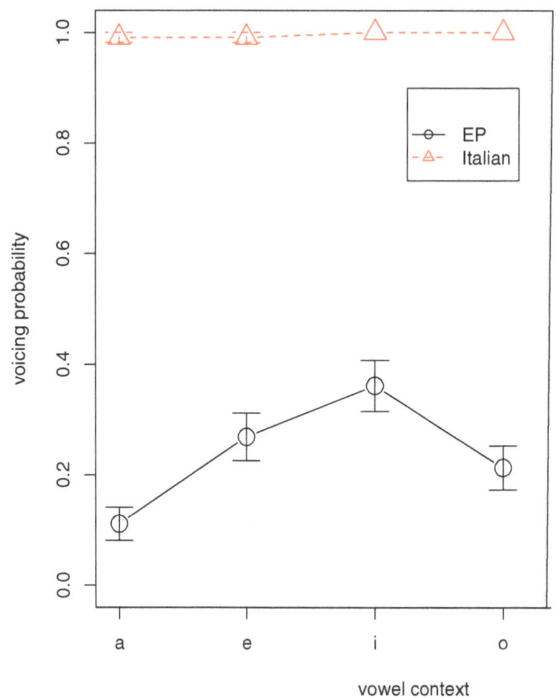

Figure 3 Interaction of language and vowel context for the voicing probability. At the acoustic midpoint of the velar stop closure (medial position) for the production study. Shown are the means ±1 standard error.

x-axis) on the mean percentage of listeners' voiced /g/ responses over all listeners are shown.

For EP, two effects can be observed for the factor voicing maintenance. First, there is a clear bias for listeners to prefer /g/ responses, with a lack of complete /k/ responses (i.e. a 0% to 20% *voiced* response probability), but a lack of 100% /g/ responses can also be observed. Second, increasing voicing maintenance leads to an increase of voiced responses for all vowel duration and stop duration conditions. Furthermore, increasing stop duration leads to a decrease in the probability of voiced responses but mainly for low voicing maintenance values (i.e. for highly devoiced or voiceless items). Increasing vowel duration (panels from left to right) results in an increase in the /g/ response probability, except for high voicing maintenance values. Increasing the voicing maintenance thus increases the /g/ response probability for all vowel durations and stop durations, but with different magnitudes. Stop duration and vowel duration have a strong influence on voicing decisions, but this effect is limited to low voicing maintenance values. For higher voicing maintenance percentages, a ceiling effect for all stop durations and vowel durations is observed. It seems that the voicing maintenance cue is very strong here and overrides the other two acoustic cues.

For Italian, there seems to be a complete lack of effect for the two factors vowel duration and stop duration on the listeners' voicing decisions. Instead, listeners show more or less a stable voicing probability of 0.7 (voiced /g/ responses), independent of the presented vowel duration and stop duration. The factor voicing maintenance has a small influence, much less than in EP and only clearly present for the second panel (vowel duration 100 ms). It is unlikely that the reason for the absence of the voicing maintenance effect is the longer preceding vowel duration in Italian (see Table 2), since in this case, the effect should be present and stronger in the third panel (vowel duration 130 ms).

In summary, Italian listeners are not influenced by the three examined factors and show a robust voicing decision towards a voiced /g/ response across all three varying factors. In contrast, EP listeners are strongly influenced by both vowel duration and voicing maintenance but less influenced by varying stop duration.

With the aim to isolate the effect of voicing maintenance from the other two acoustic cues, in the following, we define - for both languages separately - an *ambiguous prototype*, described by intermediate durational values between voiced and voiceless velar stops. Based on the values in Table 2, for the language EP, this prototype would be characterised by contextual vowel duration of 100 ms and stop duration of 125 ms. For Italian, the stop duration would be identical (125 ms), but the preceding vowel duration increases to the maximum

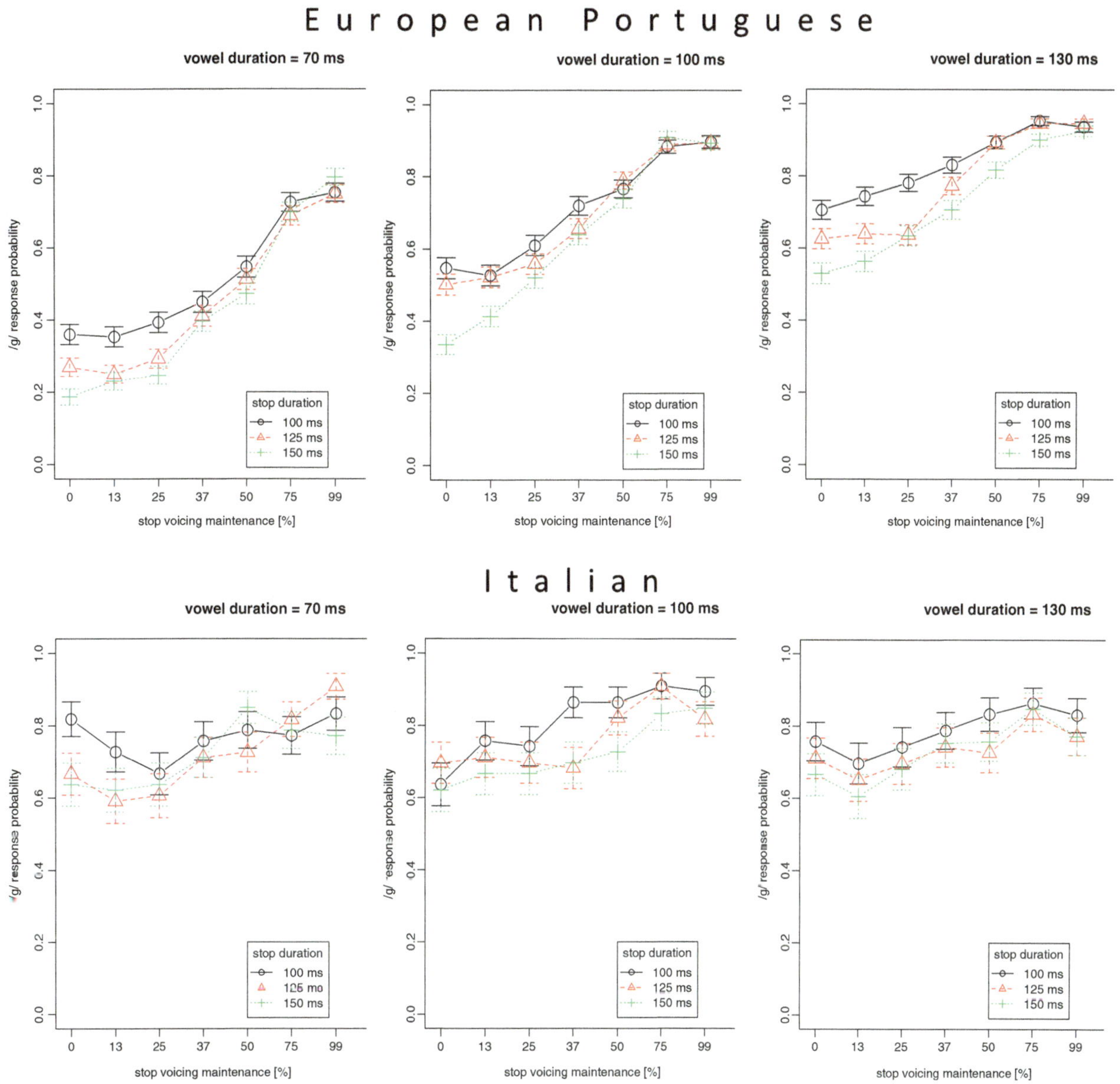

Figure 4 Means and standard errors for the perception experiment. For the EP listeners (top panels) and the Italian listeners (bottom panels). Shown are the percentages of /g/ responses (*y*-axis) over all listeners with respect to the factor voicing maintenance (in percent of the stop closure duration, *x*-axis). The three panels split the data by vowel duration (from left to right, 70, 100 and 130 ms, respectively) and within each panel the different lines correspond to the three stop durations (100, 125 and 150 ms, respectively).

duration examined in this study (i.e. 130 ms). For these ambiguous prototypes, the perceptual system is not able to use the acoustic cues vowel duration or stop duration to obtain a robust voiced/voiceless distinction and therefore has to rely only on the voicing maintenance cue (burst is not present in the stimuli). If we compare the two corresponding curves in Figure 4 (top middle panel dashed line for EP, bottom right panel dashed line for Italian), it can be seen that for stimuli with voicing maintenance higher than 50% of the stop closure, the two curves are nearly identical for the two languages, showing robust /g/ responses (with a probability of 0.8). However, with increasing devoicing of the presented stimulus (i.e. for voicing maintenance below 50%), only EP participants show an effect on listener responses, while Italian listeners maintain their stable /g/ responses. Thus, in the absence of vowel duration and stop duration cues, only EP shows an influence of voicing maintenance, while Italian listeners decisions listeners are not affected. In other words, the two languages show substantial response differences but only for higher percentages of devoicing: the more voiced the stimuli are, the more similar the resulting voicing probabilities are across the two languages.

Statistical validation of the observed differences by means of GLMM showed that all main factors language ($z = 12.1$, $p < 0.001$), with Italians reporting more /g/s than EP listeners, contextual vowel duration ($z = 37.8$, $p < 0.001$), stop duration ($z = -10.0$, $p < 0.001$) and voicing maintenance ($z = 43.6$, $p < 0.001$) had a significant effect on the listeners' voicing decision, but not the factors contextual vowel identity and *repetition* number. The following interactions were significant: voicing maintenance with language ($z = -12.2$, $p < 0.001$); voicing maintenance with stop duration ($z = 5.7$, $p < 0.001$) with higher voicing maintenance differences for short stop durations; vowel duration with language ($z = -16.2$, $p < 0.001$); vowel duration with stop duration ($z = -2.34$, $p < 0.001$).

Conclusions

Production study

The Italian voicing profiles generated from velar stop data are similar to those published in [12]. Data presented in this paper shows significantly higher stop devoicing for EP compared to Italian. Comparing the voicing profiles and statistical results of EP not only against those of Shih et al. [12] for Italian and Spanish but also with our own Italian data, one can observe that EP does not behave like the other Romance languages Spanish and Italian with regard to devoicing of phonologically voiced stops. This result, differentiating EP from Italian stop voicing, is backed up by the linear mixed models analysis of voicing status at the central part of the stop closure, with significant differences between Italian and EP.

A possible explanation for the devoicing differences in production could be related to the fact that, contrary to other Romance languages (e.g. Spanish and Italian), EP is classically reported [34] as being stress-timed (see however [35] for evidence that EP shows shared properties of both stress-timed and syllable-timed languages), with reduction and neutralisation characteristics similar to other stress-timed languages (e.g. German and English).

In stress-timed languages, segments between stresses have the tendency to undergo substantial changes, with strong effects on segments themselves (e.g. vowel centralisation, devoicing and reduction). In the light of speech economy theories, certain segments or acoustic cues are thus enhanced in stressed parts, thus forcing other acoustic cues to weaken. When devoicing of phonologically voiced stops occurs, it could be the case that the strengthening of other segments or cues is more important, thus leading to extinction of the (cost intensive) voicing during stop closure for phonologically voiced stops. As a result, the devoicing characteristics for the two languages examined in this study (stress-timed EP and syllable-timed Italian) differ, with EP behaving more like other stress-timed languages (e.g. German). This similarity of devoicing patterns comparing EP and German was shown for all of the phonologically voiced stops and fricatives in a cross-linguistic comparison study of devoicing [14].

Perception study

The perception experiment provides evidence for the significant role of the acoustic cues voicing maintenance, preceding vowel duration and stop duration for voiced versus voiceless identification for EP but not for Italian. Italian listeners are more or less insensitive to these acoustic cues. They show a stable voiced response for all of the presented stimuli in the fully crossed design. Thus, it seems that EP listeners are strongly influenced by the three acoustic cues, but not the Italian participants.

In this regard, the perceptual results for EP are comparable to those obtained for other languages, for example English [6]. However, the phonetic realisations of phonologically voiced stops in EP are often highly devoiced throughout the complete duration of their stop closure, as seen in the production study (Figure 2). From these production results, one could assume that for EP stop voicing perception, the acoustic cues vowel duration and/or stop durations are more important than voicing maintenance, or that all these factors interact in a complex manner that does not give voicing maintenance the strongest weight in the identification process. However, the perception results in Figure 4 show that voicing maintenance is the dominant cue *if* the presented stimuli are fully voiced and therefore the major cue triggering the voicing decision of EP listeners. If the presented stimuli are devoiced, however, other acoustic cues (vowel duration and stop duration) take over by triggering the voiced/voiceless listener responses based on the extracted phoneme durations. In other words, we encounter cue weighting among different acoustic cues in the perception of EP stop voicing. In summary, these observations support the hypothesis that voicing maintenance is a major but not a required cue for stop voicing perception in EP, indicating that there are strong interactions between the three observed acoustic cues. For Italian, no such cue weighting was found.

Another finding is the bias of all listeners towards voiced responses, as can be seen in Figure 4 even for the fully devoiced/voiceless condition and durational values prototypical of a voiceless stop. For example, over all listeners, the response probability never reaches a probability lower than 0.2 for EP, and for Italian, the probability does not drop below the 0.5 threshold, thus resulting in voiced responses from the Italian listeners only. In other words, none of the two languages showed a stable voiceless (/k/) response floor effect. This bias could be due to the missing burst in the presented stimuli and thus to the problem of extracting a stable VOT cue.

For Italian, this could explain the absence of voicing differences for all of the acoustic cues. Listeners actively use the burst, thus obtaining the pre-voicing cue (voicing with reference to the stop release). Since the burst was deliberately missing, it could be the case that Italian listeners were not able to extract this cue, thus leading to their consistent voiced response (overall 80% /g/ perception) throughout all of the three continua. VOT could be important to extract a voiceless response, thus pointing to the *short lag* versus *long lag* VOT used as a distinctive cue. However, examining the interplay and cue weighting of other acoustic cues was not the focus of this study; but based on other languages, burst and VOT have an important effect on stop voicing distinction [2,6,36]. Therefore, clearly, the results would have been different if the burst had been included among the perceptual constructs.

For EP however, the missing burst cannot explain the perception results. A large amount of EP voiced stops, and even voiceless stops, do not show a discernible burst [13,14], so in this case, it is not clear how the listeners would rely on a VOT cue in ambiguous conditions. The stimuli used in our perception experiments were designed to exclude the burst cue (i.e. to exclude both burst and voicing onset time cues) to examine the influence of voicing maintenance and durational cues on stop voicing distinction.

For EP, with the constraint of the missing burst, the results of the perception experiment show that the voicing maintenance cue is strongly used to distinguish voicing, in addition to and in combination with the vowel duration and stop duration cue. Even with the missing burst and in the absence of voicing during stop closure, EP listeners are able to make stable voiced/voiceless decisions based on vowel durations and stop durations only. However, if the stimuli to be judged are more or less fully voiced, then voicing maintenance is found to be the major cue. In this case, it overrides the vowel duration and stop duration cues and guarantees a stable voiced response of all EP listeners, even in the absence of a facilitating burst and with contradicting duration values. In summary, this study constitutes new evidence that, in the absence of a facilitating burst, multiple acoustic cues are used and combined with different cue weighting to obtain a stable stop voicing distinction for EP.

Competing interests

The authors declare that they have no competing interests.

Acknowledgements

This work was partially funded by National Funds through the Foundation for Science and Technology (FCT), in the context of the project PEst-OE/EEI/UI0127/2014 to IEETA, and the post-doctoral fellowship from FCT (Portugal) SFRH/BPD/48002/2008 to Daniel Pape. We would like to thank the Instituto di Scienze e Tecnologie della Cognizione (ISTC) in Padova, especially Claudio Zmarich, for enabling us to record our Italian production and perception data there and use the soundproof room. We also thank all our subjects for their participation in our experiments.

Author details

[1]Institute of Electronics and Informatics Engineering of Aveiro (IEETA), University of Aveiro, 3810-193 Aveiro, Portugal. [2]School of Health Sciences (ESSUA), University of Aveiro, 3810-193 Aveiro, Portugal.

References

1. N Cuartero, *Voicing Assimilation in Catalan and English*. Ph.D. Thesis (Universitat Autonoma de Barcelona, 2002)
2. M Jessen, *Phonetics and Phonology of Tense and Lax Obstruents in German* (John Benjamins Publishing, Amsterdam, 1999)
3. L Lisker, AS Abramson, The voicing dimension: Some experiments in comparative phonetics, in *Proceedings of the 6th International Congress of Phonetic Sciences* (Prague, 1967), pp. 563–567
4. PA Luce, J Charles-Luce, Contextual effects on vowel duration, closure duration, and the consonant vowel ratio in speech production. J Acoust Soc Am. **78**, 1949–1957 (1985)
5. MC Viana, *Étude de deux aspects du consonantisme du Portugais: fricatisation et dévoisement*. Ph.D. Thesis (Université des Sciences Humaines de Strasbourg, 1984)
6. L Lisker, "Voicing" in English: a catalogue of acoustic features signaling /b/ versus /p/ in trochees. Lang Speech. **29**(1), 3–11 (1986)
7. A Francis, K Baldwin, C Nusbaum, Effects of training on attention to acoustic cues. Percept Psychophys. **62**, 1668–1680 (2000)
8. LJ Raphael, Preceding vowel duration as a cue to the perception of the voicing characteristic of word-final consonants in American English. J Acoust Soc Am. **51**(4B), 1276–1303 (1972)
9. F Li, A Mennon, J Allen, A psychoacoustic method to find the perceptual cues of stop consonants in natural speech. J Acoust Soc Am. **127**(4), 2599–2610 (2010)
10. PJ Bailey, Q Summerfield, Information in speech: observations on the perception of [s]-stop clusters. J Exp Psychol Hum Percept Perform. **6**, 536–563 (1980)
11. EN Oglesbee, *Multidimensional stop categorization in English, Spanish, Korean, Japanese and Canadian French*. Ph.D. Thesis (Indiana University, 2008)
12. C Shih, B Möbius, B Narasimhan, Contextual effects on consonantal voicing profiles: a cross-linguistic study, in *Proceedings of the 14th International Congress of the Phonetic Sciences* (San Francisco, 1999), pp. 989–992
13. M Lousada, LMT Jesus, A Hall, Temporal acoustic correlates of the voicing contrast in European Portuguese stops. J Int Phon Assoc. **40**, 261–275 (2010)
14. D Pape, LMT Jesus, Devoicing of phonologically voiced obstruents: Is European Portuguese different from other Romance languages? in *Proceedings of the 17th International Congress of Phonetic Sciences* (Hong Kong, 2011), pp. 1566–1569
15. J Veloso, The role of consonantal duration and tenseness in the perception of voicing distinctions of Portuguese stops, in *Proceedings of the 6th International Congress of Phonetic Sciences* (Stockholm, 1995), pp. 266–269
16. MRD Martins, Vogais e consoantes do Português: Estatística de ocorrência, duração e intensidade. Boletim da Filologia **XXIV**(1–4), 1–11 (1975)
17. P Boersma, Praat, a system for doing phonetics by computer. Glot International. **5**(9/10), 341–345 (2001)
18. D Pape, LMT Jesus, P Perrier, Constructing physically realistic VCV stimuli for the perception of stop voicing in European Portuguese, in *Computational Processing of the Portuguese Language*, ed. by H Caseli, A Teixeira, A Villavicencio (Springer, Heidleberg, 2012), pp. 328–349
19. P Perrier, Y Payan, M Zandipour, J Perkell, Influences of tongue biomechanics on speech movements during the production of velar stop consonants: a modeling study. J Acoust Soc Am. **114**, 1582–1599 (2003)
20. BH Story, A parametric model of the vocal tract area function for vowel and consonant simulation. J Acoust Soc Am. **117**(5), 3231–3254 (2005)
21. BH Story, Vowel acoustics for speaking and singing. Acta Acustica. **90**(4), 629–640 (2004)
22. JR Flanagan, DJ Ostry, AG Feldman, Control of trajectory modifications in target-directed reaching. Journal of Motor Behaviour **25**, 140–152 (1993)

23. PL Gribble, DJ Ostry, Are complex control signals required for human arm movements? J Neurophysiol. **79**(3), 1409–1424 (1998)
24. PL Gribble, DJ Ostry, Compensation for loads during arm movements using equilibrium-point control. Exp Brain Res. **135**, 474–482 (2000)
25. P Perrier, S Fuchs, Speed-curvature relations in speech production challenge the one-third power law. J Neurophysiol. **100**, 1171–1183 (2008)
26. S Tasko, J Westbury, Defining and measuring speech movement events. J Speech Hear Res. **45**, 127–142 (2002)
27. LMT Jesus, C Shadle, A parametric study of the spectral characteristics of fricatives. J Phon. **30**, 437–464 (2002)
28. LMT Jesus, C Shadle et al., Devoicing measures of European Portuguese fricatives, in *Computational Processing of the Portuguese Language*, ed. by NJ Mamed, J Baptista, I Trancoso (Springer, Berlin, 2003), pp. 1–8
29. JM Hillenbrand, RT Gayvert, Open source software for experiment design and control. J Speech Lang Hear Res. **48**(1), 45–60 (2005)
30. D Bates, M Maechler, B Bolker, *lme4: Linear mixed-effects models using S4 classes, R package version 0.999375-42*, 2011
31. RDCT, *"R: A Language and Environment for Statistical Computing," R Development Core Team (RDCT), R Foundation for Statistical Computing*, 2010
32. H Quené, HVD Berg, Examples of mixed effects modelling with crossed random effects and with binomial data. J Mem Lang. **59**, 413–425 (2008)
33. HR Baayen, *Analyzing linguistic data: A practical introduction to statistics using R* (Cambridge University Press, Cambridge, 2008)
34. M Cruz-Ferreira, *Handbook of the International Phonetic Association: A Guide to the Use of the International Phonetic Alphabet* (Cambridge University Press, Cambridge, 1999), pp. 126–130
35. S Frota, M Vigário, On the correlates of rhythmic distinctions: The European/Brazilian Portuguese case. Probus. **13**, 247–275 (2001)
36. R Lasky, A Srydal-Lasky, R Klien, VOT discrimination by four to six and a half month old infants from Spanish environments. J Exp Child Psychol. **20**, 215–225 (1975)

3

Single-channel acoustic echo cancellation in noise based on gradient-based adaptive filtering

Upal Mahbub[1*], Shaikh Anowarul Fattah[1], Wei-Ping Zhu[2] and M Omair Ahmad[2]

Abstract

In this paper, a two-stage scheme is proposed to deal with the difficult problem of acoustic echo cancellation (AEC) in single-channel scenario in the presence of noise. In order to overcome the major challenge of getting a separate reference signal in adaptive filter-based AEC problem, the delayed version of the echo and noise suppressed signal is proposed to use as reference. A modified objective function is thereby derived for a gradient-based adaptive filter algorithm, and proof of its convergence to the optimum Wiener-Hopf solution is established. The output of the AEC block is fed to an acoustic noise cancellation (ANC) block where a spectral subtraction-based algorithm with an adaptive spectral floor estimation is employed. In order to obtain fast but smooth convergence with maximum possible echo and noise suppression, a set of updating constraints is proposed based on various speech characteristics (e.g., energy and correlation) of reference and current frames considering whether they are voiced, unvoiced, or pause. Extensive experimentation is carried out on several echo and noise corrupted natural utterances taken from the TIMIT database, and it is found that the proposed scheme can significantly reduce the effect of both echo and noise in terms of objective and subjective quality measures.

Keywords: Adaptive filter; Convergence analysis; Echo cancellation; Least mean squares algorithm; Noise reduction; Spectral subtraction; Single-channel communication

1 Introduction

The phenomenon of acoustic echo occurs when the output speech signal from a loudspeaker gets reflected from different surfaces, like ceilings, walls, and floors and then fed back to the microphone. In its worst case, acoustic echo can cause howling of a significant portion of sound energy [1,2]. In real life applications, such as a lecture in a large conference hall or in the public address system of a trade fair, the presence of acoustic echo along with the environmental noise is a very common phenomenon, which degrades the speech quality even leading to complete loss of intelligibility.

In order to deal with the problem of acoustic echo cancellation (AEC), conventionally echo suppressors, earphones, and directional microphones have been used, which generally place restrictions on the talkers' movement [2]. As an alternate of such hardware-based solutions, adaptive filter algorithms are widely being applied where apart from the input channel, a separate echo-free reference channel is required [3-13]. Among different adaptive filter algorithms, the least mean squares (LMS) algorithm and its different variants are very popular for their satisfactory performances and less computational burden [4,10,12-14]. Besides these algorithms, the recursive least squares (RLS) algorithm is well-known for its fast convergence at the expense of computational complexity [13]. The adaptive filter algorithms have also been used for acoustic noise cancellation (ANC) [15].

There are some methods that deal with both acoustic echo and noise cancellation (AENC) [16-18]. The echo canceller used in [16] utilizes a sub-band noise cancellation scheme. In [17], echo cancellation is done by an adaptive LMS filter while a linear prediction error fil-

*Correspondence: omeecd@eee.buet.ac.bd
[1] Department of Electrical and Electronic Engineering, Bangladesh University of Engineering and Technology, Dhaka 1000, Bangladesh
Full list of author information is available at the end of the article

ter removes the residual echo and noise. In [18], a single Wiener filter is employed to simultaneously suppress the echo and noise. It is to be mentioned that all these AENC methods employ more than one microphone, while the solutions using single microphone are favorable in most of the real-life applications.

In this paper, an AENC scheme is proposed which can efficiently deal with the single-channel scenario. First, unlike conventional LMS algorithm, considering the delayed version of the previously echo- and noise-suppressed signal as reference, a gradient-based adaptive LMS algorithm is developed for single channel AEC. Preliminary results obtained by using this idea is reported in [19]. However, in the current paper, analytical proof of convergence towards the optimum Wiener-Hopf solution is presented. Next, a single-channel ANC algorithm based on spectral subtraction with an adaptive spectral floor estimation is developed, which reduces not only the effect of noise but also some residual echo. Finally, analyzing different speech characteristics of the reference and current frames, multiconditional updating constraints are proposed in order to obtain precise control on convergence characteristics. For performance evaluation, extensive experimentation is conducted on several real-life echo and noise corrupted speech signals at different acoustic environments.

2 Problem formulation

In order to formulate the problem of single-channel AENC, for a better understanding, first, a dual channel AENC scheme is presented in Figure 1 (according to [17]). Here, $s_1(n)$ and $s_2(n)$ are speech signals corresponding to near-end and far-end speakers, while $v_1(n)$ and $v_2(n)$ are additive noises, respectively. The noise corrupted far-end signal $(s_2(n) + v_2(n))$ is played through a loudspeaker at the near-end acoustic room environment and the echo signal $x_2(n)$ is generated. Thus, the input $y_1(n)$ to the near-end microphone is given by

$$y_1(n) = s_1(n) + v_1(n) + x_2(n). \tag{1}$$

The task of the adaptive filter-based AEC block placed at the near-end is to produce an estimate $\widehat{x}_2(n)$ of the echo $x_2(n)$ by minimizing the error

$$e_1(n) = y_1(n) - \widehat{x}_2(n). \tag{2}$$

Two major issues in dual channel system are (i) availability of a separate reference signal required for the adaptive filter, for example, here the delayed version of $(s_2(n) + v_2(n))$ and (ii) different speakers for input and echo signals. Moreover, use of the double talk detector (DTD) helps in controlling the update process. Unfortunately, these features are absent in single-channel scenario as shown in Figure 2. Instead of two speakers, in this case, the microphone receives the input $s(n)$ corrupted by noise $v(n)$ and echo generated from the same speaker.

In the presence of noise $v(n)$, the sole microphone input signal in single-channel scenario is given by

$$y(n) = s(n) + v(n) + x_s(n) + x_v(n), \tag{3}$$

where $x_s(n)$ and $x_v(n)$ denote the echo of the input speech and noise, respectively. The echo signals can be expressed as

$$x_s(n) = \mathbf{a}_n^T \mathbf{s}(n - k_0), \tag{4}$$

$$x_v(n) = \mathbf{a}_n^T \mathbf{v}(n - k_0), \tag{5}$$

where $\mathbf{s}(n-k_0) = [s(n-k_0-1), s(n-k_0-2), \ldots, s(n-k_0-p)]^T$ and $\mathbf{v}(n-k_0) = [v(n-k_0-1), v(n-k_0-2), \ldots, v(n-k_0-p)]^T$ with k_0 being a predefined flat delay and $\mathbf{a}_n = [a_n(1), a_n(2), \ldots, a_n(p)]^T$ consists of the coefficients corresponding to the acoustic room transfer function $A(z)$. The order p and coefficient values of $A(z)$ depend on the room characteristics. It is to be noted that in this case, there is no scope of obtaining a separate echo-free reference or a separate noise-only reference, which makes the single-channel AENC problem extremely difficult to handle.

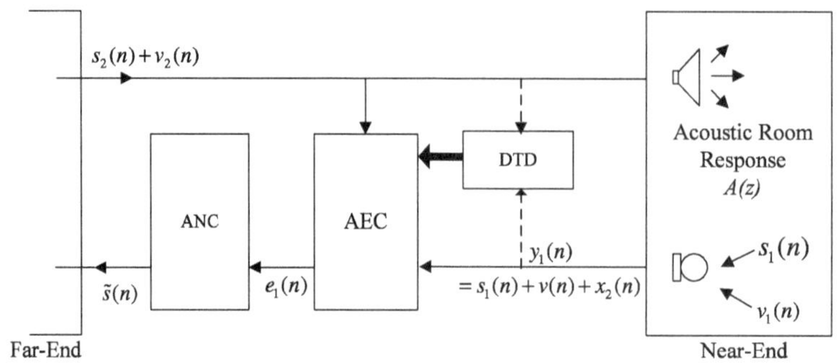

Figure 1 Adaptive filter-based echo and noise cancellation in dual channel communication system.

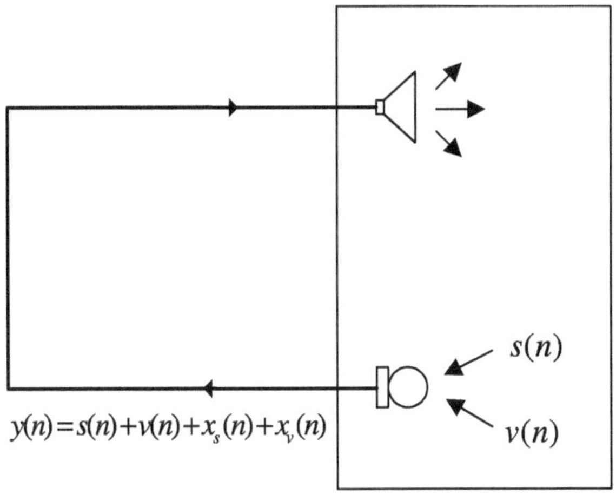

Figure 2 **Single-channel acoustic echo generation in noisy room environment.**

3 Proposed single-channel AENC scheme

3.1 Proposed two-stage setup

In Figure 3a, a simple block diagram showing two stages of the proposed AENC scheme is presented and in Figure 3b, more detail of the adaptive filter-based AEC algorithm involved in the first stage is shown. Similar to Figure 2, the input to the microphone $y(n)$ can be described by (3). For the case of single-channel AEC, for example, while delivering a lecture in a large conference hall, the microphone in front of the speaker receives input speech $s(n)$ corrupted by $v(n)$. Once this noise-corrupted speech is transmitted through loudspeaker, echo signal is generated and thus the microphone after some initial time delay will receive noise-corrupted speech and echo of previously uttered speech. The task of AEC is to cancel the echo part from this input by using adaptive filter algorithm. In order to obtain adaptively an estimate $\widehat{x}_s(n) + \widehat{x}_v(n)$ of the echo signal, we propose to utilize delayed versions of the previously echo-suppressed samples of the noisy speech as reference signal [19]. A symbol hat on the variable is used to indicate estimated value. The error signal $e(n)$ thus obtained is given by

$$e(n) = y(n) - [\widehat{x}_s(n) + \widehat{x}_v(n)] . \tag{6}$$

The estimate of the echo signal can be expressed as

$$\widehat{x}_s(n) + \widehat{x}_v(n) = \widehat{\mathbf{w}}_n^T [\widehat{\mathbf{s}}(n - k_0) + \widehat{\mathbf{v}}(n - k_0)] , \tag{7}$$

where $\widehat{\mathbf{w}}_n = [\widehat{w}_n(1), \widehat{w}_n(2) \ldots \widehat{w}_n(p)]^T$ is the estimated coefficient vector. The task of the adaptive filter is to obtain an optimum $\widehat{\mathbf{w}}_n$ by minimizing the error in (6) i.e.,

$$e(n) = s(n) + \{(v(n) + \delta_s(n)) + \delta_v(n)\}, \tag{8}$$

where $\delta_s(n) = x_s(n) - \widehat{x}_s(n)$ and $\delta_v(n) = x_v(n) - \widehat{x}_v(n)$ are the residual echo of the speech and noise portions of the input signal, respectively, and it is assumed that these signals exhibit the properties of white Gaussian noise. Next, $e(n)$ is passed through a spectral subtraction-based single-channel ANC block which produces output $\widetilde{s}(n) \approx s(n) + \Psi(n)$ that closely resembles $s(n)$ provided that the residual echo-noise portion $\Psi(n)$ becomes very small.

It is to be noted that the task of noise reduction, unlike the proposed AENC scheme, may be carried out prior

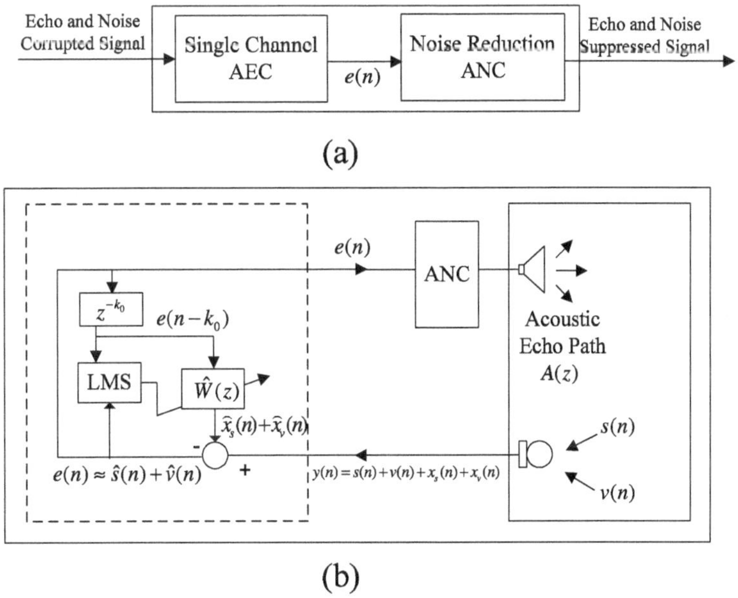

Figure 3 **Block diagram of proposed single-channel AENC scheme.** **(a)** Two stages and **(b)** details of proposed adaptive filter-based AEC algorithm.

to the AEC block. However, because of possible nonlinearities introduced by the prior noise reduction block, no proper reference would be available for the single-channel AEC block [17]. Hence, the arrangement shown in Figure 3a is adopted, in which the noise reduction block also serves as a post-processor for attenuating the residual echo.

3.2 Development of proposed gradient-based single-channel LMS AEC scheme

A delayed version of the adaptive filter output $e(n)$ is proposed to use as the reference signal, and from (8), filter output $e(n)$ can be written as

$$e(n) = \widehat{s}(n) + \widehat{v}(n), \tag{9}$$

where $\widehat{s}(n) = s(n) + \delta_s(n)$ and $\widehat{v}(n) = v(n) + \delta_v(n)$. The objective function of the adaptive filter involves minimization of the mean square estimation of the error function and using (6) it can be written as

$$\begin{aligned} E\{e^2(n)\} &= E\{(s(n)+v(n))^2\} + E\{(x_s(n)+x_v(n) \\ &\quad - \widehat{x}_s(n) - \widehat{x}_v(n))^2\} + 2E\{(s(n)+v(n)) \\ &\quad \times (x_s(n)+x_v(n) - \widehat{x}_s(n) - \widehat{x}_v(n))\}, \end{aligned} \tag{10}$$

where $E\{.\}$ denotes the expectation operator. In (10), it is intended to use the basic definition of cross-correlation operation, for example, the cross-correlation function between $s(n)$ and $v(n)$ is defined as

$$r_{sv}(m) = E\{s(n)v(n-m)\}, \tag{11}$$

where m denotes the lag. Using (4), (5), (7), and the above definition, the last term of (10) can be expressed as

$$\begin{aligned} &2E\{[\,(s(n)+v(n))(x_s(n)+x_v(n)-\widehat{x}_s(n)-\widehat{x}_v(n))]\,\} \\ &= 2\sum_{k=1}^{k=p}\{(a_n(k)-\widehat{w}_n(k))(r_{ss}(k_0+k)+r_{sv}(k_0+k) \\ &\quad + r_{vs}(k_0+k)+r_{vv}(k_0+k)) - r_{s\delta_s}(k_0+k) \\ &\quad - r_{s\delta_v}(k_0+k) - r_{v\delta_s}(k_0+k) - r_{v\delta_v}(k_0+k)\}. \end{aligned} \tag{12}$$

Here, $r_{ss}(k_0+k)$ corresponds to the (k_0+k)th lag of the cross-correlation between $s(n)$ and its previous samples $s(n-k_0-k)$, and $r_{sv}(k_0+k)$ corresponds to the (k_0+k)th lag of the cross-correlation between $s(n)$ and $v(n-k_0-k)$. In a similar way, $r_{vs}(k_0+k)$, $r_{vv}(k_0+k)$, $r_{s\delta_s}(k_0+k)$, $r_{s\delta_v}(k_0+k)$, $r_{v\delta_s}(k_0+k)$, and $r_{v\delta_v}(k_0+k)$ can be defined. It is well known that the value of cross-correlation decreases rapidly with the increasing lags when two signals are uncorrelated. In ideal case, the cross-correlation function between two random noise signals would be nonzero only at the zero lag. Since $v(n)$ is assumed to be white Gaussian noise and, generally, the value of k_0 is very large, in (12), the effect of the terms $r_{sv}(k_0+k)$, $r_{vs}(k_0+k)$, and $r_{vv}(k_0+k)$ can be neglected. Moreover, because of noise-like characteristics of $\delta_s(n)$ and $\delta_v(n)$, in (12), one can neglect $r_{s\delta_v}(k_0+k)$, $r_{v\delta_s}(k_0+k)$, and $r_{v\delta_v}(k_0+k)$ too. Hence, it can easily be comprehended that optimal filter performance occurs when $r_{ss}(n)$ is minimum, i.e., the least possible correlation between $s(n-k_0-k)$ and $s(n)$ is desired. As a result, (10) reduces to

$$\begin{aligned} E\{e^2(n)\} &= E\{(s(n)+v(n))^2\} \\ &\quad + E\{[\,x_s(n)+x_v(n)-\widehat{x}_s(n)-\widehat{x}_v(n)]^2\,\} \\ &\quad + 2\sum_{k=1}^{k=p}(a_n(k)-\widehat{w}_n(k))r_{ss}(k_0+k). \end{aligned} \tag{13}$$

Here, the magnitude of $r_{ss}(k_0+k)$ strongly depends on speech characteristics and the amount of flat delay k_0. For a reasonably large k_0, the effect of $r_{ss}(k_0+k)$ in 13 can be neglected, and minimization of (13) results in

$$\frac{\partial E\{e^2(n)\}}{\partial \widehat{\mathbf{w}}_n^T} = 0$$

$$E[\{x_s(n)+x_v(n)-\widehat{x}_s(n)-\widehat{x}_v(n)\}\{\widehat{\mathbf{s}}(n-k_0)+\widehat{\mathbf{v}}(n-k_0)\}]=0. \tag{14}$$

Hence, we obtain

$$\begin{aligned} &E\{(x_s(n)+x_v(n))(\widehat{\mathbf{s}}(n-k_0)+\widehat{\mathbf{v}}(n-k_0))\} \\ &= \widehat{\mathbf{w}}_n^T E[\,\{\widehat{\mathbf{s}}(n-k_0)+\widehat{\mathbf{v}}(n-k_0)\}\{\widehat{\mathbf{s}}(n-k_0)+\widehat{\mathbf{v}}(n-k_0)\}]\,. \end{aligned} \tag{15}$$

The above equation is similar to Wiener-Hopf equation and its solution can be written as

$$\widehat{\mathbf{w}}_n^T = \mathbf{R}_{(s+v)(s+v)}(n-k_0)^{-1}\mathbf{r}_{(x_s+x_v)(s+v)}(n-k_0), \tag{16}$$

where $\mathbf{r}_{(x_s+x_v)(s+v)}(n-k_0)$ consists of different lags of cross-correlation between the echo signal $x_s(n)+x_v(n)$ and the noisy input signal $s(n)+v(n)$, while $\mathbf{R}_{(s+v)(s+v)}$ is the auto-correlation matrix of $s(n)+v(n)$. There is no doubt that $\widehat{\mathbf{w}}_n$ is the most optimum solution possible. Hence, it is shown that even for a single-channel noise corrupted AEC problem, the most optimum solution $\widehat{\mathbf{w}}_n$ can be achieved under the assumptions stated earlier.

For iterative estimation of optimal filter coefficients, the adaptive LMS algorithm is very popular. It is fast and efficient, and it does not require any correlation measurements or matrix inversion [13]. The update equation of the LMS adaptive algorithm is generally expressed as

$$\widehat{\mathbf{w}}_{n+1}^T = \widehat{\mathbf{w}}_n^T - \mu\nabla\xi(n), \tag{17}$$

where μ is the step factor controlling the stability and rate of convergence, $\xi(n)$ is the cost function, and ∇ is the gradient operator. The LMS algorithm simply approximates the mean square error by the square of the instantaneous error, i.e., $\xi(n) = e^2(n)$, and therefore, from (6) and (7) , the gradient of $\xi(n)$ can be expressed as

$$\nabla\xi(n) = \frac{\partial\xi(n)}{\partial\widehat{\mathbf{w}}_n^T} = -2e(n)(\widehat{\mathbf{s}}(n-k_0)+\widehat{\mathbf{v}}(n-k_0)).$$

Thus, the update equation for the proposed single-channel LMS adaptive scheme can be written as

$$\widehat{\mathbf{w}}_{n+1}^T = \widehat{\mathbf{w}}_n^T + 2\mu e(n)(\widehat{\mathbf{s}}(n-k_0)+\widehat{\mathbf{v}}(n-k_0)). \quad (18)$$

3.3 Convergence analysis of the proposed AEC scheme

Considering expectation operation on both sides of the update Eq. 18, one can obtain

$$\underline{\widehat{\mathbf{w}}}_{n+1}^T = \underline{\widehat{\mathbf{w}}}_n^T + 2\mu E\{e(n)(\widehat{\mathbf{s}}(n-k_0)+\widehat{\mathbf{v}}(n-k_0))\}. \quad (19)$$

Here, an underline beneath $\widehat{\mathbf{w}}_n$ is introduced to represent the expected value $E\{\widehat{\mathbf{w}}_n\}$. For the kth unknown weight vector (where $k = 1, 2, \ldots, p$), using (6) and neglecting the effect of $r_{ss}(n)$ that has already been discussed in the previous subsection, the last term of (19) can be written as

$$\begin{aligned}&2\mu E\{e(n)(\widehat{\mathbf{s}}(n-k_0)+\widehat{\mathbf{v}}(n-k_0))\}\\&= 2\mu E\{[x_s(n)+x_v(n)-\widehat{x}_s(n)-\widehat{x}_v(n)]\times(\widehat{\mathbf{s}}(n-k_0)\\&\quad+\widehat{\mathbf{v}}(n-k_0))\}. \quad (20)\end{aligned}$$

Based on the assumptions on cross-correlation terms stated in the previous subsection, one can obtain

$$\begin{aligned}E\{e(n)(\widehat{\mathbf{s}}(n-k_0)+\widehat{\mathbf{v}}(n-k_0))\} &= \mathbf{r}_{(x_s+x_v)(s+v)}(n-k_0)\\&\quad-\mathbf{R}_{(s+v)(s+v)}(n-k_0)\underline{\widehat{\mathbf{w}}}_n^T. \quad (21)\end{aligned}$$

Using (21), the update Eq. 19 can be written as

$$\begin{aligned}\underline{\widehat{\mathbf{w}}}_{n+1}^T &= \underline{\widehat{\mathbf{w}}}_n^T - 2\mu\mathbf{R}_{(s+v)(s+v)}(n-k_0)\underline{\widehat{\mathbf{w}}}_n^T\\&\quad+2\mu\mathbf{r}_{(x_s+x_v)(s+v)}(n-k_0). \quad (22)\end{aligned}$$

Evaluating the homogeneous and particular solutions of (22), the total solution can be obtained as (see Appendix)

$$\underline{\widehat{w}}_{n+1}^U(k) = C_k(1-2\mu\lambda(k))^n + \frac{1}{\lambda(k)}r^U(n-k_0-k), \quad (23)$$

where $\lambda(k)$ is the kth diagonal element of the eigenvalue matrix obtained by eigenvalue decomposition of $\mathbf{R}_{(s+v)(s+v)}(n-k_0)$ and $r^U(n-k_0-k)$ is the kth element of $\mathbf{U}^T\mathbf{r}_{(x_s+x_v)(s+v)}(n-k_0) = \mathbf{r}^U_{(x_s+x_v)(s+v)}(n-k_0)$ with the matrix $\mathbf{U}$ consisting of eigenvectors corresponding to eigenvalues. Since in the iterative update procedure, the homogeneous part $(1-2\mu\lambda(k))^n$ diminishes with iterations, (23) in a matrix form can be expressed as

$$\begin{aligned}\underline{\widehat{\mathbf{w}}}^T &= \mathbf{U}\Lambda^{-1}\mathbf{U}^T\mathbf{r}_{(x_s+x_v)(s+v)}(n-k_0)\\&= \mathbf{R}^{-1}_{(s+v)(s+v)}(n-k_0)\mathbf{r}_{(x_s+x_v)(s+v)}(n-k_0). \quad (24)\end{aligned}$$

Thus, it is found that the average value of the weight vector converges to the Wiener-Hopf solution, which is the optimum solution with increasing number of iteration.

3.4 Noise reduction in spectral domain

In the proposed AENC scheme, the operation of the ANC block is processed frame by frame for noise reduction based on single-channel spectral subtraction algorithm [20-22]. According to (9), for the ith frame, the error signal for the duration of a frame length can be written as

$$e_i(n) = \widehat{s}_i(n) + \widehat{v}_i(n). \quad (25)$$

Corresponding frequency domain representation is given by

$$E_i(\omega) = \widehat{S}_i(\omega) + \widehat{V}_i(\omega). \quad (26)$$

The magnitude squared spectrum of $\widehat{s}_i(n)$ can be written as

$$|\widehat{S}_i(\omega)|^2 = |E_i(\omega)|^2 - |\widehat{V}_i(\omega)|^2 - \widehat{V}_i(\omega)\widehat{S}_i^*(\omega) - \widehat{S}_i(\omega)\widehat{V}_i^*(\omega). \quad (27)$$

It is desired to choose an estimate $\widetilde{S}_i(\omega)$ that will minimize

$$Err_i(\omega) = ||\widetilde{S}_i(\omega)|^2 - |\widehat{S}_i(\omega)|^2|. \quad (28)$$

Since the noise is assumed to be zero mean and uncorrelated with the signal, the expected values of the last two terms of (27) can be neglected. Thus, (28) can be expressed as

$$Err_i(\omega) = |\widetilde{S}_i(\omega)|^2 - |E_i(\omega)|^2 + E\{|\widehat{V}_i(\omega)|^2\}. \quad (29)$$

This expression of $Err_i(\omega)$ can be minimized by choosing

$$|\widetilde{S}_i(\omega)|^2 = |E_i(\omega)|^2 - E\{|\widehat{V}_i(\omega)|^2\}. \quad (30)$$

With an estimate of noise spectrum $E\{|\widehat{V}_i(\omega)|^2\}$, signal spectrum $\widetilde{S}_i(\omega)$ can be computed as

$$\widetilde{S}_i(\omega) = |\widetilde{S}_i(\omega)|\, e^{j\arg[E_i(\omega)]}, \quad (31)$$

where the phase ($\arg[E_i(\omega)]$) is generally assumed to be the phase of the noise corrupted signal without causing significant degradation in terms of loss of intelligibility of the speech signal [20]. It can be seen that an estimate of the magnitude spectrum $|\widetilde{S}_i(\omega)|$ of the signal can be obtained provided an estimate of noise spectrum $E\{|\widehat{V}_i(\omega)|^2\}$ is available, which is generally computed during the periods when speech is known *a priori* not to be present.

Final output of the AENC system is the speech frame ($\widetilde{s}_i(n)$), which consists of the original speech $s_i(n)$ and a negligible amount of noise-like signal $\Psi_i(n)$. The signal $\Psi_i(n)$, although very weak, may contain some signature of the input noise $v(n)$, the residual echo $\delta_s(n)$, and the residual noise $\delta_v(n)$. In order to overcome the problem of musical noise and to avoid the speech distortion caused by speech subtraction, in (31), an over estimate of the noise power spectrum can be subtracted carefully such that the spectral floor is preserved [21]. Thus, (31) can be modified as

$$\begin{aligned} |\, \widetilde{S}_i(\omega)\,|^2 &= |\, \widehat{E}_i(\omega)\,|^2 - \alpha_{ss} E\{|\, \widehat{V}_i(\omega)\,|^2\}, \\ &\quad \text{if } |\, \widetilde{S}_i(\omega)\,|^2 > \beta_{ss}\{|\, \widehat{V}_i(\omega)\,|^2\} \\ &= \beta_{ss}\{|\, \widehat{V}_i(\omega)\,|^2\}, \text{ otherwise.} \end{aligned} \tag{32}$$

Here, α_{ss} is the subtraction factor and β_{ss} is the spectral floor parameter with $\alpha_{ss} \geq 1$ and $0 \leq \beta_{ss} \leq 1$. The task of noise power spectral density estimation is carried out based on the minimum statistics noise estimator proposed in [23] which can handle the time-varying nature of the noise.

4 Development of adaptive update constraints

The AEC part of the proposed AENC scheme may suffer from some common problems of adaptive filter-based algorithms, such as slow convergence rate and fluctuation around the desired estimates, especially in practical cases where the assumption on negligibility of cross-correlation terms (stated in the previous section) may not strictly hold. In order to overcome such problems, some updating constraints are proposed based on the following speech characteristics:

(i) The level of cross-correlation
(ii) The amount of signal power
(iii) The mean square error (MSE) between consecutive estimates of the unknown filter coefficients.

Through extensive experimentation on different speech frames, it is found that the negligibility of the cross-correlation terms $r_{ss}(n)$, $r_{s\delta_v}(n)$, $r_{v\delta_s}(n)$, and $r_{v\delta_v}(n)$ (as described after (12)) strongly depends on the voicing characteristics of speech frames and the input noise. Because of inherent periodicity of the voiced speech frame, the degree of cross-correlation between two voiced speech frames of a person becomes higher in comparison to that between two unvoiced speech frames which are random in nature. Regarding signal power, the ratio of power of a voiced speech frame and an unvoiced speech frame is found to be higher in comparison to that of the two voiced speech frames. As white Gaussian noise is considered, the degree of cross-correlation between the speech and noise is found to be negligible and the noise powers in two different frames may not differ significantly. As a result, the effect of input noise is found to be negligible on the power ratio.

For a flat delay of k_0 samples, the initial k_0 samples of the utterance $s(n) + v(n)$ can be treated as a reference signal (echo-free signal) responsible for the generation of echo signal that corrupts the current samples at or after k_0 samples. Considering a window of M samples with $M \ll K_0$, power of the reference signal $(\widehat{s}(n - k_0) + \widehat{v}(n - k_0))$ can be computed as

$$P_{\text{ref}}(n) = \frac{1}{M} \sum_{i=-\frac{M}{2}}^{\frac{M}{2}-1} [\widehat{s}(n - k_0 + i) + \widehat{v}(n - k_0 + i)]^2 . \tag{33}$$

For a window of last M samples of the echo-suppressed speech signal $\widehat{s}(n)$, the average power $P_{\text{sup}}(n)$ can be computed as

$$P_{\text{sup}}(n) = \frac{1}{M} \sum_{j=0}^{M-1} [\widehat{s}(n - j) + \widehat{v}(n - j)]^2 . \tag{34}$$

The ratio of $P_{\text{ref}}(n)$ and $P_{\text{sup}}(n)$ is denoted as the power ratio $P_{\text{rs}}(n)$ and considered as one of the control characteristics.

Another important characteristic criterion is the correlation coefficient $C_{\text{rs}}(n)$ between a frame of the noisy reference signal $(\widehat{s}(n - k_0) + \widehat{v}(n - k_0))$ and a frame of the current noisy signal $(\widehat{s}(n) + \widehat{v}(n))$. For a frame length of M samples, correlation coefficient $C_{\text{rs}}(n)$ is defined as

$$\begin{aligned} C_{\text{rs}}(n) &= \frac{1}{\sigma_{\widehat{s}(n-k_0+i)+\widehat{v}(n-k_0+i)}\sigma_{\widehat{s}(n-j)+\widehat{v}(n-j)}} \\ &\quad \times \{\text{cov}((\widehat{s}(n - k_0 + i) + \widehat{v}(n - k_0 + i)) \\ &\quad \times (\widehat{s}(n - j) + \widehat{v}(n - j)))\} \end{aligned} \tag{35}$$

where $-M/2 \leq i \leq M/2 - 1$ and $0 \leq j \leq (M - 1)$.

Finally, the parameter estimation accuracy is also considered for the purpose of analyzing the convergence property. In this regard, the mean square error $\text{MSE}_{\text{ideal}}(n)$ between the values of estimated coefficients $\widehat{w}_n$ and those of true coefficients a_n is computed as

$$\text{MSE}_{\text{ideal}}(n) = \frac{1}{p} \sum_{k=1}^{p} [\widehat{w}_n(k) - a_n(k)]^2 . \tag{36}$$

In Figure 4, considering a real-life speech utterance of 250 ms corrupted by echo and noise, behavior of the control parameters obtained by using (33), (34), (35), and (36) is shown. The speech utterance (/iy/ – /ix/) contains a voiced phoneme followed by another voiced phoneme [24]. Here $k_0 = 1{,}000$, $M = 100$, $N_f = 1002$, sampling frequency 16 kHz and SNR = 15 db is used.

In a similar fashion, in Figure 5, a speech utterance consisting of a voiced phoneme /ih/ followed by an unvoiced phoneme /sh/ and, in Figure 6, a voiced phoneme /ih/ followed by pause are considered. It is observed that the characteristic parameters vary depending on the nature of reference and current frames. When the current frame

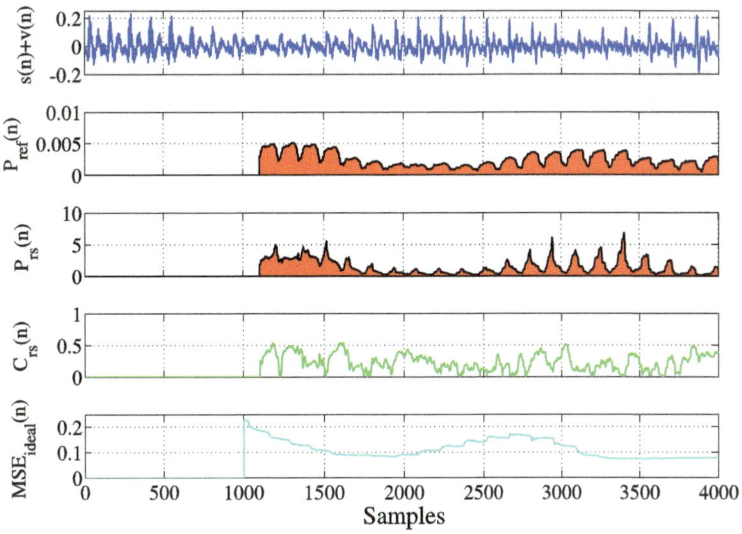

Figure 4 **Characteristics of controlling factors - a voiced phoneme followed by another voiced frame.**

is a pause or weakly unvoiced, the power ratio becomes higher in comparison to the case when the current frame is a voiced one. On the contrary, the correlation coefficient becomes smaller when measured between a voiced and an unvoiced frame, but it becomes quite larger when measured between two voiced frames. It is also found that the presence of voiced frame as a reference strongly governs the rate of convergence and the estimation error of the proposed LMS algorithm. In Figure 4, because of all through presence of the voiced frame as the reference as well as the current frame, it is found that the convergence performance is not very satisfactory and the estimation error is relatively higher. On the other hand, in Figure 6, it is observed that when the current frame is pause, even in the presence of voiced reference frame, a very fast convergence is obtained with a little estimation error. In Figure 5, as the current frame is unvoiced instead of pause, a comparatively slower convergence is observed with higher estimation error.

Next, in Figures 7, 8, 9, the reference frame is considered unvoiced, and in Figures 10, 11, 12, it is considered pause. When the reference frame is considered unvoiced because of the existence of a little correlation between the current and reference frames, the convergence performance of the proposed LMS algorithm is found quite satisfactory irrespective of the power of the reference

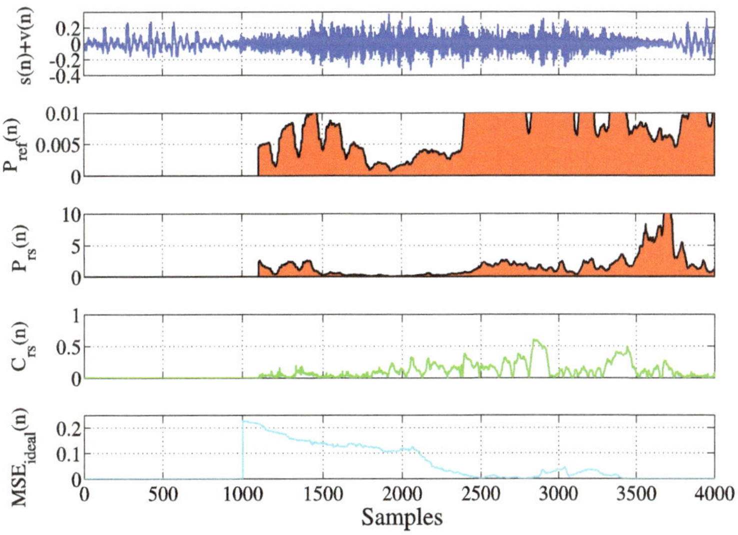

Figure 5 **Characteristics of controlling factors - a voiced phoneme followed by an unvoiced phoneme.**

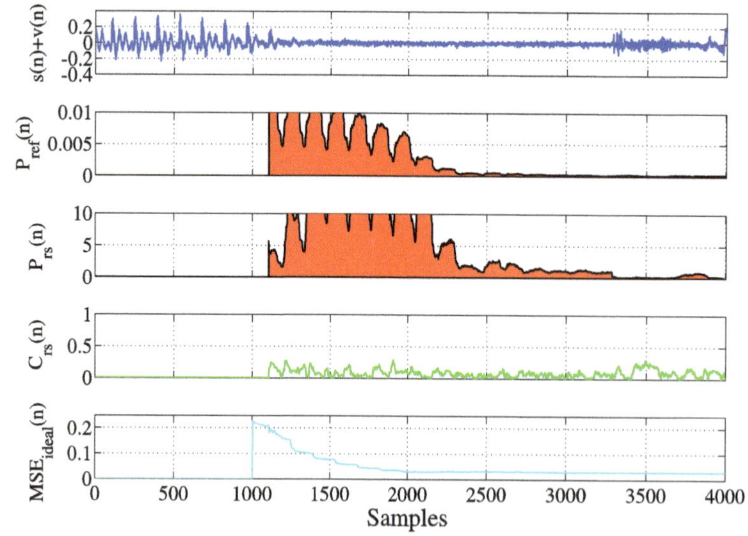

Figure 6 Characteristics of controlling factors - a voiced phoneme followed by pause frame.

signal (strong unvoiced or weakly unvoiced). In the case when the current frame is pause, no matter whether the reference frame is voiced or unvoiced, a fast convergence with high estimation accuracy is achieved using the proposed LMS algorithm. The reasons behind are (i) negligible cross-correlation between reference frame and current frame and (ii) a comparatively higher power ratio. In Figures 10, 11, 12, it is observed that even the reference frame is a pause or stop because of the presence of additive white noise, the reference frame may contain significant energy. In these cases, a reasonable estimation of the room response can be obtained given that the noise power is quite high. Findings in the above cases are summarized in Table 1.

First of all, it is observed that a better convergence in terms of iterations and estimation error is obtained when the current frame is a pause (P) or stop and the reference frame is either voiced (V) or unvoiced (U), namely, V-P and U-P. This fact leads to a decision that the updating needs to be carried out at high level of power ratio, i.e.,

$$P_{rs}(n) = \frac{P_{ref}(n)}{P_{sup}(n)} \geq \zeta, \tag{37}$$

where $P_{ref}(n)$ and $P_{sup}(n)$ are defined in (33) and (34), respectively. If the value of the lower bound ζ is chosen too large, the updating would be postponed for most of the instances resulting in very slow convergence. On the other hand, a very small value of ζ may cause more frequent

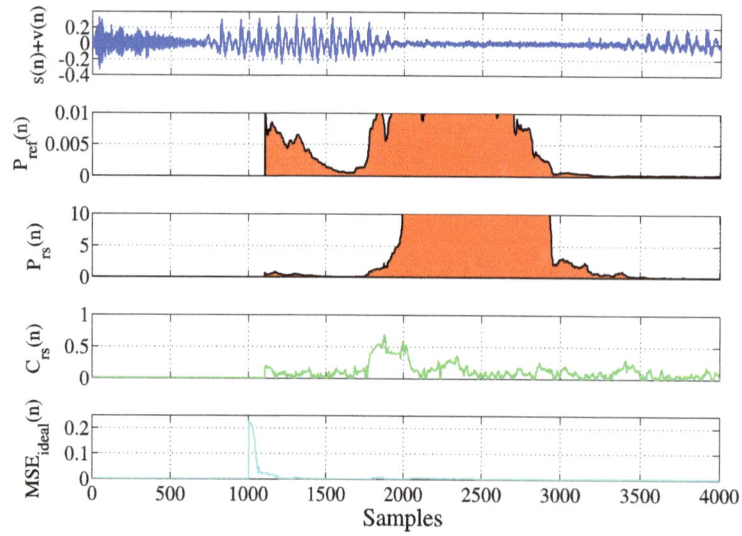

Figure 7 Characteristics of controlling factors - an unvoiced frame followed by a voiced frame.

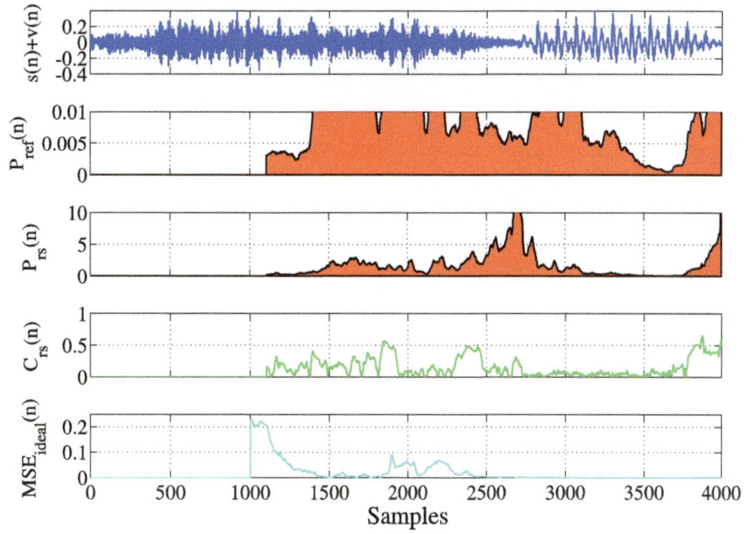

Figure 8 Characteristics of controlling factors - an unvoiced frame followed by another unvoiced frame.

updates where possibility of wrong estimations of filter coefficients would be higher, especially in V-P, U-P, and P-P cases. It is to be noted that considering only a lower bound of $P_{rs}(n)$ may not always be sufficient to ensure that the reference frame possesses significant energy. For example in Figure 13, it is shown that high value of $P_{rs}(n)$ may arise (marked block in the figure) from an initial silence frame where only a very little amount of noise is present. In order to prevent the updating in these situations, a lower bound β on the power of the reference frame is employed, i.e., $P_{ref}(n) \geq \beta$. The value of β should surpass the power of speech pauses and ensure that the LMS update is postponed even if a frame of speech containing a partial pause is available as the reference. Hence, the first constraint for updating the algorithm is proposed as

Condition I: $P_{rs}(n) \geq \zeta$ and $P_{ref}(n) \geq \beta$.

In some cases, it is observed that though the power ratio is very small, quite satisfactory updating is obtained, such as the U-V case shown in Figure 7. Another characteristic observed here is lower value of correlation coefficient $C_{rs}(n)$ with higher value of $P_{ref}(n)$. It is to be mentioned that the proposed AEC algorithm is developed on the assumption of negligibility of the cross correlation between current frame and reference frame. However,

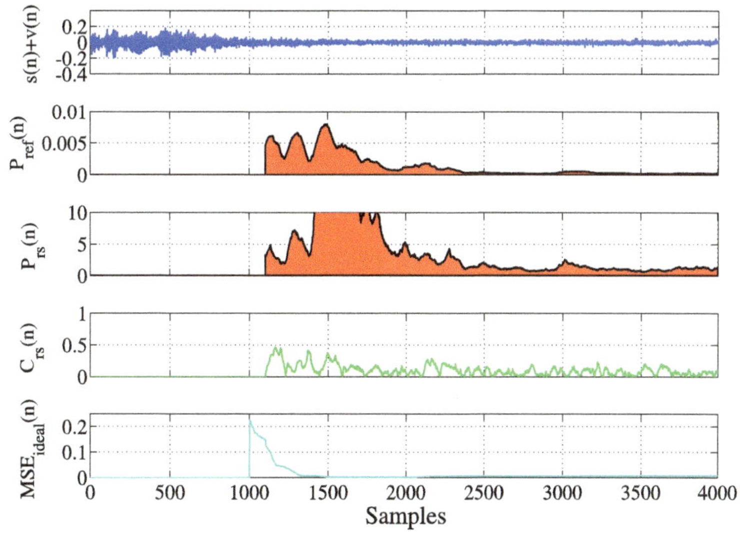

Figure 9 Characteristics of controlling factors - an unvoiced frame followed by a pause.

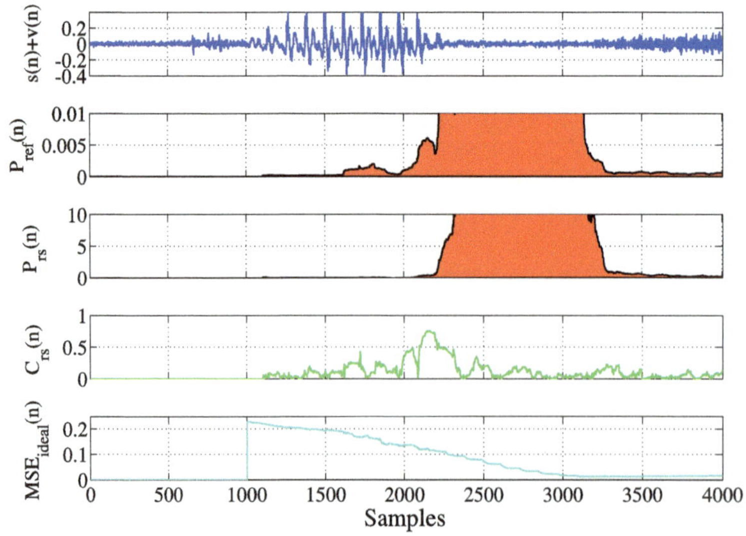

Figure 10 Characteristics of controlling factors - pause followed by a voiced frame.

since both reference and current frame may belong to the same person, in case of high degree of correlation, the adaptive algorithm would try to suppress portion from the echo-corrupted signal resulting in unusual degradation = in convergence performance. Hence, introducing an upper bound on $C_{rs}(n)$, the second condition is proposed as

Condition II: $C_{rs}(n) \leq \Upsilon 1$ and $P_{ref} \geq \beta$.

The presence of a certain level of noise can be utilized as an advantage in pause instances where generally the updating is not performed. Since noise is considered uncorrelated to itself, updating at frames where only noise is present would be quite satisfactory. In this case, the value of $C_{rs}(n)$ must be very small and thus another condition on updating is proposed as

Condition III: $C_{rs}(n) \leq \Upsilon 2 \leq \Upsilon 1$.

Another important factor is the MSE of the estimations of successive iterations, which is defined as

$$e_{coeff}(n) = \sum_{K=1}^{p} (\widehat{w}_n(k) - \widehat{w}_{n-1}(k))^2/p. \tag{38}$$

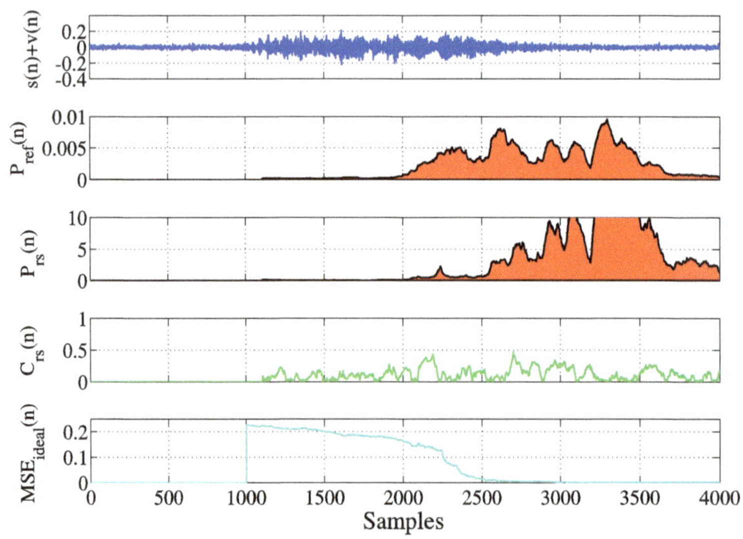

Figure 11 Characteristics of controlling factors - pause followed by an unvoiced frame.

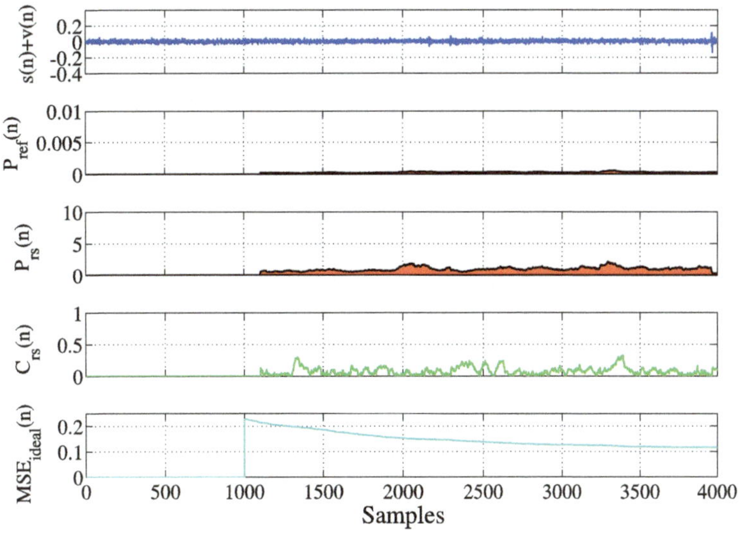

Figure 12 Characteristics of controlling factors - pause followed by another pause.

In order to continue the updating, an upper bound on the variation of successive estimates is set as following condition:

Condition IV: $e_{coeff}(n) \leq \aleph.$

Considering smaller values of $e_{\text{coeff}}(n)$ allows to avoid updating at those instances where abrupt and significant changes occur in the estimated coefficients. In the proposed method, in order to carry out the LMS update, at least one of the above four conditions must be fulfilled.

5 Simulation results and comments

Performance of the proposed algorithm is investigated in different echo-generating environments at various input noise levels considering several male and female utterances available in the TIMIT database [24]. An acoustic room environment is simulated using an FIR filter of length N_f, where as per conventional approaches, filter coefficients during the flat delay portion are assumed to be zero. The flat delay time (k_0) can be pre-calculated based on the distance between the microphone and the speaker [25]. Because of the implicit zeros corresponding to the flat delay, it is evident that a few number ($N_f - k_0$) of unknown coefficients has to be determined. In the proposed method, a smaller step size is used to obtain a smooth convergence.

Table 1 Variation of LMS updating performance due to various characteristics of reference and current speech frame

Reference speech sample	Current noise- and echo-corrupted speech sample	LMS update performance
Voiced	Voiced	Poor
Voiced	Unvoiced	Unsatisfactory
Voiced	Pause	Satisfactory/Excellent
Unvoiced	Voiced	Excellent
Unvoiced	Unvoiced	Excellent
Unvoiced	Pause	Excellent
Pause	Voiced	Poor
Pause	Unvoiced	Poor
Pause	Pause	Poor

First, a subjective evaluation is carried out based on the feedback about the quality of the echo- and noise-suppressed signal provided by five individual listeners at different noisy echo-generating environments. From the overall response of the listeners in terms of mean objective score (MOS), a very satisfactory performance of the proposed method is obtained even under severe echo-generating conditions in noise.

Next, two objective measures, namely, echo return loss enhancement (ERLE) and signal-to-distortion ratio (SDR) are employed. The ERLE is defined as the ratio of the instantaneous power of the residual echo signal $\eta_\varsigma(n)$ and that of the input echo signal $\eta_x(n)$ and expressed in dB as [1]

$$\text{ERLE}(n) = -10\log\frac{\eta_\varsigma(n)}{\eta_x(n)}. \quad (39)$$

The average value of ERLE(n) over time is considered. The input and output SDRs in dB are respectively defined as

$$\text{SDR}_{\text{in}} = 10\log\frac{P_s}{P_{x+v}} \quad (40)$$

$$\text{SDR}_{\text{out}} = 10\log\frac{P_s}{P_{\hat{s}+\hat{v}-s}}, \quad (41)$$

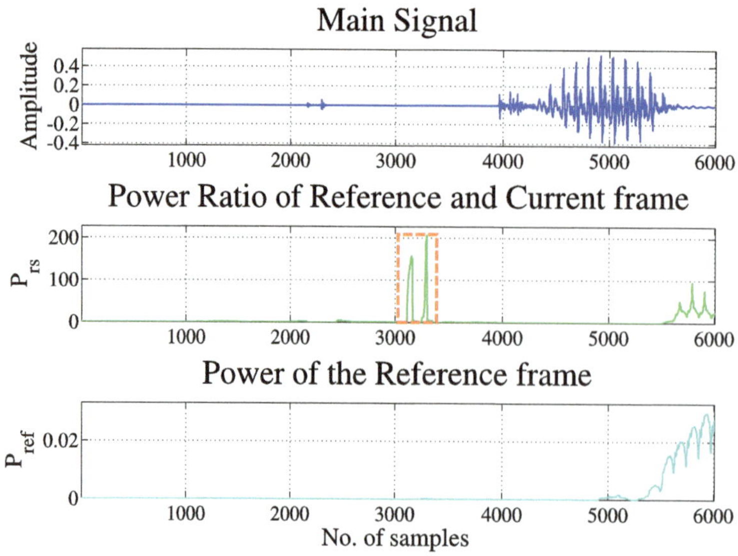

Figure 13 Example of high power ratio during initial silence frame.

where P_s is the power of original signal $s(n)$, P_{x+v} is the power of microphone input, and $P_{\widehat{s}+\widehat{v}-s}(n)$ is the power of distortion present in the echo-suppressed output signal. The SDR improvement is given by

$$\mathrm{SDRI} = \mathrm{SDR}_{\mathrm{out}} - \mathrm{SDR}_{\mathrm{in}}, \tag{42}$$

which indicates the overall distortion removal.

The proposed algorithm has been tested on several different sentences taken from the TIMIT database. In order to demonstrate the principle of selecting different threshold values required in the proposed updating constraints, as a typical example, a sample utterance 'Good service should be rewarded by big tips' is shown in Figure 14 [24]. Voicing decisions are marked in the figure as 'P' for pause, 'V' for voiced, and 'U' for unvoiced. Considering white Gaussian noise with SNR = 15 dB, $N_f = 1,002$, $k_0 = 1,000$, and $M = 100$ in Figure 14b,c,d,e, $P_{\mathrm{rs}}(n)$, $P_{\mathrm{ref}}(n)$, $C_{\mathrm{rs}}(n)$, and $\mathrm{MSE}_{\mathrm{ideal}}(n)$ are shown, respectively. Note that

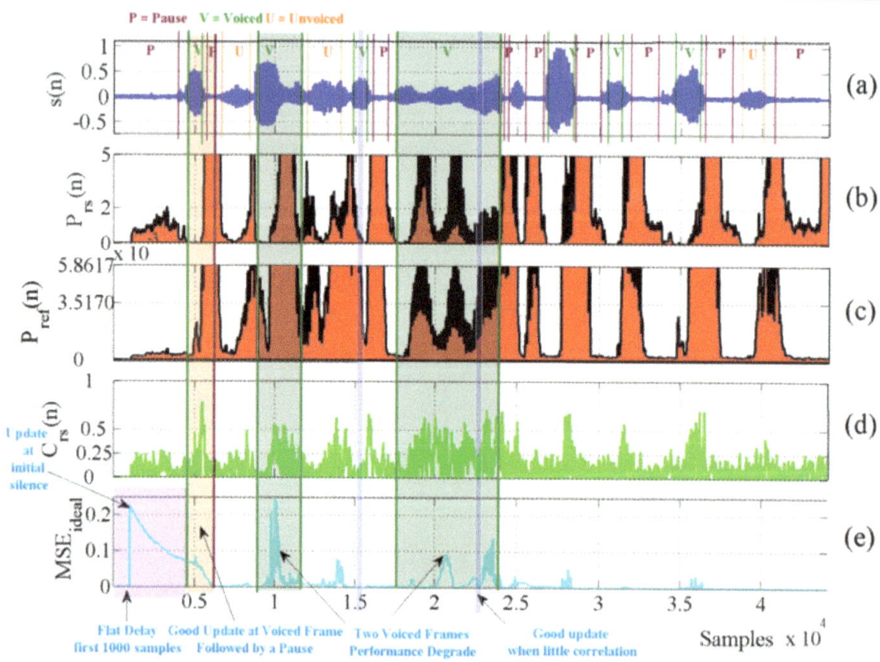

Figure 14 Plots of (a) utterance $s(n)$ and update parameters (b) $P_{rs}(n)$, (c) $P_{ref}(n)$, (d) $C_{rs}(n)$, and (e) MSE (without using constraint).

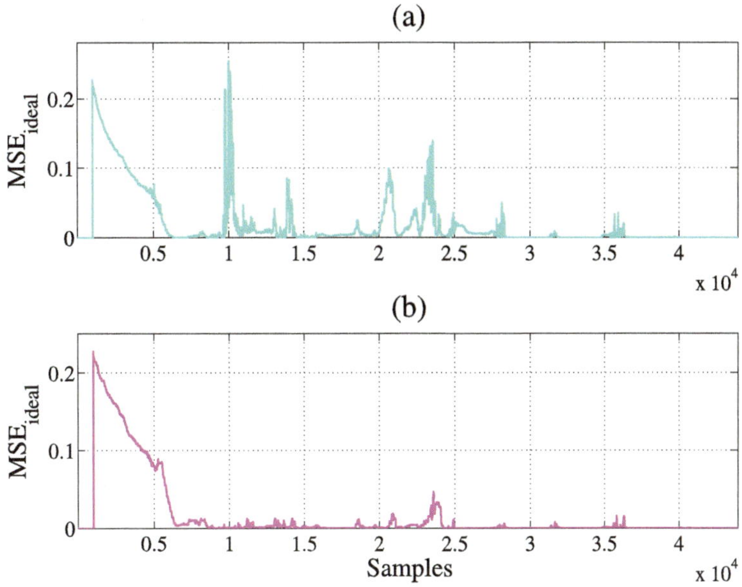

Figure 15 MSEs for the utterance shown in Figure 14. (a) Without conditions and **(b)** with conditions.

in this case, the proposed algorithm is used without the update constraints, and thus, the $\text{MSE}_{\text{ideal}}(n)$ exhibits some higher values. The comments provided in Table 1 can be better visualized from different marked zones of this figure. From extensive experimentations, it is found that a better update requires $P_{\text{ref}}(n)$ to be at least twice of $P_{\text{supp}}(n)$ and a small percentage (1% to 5%) of the power of a regular voiced frame can be chosen as the lower bound of β for $P_{\text{ref}}(n)$. Analyzing $C_{\text{rs}}(n)$ in different speech frames, $\Upsilon 1$ in condition 2 is chosen as 0.25 to ensure that no speech is being suppressed during the update procedure by confusing it with the echo and $\Upsilon 2$ is kept very small, i.e, $\Upsilon 2 \approx 0.1$ to allow updating for cases where there exists no correlation or extremely low correlation

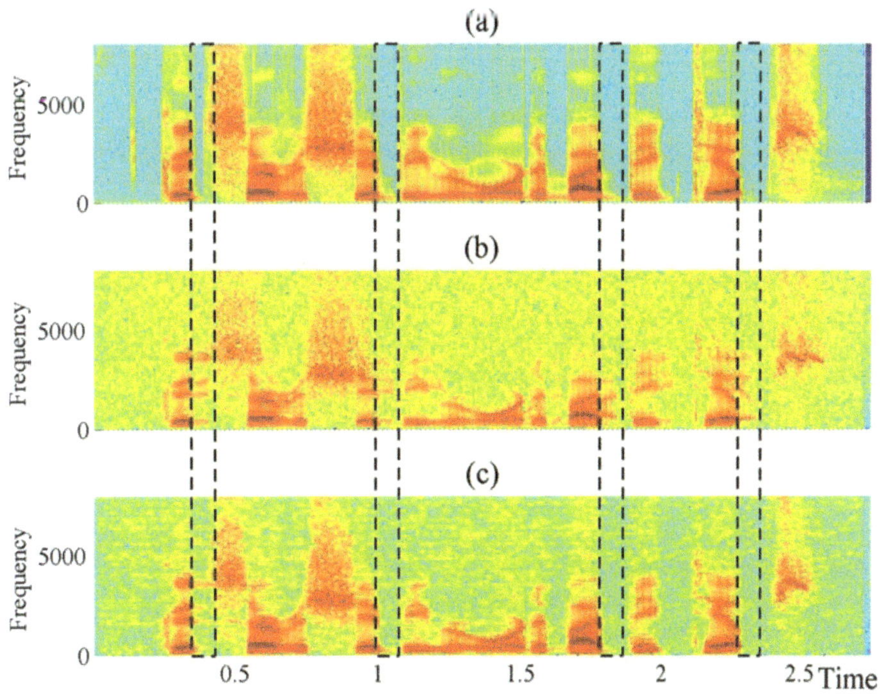

Figure 16 Spectrograms of (a) original signal, (b) echo- and noise-corrupted input and (c) enhanced output.

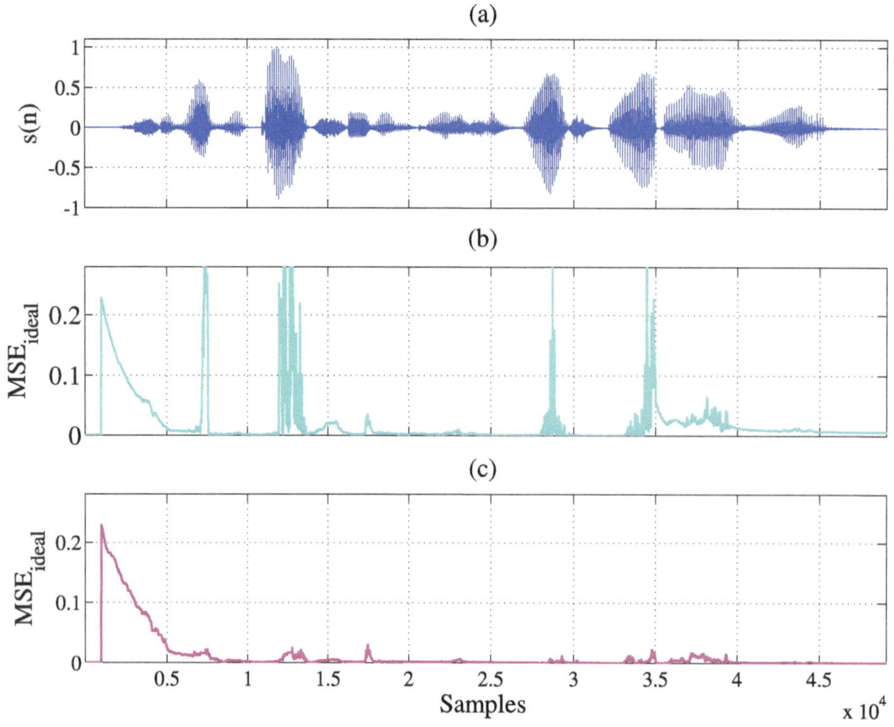

Figure 17 Another TIMIT utterance (a). MSEs of LMS estimations: **(b)** without conditions and **(c)** with conditions.

between the reference signal and echo-suppressed signal. The value of the threshold $\aleph$ for $e_{\text{coeff}}(n)$ in condition IV is chosen to be very small (0.7×10^{-4}) such that there will be no update of the LMS algorithm when the magnitude of $e_{\text{coeff}}(n)$ is comparatively much larger.

In Figure 15, the effect of incorporating the proposed conditions is shown. It is vividly observed from Figure 15 that by employing the proposed conditions, the convergence is improved to a greater extent. Moreover, in order to demonstrate the performance in frequency domain, spectrograms of the original signal, echo- and noise-corrupted signal, and the output of the proposed AENC block are depicted in Figure 16a,b, respectively. For convenience, some zones are marked on the spectrograms where significant reduction in echo and noise can easily be observed

In order to show the effectiveness of the proposed conditions, the $\text{MSE}_{\text{ideal}}(n)$ obtained in Figure 14e is redrawn in Figure 15. In Figure 15, the effect of incorporating the conditions is shown. It is vividly observed from Figure 15 that by employing the proposed conditions, the convergence is improved to a greater extent. Moreover, in order to demonstrate the performance in frequency domain, spectrograms of the original signal, echo- and noise-corrupted signal, and the output of the proposed AENC block are depicted in Figure 16a,b, respectively. For convenience, some zones are marked on the spectrograms where significant reduction in echo and noise can easily be observed. For a better understanding, another TIMIT utterance 'She had your dark suit in greasy wash water all year', under similar acoustic environment as used in Figure 14, is considered and corresponding echo- and noise-corrupted speech signal is shown in Figure 17a. The MSEs obtained by using the proposed method with and without the conditions are presented in Figure 17b,c, which clearly demonstrate the performance improvement in the later case.

In Table 2, the performance of the proposed algorithm with and without applying the conditions is shown in

Table 2 Performance comparison with varying room acoustics

	No condition		With conditions	
$N_f - k_0$	SDRI (dB)	Avg. ERLE (dB)	SDR (dB)	Avg. ERLE (dB)
2	4.9921	8.8496	6.9848	10.6772
4	4.9027	2.0696	5.7731	2.2787
6	8.391	4.6507	9.2744	5.0313
8	6.4551	2.4214	6.5558	2.6797
10	6.0507	2.6341	6.1730	2.854
12	6.7127	3.0277	7.0978	3.2048
14	7.8763	3.7481	8.2515	3.8909

Table 3 Performance comparison with noise level variation

Input noise Level (dB)	No condition		With conditions	
	SDRI (dB)	Avg. ERLE (dB)	SDR (dB)	Avg. ERLE (dB)
25	7.4065	3.183	7.8189	3.2759
20	7.613	3.5382	7.9346	3.6171
15	7.8763	3.7481	8.2515	3.8909
10	8.2085	3.5999	8.386	3.6064
5	8.2434	3.0533	8.8839	3.0765
0	8.7968	2.4493	9.4557	2.542
-5	8.2259	2.0032	10.5136	2.2912

terms of the SDR improvement (dB) and ERLE (dB) for utterance 1. In order to evaluate the performance under different room environments, length (N_f) and parameter values of the room response filter are varied while keeping the input SNR constant to 15 dB. Considering $k_0 = 1,000$, $N_f - k_0$ is varied from 2 to 14. Results shown in the table clearly demonstrate the effectiveness of using the conditions on performance measures; in all cases, higher values of SDR and ERLE are obtained.

In Table 3, the performance of the proposed algorithm with and without applying the conditions is evaluated for different levels of input SNR ranging from 25 to -5 dB for the first utterance considering white Gaussian noise and $N_f = 1014$. It can be seen that the proposed method provides satisfactory performance at all SNR levels. Especially, the use of proposed conditions exhibits comparatively better performance.

6 Conclusion

The problem of echo cancellation in the presence of noise, especially in single-channel environment, is a very challenging task, which has been efficiently tackled in this paper. First, the single-channel AEC block is designed based on the gradient-based adaptive LMS filter where to overcome the problem of getting a separate reference signal, we propose to use the delayed version of the echo-suppressed signal. Such a unique proposal of getting the reference signal is justified by presenting a detailed mathematical proof of achieving the most optimum Wiener-Hopf solution of the estimated filter coefficients, and a convergence analysis is carried out. Moreover, in order to achieve fast and smooth convergence, a set of updating constraints is proposed by analyzing the speech characteristics of different types of speech frames, such as voiced, unvoiced, and pause. In the ANC block, a modified single-channel spectral subtraction method is considered for its robust performance. It is shown that the proposed AENC scheme with updating constraints provides a very satisfactory performance in different echo-generating conditions and various levels of SNR in terms of SDR and ERLE.

Appendix

Derivation of the solution of the LMS update

In order to obtain a homogeneous solution of the update Eq. 22, one may consider

$$\underline{\widehat{\mathbf{w}}}_{n+1}^{T} = \underline{\widehat{\mathbf{w}}}_{n}^{T} - 2\mu \mathbf{R}_{(s+v)(s+v)}(n-k_0)\underline{\widehat{\mathbf{w}}}_{n}^{T}. \tag{43}$$

Eigenvalue decomposition of the correlation matrix $\mathbf{R}_{(s+v)(s+v)}(n-k_0)$ results in

$$\mathbf{R}_{(s+v)(s+v)}(n-k_0) = \mathbf{U}\Lambda\mathbf{U}^T, \tag{44}$$

where each column of the matrix $\mathbf{U}$ consists of eigenvectors corresponding to eigenvalues constituting the diagonal elements of the matrix Λ and $\mathbf{U}^T\mathbf{U} = \mathbf{I}$. Forward multiplication by $\mathbf{U}^T$ on both sides of (43) results in

$$\underline{\widehat{\mathbf{w}}}_{n+1}^{T^{\mathcal{U}}} = \underline{\widehat{\mathbf{w}}}_{n}^{T^{\mathcal{U}}} - 2\mu\Lambda\underline{\widehat{\mathbf{w}}}_{n}^{T^{\mathcal{U}}}, \tag{45}$$

where $\mathbf{U}^T\underline{\widehat{\mathbf{w}}}_{n}^{T} = \underline{\widehat{\mathbf{w}}}_{n}^{T^{\mathcal{U}}}$. The kth coefficient of the weight vector can be expressed as

$$\underline{\widehat{w}}_{n+1}^{\mathcal{U}}(k) = (1 - 2\mu\lambda(k))\underline{\widehat{w}}_{n}^{\mathcal{U}}(k), \tag{46}$$

where $\lambda(k)$ is the kth diagonal element of the eigenvalue matrix obtained by eigenvalue decomposition of $\mathbf{R}_{(s+v)(s+v)}(n-k_0)$. Hence, the homogeneous solution can be obtained as

$$\widehat{w}_{h.s} = C_k(1 - 2\mu\lambda(k))^n, \tag{47}$$

where C_k is a constant. Next, in order to obtain the particular solution for the kth coefficient, based on (22) one can get

$$\widehat{w}_{p.s} = \widehat{w}_{p.s} - 2\mu\lambda(k)\widehat{w}_{p.s} + 2\mu r^{\mathcal{U}}(n-k_0-k). \tag{48}$$

Here, $r^{\mathcal{U}}(n-k_0-k)$ is the kth element of $\mathbf{U}^T$ $\mathbf{r}_{(x_s+x_v)(s+v)}(n-k_0) = \mathbf{r}^{\mathcal{U}}_{(x_s+x_v)(s+v)}(n-k_0)$. For a particular solution $\widehat{w}_{p.s} = K_p r^{\mathcal{U}}(n-k_0-k)$, (48) can be written as

$$\begin{aligned} K_p r^{\mathcal{U}}(n-k_0-k) &= K_p r^{\mathcal{U}}(n-k_0-k) \\ &\quad - 2\mu\lambda(k)K_p r^{\mathcal{U}}(n-k_0-k) \\ &\quad + 2\mu r^{\mathcal{U}}(n-k_0-k), \end{aligned} \tag{49}$$

which leads to $K_p = \frac{1}{\lambda(k)}$ and the particular solution

$$\widehat{w}_{p.s} = \frac{1}{\lambda(k)} r^{\mathcal{U}}(n-k_0-k). \tag{50}$$

Competing interests

The authors declare that they have no competing interests.

Author details
[1]Department of Electrical and Electronic Engineering, Bangladesh University of Engineering and Technology, Dhaka 1000, Bangladesh. [2]Department of Electrical and Computer Engineering, Concordia University, Montreal, Quebec H3G 1M8, Canada.

References

1. SV Vaseghi, *Advanced Digital Signal Processing and Noise Reduction*, 2nd edn. (Wiley, Chichester, 2000)
2. SM Kuo, BH Lee, *Real-Time Digital Signal Processing*. (Wiley, 2001)
3. C Breining, P Dreiseitel, E Hänsler, A Mader, B Nitsch, H Puder, T Schertler, G Schmidt, J Tilp, Acoustic echo control - an application of very-high-order adaptive filters. IEEE Signal Process. Mag. **16**(4), 42–69 (1999)
4. E Hänsler, The hands-free telephone problem: an annotated bibliography. Signal Process. **27**(3), 259–271 (1992)
5. AWH Khong, PA Naylor, Stereophonic acoustic echo cancellation employing selective-tap adaptive algorithms. IEEE Trans. Audio, Speech, Lang. Process. **14**(3), 785–796 (2006)
6. F Lindstrom, C Schuldt, I Claesson, An improvement of the two-path algorithm transfer logic for acoustic echo cancellation. IEEE Trans. Audio, Speech, Lang. Process. **15**(4), 1320–1326 (2007)
7. S Wu, X Qiu, M Wu, Stereo acoustic echo cancellation employing frequency-domain preprocessing and adaptive filter. IEEE Trans. Audio, Speech, Lang. Process. **19**(3), 614–623 (2011)
8. R Nath, Adaptive echo cancellation based on a multipath model of acoustic channel. Circuits, Syst. Signal Process., Springer US. **32**(4), 1673–1698 (2013)
9. M Yukawa, RC de Lamare, R Sampaio-Neto, Efficient acoustic echo cancellation with reduced-rank adaptive filtering based on selective decimation and adaptive interpolation. IEEE Trans. Audio, Speech, Lang. Process. **16**(4), 696–710 (2008)
10. E Hänsler, G Schmidt, *Acoustic Echo and Noise Control: a Practical Approach*. (Wiley, New York, 2004)
11. V Myllylä, Residual echo filter for enhanced acoustic echo control. Signal Process. **86**(6), 1193–1205 (2006)
12. R Topa, I Muresan, BS Kirei, I Homana, A digital adaptive echo-canceller for room acoustics improvement. Adv. Electrical Comput. Eng. **10**, 450–453 (2004)
13. S Haykin, *Adaptive Filter Theory*, 3rd edn. (Prentice-Hall, Inc., Upper Saddle River, NJ, 1996)
14. G Schmidt, Applications of acoustic echo control: an overview, in *Proc. Eur. Signal Process. Conf.* (EUSIPCO Vienna, 2004), pp. 9–16
15. B Widrow, JRJ Glover, JM McCool, J Kaunitz, CS Williams, RH Hearn, JR Zeidler, JE Dong, RC Goodlin, Adaptive noise cancelling: principles and applications. Proc. IEEE. **63**(12), 1692–1716 (1975)
16. H Yasukawa, An acoustic echo canceller with sub-band noise cancelling. IEICE Trans. Fundamentals Electron. Commun. Comput. Sci. **E75–A**(11), 1516–1523 (1992)
17. SJ Park, CG Cho, C Lee, DH Youn, Integrated echo and noise canceller for hands-free applications. IEEE Trans. Circuits Syst.-II: Analog Digital Signal Process. **49**(3) (2002)
18. C Beaugeant, V Turbin, P Scalart, A Gilloire, New optimal filtering approaches for hands-free telecommunication terminals. Signal Process. **64**(1), 33–47 (1998)
19. U Mahbub, SA Fattah, Gradient based adaptive filter algorithm for single channel acoustic echo cancellation in noise, in *Proc. Int. Conf. Electrical Computer Engineering (ICECE), 2012 7th International Conference On* (Dhaka, 688 Bangladesh, 2012), pp. 880–883
20. S Boll, A spectral subtraction algorithm for suppression of acoustic noise in speech, in *Proc. IEEE Int. Conf. Acoust. Speech, Signal Process. (ICASSP) '79*, (1979), pp. 200–203
21. M Berouti, R Schwartz, J Makhoul, Enhancement of speech corrupted by acoustic noise. IEEE Conf. Acoust. Speech Signal Process. (ICASSP), 208–211 (1979)
22. JS Lim, Evaluation of a correlation subtraction method for enhancing speech degraded by additive white noise. IEEE Trans. Acoust. Speech Signal Process. **26**(5), 471–472 (1978)
23. R Martin, Noise power spectral density estimation based on optimal smoothing and minimum statistics. IEEE Trans. Speech Audio Process. **9**(5), 504–512 (2001)
24. JS Garofolo, LF Lamel, WM Fisher, JG Fiscus, DS Pallett, NL Dahlgren, V Zue, *Timit acoustic-phonetic continuous speech corpus*. (Linguistic Data Consortium, Philadelphia, 1993)
25. F Guangzeng, L Feng, A new echo caneller with the estimation of flat delay, in *IEEE Region Ten Conf. TENCON 92* (Melbourne, Australia, 1992). vol. 1, pp. 1–5, Print ISBN 0-7803-0849-2, DOI- 10.1109/TENCON.1992.271995

A preliminary demonstration of exemplar-based voice conversion for articulation disorders using an individuality-preserving dictionary

Ryo Aihara[1*], Ryoichi Takashima[1], Tetsuya Takiguchi[2] and Yasuo Ariki[2]

Abstract

We present in this paper a voice conversion (VC) method for a person with an articulation disorder resulting from athetoid cerebral palsy. The movement of such speakers is limited by their athetoid symptoms, and their consonants are often unstable or unclear, which makes it difficult for them to communicate. In this paper, exemplar-based spectral conversion using nonnegative matrix factorization (NMF) is applied to a voice with an articulation disorder. To preserve the speaker's individuality, we used an individuality-preserving dictionary that is constructed from the source speaker's vowels and target speaker's consonants. Using this dictionary, we can create a natural and clear voice preserving their voice's individuality. Experimental results indicate that the performance of NMF-based VC is considerably better than conventional GMM-based VC.

Keywords: Voice conversion; NMF; Articulation disorders; Voice reconstruction; Assistive technologies

1 Introduction

In recent years, a number of assistive technologies using information processing have been proposed, for example, sign language recognition using image recognition technology [1-3], text reading systems from natural scene images [4-6], and the design of wearable speech synthesizers [7]. In this study, we focused on a person with an articulation disorder resulting from athetoid cerebral palsy. There are about 34,000 people with speech impediments associated with an articulation disorder in Japan alone, and one of the causes of speech impediments is cerebral palsy.

Cerebral palsy is a result of damage to the central nervous system, and the damage causes movement disorders. Three general times are given for the onset of the disorder: before birth, at the time of delivery, and after birth. Cerebral palsy is classified into the following types: (1) spastic, (2) athetoid, (3) ataxic, (4) atonic, (5) rigid, and (6) a mixture of these types [8].

Athetoid symptoms develop in about 10% to 15% of cerebral palsy sufferers. In the case of a person with this type of articulation disorder, his/her movements are sometimes more unstable than usual. That means their utterances (especially their consonants) are often unstable or unclear due to the athetoid symptoms. Athetoid symptoms also restrict the movement of their arms and legs. Most people suffering from athetoid cerebral palsy cannot communicate by sign language or writing, so there is a great need for voice systems for them.

In [9], we proposed robust feature extraction based on principal component analysis (PCA) with more stable utterance data instead of DCT. In [10], we used multiple acoustic frames (MAF) as an acoustic dynamic feature to improve the recognition rate of a person with an articulation disorder, especially in speech recognition using dynamic features only. In spite of these efforts, the recognition rate for articulation disorders is still lower than that of physically unimpaired persons. Maier et. al. [11] proposed automatic speech recognition systems for the evaluation of speech disorders that resulted from head and neck cancer.

In this paper, we propose a voice conversion (VC) method for articulation disorders. Regarding speech recognition for articulation disorders, the recognition rate

*Correspondence: aihara@me.cs.scitec.kobe-u.ac.jp
[1]Graduate School of System Informatics, Kobe University, 1-1 Rokkodai, Nada-ku, Kobe 657-8501, Japan
Full list of author information is available at the end of the article

using a speaker-independent model, which is trained by well-ordered speech, is 3.5% [9]. This result implies that the utterance of a person with an articulation disorder is difficult to understand for people who have not communicated with them before. In recent years, people with an articulation disorder may use slide shows and a previously synthesized voice when they give a lecture. However, because their movement is restricted by their athetoid symptoms, to make slides or synthesize their voice in advance is hard for them. People with articulation disorders desire a VC system that converts their voice into a clear voice that preserves their voice's individuality. However, a speech conversion method for people with articulation disorders resulting from athetoid cerebral palsy has not been successfully developed.

In the research discussed in this paper, we conducted VC for articulation disorders using nonnegative matrix factorization (NMF) [12]. NMF is a matrix decomposition method with nonnegativity constraint. In the field of speech processing, NMF is a well-known approach for source separation and speech enhancement [13]. In these approaches, the observed vector is represented by a linear combination of a small number of elementary vectors, referred to as the basis, and its weights. The collection of the basis is called a 'dictionary', and the joint matrix of weights is called an 'activity':

$$\mathbf{x}_l = \sum\nolimits_{j=1}^{J} \mathbf{a}_j h_{j,l} = \mathbf{A}\mathbf{h}_l \quad (1)$$

where $\mathbf{x}_l$ is the lth frame of the observation, and $\mathbf{a}_j$ and $h_{j,l}$ are the jth basis and the weight, respectively. $\mathbf{A}$ and $\mathbf{h}_l$ are the dictionary and the activity of frame l, respectively. In some separation approaches, a dictionary is constructed for each source, and the mixed signals are expressed with a sparse representation of these dictionaries. Using only the weights (called activity in this paper) of basis in the target dictionary, the target signal can be reconstructed. Gemmeke et al. also used the activity of the speech dictionary as phonetic scores instead of likelihoods of hidden Markov models (HMMs) for speech recognition [14].

In our study, we adopt the supervised NMF approach [15], with a focus on VC from poorly articulated speech resulting from articulation disorders into well-ordered articulation. An input spectrum with an articulation disorder is represented by a linear combination of an articulation disorder basis and its weights using NMF. By replacing the basis produced by someone with an articulation disorder with a well-ordered basis, the original speech spectrum is replaced with a well-ordered spectrum. In the voice of a person with an articulation disorder, their consonants are often unstable and that makes their voices unclear. Hence, by replacing the articulation disorder basis of consonants only, a voice with an articulation disorder is converted into a clear voice that preserves the individuality of the speaker's voice.

The rest of this paper is organized as follows. Section 2 discusses related works, while Section 3 describes the NMF-based VC method. Section 4 presents the experimental results, and the final section presents the conclusions.

2 Related works

VC is a technique for changing specific information in an input speech while retaining the other information in the utterance such as its linguistic information. Unlike speech synthesis, VC application does not need text input. Many statistical approaches to VC have been studied and applied to various tasks. One of the most popular VC applications is speaker conversion [16]. In speaker conversion, a source speaker's voice individuality is changed to a specified target speaker's so that the input utterance sounds as if it had been spoken by a specified target speaker. The other information, such as its linguistic information or emotional information, is retained. Emotion conversion is a technique for changing emotional information in input speech while maintaining linguistic information and speaker individuality [17,18]. With those approaches, a mapping function is trained in advance using a small amount of training data consisting of utterance pairs consisting of source and target voices.

A Gaussian mixture model (GMM)-based approach is widely used for VC because of its flexibility and good performance [19]. The conversion function is interpreted as the expectation value of the target spectral envelope. The conversion parameters are evaluated using minimum mean-square error (MMSE) using a parallel training set. A number of improvements to this approach have been proposed. Toda et al. [16] introduced the global variance (GV) of the converted spectra over time sequence. Helander et al. [20] proposed transforms based on partial least squares (PLS) in order to prevent the over-fitting problem of standard multivariate regression. However, over-smoothing and over-fitting problems in these GMM-based approaches have been reported [20] because of statistical averages and large number of parameters. These problems degrade the quality of synthesized speech. There have also been approaches that do not require parallel data using GMM adaptation techniques [21] or eigen-voice GMM (EV-GMM) [22].

However, these approaches have been developed for speaker conversion. If the person with an articulation disorder is set as a source speaker and a physically unimpaired person is set as a target speaker, an articulation disorder voice may be converted into a well-ordered voice, but the source speaker's voice individuality is also converted into the target speaker's individuality.

In the field of assistive technology, Nakamura et al. [23,24] proposed GMM-based VC systems that reconstruct a speaker's individuality in electrolaryngeal speech and speech recorded by NAM microphones. These systems are effective for electrolaryngeal speech and speech recorded by NAM microphones, but the target speaker's individuality will be changed to the source speaker's individuality. Veaux et al. [25] used HMM-based speech synthesis to reconstruct the voice of individuals with degenerative speech disorders. HMM-based speech synthesis needs text input to synthesize speech. In the case of people with an articulation disorder resulting from athetoid cerebral palsy, it is difficult for them to input text because of their athetoid symptoms.

The goals of this study can be divided into in following three points: (1) convert the voice uttered by a person with an articulation disorder so that everyone can understand what he/she said, (2) preserve the individuality of the speaker's voice, and (3) output a natural-sounding voice. Our proposed exemplar-based VC can create a natural-sounding voice because there is no statistical model in our approach, and the source speaker's individuality can be preserved using our individuality-preserving dictionary.

3 Voice conversion based on NMF

3.1 Exemplar-based voice conversion

Figure 1 shows the basic approach of our exemplar-based VC using NMF. L and J represent frames of dictionary and basis of dictionary, respectively. D represents the number of dimensions of the source feature, which consists of a segment feature of the source speaker's spectrum, and d represents the number of dimensions of a target feature, which consists a single feature of the target speaker's spectrum.

A dictionary is a collection of source or target basis. Our VC method needs two dictionaries that are phonemically parallel. One dictionary is a source dictionary, which is constructed from a source feature. The source feature is constructed from an articulation-disordered spectrum and its segment feature which consists of some consecutive frames. We use a segment feature in order to consider temporal information in activity estimation. The other dictionary is a target dictionary, which is constructed from a target feature. The target feature is mainly constructed from a well-ordered spectrum. These two dictionaries consist of the same words and are aligned with dynamic time warping (DTW). Hence, these dictionaries have the same number of bases.

Input source features, $\mathbf{X}^s$, which consist of an articulation-disordered spectrum and its segment features, are decomposed into a linear combination of bases from the source dictionary $\mathbf{A}^s$ by NMF. The weights of the basis are estimated as an activity $\mathbf{H}^s$. Therefore, the activity includes the weight information of input features for each basis.

Then, the activity is multiplied by a target dictionary in order to obtain converted spectral features $\hat{\mathbf{X}}^t$ which are represented by a linear combination of bases from the target dictionary. Because the source and target dictionaries are parallel phonemically, the basis used in the converted

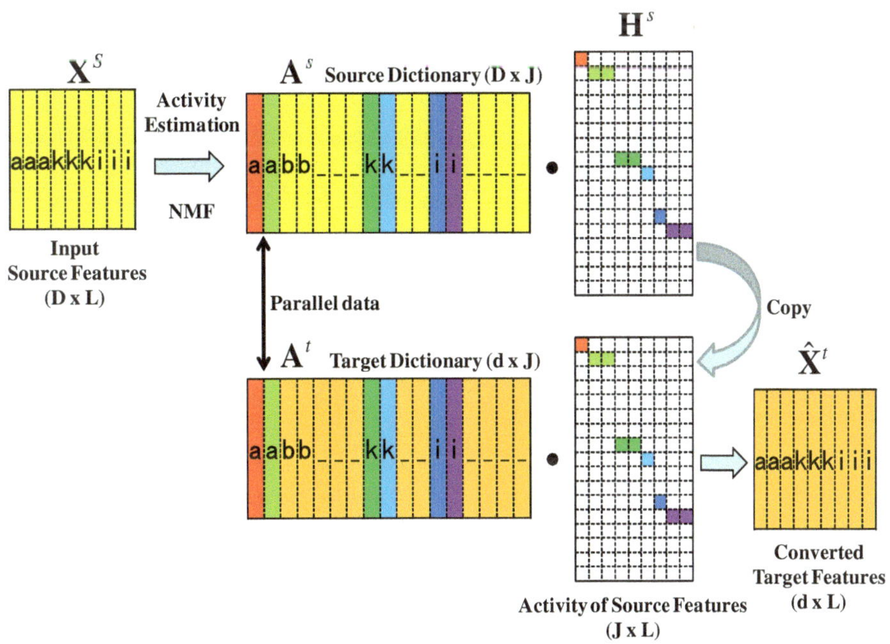

Figure 1 Basic approach of NMF-based voice conversion.

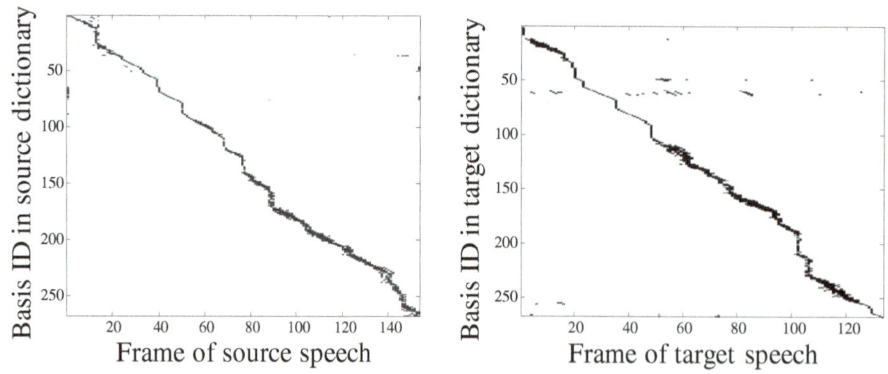

Figure 2 Activity matrices for the articulation-disordered utterance (left) and well-ordered utterance (right).

features is phonemically the same as that of the source features.

Figure 2 shows an example of the activity matrices estimated from the word *ikioi* ('vigor' in English). One was uttered by a person with an articulation disorder and the other by a physically unimpaired person. To show an intelligible example, each dictionary was structured from just the one word *ikioi* and aligned with DTW. As shown in Figure 2, these activities have high energies at similar elements. For this reason, when there are parallel dictionaries, it may be possible to substitute the activity of the source features estimated with the source dictionary with that of the target features. Therefore, the target speech can be constructed using the target dictionary and the activity of the source signal as shown in Figure 1.

Spectral envelopes extracted by STRAIGHT analysis [26] are used in the source and target features. The other features extracted by STRAIGHT analysis, such as F0 and the aperiodic components, are used to synthesize the converted signal without any conversion.

3.2 Preserving the individuality of the speaker's voice

In order to make a parallel dictionary, some pairs of parallel utterances are needed, where each pair consists of the same text. One is spoken by a person with an articulation disorder (source speaker), and the other is spoken by a physically unimpaired person (target speaker).

The left side of Figure 3 shows the process for constructing a parallel dictionary. Spectrum envelopes, which are extracted from parallel utterances, are phonemically aligned. In order to estimate activities of source features precisely, segment features, which consist of some consecutive frames, are constructed. Target features are constructed from consonant frames of the target's aligned spectrum and vowel frames of the source's aligned spectrum. Source and target dictionaries are constructed by

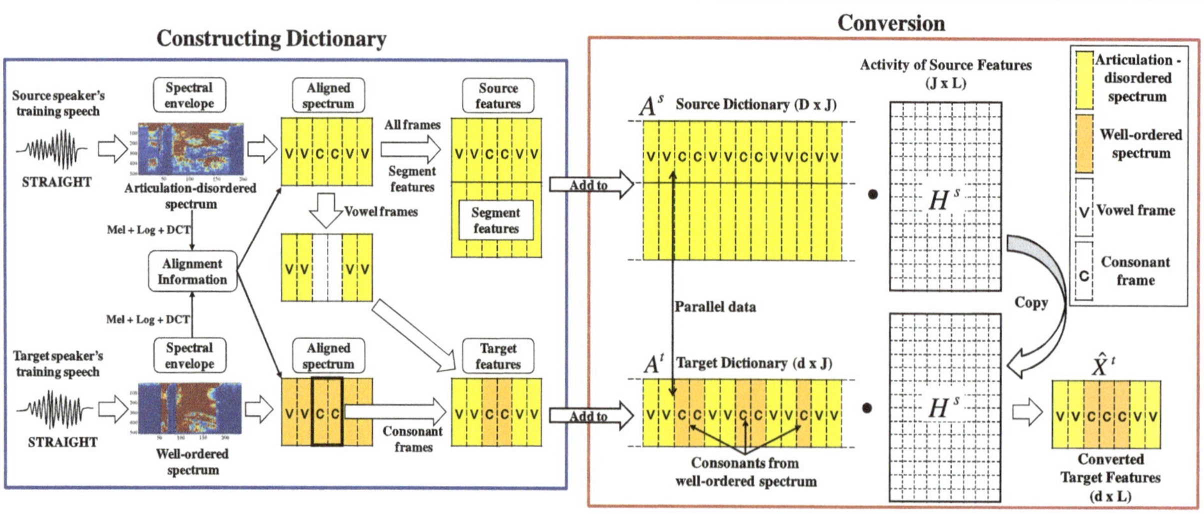

Figure 3 Individuality-preserving voice conversion.

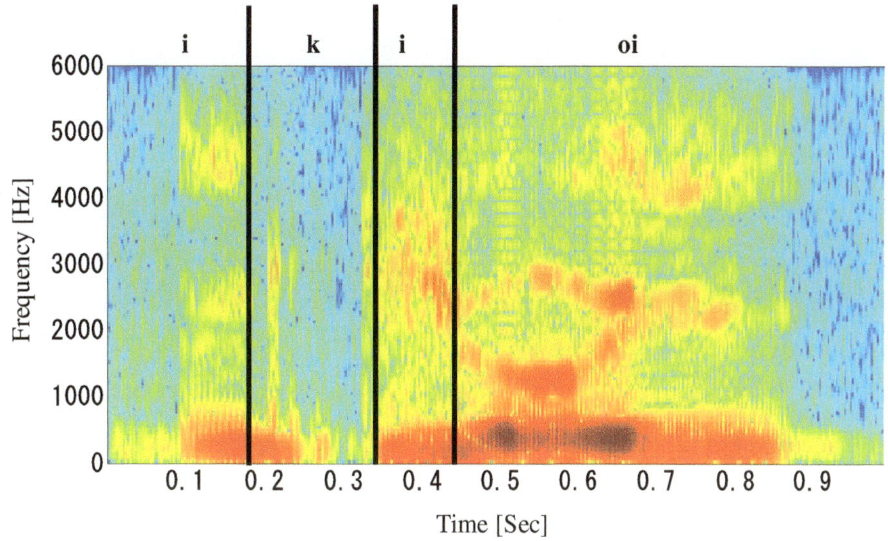

Figure 4 **Example of a spectrogram spoken by a person with an articulation disorder i k i oi.**

lining up each of the features extracted from parallel utterances.

Figure 3 shows how to preserve a source speaker's voice individuality in our VC. The vowels voiced by the speaker strongly indicate the speaker's individuality. On the other hand, consonants of people with articulation disorders are often unstable. Figure 4 shows an example of the spectrogram for the word *ikioi* (vigor in English) of a person with an articulation disorder. The spectrogram of a physically unimpaired person speaking the same word is shown in Figure 5. In Figure 4, the area labeled 'k' is not clear, compared to the same region in Figure 5. By combining a source speaker's vowels and target speaker's consonants in the target dictionary, the individuality of the source speaker's voice can be preserved.

3.3 Estimation of activity

In the NMF-based approach, the spectrum source signal at frame l is approximately expressed by a nonnegative linear combination of the source dictionary and the activities:

$$\begin{aligned}\mathbf{x}_l &= \mathbf{x}_l^s \\ &\approx \sum_{j=1}^{J} \mathbf{a}_j^s h_{j,l}^s \\ &= \mathbf{A}\mathbf{h}_l \quad \text{s.t.} \quad \mathbf{h}_l \geq 0\end{aligned} \tag{2}$$

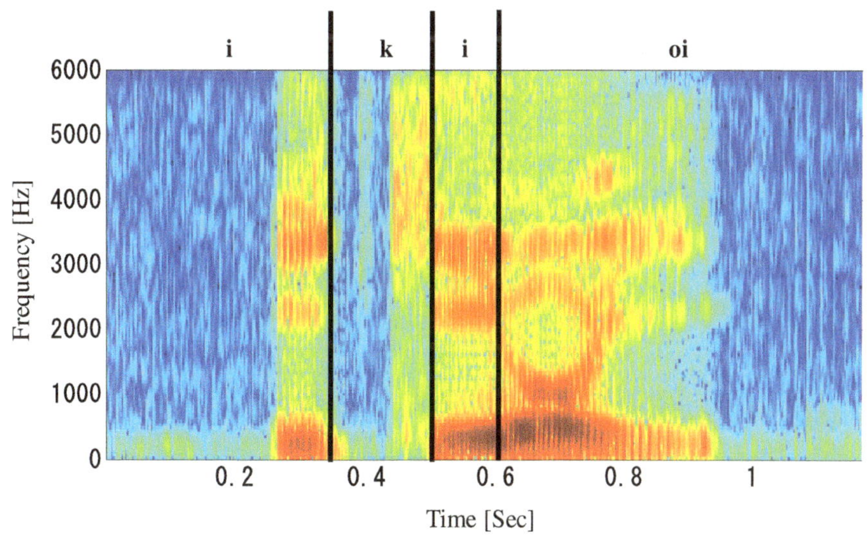

Figure 5 **Example of a spectrogram spoken by a physically unimpaired person i k i oi.**

where $\mathbf{x}_l^s \in \mathbf{X}^s$ is the magnitude spectra of the source signal. Given the spectrogram, Equation 2 can be written as follows:

$$\mathbf{X}^s \approx \mathbf{A}^s\mathbf{H}^s \quad \text{s.t.} \quad \mathbf{H}^s \geq 0. \tag{3}$$

The joint matrix $\mathbf{H}^s$ is estimated based on NMF with the sparse constraint that minimizes the following cost function:

$$d(\mathbf{X}^s, \mathbf{A}^s\mathbf{H}^s) + ||(\lambda\mathbf{1}^{1\times L}).*\mathbf{H}^s||_1 \quad \text{s.t.} \quad \mathbf{H}^s \geq 0 \tag{4}$$

where $\mathbf{1}$ is an all-one matrix and $.*$ denotes element-wise multiplication, respectively. The first term is the Kullback-Leibler (KL) divergence between $\mathbf{X}^s$ and $\mathbf{A}^s\mathbf{H}^s$. The second term is the sparse constraint with the L1-norm regularization term that causes $\mathbf{H}^s$ to be sparse. The weights of the sparsity constraints can be defined for each exemplar by defining $\lambda^T = [\lambda_1 \ldots \lambda_J]$.

$\mathbf{H}^s$ minimizing Equation 4 is estimated iteratively applying the following update rule [12,27]:

$$\mathbf{H}_{n+1}^s = \mathbf{H}_n^s.*(\mathbf{A}^{sT}(\mathbf{X}^s./(\mathbf{A}^s\mathbf{H}^s))) \\ ./(\mathbf{A}^{sT}\mathbf{1}^{D\times L} + \lambda\mathbf{1}^{1\times L}). \tag{5}$$

To increase the sparseness of $\mathbf{H}^s$, elements of $\mathbf{H}^s$, which are less than threshold, are rounded to zero.

Using the activity and the target dictionary, the converted spectral features are constructed:

$$\hat{\mathbf{X}}^t = (\mathbf{A}^t\mathbf{H}^s). \tag{6}$$

4 Experiments

4.1 Experimental conditions

The proposed method was evaluated on word-based VC for one person with an articulation disorder. We recorded 432 utterances (216 words, repeating each two times) included in the ATR Japanese speech database [28]. The speech signals were sampled at 16 kHz and windowed with a 25-ms Hamming window every 10 ms. A physically unimpaired Japanese male in the ATR Japanese speech database was chosen as a target speaker.

The source feature is a 2,565-dimensional segment spectrum, and the target feature is a 513-dimensional single spectrum. Those spectra are extracted by STRAIGHT analysis. The mel-cepstral coefficient, which is converted from the STRAIGHT spectrum, is used for DTW in order to align the temporal fluctuation.

We compared our NMF-based VC to conventional GMM-based VC. In GMM-based VC, the first through 24th cepstral coefficients extracted by STRAIGHT are used as source and target features.

4.2 Objective evaluation of exemplar-based VC

Mel-cepstral distortion between the target and converted mel-cepstra given by the following equation [29] was used

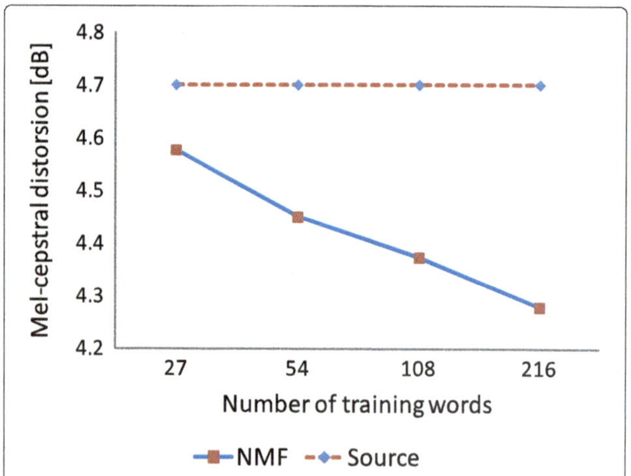

Figure 6 Mel-cepstral distortion as a function of the number of training words.

in order to evaluate the size of training words and the weights of the sparsity constraint:

$$\text{MCD}(v^{\text{conv}}, v^{\text{ref}}) = (\alpha/T)\sum_{t=0}^{T-1}\sqrt{\sum_{d=1}^{24}(v_d^{\text{conv}}(t) - v_d^{\text{ref}}(t))^2} \tag{7}$$

$$\alpha = 10\sqrt{2}/\ln 10 \tag{8}$$

where $v_d^{\text{conv}}(t)$ and $v_d^{\text{ref}}(t)$ are the dth coefficients on frame t of the converted and target mel-cepstra, T is a total number of frames, and α is a scaling factor, respectively.

We selected 50 words at random as an evaluation set and used the target speaker's utterance as a reference. An individuality-preserving dictionary is not suitable for mel-cepstral distortion because vowel frames in converted voices are different from the reference. Therefore, in this

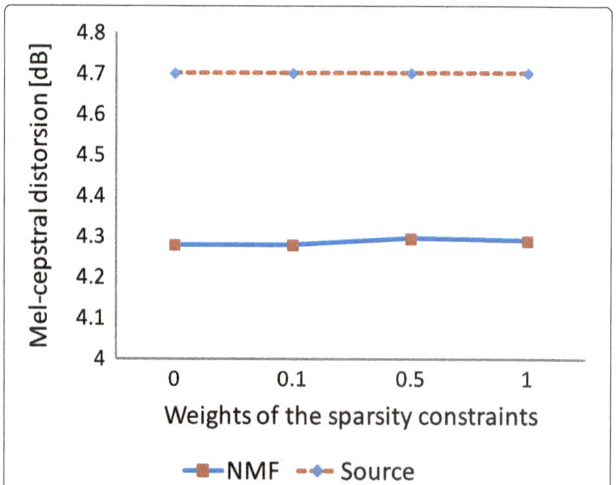

Figure 7 Mel-cepstral distortion as a function of the weights of the sparsity constraints.

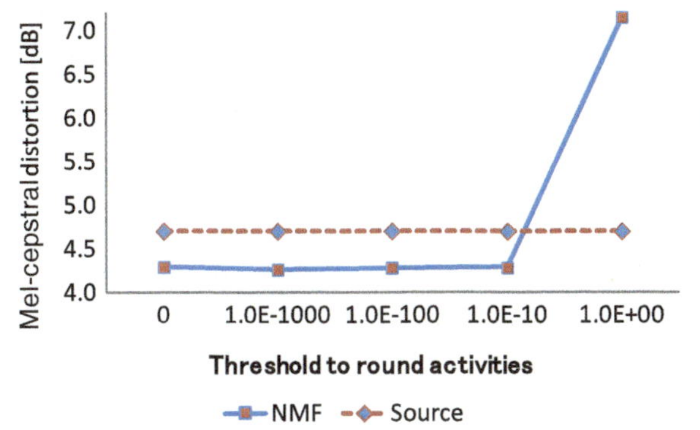

Figure 8 Mel-cepstral distortion as a function of the threshold to round activities.

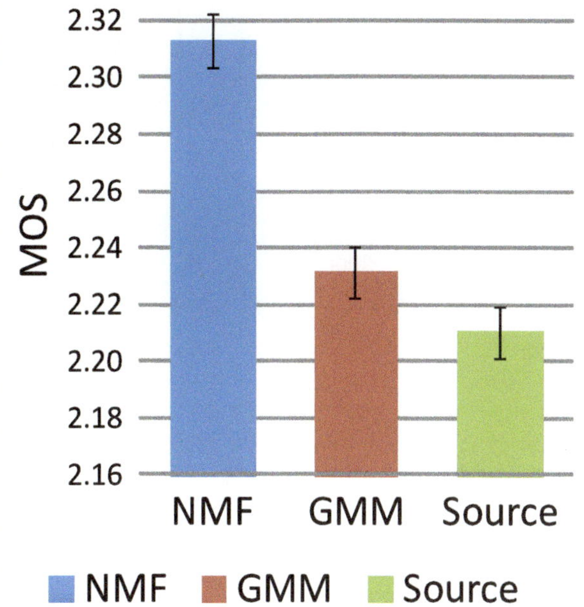

Figure 10 Results of MOS test on speech quality.

subsection, we used a target dictionary that is constructed from the target speaker's vowels and consonants.

Figure 6 shows the mel-cepstral distortions in the evaluation set as a function of the number of training words. The distortion before conversion (shown by the dotted line) is 4.70 dB. With NMF-based VC, there is less distortion as the number of training words increases.

Figure 7 shows the mel-cepstral distortions in the evaluation set as a function of the weights of the sparsity constraint (λ in Equation 4). The distortion between the converted voice and the reference is almost the same despite the increase in the weights of the sparsity constraint.

Figure 8 shows the mel-cepstral distortions in the evaluation set as a function of the threshold to round activities to zero. From Figure 8, the threshold was set to 0.1.

4.3 Subjective evaluation

We used 216 utterances for training and used the other 216 utterances for the test. We conducted subjective evaluation on three topics. A total of ten Japanese speakers took part in the test using headphones. For the 'speech quality' evaluation, we performed a mean opinion score (MOS) test [30]. The opinion score was set to a five-point scale (5 excellent, 4 good, 3 fair, 2 poor, 1 bad). Thirty-two words, which are difficult for a person with an articulation disorder to utter, were evaluated. The subjects were asked about the speech quality in the articulation disordered voice, the NMF-based converted voice, and the GMM-based converted voice. Each voice uttered by a physically

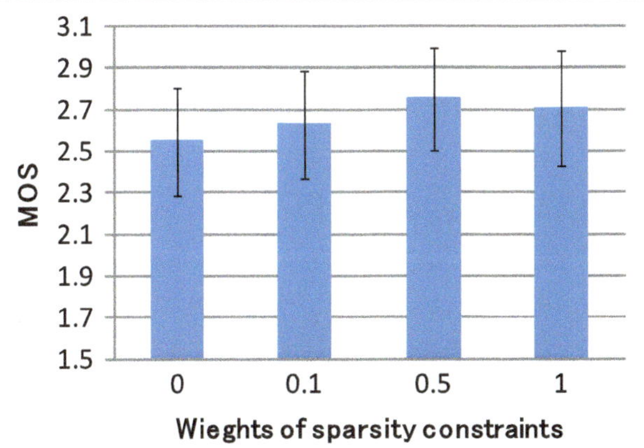

Figure 9 Results of MOS test on speech quality as a function of the weights of the sparsity constraints.

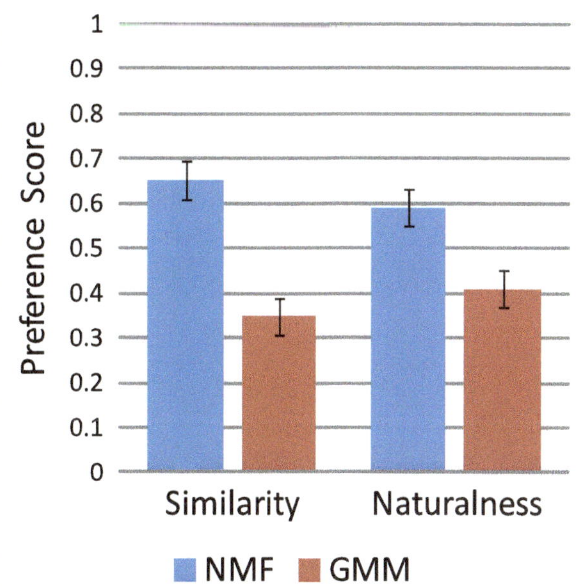

Figure 11 Results of MOS test on the similarity to the source speaker and naturalness.

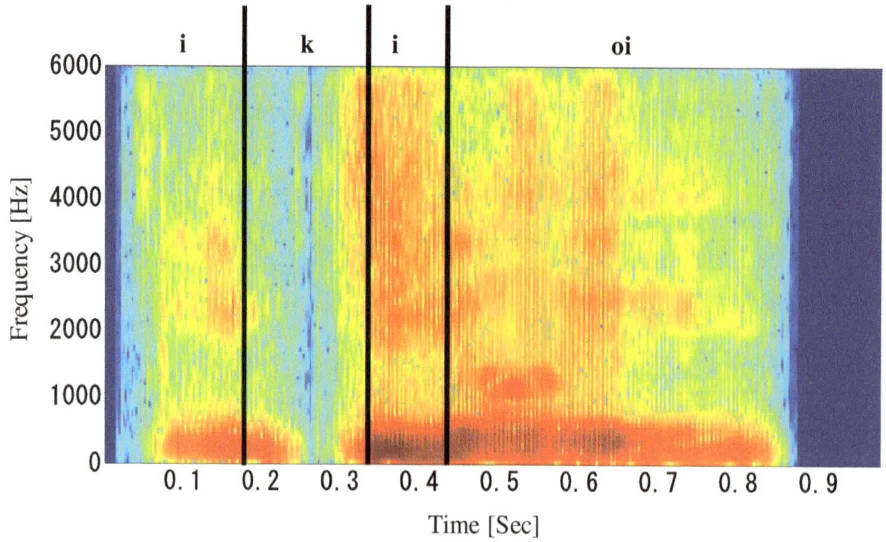

Figure 12 Example of a spectrogram converted by NMF-based VC i k i oi.

unimpaired person was presented as a reference of five points on the MOS test.

Fifty words were converted using NMF-based VC and GMM-based VC for the following evaluations. On the 'similarity' evaluation, the XAB test was carried out. In the XAB test, each subject listened to the articulation-disordered voice. Then the subject listened to the voice converted by the two methods and selected which sample sounded most similar to the articulation-disordered voice. On the 'naturalness' evaluation, a paired comparison test was carried out, where each subject listened to pairs of speech converted by the two methods and selected which sample sounded more natural.

4.4 Results and discussion

Figure 9 shows the results of the MOS test on the speech quality of the words converted by NMF-based VC, where the weights of the sparsity constraints are marked on the horizontal axis. The error bars show a 95% confidence score. The p value test of 0.05 showed that there are no significant differences in these scores. In this paper, all elements for λ in Equation 5 were set to 0.5 from Figure 9.

Figure 10 shows the results on the MOS test on speech quality. The error bars show a 95% confidence score. The difference of MOS between NMF-based VC and GMM-based VC was confirmed by a p value test of 0.05. Moreover, the p value test of 0.05 showed that there are

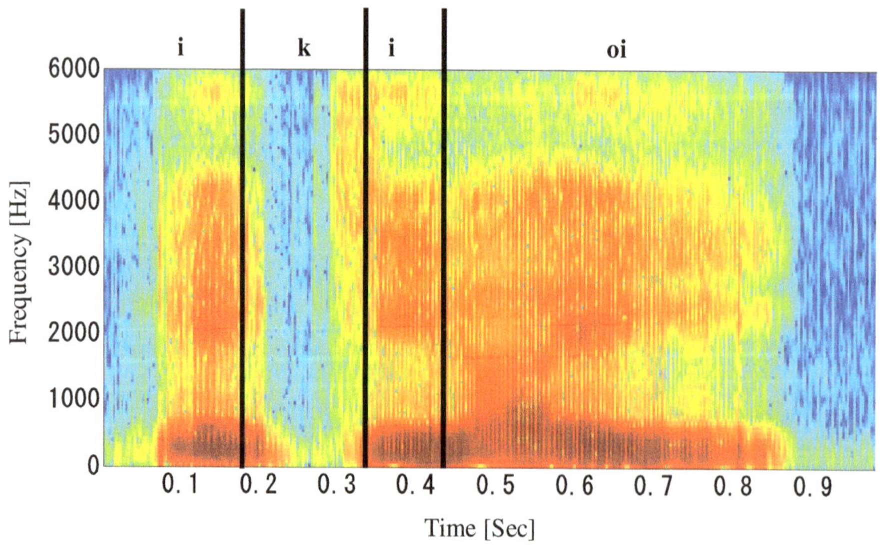

Figure 13 Example of a spectrogram converted by GMM-based VC i k i oi.

no significant differences between GMM-based VC and utterances of a person with an articulation disorder. The difference between NMF-based VC and an articulation-disordered utterance was significant at the 0.1 level. As shown in Figure 10, NMF-based VC obtained a higher score than GMM-based VC on the speech quality test. This is because GMM-based VC creates conversion noise. NMF-based VC also creates some conversion noise, but it is less than that created by GMM-based VC. Using the individuality-preserving dictionary, we can reduce the effect of noise in the vowel portion.

Figure 11 shows the results of the XAB test on the similarity to the source speaker and naturalness of the converted voice. The error bars show a 95% confidence score. The NMF-based VC obtained a higher score than the GMM-based conversion on similarity because NMF-based conversion uses an individuality-preserving dictionary. The NMF-based VC also obtained a higher score than the GMM-based conversion on naturalness although NMF-based conversion mixed the source speaker's vowels and target speaker's consonants. The results of this test were confirmed by a p value test of 0.05.

Figure 12 shows an example of an NMF-based converted spectrogram of *ikioi*. The same word converted by GMM-based VC is also shown in Figure 13. In Figure 12, the 'oi' area is similar to the same area in Figure 4, as compared to Figure 13. Using the individuality-preserving dictionary, NMF-based VC can keep naturalness in the vowel portion because our method converts consonants only.

5 Conclusions

We proposed a spectral conversion method based on NMF for a voice with an articulation disorder. By combining articulation-disordered vowels and well-ordered consonants, we constructed an individuality-preserving dictionary. Experimental results demonstrated that our VC method can improve the listening speech quality of the words uttered by a person with an articulation disorder.

Our proposed method has the following benefits compared to conventional GMM-based VC: (1) NMF-based VC can preserve the individuality of the source speaker's voice using an individuality-preserving dictionary, and (2) our NMF-based VC can create a natural voice because the individuality-preserving dictionary keeps the source speaker's vowels.

Some problems remain with this method. Articulation-disordered speech has a co-articulation effect between phonemes. As shown in Figure 4, the mispronunciation of consonants also affects the vowels that follow. Solving this problem will be the focus of our future work. In this study, there was only one subject person, so in future experiments, we will increase the number of subjects and further examine the effectiveness of our method.

Competing interests

The authors declare that they have no competing interests.

Acknowledgements

This research was supported in part by MIC SCOPE.

Author details

[1]Graduate School of System Informatics, Kobe University, 1-1 Rokkodai, Nada-ku, Kobe 657-8501, Japan. [2]Organization of Advanced Science and Technology, Kobe University, 1-1 Rokkodai, Nada-ku, Kobe 657-8501, Japan.

References

1. J Lin, W Ying, TS Huang, Capturing human hand motion in image sequences, in *IEEE Workshop on Motion and Video Computing*, Orlando, 5-6 Dec 2002 (IEEE, Piscataway, 2002), pp. 99–104
2. T Starner, J Weaver, A Pentland, Real-time American sign language recognition using desk and wearable computer based video. IEEE Trans. Pattern Anal. Mach. Intell. **20**(12), 1371–1375 (1998)
3. G Fang, W Gao, D Zhao, Large vocabulary sign language recognition based on hierarchical decision trees. 5th International Conference on Multimodal Interfaces. **34**(3), 125–131 (2004)
4. N Ezaki, M Bulacu, L Schomaker, Text detection from natural scene images: towards a system for visually impaired persons. Proceedings of the 17th International Conference on Pattern Recognition (ICPR 2004). **2**, 683–686 (2004)
5. MK Bashar, T Matsumoto, Y Takeuchi, H Kudo, N Ohnishi, Unsupervised texture segmentation via wavelet-based locally orderless images (WLOIs) and SOM, in *Computer Graphics and Imaging* (IASTED/ACTA Press, Calgary, 2003), pp. 279–284
6. V Wu, R Manmatha, EM Riseman, Textfinder: an automatic system to detect and recognize text in images. IEEE Trans. Pattern Anal. Mach. Intell. **21**(11), 1224–1229 (1999)
7. K Yabu, T Ifukube, S Aomura, A basic design of wearable speech synthesizer for voice disorders [Japanese]. EIC Technical report (Institute Electron. Inf. Commun. Eng). **105**(686), 59–64 (2006)
8. ST Canale, WC Campbell, *Campbell's Operative Orthopaedics*, vol. 4 (Mosby-Year Book, St. Louis, 2002)
9. H Matsumasa, T Takiguchi, Y Ariki, I Li, T Nakabayachi, Integration of metamodel and acoustic model for dysarthric speech recognition. J. Multimedia. **4**(4), 254–261 (2009)
10. C Miyamoto, Y Komai, T Takiguchi, Y Ariki, I Li, Multimodal speech recognition of a person with articulation disorders using AAM and MAF, in *IEEE International Workshop on Multimedia Signal Processing (MMSP'10)*, St. Malo, 4–6 Oct 2010 (IEEE, Piscataway, 2010), pp. 517–520
11. A Maier, T Haderlein, F Stelzle, E Noth, E Nkenke, F Rosanowski, A Schutzenberger, M Schuster, Automatic speech recognition systems for the evaluation of voice and speech disorders in head and neck cancer. EURASIP J. Audio Speech Music Process. **2010**, 926951 (2010)
12. D Lee, HS Seung, Algorithms for non-negative matrix factorization, in *Advances in Neural Information Processing 13 (NIPS 2000)* (MIT Press, Massachusetts, 2001), pp. 556–562
13. T Virtanen, Monaural sound source separation by non-negative matrix factorization with temporal continuity and sparseness criteria. IEEE Trans. Audio Speech Lang. Process. **15**(3), 1066–1074 (2007)
14. JF Gemmeke, T Virtanen, Noise robust exemplar-based connected digit recognition, in *2010 IEEE International Conference on Acoustics Speech and Signal Processing (ICASSP)*, Dallas, 14–19 March 2010 (IEEE, Piscataway, 2010), pp. 4546–4549
15. MN Schmidt, RK Olsson, Single-channel speech separation using sparse non-negative matrix factorization, in *Interspeech 2006 - ICSLP, Ninth International Conference on Spoken Language Processing*, Pittsburgh, 17–21 Sept 2006 (Curran Associates, Inc., New York, 2006), pp. 2614–2617
16. T Toda, A Black, K Tokuda, Voice conversion based on maximum likelihood estimation of spectral parameter trajectory. IEEE Trans. Audio Speech Lang. Process. **15**(8), 2222–2235 (2007)
17. Y Iwami, T Toda, H Saruwatari, K Shikano, GMM-based voice conversion applied to emotional speech synthesis. IEEE Trans. Speech Audio Process. **7**, 2401–2404 (1999)

18. R Aihara, R Takashima, T Takiguchi, Y Ariki, GMM-Based emotional voice conversion using spectrum and prosody features. Am. J. Signal Process. **2**(5), 135–138 (2012)
19. Y Stylianou, O Cappe, E Moulines, Continuous probabilistic transform for voice conversion. IEEE Trans. Speech Audio Process. **6**(2), 131–142 (1998)
20. E Helander, T Virtanen, J Nurminen, M Gabbouj, Voice conversion using partial least squares regression. IEEE Trans. Audio Speech Lang. Process. **18**(5), 912–921 (2010)
21. CH Lee, CH Wu, Map-based adaptation for speech conversion using adaptation data selection and non-parallel training, in *Interspeech 2006 - ICSLP, Ninth International Conference on Spoken Language Processing,* Pittsburgh, 17–21 Sept 2006 (Curran Associates, Inc., New York, 2006), pp. 2254–2257
22. T Toda, Y Ohtani, K Shikano, Eigenvoice conversion based on Gaussian mixture model, in *Interspeech 2006 - ICSLP, Ninth International Conference on Spoken Language Processing,* Pittsburgh, 17–21 Sept 2006 (Curran Associates, Inc., New York, 2006), pp. 2446–2449
23. K Nakamura, T Toda, H Saruwatari, K Shikano, Speaking-aid systems using GMM-based voice conversion for electrolaryngeal speech. Speech Commun. **54**(1), 134–146 (2012)
24. K Nakamura, T Toda, H Saruwatari, K Shikano, Speaking aid system for total laryngectomees using voice conversion of body transmitted artificial speech, in *Interspeech 2006 - ICSLP, Ninth International Conference on Spoken Language Processing,* Pittsburgh, 17–21 Sept 2006 (Curran Associates, Inc., New York, 2006), pp. 1395–1398
25. C Veaux, J Yamagishi, S King, Using HMM-based speech synthesis to reconstruct the voice of individuals with degenerative speech disorders, in *13th Annual Conference of the International Speech Communication Association 2012 (INTERSPEECH 2012),* Portland, 9–13 September 2012 (Curran Associates, Inc. New York, 2012), pp. 966–969
26. H Kawahara, I Masuda-Katsuse, A Cheveigne, Restructuring speech representations using a pitch-adaptive time-frequency smoothing and an instantaneous-frequency based F0 extraction: possible role of a repetitive structure in sounds. Speech Commun. **27**(3–4), 187–207 (1999)
27. JF Gemmeke, T Viratnen, A Hurmalainen, Exemplar-based sparse representations for noise robust automatic speech recognition. IEEE Trans. Audio Speech Lang. Process. **19**(7), 2067–2080 (2011)
28. A Kurematsu, K Takeda, Y Sagisaka, S Katagiri, H Kuwabara, K Shikano, ATR Japanese speech database as a tool of speech recognition and synthesis. Speech Commun. **9**, 357–363 (1990)
29. J Kominek, T Schultz, AW Black, Synthesizer voice quality of nwe languages calibrated with mean mel cepstral distortion, in *The International Workshop on Spoken Language Technology for Under-Resourced Languages (SLTU)* (Hanoi University of Technology, Hanoi, 5–7 May 2008)
30. International Telecommunication Union, *ITU-T Recommendation P.800-P.899: Methods for Objective and Subjective Assessment of Quality* (ITU, Geneva, 2003)

5

Musical note analysis of solo violin recordings using recursive regularization

Yi-Ju Lin, Tien-Ming Wang*, Ta-Chun Chen, Yin-Lin Chen, Wei-Chen Chang and Alvin WY Su

Abstract

Composers may not provide instructions for playing their works, especially for instrument solos, and therefore, different musicians may give very different interpretations of the same work. Such differences usually lead to time, amplitude, or frequency variations of musical notes in a phrase in the signal point of view. This paper proposes a frame-based recursive regularization method for time-dependent analysis of each note presenting in solo violin recordings. The system of equations evolves when a new frame is added and an old frame is dropped to track the varying characteristics of violin playing. This method is compared with a time-dependent non-negative matrix factorization method. The complete recordings of both BWV 1005 No. 3 played by Kuijken and 24 Caprices op. 1 no. 24 in A minor played by Paganini are used for the transcription experiment, where the proposed method performs strongly. The analysis results of a short passage extracted from BWV 1005 No. 3 performed by three famous violinists reveal numerous differences in the styles and performances of these violinists.

Introduction

Analyses of performances are mostly subjective in the domain of musicology. Objective analysis has become possible with advances in information technologies and sound/music analysis tools, such as pitch/partial tracking, score alignment/following, melody tracking, and extraction. Non-negative matrix factorization (NMF) [1] is a popular tool for musical signal analysis such as pitch estimation, chord recognition, and automatic transcription [2-4]. In NMF, the matrix of the input magnitude spectrum is decomposed into the product of two matrices. One matrix is formed by a certain number of magnitude spectra and is called the template matrix or dictionary matrix. The other matrix is the intensity information of the notes and is called the intensity matrix or activation matrix. When considering the decomposition of audio recordings, these matrices are, for both NMF and the proposed method, related to several notes, each with a quasi-harmonic spectrum activating during a specific time period. Furthermore, NMF is usually used on the Fourier spectrogram which is easy to apply time-frequency masking but hard to extract time-varying sources. Some additional models are needed to enforce the procedure of decomposition, such as time-dependent parametric and harmonic templates [5] and Markov-chained base [6]. On the other hand, decomposing constant-Q spectrograms is difficult to apply time-frequency masking but shows good potential to deal with spreading of higher harmonic frequencies when the pitch is getting higher, such like scale invariance across linear frequency [7] and shift invariance across log-frequency [8]. State-of-the-art methods in these areas can be found in the annual Music Information Retrieval Evaluation eXchange (MIREX) [9].

Good results can be achieved when the number of notes and/or the spectra information of notes is known *a priori*. In the analysis of polyphonic recordings, to determine the number of notes appearing in a single time frame is firstly discussed. A harmonic structure is generally desirable, and the spectral bases are usually constrained to be harmonic in the applications [10]. A fixed number of templates are usually set in previous works according to the note range of interest. Pitches of a violin can, however, vary continually, and fixed pitch templates are unsuitable in the analysis of bowed string instruments. Two issues are then welcome to be discussed in this work: (a) how to determine the exact number of notes and (b) how to model the time-varying notes with suitable templates.

Methods to estimate the possible number of notes have been discussed in [11,12]. In [13], a dynamic note

*Correspondence: showmin@csie.ncku.edu.tw
SCREAM Laboratory, Department of Computer Science Information Engineering, National Cheng-Kung University, Tainan 701, Taiwan

number detection for NMF is proposed to analyze solo bowed string instrument recordings. Since fixed template is employed in [13], a note with a time-varying spectrum which resulted from performing skills such as vibrato and portamento is encouraged to be obtained by using multiple templates. In [5], the time-dependent parametric and harmonic templates are applied to NMF when the pitch of a note varies. The method provides a parametric representation of the harmonic atoms, which can depend on a fundamental frequency parameter, a chirp parameter, and so on with respect to time. It can therefore represent a time-varying note by using only one single template.

Recursive regularization [14] has been widely applied in the areas of system identification, image restoration, noise reduction, echo cancellation, and blind deconvolution [15,16]. The proposed method decomposes the magnitude spectrogram into the product of a template matrix and an intensity matrix based on the modified version of the previous work in high-resolution image reconstruction [16]. In this work, a new algorithm is developed such that the two matrices are updated whenever a new frame is added and an old frame is dropped. This online scheme is similar to the so-called online dictionary learning but does not keep a global dictionary for identified patterns, i.e., musical notes of the same pitch. To analyze violin solo recordings, we regard each note as one single source in this paper. Some works have been proposed for online dictionary learning using L2 norms [17,18], KL divergence [19], and IS divergence [20]. Here, we considered L2 norms to simplify the derivations of the proposed recursive algorithm. Similar to [16], the new iterative update procedure also eliminates the matrix inversion operation to reduce the computational complexity. Because the convergence of the recursive regularization has been well addressed, those who are interested could find related materials in [16]. The systematic flow proposed in the previous work [13] to find a new note template is modified for the application of this work. The concepts of harmonic and sparseness constraint [21,22] are also adopted. The proposed method is compared with the time-dependent NMF method [5] by using the complete recordings of BWV 1005 No. 3 played by Kuijken [23] and 24 Caprices op. 1 no. 24 in A minor played by Paganini [24]. Finally, the proposed method is tested using Bach solo violin recordings by three violinists, that is, Arthur Grumiaux, Sigiswald Kuijken, and Hilary Hahn [23,25,26]. It is easy to identify the differences in their playing styles when note-by-note spectral and intensity information is available. The insightful discussions will be discussed in the 'Results' section.

The remainder of this paper is organized as follows: The 'Formulation of regularized analysis system' section presents the basic formulation of the decomposition problem using the regularization method. The 'Frame-based recursive regularization analysis' section presents the frame-based recursive regularization analysis method. We then present some experiments and corresponding results in the 'Experiments' and 'Results' sections. Lastly, the 'Conclusions' section offers the conclusion and the discussion of future works.

Formulation of regularized analysis system

Let m be the number of consecutive frames. Each frame has $2n$ samples. Discrete Fourier transform (DFT) is performed for each frame to obtain its magnitude information. We obtain an m-by-n matrix, $\mathbf{V}$, where $\mathbf{V}_{ji}$ represents the magnitude of the ith Fourier coefficient of the jth frame. Let r be the number of pitches present in these frames. We obtain an r-by-n template matrix, $\mathbf{W}$, where $\mathbf{W}_{ki}$ represents the magnitude of the ith Fourier coefficient of the kth template. Finally, we obtain an m-by-r intensity matrix, $\mathbf{H}$, where $\mathbf{H}_{jk}$ represents the intensity of the kth template of the jth frame. Hence, $\mathbf{W}$ and $\mathbf{H}$ are used to construct $\mathbf{V}$, as follows:

$$\mathbf{V} = \mathbf{HW}. \tag{1}$$

A cost function can be set as

$$D = \|\mathbf{V} - \mathbf{HW}\|^2. \tag{2}$$

Subsequently, the template matrix and the intensity matrix can be obtained easily as

$$\mathbf{W} = (\mathbf{H}^\top\mathbf{H})^{-1}(\mathbf{H}^\top\mathbf{V}) \tag{3}$$

$$\mathbf{H}^\top = (\mathbf{W}\mathbf{W}^\top)^{-1}(\mathbf{W}\mathbf{V}^\top). \tag{4}$$

The result can be obtained by evaluating (3) and (4) iteratively. Although (1) is similar to NMF in its formulation, (3) and (4) do not enforce the factorization of a non-negative matrix into two non-negative matrices, in comparison to NMF. Since the goal is to get a reasonable distribution of frequency energies, the negative elements of $\mathbf{W}$ and $\mathbf{H}$ can be set to zeros to re-evaluate the equations and obtain a non-negative result in every iteration. Notice that the matrix $\mathbf{V}$ is represented as (1) rather than $\mathbf{V} = \mathbf{WH}$ commonly used in NMF-related literatures to make the derivatives of following formulations more readable without loss of generality.

To improve (3) and (4), a penalty term is added to support a temporal smoothness mechanism in the context of the proposed online scheme. For example, (2) can be modified to

$$J = \|\mathbf{V} - \mathbf{HW}\|^2 + \lambda\|\mathbf{W} - \mathbf{C_W}\|^2 + \gamma\|\mathbf{H} - \mathbf{C_H}\|^2, \tag{5}$$

where $\lambda > 0$ and $\gamma > 0$ are carefully chosen to ensure the stability of the solution. Such a process is called regularization [14]. Since both frequency response and intensity of a note evolve slowly in a short time, $\mathbf{C_W}$ is determined as a template matrix in the previous update iteration, where λ is a corresponding penalty factor to achieve spectral smoothness. Similarly, $\mathbf{C_H}$ and γ are the terms related to temporal smoothness. The solution becomes

$$\mathbf{W} = (\mathbf{H}^\top\mathbf{H} + \lambda\mathbf{I})^{-1}(\mathbf{H}^\top\mathbf{V} + \lambda\mathbf{C_W}), \quad (6)$$

$$\mathbf{H}^\top = (\mathbf{W}\mathbf{W}^\top + \gamma\mathbf{I})^{-1}(\mathbf{W}\mathbf{V}^\top + \gamma\mathbf{C}_\mathbf{H}^\top). \quad (7)$$

In our experience, the system described by (6) and (7) requires a smaller number of iterations than NMF to converge. Furthermore, the system also requires a smaller number of frames than NMF to obtain reasonably good results. The 'Experiments' section presents the simulation results of the proposed method and a comparison to other methods.

The proposed method is designed by considering the following issues. Firstly, it is crucial to determine the exact number of templates to obtain reasonably good results. Such an issue is widely discussed in [13] for conventional NMF-based methods. Secondly, it is crucial to determine in which manner the penalty term is set in (5). Finally, matrix inversion consumes substantial computing power, compared to the gradient descent algorithms used in NMF. These problems are discussed in the following section.

Frame-based recursive regularization analysis

Refined update rules

This section presents the high computation complexity problem caused by matrix inversions. The template and intensity matrices can be adaptively learned from the current input frame and some previous input frames. For brevity, we present only the derivation to recursively evaluate the time-varying template matrix $\mathbf{W}$. The derivation of the update rule for the intensity matrix $\mathbf{H}$ is omitted. Let the system start with the first m input frames. m must not be large when the signal varies rapidly. Let the r-by-n template matrix for the lth input frame be denoted by $\mathbf{W}(l)$, which is obtained using frame-l, frame-$(l-1), \ldots$, and frame-$(l-m+1)$. $\mathbf{V}(l)$ and $\mathbf{H}(l)$ are subsequently defined as

$$\mathbf{V}(l) = [\mathbf{v}(l-m+1)\mathbf{v}(l-m+2)\mathbf{v}(l-m+3)\ldots\mathbf{v}(l)]^\top \quad (8)$$

$$\mathbf{H}(l) = [\mathbf{h}(l-m+1)\mathbf{h}(l-m+2)\mathbf{h}(l-m+3)\ldots\mathbf{h}(l)]^\top, \quad (9)$$

where

$$\mathbf{v}(q) = [v_{q1}v_{q2}v_{q3}\ldots v_{qr}]^\top \quad (10)$$

$$\mathbf{h}(q) = [h_{q1}h_{q2}h_{q3}\ldots h_{qr}]^\top, \quad (11)$$

$l-m+1 \leq q \leq l$. Hence, the cost function in (5) for frame-l is rewritten as

$$J = \|\mathbf{V}(l) - \mathbf{H}(l)\mathbf{W}(l)\|^2 + \lambda\|\mathbf{W}(l) - \mathbf{C_W}(l)\|^2 + \gamma\|\mathbf{H}(l) - \mathbf{C_H}(l)\|^2. \quad (12)$$

Subsequently, the template matrix is obtained as

$$\mathbf{W}(l) = (\mathbf{H}^\top(l)\mathbf{H}(l) + \lambda I)^{-1}(\mathbf{H}^\top(l)\mathbf{V} + \lambda\mathbf{C_W}(l)). \quad (13)$$

Therefore, to obtain the template matrix and the corresponding intensity values for every new input frame, (6) and (7) must be re-evaluated when a new input frame is added. A recursive procedure is proposed to reduce the number of matrix inversions. First, two new matrices are defined as

$$\mathbf{P}(l) = (\mathbf{H}(l)^\top\mathbf{H}(l) + \lambda I)^{-1} \quad (14)$$

$$\mathbf{R}(l) = \mathbf{H}(l)^\top\mathbf{V}(l) + \lambda\mathbf{C_W}(l), \quad (15)$$

where

$$\mathbf{P}(l) = \left(\left[\mathbf{h}^\top(l-m+1)\ldots\mathbf{h}^\top(l)\right] \times \begin{bmatrix}\mathbf{h}^\top(l-m+1)\\ \mathbf{h}^\top(l-m+2)\\ \vdots\\ \mathbf{h}^\top(l)\end{bmatrix} + \lambda I\right)^{-1} = \left(\sum_{k=l-m+1}^{l}\mathbf{h}^\top(k)\mathbf{h}(k) + \lambda I\right)^{-1} \quad (16)$$

$$\mathbf{R}(l) = \left[\mathbf{h}^\top(l-m+1)\ldots\mathbf{h}^\top(l)\right] \times \begin{bmatrix}\mathbf{v}(l-m+1)\\ \mathbf{v}(l-m+2)\\ \vdots\\ \mathbf{h}(l)\end{bmatrix} + \lambda\mathbf{C_W}(l) = \sum_{k=l-m+1}^{l}\mathbf{h}^\top(k)\mathbf{v}(k) + \lambda\mathbf{C_W}(l). \quad (17)$$

Therefore, $\mathbf{W}(l) = \mathbf{P}(l)\mathbf{R}(l)$ and the template matrix for frame-$(l+1)$ can also be calculated by $\mathbf{W}(l+1) = \mathbf{P}(l+1)\mathbf{R}(l+1)$.

Similar to (16) and (17), we obtain

$$\mathbf{P}(l+1) = \left(\left[\mathbf{h}^\top(l-m+2)\ldots\mathbf{h}^\top(l+1)\right] \times \begin{bmatrix}\mathbf{h}(l-m+2)\\ \mathbf{h}(l-m+3)\\ \vdots\\ \mathbf{h}(l+1)\end{bmatrix} + \lambda I\right)^{-1} = \left(\sum_{k=l-m+2}^{l+1}\mathbf{h}^\top(k)\mathbf{h}(k) + \lambda I\right)^{-1} \quad (18)$$

$$\mathbf{R}(l+1) = \left[\mathbf{h}^\top(l-m+2)\ldots\mathbf{h}^\top(l+1)\right] \times \begin{bmatrix} \mathbf{v}(l-m+2) \\ \mathbf{v}(l-m+3) \\ \vdots \\ \mathbf{v}(l+1) \end{bmatrix} + \lambda\mathbf{C_W}(l+1) = \sum_{k=l-m+2}^{l+1} \mathbf{h}^\top(k)\mathbf{v}(k) + \lambda\mathbf{C_W}(l+1). \tag{19}$$

We define

$$\tilde{\mathbf{P}}(l) = \left(\left[0\ \mathbf{h}^\top(l-m+2)\ldots\mathbf{h}^\top(l)\right] \times \begin{bmatrix} 0 \\ \mathbf{h}(l-m+2) \\ \vdots \\ \mathbf{h}(l) \end{bmatrix} + \lambda I \right)^{-1} = \left(\sum_{k=l-m+1}^{l} \mathbf{h}^\top(k)\mathbf{h}(k) + \lambda I - \mathbf{h}^\top(l-m+1)\mathbf{h}(l-m+1) \right)^{-1}. \tag{20}$$

By using the Woodbury matrix identity [27],

$$(\mathbf{A}+\mathbf{BCD})^{-1} = \mathbf{A}^{-1} - \mathbf{A}^{-1}\mathbf{B}(\mathbf{DA}^{-1}\mathbf{B}+\mathbf{C}^{-1})^{-1}\mathbf{DA}^{-1} \tag{21}$$

holds with $\mathbf{A} = \mathbf{P}(l)^{-1} = \sum_{k=l-m+1}^{l} \mathbf{h}^\top(k)\mathbf{h}(k) + \lambda I$, $\mathbf{B} = \mathbf{h}^\top(l-m+1)$, $\mathbf{C} = -1$, and $\mathbf{D} = \mathbf{h}(l-m+1)$. Therefore, we obtain $\tilde{\mathbf{P}}(l)$ without matrix inversion by

$$\tilde{\mathbf{P}}(l) = \mathbf{P}(l) - \frac{\mathbf{P}(l)\mathbf{h}^\top(l-m+1)\mathbf{h}(l-m+1)\mathbf{P}(l)}{\mathbf{h}(l-m+1)\mathbf{P}(l)\mathbf{h}^\top(l-m+1)-1}. \tag{22}$$

Next,

$$\mathbf{P}(l+1) = \left(\sum_{k=l-m+2}^{l} \mathbf{h}^\top(k)\mathbf{h}(k) + \lambda I + \mathbf{h}^\top(l+1)\mathbf{h}(l+1) \right)^{-1}. \tag{23}$$

By using (21) with $\mathbf{A} = \tilde{\mathbf{P}}(l)^{-1} = \sum_{k=l-m+2}^{l} \mathbf{h}^\top(k)\mathbf{h}(k) + \lambda I$, $\mathbf{B} = \mathbf{h}^\top(l+1)$, $\mathbf{C} = +1$, and $\mathbf{D} = \mathbf{h}(l+1)$, we obtain

$$\mathbf{P}(l+1) = \tilde{\mathbf{P}}(l) - \frac{\tilde{\mathbf{P}}(l)\mathbf{h}^\top(l+1)\mathbf{h}(l+1)\tilde{\mathbf{P}}(l)}{\mathbf{h}(l+1)\tilde{\mathbf{P}}(l)\mathbf{h}^\top(l+1)+1}. \tag{24}$$

A matrix inversion is unnecessary when $\mathbf{R}(l+1)$ is achieved by

$$\tilde{\mathbf{R}}(l) = \mathbf{R}(l) - \mathbf{h}^\top(l-m+1)\mathbf{h}(l-m+1) - \lambda\mathbf{C_W}(l) \tag{25}$$

$$\mathbf{R}(l+1) = \tilde{\mathbf{R}}(l) + \mathbf{h}^\top(l+1)\mathbf{h}(l+1) + \lambda\mathbf{C_W}(l+1). \tag{26}$$

The oldest frame is removed by using (22) and (25), and the new input frame is added by using (24) and (26). Hence, the template matrix for each new input frame can be computed recursively without matrix inversion by using the results generated by previous input frames. Since both frequency response and intensity of a note evolve slowly in a short time, $\mathbf{C_W}$ and $\mathbf{C_H}$ are determined

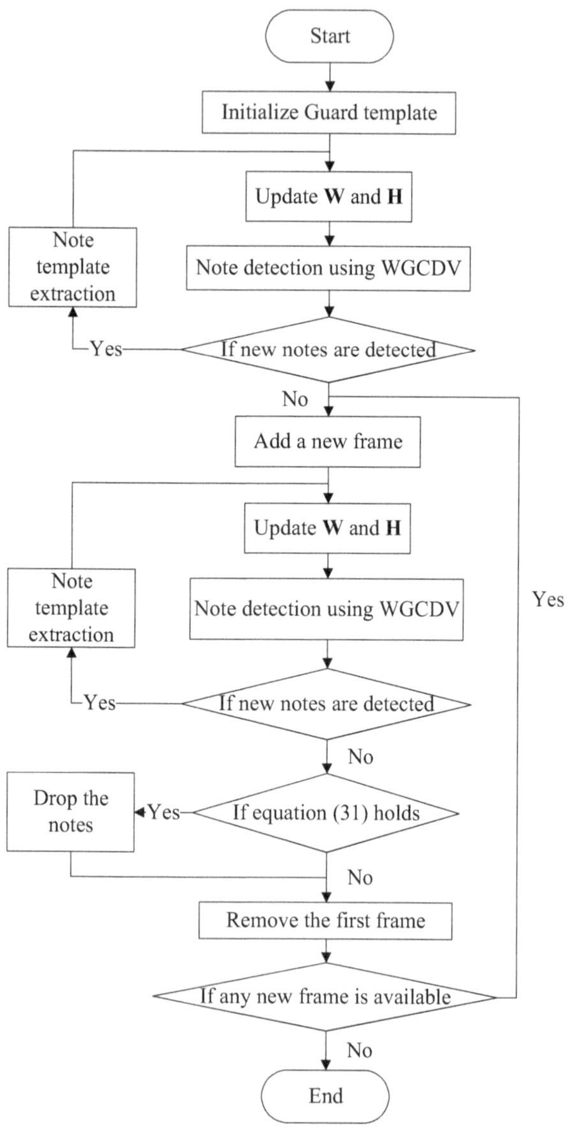

Figure 1 The procedure of the proposed system. The block diagram of the proposed system with the recursive regularization procedure.

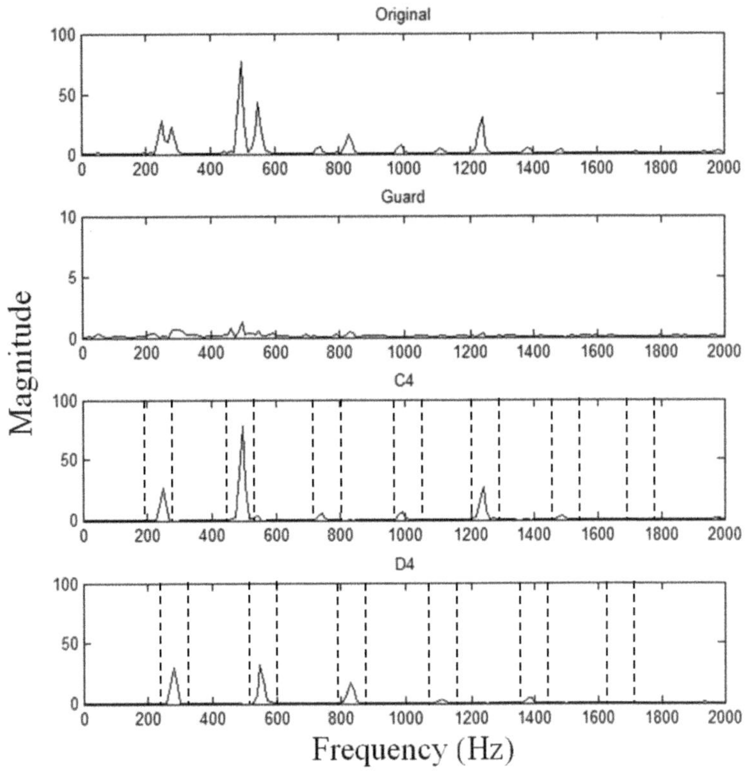

Figure 2 An example of extracted harmonic structures. The original Guard template is separated into C4 and D4 constrained templates and a new Guard template. Sound source: Kuijken's recording on BWV 1005 No. 3 [23].

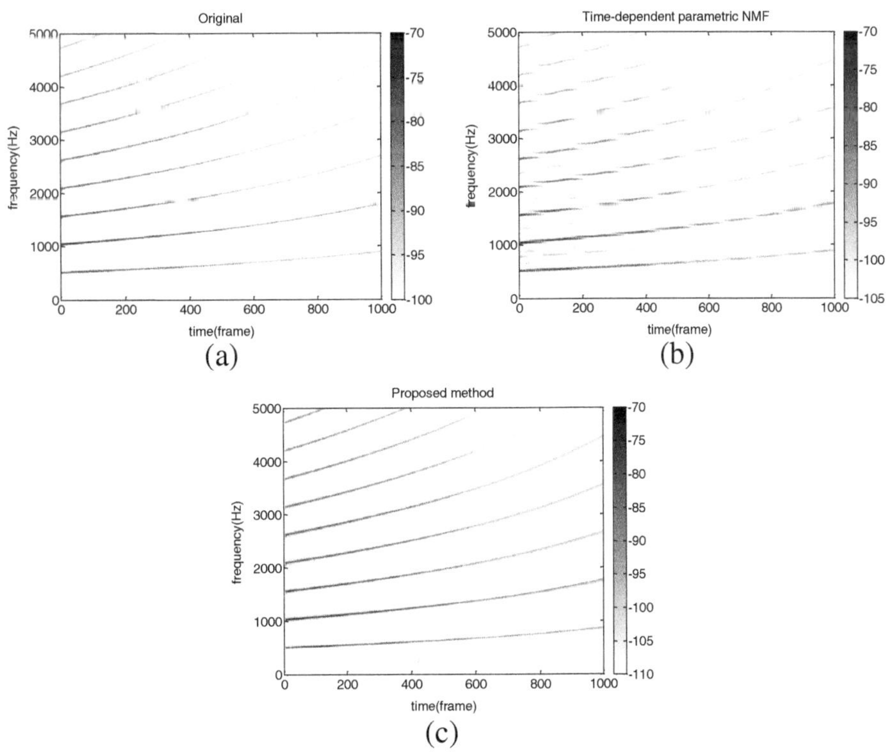

Figure 3 A chirp signal varies from C5 to A5 in 1,000 time frames. (a) The original spectrogram. **(b)** The reconstructed spectrogram of [5]. **(c)** The reconstructed spectrogram of the proposed method.

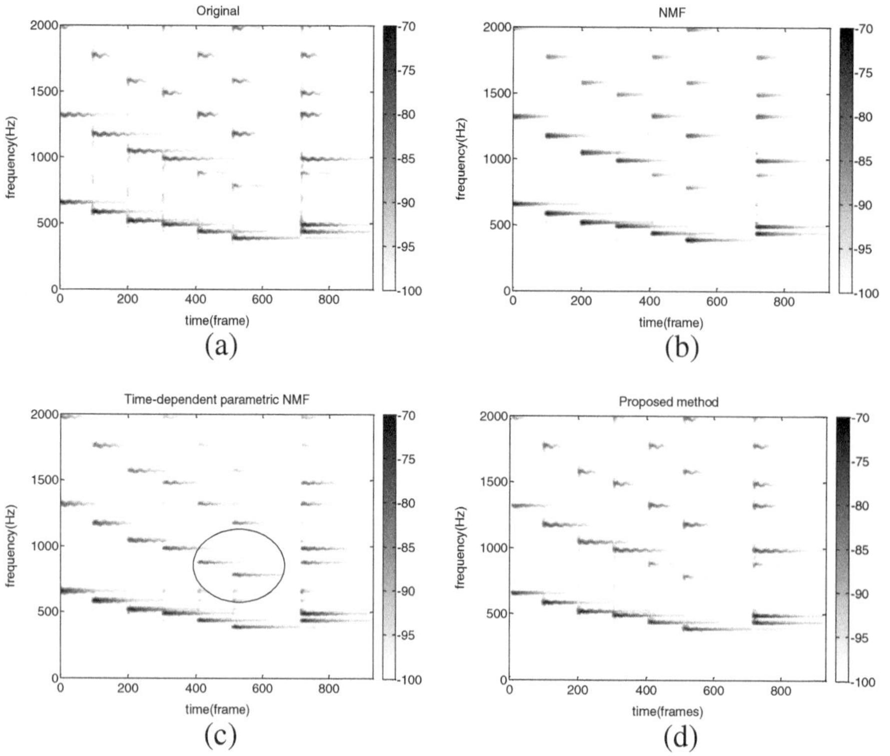

Figure 4 Six vibrating notes presented in the following order: E5, D5, C5, B4, A4, G4, and A4+B4. The spectrograms of **(a)** the original, **(b)** the harmonic-constrained NMF [13], **(c)** time-depend parametric NMF [5], and **(d)** the proposed method of G4, A4, B4, C5, D5, and E5 notes that include vibrato.

by the the results of $\mathbf{W}$ and $\mathbf{H}$ obtained in the previous update iteration, i.e., $\mathbf{C_W}$ is updated by $\mathbf{C_W}(l+1) = \mathbf{W}(l)$.

The intensity matrix is subsequently calculated. Similar to (13), the intensity matrix for the $(l-m+1)$th frame, the $(l-m+2)$th frame, ..., and the lth frame is obtained as

$$\mathbf{H}^\top(l) = \left(\mathbf{W}(l)\mathbf{W}^\top(l) + \gamma I\right)^{-1} (\mathbf{W}(l)\mathbf{V}^\top(l) + \gamma \mathbf{C_H}^\top). \tag{27}$$

Subsequently, the intensity information corresponding to the template matrix of the $(l+1)$th frame is computed as

$$\mathbf{h}^\top(l+1) = \left(\mathbf{W}(l+1)\mathbf{W}^\top(l+1) + \gamma I\right)^{-1} \times \left(\mathbf{W}(l+1)\mathbf{v}^\top(l+1) + \gamma \mathbf{C_H}^\top(l+1)\right). \tag{28}$$

In (28), $\mathbf{C_H}^\top(l+1)$ can be set to $\mathbf{H}^\top(l)$ because it is assumed that the intensity cannot change abruptly. The forgetting factors λ and γ can determine the effects of old frames. They are both set to 100 in this work. The time-varying template matrix and the corresponding intensity matrix can be calculated alternatively when a new input frame is added. Further details and the overall procedure are presented in the following section.

Analysis procedure

As shown in Figure 1, the analysis system started from only one template initialized with random values, which is called Guard template. An initialization loop determines all possible note templates and evaluates the template matrix and the intensity matrix in the same time. After the initialization loop, the main loop takes care of the new frame addition, the new note template detection, the offset note template evaluation, and the old frame removal. For the new note template detection, we used the weighted greatest common divisor and vote (WGCDV) [28] method to detect new tones in Guard template by using a floating point GCD lookup table and a frame-based correction method. WGCDV estimates F0 in three steps: (a) locates the peaks of the frequency response of

Table 1 Objective performance comparison

	SIR	SAR	SDR
NMF with harmonic constraints [13]	26.08	6.19	6.13
Time-dependent parametric NMF [5]	23.06	6.78	6.56
Proposed method	29.08	6.74	6.70

Objective performance comparison of NMF with harmonic constraints [13], time-dependent parametric NMF [5], and the proposed method by using SIR, SAR, and SDR.

Table 2 Transcription results 1

	Accuracy (Acc), %	Precision (*P*), %	Recall (*R*), %	*F*-measure (*F*), %
NMF with harmonic constraints [13]	79.64	86.27	90.97	88.56
Time-dependent parametric NMF [5]	81.94	86.87	93.53	90.08
RR without online scheme [14]	77.45	78.81	97.82	87.29
Proposed method	83.36	97.28	85.35	90.93

Transcription report of the proposed work in the analysis of Kuijken's recording on BWV 1005 No. 3.

the current frame and regards them as possible partials, (b) finds a likely GCD value for each partial pair using a lookup table method, (c) weights the likely GCD values and voting the deterministic GCD according to the spectral energy. Once a harmonic structure is recognized within Guard template at the current frame, these harmonic peaks will be extracted from Guard template and a corresponding new template is added to the original template matrix.

Figure 2 shows an example of extracted harmonic structures of C4 and D4 notes from the original Guard template by using a mask function, S. As shown in Figure 2, each mask function represented in a dashed line can be defined by a harmonic set corresponding to the detected note as follows:

$$S_j = \sum_{p=1}^{N} I(pf_j - \epsilon, pf_j + \epsilon), \tag{29}$$

where $I(\alpha, \beta) = 1$ in the interval $[\alpha, \beta]$; otherwise, it is 0. f_j is the fundamental frequency of the jth recognized tone, and p is the partial index. ϵ is set at 3% of the partial frequency, pf_j.

Subsequently, the new template is computed by

$$\mathbf{W}_j^l = \mathbf{S}_j^l \otimes \mathbf{W}_0^l, \tag{30}$$

where $\mathbf{S}_j^l$ is the mask function of frame-l. In (30), $\mathbf{W}_0^l = [\mathbf{W}_{01}^l \mathbf{W}_{02}^l \ldots \mathbf{W}_{0n}^l]^T$ represents the original Guard template of frame-l, and $\otimes$ is the element-wise multiplication. The number of templates, r, is equal to $j+1$. The procedure described in the previous section is performed again for frame-$(l+1)$ to obtain the new template matrix and intensity matrix.

A re-estimation of the pitch of each note based on the updated template matrix is necessary because all templates, as well as pitches, can vary by frame. Consequently, the mask functions of all templates must be updated. Because each template contains only one harmonic set, $\mathbf{C_W}(l+1)$ in Equation 13 is computed by $\mathbf{S}(l) \otimes \mathbf{W}(l)$, where $\mathbf{S}(l) = \left[\mathbf{S}_0^l \mathbf{S}_1^l \ldots \mathbf{S}_r^l\right]^T$. Based on Equation 5, the iterative update procedure forces $\mathbf{W}(l+1)$ to retain harmonic structures for all the notes as much as possible, depending on the regularization parameter, λ.

Finally, a musical note is muted eventually after it is played for a while. In this study, a note is removed from the analysis process by removing its template and the corresponding intensity information. A note can be removed if

$$\sum_{q=l}^{l-k} |h_{qi}| \le T, 1 \le i \le r. \tag{31}$$

That is, the ith note is removed after frame-l if Equation 31 holds. T is empirically set to 0.1 in this work. By removing such notes, the computation complexity is also reduced.

Experiments

Database

Two excerpts are generated by a MIDI synthesizer for preliminary tests. Firstly, a synthetic chirp signal is generated, and its pitch varies from C5 to A5 in 1,000 time frames. The second test uses a signal with vibrating notes including six notes in the following order: E5, D5, C5, B4, A4, G4, and A4+B4. A vibrato effect is generated by setting proper MIDI commands. All parameter sets are the same as those in the previous test. Moreover, two recordings, the BWV 1005 No. 3 performed by Kuijken [23] and RWC database

Table 3 Transcription results 2

	Accuracy (Acc), %	Precision (*P*), %	Recall (*R*), %	*F*-measure (*F*), %
NMF with harmonic constraints [13]	56.22	57.09	97.64	71.98
Time-dependent parametric NMF [5]	74.11	75.29	97.94	85.13
RR without online scheme [14]	49.27	50.71	94.93	66.01
Proposed method	76.26	85.99	87.08	86.53

Transcription report of the proposed work in the analysis of RWC database C038.

C038 [24], are used to evaluate the accuracy of all methods, as proposed in [29]. The former contains 587 notes that are manually annotated as the ground truth. The latter contains 1,745 notes whose ground truth is provided from the syncRWC annotations [30].

Parameters

The window size is 4,096 samples, the hop size is 256 samples, and the sampling rate is 44.1 kHz. A Hamming window and 4096-FFT are subsequently applied.

Performance evaluation

An objective measure for evaluating the performance of a source separation method proposed in [31] is adopted for the following discussion. To compare different approaches, the signal-to-distortion ratio (SDR), the signal-to-artifact ratio (SAR), and the signal-to-interference ratio (SIR) are computed with each note as the target. In this work, we considered each note as a separate source. The SIR, SAR, and SDR values of these notes are averaged respectively.

To evaluate the performance of music transcription, we chose note-level metrics with a tolerance of one window before and after the reference onset time. Overall accuracy is defined as

$$\text{Acc} = \frac{\text{TP}}{\text{FP} + \text{FN} + \text{TP}}, \tag{32}$$

where TP (true positives) is the number of correctly transcribed voiced notes (within a quarter tone distance in frequency and 50-ms onset distance in time), FP (false positives) is the number of unvoiced notes transcribed as voiced, and FN (false negatives) is the number of voiced notes transcribed as unvoiced. This measure is defined for note-based onset evaluation in this case and is bounded by 0 and 1, with 1 corresponding to perfect transcription. Another metric, called F-measure, is defined as

$$F = 2\frac{P \cdot R}{P + R}, \tag{33}$$

whereas

$$P = \frac{\text{TP}}{\text{TP} + \text{FP}} \tag{34}$$

$$R = \frac{\text{TP}}{\text{TP} + \text{FN}}, \tag{35}$$

where P and R represent the precision rate and recall rate, respectively.

Results

The first preliminary test: glissando

The first part of this section presents a comparison of the proposed method by using a glissando signal to the conventional NMF method with harmonic constraints on matrix $\mathbf{W}$ [32] and the time-dependent parametric

Table 4 Complexity comparison

	Numbers of matrix multiplications	Big-O
Update rules of NMF	$(m-k+1)(2nrk+2r^2n+2r^2k+2rn+2rk)$	$O(mkrn)$
Proposed update rules	$(m-k)(2r^2n+9rn+2r^3+15r^2+4r)$ $+(2r^2n+2krn+4r^3+(2k+2)r^2)$	$O(mr^2n)$

The complexity of the proposed update rule is compared with the NMF update method.

NMF method [5] that uses the same harmonic structure for different notes of the same instrument. For the time-dependent parametric NMF, 10 templates are used to cover the possible pitch range of the signals. For the proposed method, it started with the initialized Guard template, and the required number of templates is determined adaptively during the analysis process. Figure 3a shows the original spectrogram of the note, and Figure 3b,c shows the reconstructed spectrograms that resulted from the time-dependent NMF method and the proposed method, respectively. In [5], the pitch of a template is constrained to vary within a fixed semitone; therefore, a number of discontinuous effects occurred among the templates, as shown in Figure 3b. Moreover, 10 note templates are required for the time-dependent NMF approach in this case. Because the pitch of a template is allowed to vary freely in the proposed method, only one template is required to represent the chirp signal. This accounts for the superior performance of the proposed method in which the signal varies more than one semitone.

The second preliminary test: vibrato

In this case, a synthetic signal with six vibrating notes appearing in the following order, E5, D5, C5, B4, A4, G4, and A4+B4, is used. Figure 4 shows the simulation results. Figure 4b shows that the harmonic-constrained NMF is limited in presenting the signal with frequency modulation because of the consistency of the templates. As discussed in [13], separation performance can be improved by increasing the number of templates for each note. The time-dependent parametric NMF performs efficiently in signal that includes vibrato effects; the result is shown in Figure 4c. The only disadvantage is that the amplitudes of the partials are not determined independently

Figure 5 BWV 1005 No. 3. Five notes in the bracket are analyzed.

from time to time, as shown in the circled region of Figure 4c. The amplitude set trends toward the average among all played notes. The spectrogram obtained from the proposed method is closer to the original spectrogram than those obtained from NMF methods. Table 1 shows the separation results of all methods. The time-dependent NMF experiences a number of interference errors from the additional partial energy shown in the circled region of Figure 4c. By considering the SAR, the proposed method preserves most of the vibrato effects in all cases, whereas the harmonic-constrained NMF preserves only steady partial energies. Based on the success of both measures, the proposed method scores the highest SDR as well.

Main experiments

The system is tested on the complete recordings of BWV 1005 No. 3 played by Kuijken [23] and 24 Caprices op. 1 no. 24 in A minor played by Paganini from RWC database [24]. A total of 587 + 1,745 notes are annotated as the ground truth. The analysis is performed blindly; however, pitches outside the possible range are excluded. Additional score information is also excluded from the process.

Table 2 shows the results of analyzing Kuijken's recording. It indicates that 501 notes are correctly detected (TP), 14 detected notes are not in the score (FP), 86 of 587 notes are not detected (FN), 18 notes are missed because their octave notes are too strong, 4 notes are missed because

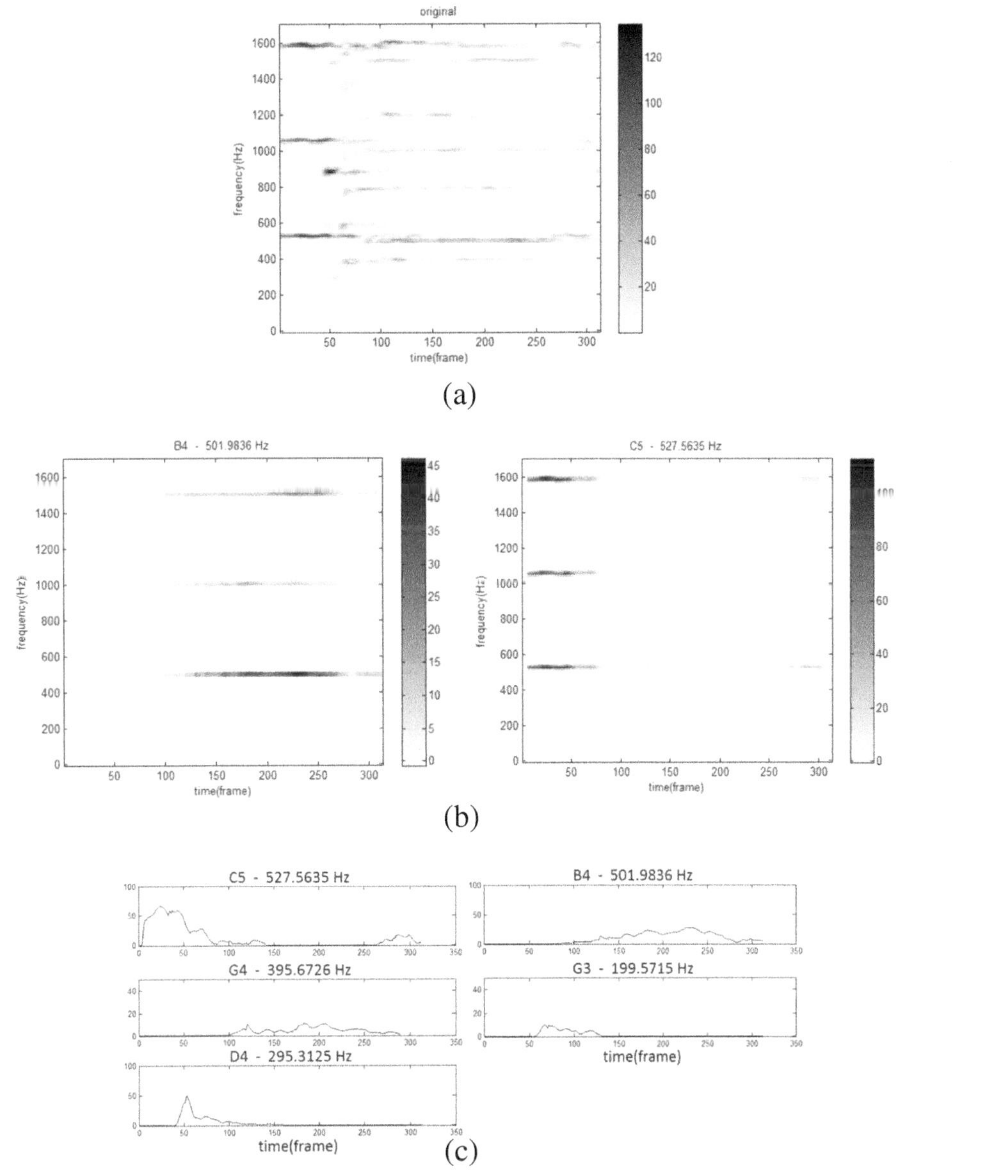

Figure 6 Analysis of Grumiaux's recording. (a) Original spectrogram, **(b)** spectrogram of B4 and C5 notes, and **(c)** intensity matrices.

their perfect fifth notes are too strong, 56 notes remain in Guard templates because the automatic detection method fails to extract them from Guard template, and 8 notes are mixed with other detected notes because their pitches are close to those of certain high intensity notes. Hence, the following results are obtained: precision rate $= 501/(501+14) = 97.28\%$, recall rate $= 501/(501+86) = 85.35\%$, accuracy $= 501/(501+14+86) = 83.36\%$, and F-measure $= 90.93\%$.

The second recording contains more complicated performing styles with larger amount of notes than the first one. False alarms are enormously increased compared with the first recording especially in NMF case, since the energies of overlapping partials of voiced notes and unvoiced notes interfere each other in the intensity matrix **H**. The performance of the proposed method maintains balance between precision and recall rate. Its number of unvoiced notes transcribed as voiced and voiced notes transcribed as unvoiced is lower than NMF and TD-NMF. That shows our approach is more stable than NMF and TD-NMF.

In Tables 2 and 3, the recursive regularization method without the online scheme means that the sliding frame size is maximized to the whole signal duration. It is

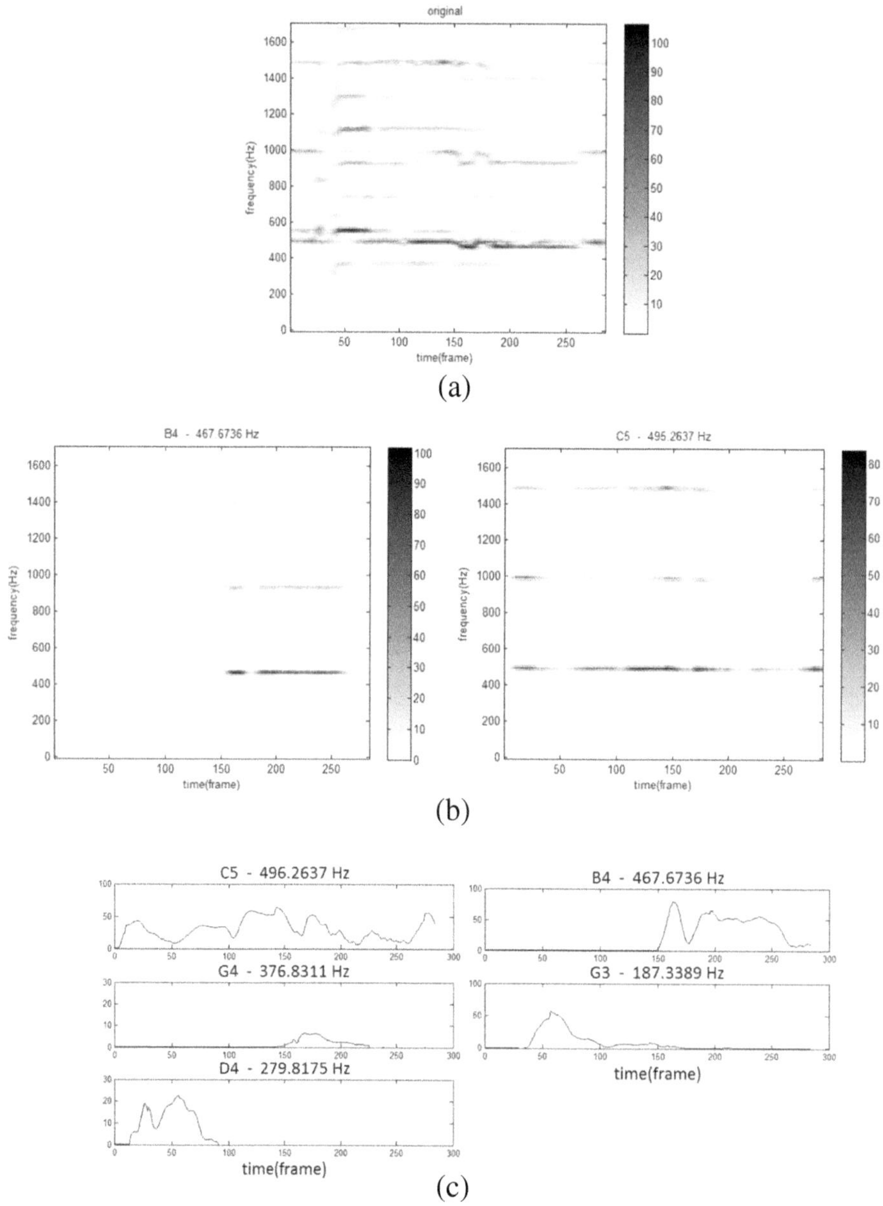

Figure 7 Analysis of Kuijken's recording. (a) Original spectrogram, **(b)** spectrogram of B4 and C5 notes, and **(c)** intensity matrices of all notes.

notable that, according to accuracy and F-measure of the transcription report, the results of this conventional recursive regularization method provides a baseline performance of the proposed algorithm. This work takes advantages of high-speed property of recursive regularization and improves the performance by imposing the proposed online scheme.

Complexity

In addition, the complexity of the proposed update rules is compared with the NMF as shown in Table 4. Let n be the number of frequency bins, m be the number of frames, and r be the number of templates. Suppose m input frames are set to be processed and only k frames are analyzed for each iteration, where $k < m$. For each iteration, the complexities of two update rules are $O(krn)$ and $O(r^2n)$. The complexity of the proposed update rules is similar to traditional NMF update rules in the case of regarding Euclidean distance as the cost function. Under the similar computational complexity, the proposed method is able to handle the time-varying music features and offer better results.

Performance analysis

Here, the proposed method attempts to analyze the performances of three violinists of different generations: Arthur Grumiaux, Sigiswald Kuijken, and Hilary Hahn. The signals are extracted from their CDs [23,25,26]. The following passages excerpted from BWV 1005 No. 3 are used to compare their performances. The musical score is shown in Figure 5. This study analyzes the signals of the short period during five notes in the bracket. The

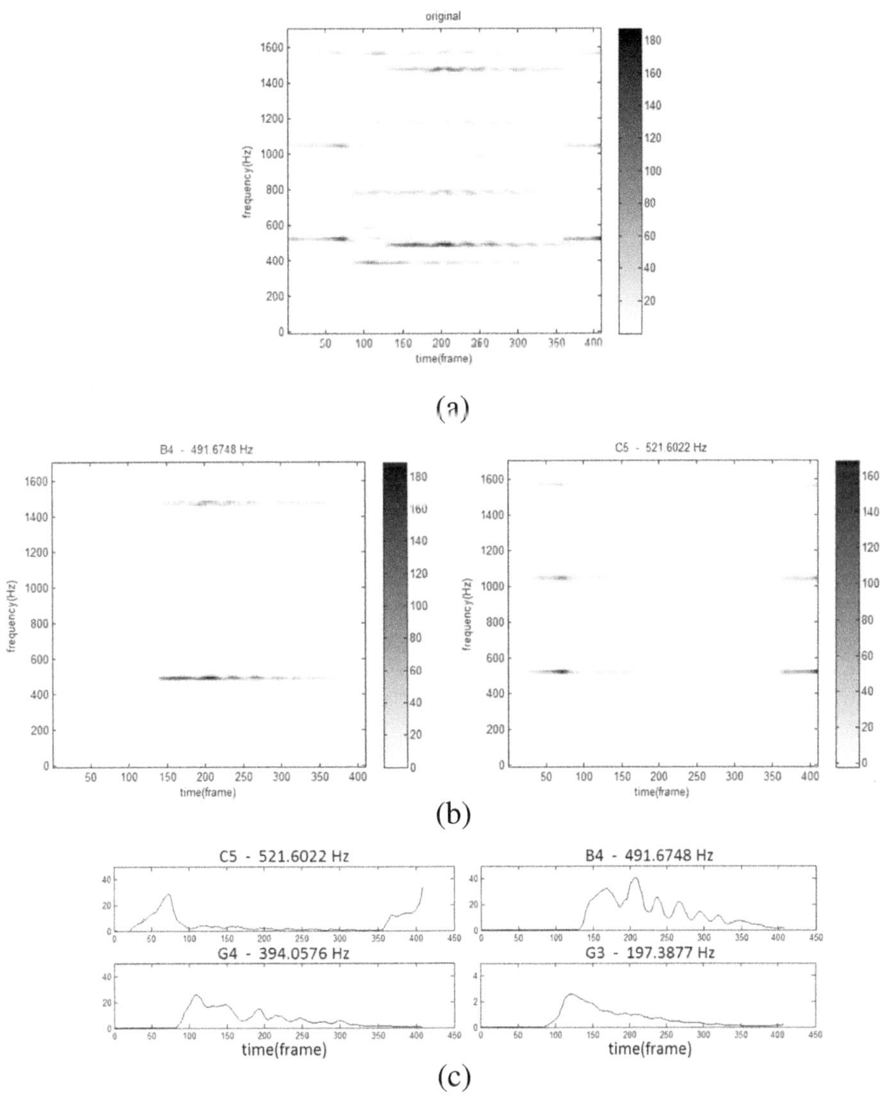

Figure 8 Analysis of Hahn's recording. (a) Original spectrogram, **(b)** spectrogram of B4 and C5 notes, and **(c)** intensity matrices of all notes.

original spectrograms and the analysis results are shown in Figures 6, 7, and 8.

According to the figures, a notable difference is that the pitches of Kuijken's performance are one semitone lower than those of the other violinists' since the instruments are usually used in historically informed performances (HIP), and musicians usually follow the tradition of the period during which the music is composed. Secondly, violinists use vibrato techniques frequently. We can observe the vibrato effects of B4 note in the performance played by Hahn, as shown in Figure 8, compared to those played by Grumiaux and Kuijken, as shown in Figures 6 and 7, respectively. Thirdly, Kuijken's style is distinct. He freely used numerous trills, which were not indicated in the score of Bach's solo violin works. This can be viewed in Figure 7b,c. As shown in Figures 6c and 8c, C5 is off and B4 is on. After B4 continues for a period of time, it is off and C5 is on again. Moreover, C5 is the strongest note in Grumiaux's playing, whereas B4 is the most prominent note in Hahn's playing. Comparatively, Kuijken played equal intensity on these two notes. Hahn's recording sounded brighter than the other two recordings since the spectral energy of the G3 note in Hahn's recording is larger than the others'. This may be caused by the decision of her balance engineer since the lower notes are weak in amplitude in all of her recordings. A notable mistake is also observed, that is, both Kuijken and Grumiaux played an extra D4 note, which is not in the original score. This may be a coincidence, or it is possible that they used a different score edition. Finally, their tempi also differ considerably. Kuijken used 1.68 s (290 frames) to finish the short passage, whereas Grumiaux used 1.83 s, and Hahn used 2.37 s. Hahn's tempo is 40% slower than Kuijken's. As an HIP musician, Kuijken played faster than the other violinists.

Conclusions

A recursive regularization analysis method is proposed to analyze acoustic recordings of solo violin works. Similar to NMF, the proposed method factorizes the matrix formed with the Fourier magnitude coefficients of multiple frames into a template matrix and an intensity matrix. The frame-by-frame-based procedure is designed for time-varying musical signals, such as solo violin recordings. The system of equations is updated by adding a new frame and dropping an old frame to avoid the problems of most NMF methods when the signal varies substantially. The proposed method is compared to the time-dependent NMF method by using two synthesized signals and exhibited superior SDR performances. The objective performance of the proposed method is also verified. For Kuijken's recording of BWV 1005 No. 3, the precision rate is 97.28%, the recall rate is 85.35%, and the F-measure is 90.93%. For a larger recording database from RWC C038, the precision rate is 85.99%, the recall rate is 87.08%, and the F-measure is 86.53%. It shows the stability of our approach. Finally, the proposed method is used to analyze the recordings of J.S. Bach's BWV 1005 No. 3 by three violinists, that is, Arthur Grumiaux, Sigiswald Kuijken, and Hilary Hahn. The results show that the time-varying characteristics of most notes appearing in the recordings can be tracked efficiently. The styles of the three violinists are easily distinguished through the separated results.

We are currently investigating possible approaches to improve the extraction of new notes from Guard template. An octave error may occur in our case because of the overlapping partials of the octave notes. In addition, because of the basis of the least-squares method, the performance of the proposed method with respect to signals of small amplitude, such as higher partials, is not as effective as NMF using other types of cost functions such as KL and IS divergences. The derivation of other cost functions into the proposed method may improve performance. Moreover, a supervised learning procedure can be introduced if note activations are available. Note activations can not only eliminate pitch detection errors but also constrain the intensity matrix for each note. As our approach preserves more musical characteristic details in the note level, nearly perfect decomposition is possible if it incorporates with more constraints, such as timbre, inharmonic bias, and phase. Therefore, many music information retrieval tasks are suitable to take our approach as a preprocessing, for example, player identification or expressive remix.

Competing interests

The authors declare that they have no competing interests.

Acknowledgements

The authors would like to thank the National Science Council, ROC, for its financial support of this work, under contract no. NSC98-2221-E-006-158-MY3.

References

1. DD Lee, HS Seung, Learning the parts of objects by non-negative matrix factorization. Nature. **401**(6755), 788–791 (1999)
2. P Smaragdis, JC Brown, Non-negative matrix factorization for polyphonic music transcription, in *Proc. IEEE Workshop on Applications of Signal Processing to Audio and Acoustics* (New Paltz, NY, 19–22 Oct. 2003), pp. 177–180
3. T Virtanen, Monaural sound source separation by nonnegative matrix factorization with temporal continuity and sparseness criteria. Audio Speech Lang. Process. IEEE Trans. on. **15**(3), 1066–1074 (2007)
4. CT Lee, YH Yang, H Chen, Automatic transcription of piano music by sparse representation of magnitude spectra, in *IEEE International Conference on Multimedia and Expo (ICME)* (Taipei, Taiwan, 11–15 July 2011), pp. 1–6
5. R Hennequin, R Badeau, B David, Time-dependent parametric and harmonic templates in non-negative matrix factorization, in *Proc. of the 13th Int. Conference on Digital Audio Effects* (Graz, Austria, 6–10 Sept. 2010)
6. M Nakano, JL Roux, H Kameoka, Nonnegative matrix factorization with Markov-chained bases for modeling time-varying patterns in music spectrograms, in *LVA/ICA'10 Proceedings of the 9th International Conference*

on Latent Variable Analysis and Signal Separation (St. Malo, France, 27–30 Sept. 2010), pp. 149–156
7. R Hennequin, Scale-invariant probabilistic latent component analysis, in *Proc. IEEE Workshop on Applications of Signal Processing to Audio and Acoustics* (New Paltz, NY, 16–19 Oct. 2011)
8. P Smaragdis, Relative-pitch tracking of multiple arbitrary sounds. J. Acoust. Soc. Am. **125**, 3406–3413 (2009)
9. MIREX, Music Information Retrieval Evaluation eXchange (MIREX.) http://www.music-ir.org/mirex/wiki/MIREX_HOME. Accessed 19 May 2011
10. N Bertin, R Badeau, E Vincent, Enforcing harmonicity and smoothness in Bayesian non-negative matrix factorization applied to polyphonic music transcription. Audio Speech Lang. Process. IEEE Trans. on. **18**(3), 538–549 (2010)
11. C Yeh, A Roebel, X Rodet, Multiple fundamental frequency estimation and polyphony inference of polyphonic music signals. Audio Speech Lang. Process. IEEE Trans. on. **18**(6), 1116–1126 (2010)
12. WC Chang, WY Su, C Yeh, A Roebel, X Rodet, Multiple-F0 tracking based on a high-order HMM model, in *Proc. of the 11th Int. Conference on Digital Audio Effects (DAFx-08)* (Espoo, Finland, 1–4 Sept. 2008)
13. TM Wang, YL Chen, WH Liao, A Su, Analysis and trans-synthesis of acoustic bowed-string instrument recordings–a case study using Bach cello suites, in *International Conference on Digital Audio Effects (DAFX)* (IRCAM, Paris, France, 19–23 Sept. 2011)
14. M Unser, A Aldroubi, M Eden, Recursive regularization filters: design, properties, and applications. IEEE Trans. on Pattern Anal. Mach. Intell. **13**(3), 272–277 (1991)
15. F Nesta, P Svaizer, M Omologo, Convolutive BSS of short mixtures by ICA recursively regularized across frequencies. Audio Speech Lang. Process. IEEE Trans. on. **19**(3), 624–639 (2011)
16. SP Kim, WY Su, Recursive high-resolution reconstruction of blurred multiframe images. Image Process. IEEE Trans. on. **2**(4), 534–539 (1993)
17. J Mairal, F Bach, J Ponce, G Sapiro, Online learning for matrix factorization and sparse coding. J. Mach. Learn. Res. **11**, 19–60 (2010)
18. F Wang, C Tan, AC König, P Li, Efficient document clustering via online nonnegative matrix factorizations, in *Proc. SIAM , Hilton Phoenix East/Mesa* (Mesa, USA, 28–30 Apr 2011)
19. Z Duan, G Mysore, P Smaragdis, Online PLCA for real-time semi-supervised source separation. Latent Variable Anal. Signal Sep. **7191**, 34–41 (2012)
20. A Lefevre, F Bach, C Févotte, Online algorithms for nonnegative matrix factorization with the Itakura-Saito divergence, in *Proc. WASPAA* (IEEE, New Paltz, NY, 16–19 Oct. 2011), pp. 313–316
21. E Vincent, N Bertin, R Badeau, Harmonic and Inharmonic nonnegative matrix factorization for polyphonic pitch transcription, in *Proc. of International Conference on Acoustics, Speech and Signal Processing* (IEEE, Las Vegas, Nevada, USA, 2008), pp. 109–112
22. PO Hoyer, Non-negative matrix factorization with sparseness constraints. J. Mach. Learn. Res. **5**, 1457–1469 (2004)
23. S Kuijken, *Bach: Sonatas & Partitas, BWV 1001-1006, CD 1* (Deutsche Harmonia Mundi, 1990)
24. M GOTO, RWC Music Database. http://staff.aist.go.jp/m.goto/RWC-MDB/. Accessed 13–17 Oct. 2002
25. A Grumiaux, *Bach: Sonatas & Partitas (BWV 1001-1006), CD 1* (Philips, 2006)
26. H Hahn, *Hilary Hahn plays Bach* (Sony, 1997)
27. MA Woodbury, Inverting modified matrices. Memorandum Rep. **42**, 106 (1950)
28. YS Siao, WC Chang, WY Su, Pitch detection/tracking strategy for musical recordings of solo bowed-string and wind instruments. J. Inf. Sci. Eng. **25**(4), 1239–1253 (2009)
29. S Dixon, On the computer recognition of solo piano music, in *Proceedings of Australasian Computer Music Conference* (Brisbane, Australia, 17 Jul 2011), pp. 31–37
30. M GOTO, Music synchronization for RWC Music Database (classical music). http://staff.aist.go.jp/m.goto/RWC-MDB/AIST-Annotation/SyncRWC/. Accessed 24 Sept. 2010
31. E Vincent, R Gribonval, C Févotte, Performance measurement in blind audio source separation. IEEE Trans. on Audio Speech Lang. Process. **14**(4), 1462–1469 (2006)
32. S Ewert, M Müller, Using score-informed constraints for NMF-based source separation, in *Proc. IEEE International Conference on Acoustics, Speech and Signal Processing (ICASSP)* (Kyoto, Japan, 25–30 Mar 2012)

An efficient algebraic codebook search for ACELP speech coder

Ning-Yun Ku[1], Cheng-Yu Yeh[2*] and Shaw-Hwa Hwang[1]

Abstract

In a bid to enhance the search performance, this paper presents an improved version of reduced candidate mechanism (RCM), an algebraic codebook search conducted on an algebraic code-excited linear prediction (ACELP) speech coder. This improvement is made based on two findings in a piece of our prior work. The first finding is that a pulse with a high contribution in the associated track is more likely to serve as an optimal pulse in the optimal codevector and the second is that the speech quality can be well maintained at a search accuracy above 50% approximately. Subsequently, a new finding in this study concerning a structured algebraic codebook in G.729 indicates that there is a 0.8321 probability that the number 1 ranked pulse in a global sorting by pulse contribution is indeed one of the optimal pulses. Hence, the number 1 pulse in the global sorting is labeled as one of the optimal pulses, following which a sequence of search tasks are fulfilled through RCM. This proposed complexity reduction algorithm, implemented on a G.729A speech codec, takes as few as eight searches, a search load tantamount to 2.5% of G.729A, 12.5% of global pulse replacement method (iteration = 2), 16.7% of iteration-free pulse replacement method, and 50% of RCM ($N = 2$). This proposal is thus found to successfully reduce the required computational complexity to a great extent as intended.

Keywords: Speech codec; VoIP; Algebraic code-excited linear-prediction (ACELP); Algebraic codebook search; Reduced candidate mechanism (RCM)

1 Introduction

An algebraic code-excited linear prediction (ACELP)-based speech coding technique [1-3] is a type of technique most widely applied to digital speech communication systems and serves as a mainstream technique adopted in a great number of speech coding standards due to a double advantage of low bit rates and high speech quality. The main coding flow for an ACELP coding technique is to perform a linear predictive coding (LPC) on an input speech signal and then perform an adaptive as well as an algebraic codebook search on an LPC residual signal.

Yet, the price paid is a high computational complexity requirement, particularly in an algebraic codebook search, due to the reason that it necessitates a tremendous computational load when conducting a full search over the algebraic codebook to locate the optimal pulses. As suggested in [4], the computational load required in a G.729 algebraic codebook search is dominated by two parts, namely, the load in a search process and the load during the algorithm initialization phase. The former and the latter respectively account for 74.9% and the remaining 25.1% of the entire computational load. Provided that there is a way to reduce the computational load to a great extent, an ACELP-based coding technique can be extensively applied to an embedded system on a handheld device. In this way, a high-performance embedded system is not seen as required, making electronic devices cost-competitive. Moreover, due to a computational load reduction, the aim of complexity saving is reached for an extended operation time period.

For this sake, full search scheme is hardly adopted in most prominent speech coding standards. There have been a great number of studies proposed on search load reduction, say, the focus search in G.729 [3] and the depth-first search in G.729A [5], among other approaches. In recent times, to further reduce the computational complexity, there are many effective algebraic codebook search algorithms [6-14] proposed to reduce

* Correspondence: cy.yeh@ncut.edu.tw
[2]Department of Electrical Engineering, National Chin-Yi University of Technology, 57, Sec. 2, Zhongshan Rd., Taiping Dist, Taichung 41170, Taiwan
Full list of author information is available at the end of the article

the pulse combinations. For example, the candidate scheme [6] attempts to decrease the number of the candidate positions. The least important pulse replacement [7] is presented to repeatedly measure the contribution of each pulse and replace the least important pulse with a new one. Analogous to [7], the global pulse replacement (GPR) [8], adopted by G.729.1 [9], and the iteration-free pulse replacement (IFPR) [10,11] methods are proposed to further efficiently reduce the search complexity.

Alternatively, the reduced candidate mechanism (RCM) approach [12], a piece of our prior work, has been presented. In RCM, individual pulse contribution is evaluated in the associated track and sorted in descending order. Subsequently, a full search is performed on the sorted top N pulses treated as candidates. In this way, the optimal pulse combination is acquired following N^4 searches, that is, a significant reduction in search complexity.

It is a significant finding in RCM that a pulse with a high contribution in the associated track is more likely to serve as the optimal pulse in the optimal codevector. Thus, a further investigation is conducted into the RCM approach, and an improved version of RCM is proposed for further load reduction while comparable speech quality is well maintained. In contrast to existing approaches, this proposal, implemented in a G.729A speech codec, is presented as a superior choice in the aspect of search complexity and as a competitive candidate in terms of speech quality.

The rest of the paper is organized as follows. The coding criterion of an algebraic codebook in G.729 and various search methods are briefly reviewed in Section 2. Section 3 proposes an efficient approach for the purpose of search complexity reduction. Experimental results are demonstrated and discussed in Section 4. This work is concluded at the end of this paper.

2 Algebraic codebook search

With the determination of an optimal codevector as the goal of the algebraic codebook search, the codebook in G.729 is configured as tabulated in Table 1, on the basis of which each codevector contains four nonzero pulses extracted out of associated track. Each pulse's amplitude can be either +1 or −1.

Table 1 A structured algebraic codebook in G.729

Track	Pulse	Sign	Positions
T_0	i_0	s_0	m_0: 0, 5, 10, 15, 20, 25, 30, 35
T_1	i_1	s_1	m_1: 1, 6, 11, 16, 21, 26, 31, 36
T_2	i_2	s_2	m_2: 2, 7, 12, 17, 22, 27, 32, 37
T_3	i_3	s_3	m_3: 3, 8, 13, 18, 23, 28, 33, 38
			4, 9, 14, 19, 24, 29, 34, 39

The optimal codevector $\mathbf{c}_k = \{c_k(n)\}$ is thus found by minimizing the mean squared weighted error between the original and the synthesized speeches [2,3], defined as

$$\varepsilon_k = \|\mathbf{x}-g\mathbf{H}\mathbf{c}_k\|^2 \tag{1}$$

where $\mathbf{x}$ denotes the target vector, g a scaling gain factor, and $\mathbf{H}$ a lower triangular convolution matrix. It can be shown that the optimal codevector is the one maximizing the term Q_k:

$$Q_k = \frac{(\mathbf{x}^{\mathrm{T}}\mathbf{H}\mathbf{c}_k)^2}{\mathbf{c}_k^{\mathrm{T}}\mathbf{H}^{\mathrm{T}}\mathbf{H}\mathbf{c}_k} = \frac{(\mathbf{d}\mathbf{c}_k)^2}{\mathbf{c}_k^{\mathrm{T}}\mathbf{\Phi}\mathbf{c}_k} \tag{2}$$

where $\mathbf{d} = \mathbf{x}^{\mathrm{T}}\mathbf{H}$, the correlation function, is expressed as

$$d(n) = \sum_{i=n}^{L-1} x(i)h(i-n), \quad 0 \le n \le L-1 \tag{3}$$

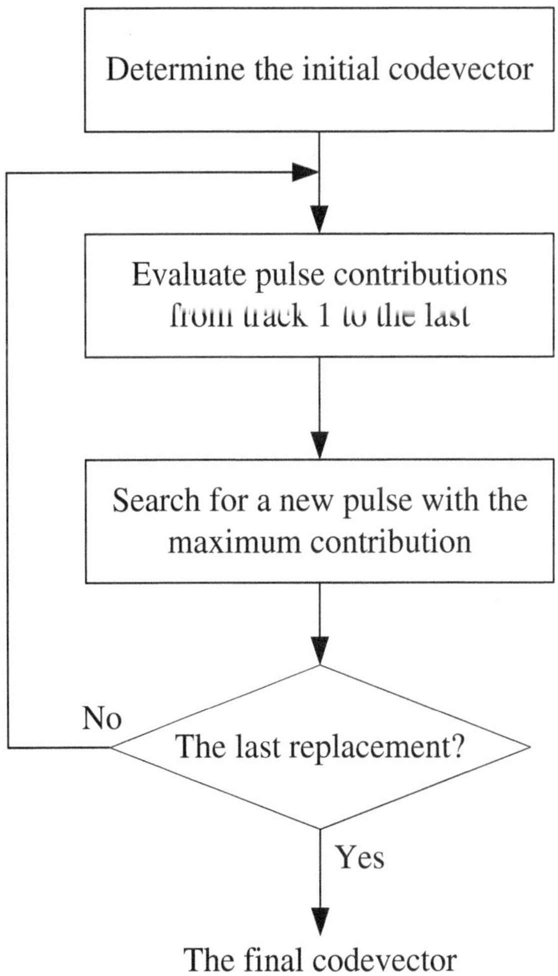

Figure 1 A system flowchart of the GPR search.

where L is the speech subframe size. The correlations of $h(n)$, are contained in the symmetric matrix $\mathbf{\Phi} = \mathbf{H}^T\mathbf{H}$, where the entries are given by

$$\varphi(i,j) = \sum_{n=j}^{L-1} h(n-i)h(n-j), \quad 0 \le i \le L-1 \; ; \; i \le j \le L-1 \tag{4}$$

It takes a total of 8,192 (8 × 8 × 8 × 16) searches, a tremendous computational load, to conduct a full search, i.e., repeated computations and comparisons in (2), for the identification of the optimal codevector. Therefore, a focused search method is adopted in G.729 to reduce the search times to below 1,440. However, the search number is further reduced to 320, adopting a depth-first tree search method in G.729A. Besides, three existing methods, the GPR, IFPR, and RCM methods, will be discussed in this section.

2.1 The GPR search method

The GPR method stems from the least important pulse replacement algorithm [7]. In order to prevent the termination of the pulse replacement procedure without finding the optimal codevector in the GPR algorithm, except for the only track that contains the least important pulse, all the tracks are searched for a new pulse. That is, the new pulse is sought by replacing each pulse in each track with a new one so that the Q_k associated with a new codevector is maximized. On the ground that the variation in Q_k is always maximized during the replacement procedure, the codevector approaches the optimal solution rapidly as this procedure is repeated. When the value of Q_k once reaches the upper bound, the search procedure is then terminated.

A system flowchart of the GPR method is sketched in Figure 1. Following an application of the GPR method to G.729A at the first stage, the initial Q_k is evaluated and

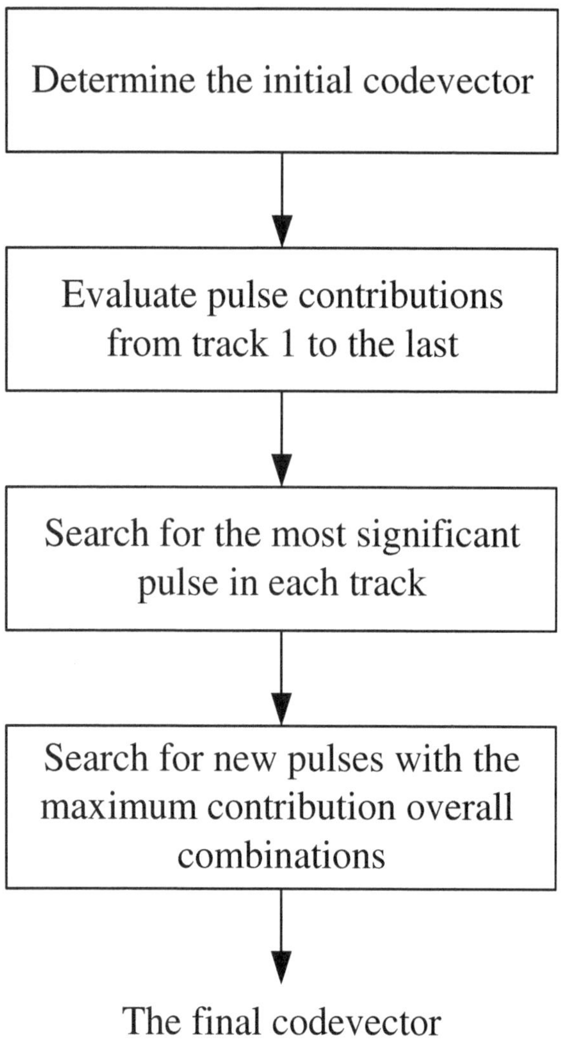

Figure 2 A system flowchart of the IFPR search.

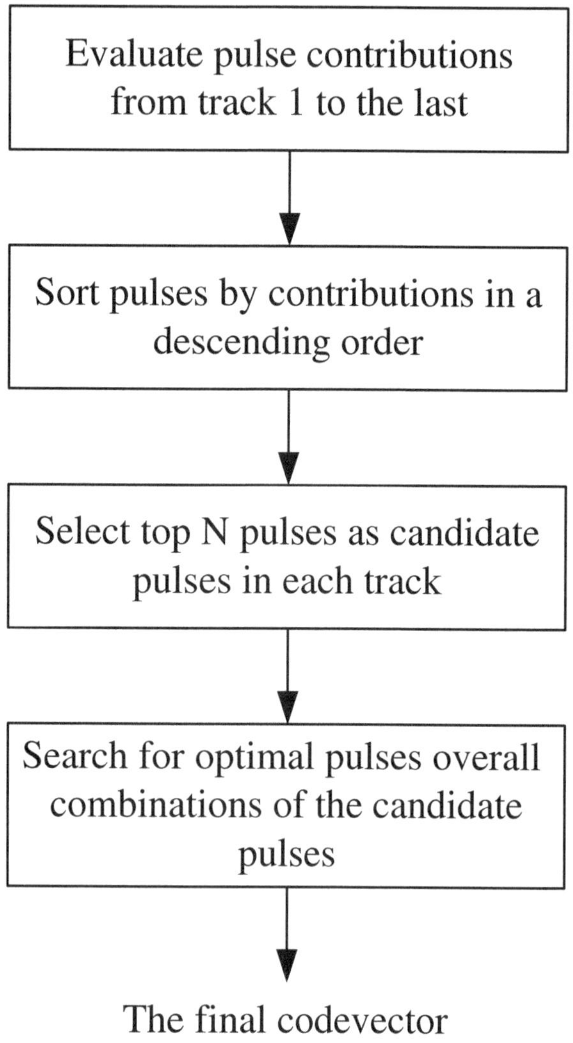

Figure 3 A system flowchart of the RCM search.

Table 2 Hit probability of each pulse contribution in a global sorting

Global sorting for pulse contribution	Hit probability	Global sorting for pulse contribution	Hit probability
1	0.8321	21	0.0189
2	0.5857	22	0.0161
3	0.4589	23	0.0144
4	0.3621	24	0.0127
5	0.2925	25	0.0116
6	0.2383	26	0.0103
7	0.1954	27	0.0096
8	0.1595	28	0.0084
9	0.1314	29	0.0079
10	0.1085	30	0.0071
11	0.0908	31	0.0063
12	0.0760	32	0.0059
13	0.0635	33	0.0054
14	0.0527	34	0.0048
15	0.0450	35	0.0047
16	0.0385	36	0.0042
17	0.0329	37	0.0040
18	0.0281	38	0.0037
19	0.0245	39	0.0032
20	0.0213	40	0.0030

the initial codevector is yielded with a single search. At the second stage, it requires 36 searches to seek the new pulse during the first pulse replacement procedure and requires an average of 27 during the second. Therefore, the overall search complexity is evaluated as $37 + 27 \times (R - 1)$ for $R \geq 1$, where R is the number of iterations of the pulse replacement procedure.

2.2 The IFPR search method

In the previously mentioned pulse replacement methods, [7] and [8], the computational load increases with the number of iterations of the replacement procedure. In the IFPR method, new pulses are sought by a number of pulse replacements at a time following pulse contributions evaluated for every track so as to maximize over all combinations a search criterion, which replaces the pulses pertaining to the initial codevector with the most significant pulses for every track.

Presented in Figure 2 is a system flowchart of the IFPR method. Applying IFPR method to G.729A at the first stage, the initial Q_k is evaluated and the initial codevector is then yielded with a single search. At the second stage, a total of 36 searches are performed to measure the pulse contribution so as to sort out the most significant pulses in each track. In order to find the final codevector in the end, it requires a total of 11 searches for all combinations, i.e., the number required from two pulses replacement to four pulses replacement, to replace the pulses of the initial codevector with the most significant pulses for every track, that is, an overall search complexity of 48.

2.3 The RCM search method

Ahead of a search task, the number of candidate pulses in each track is reduced for the purpose of search complexity reduction. This is done in this work according to the contribution of individual pulses. It is that in each track, a pulse sorting is made by the contribution thereof in descending order as the first step, and then, the top N pulses are chosen as the candidate pulses for a full search. In this way, the search process needs to be performed for merely N^4 number of times for the optimal pulse combination, and the search complexity is reduced remarkably in particular for low values of N.

The contribution made by individual pulses is given as (2), that is, a higher value of Q_k reflects a higher

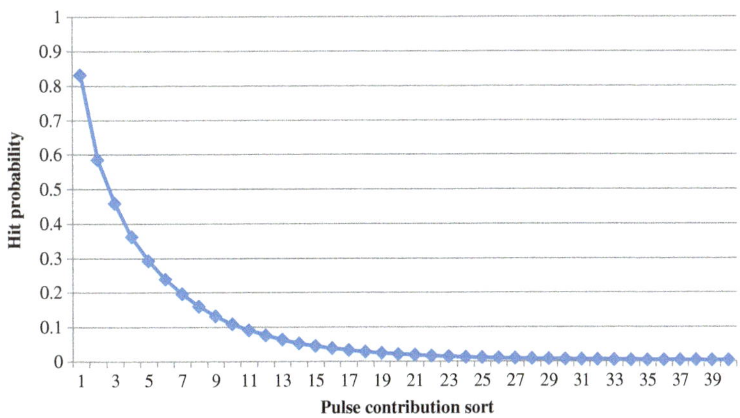

Figure 4 Plot of the hit probability of each pulse contribution in a global sorting.

contribution. In consideration of merely a single pulse contribution, the number of nonzero pulses in the codevector $\mathbf{c}_k$ of length 40 is reduced to 1 from 4. Therefore, (2) can be simplified into (5), where the numerator of (5) is derived from (2) and (3), and the denominator of (5) is derived from (2) and (4), respectively. Just as in (2), the contribution of the ith pulse is reflected by the value of Q_k^i

$$Q_k^i = \frac{d^2(i)}{\varphi(i,i)}, \quad 0 \le i \le L-1 \tag{5}$$

In [12], it is verified that a single pulse with a higher contribution within each track is more likely to be the optimal pulse out of the optimal codevector within the associated track. Thus, the RCM approach is used to reduce the search complexity by the reduction in the number of candidate pulses in each track. This approach is decomposed into two stages as follows. The first is to evaluate individual pulse contribution with (5), which indicates that a higher value of Q_k^i denotes a higher pulse contribution. Subsequently, the top N pulses, $1 \le N \le 8$, are extracted out of the Q_k^i sorting in each track as the prerequisite of the second stage. Then, in the second stage, it is proven that the pulses' combination with the highest value of Q_k, as given in (2), is indeed the optimal solution through a nest-loop search. A flowchart of the RCM approach is presented in Figure 3. The first three steps in Figure 3 are to compute the contribution of each pulse in each track and then to select the candidate pulses in each track. The last step is to compute the best combination of the candidates.

3 Proposed approach

In the aspect of RCM, the first significant finding indicates that a pulse with a high contribution is more likely to serve as one of the optimal pulses in the associated track, whereby the hit probability can be elevated when conducting a search task. The second finding reveals that the speech quality can be well maintained as long as the search accuracy exceeds a threshold, say 50% in [12]. Based on such findings, an improved version of RCM, referred to as the Fixed-G1-RCM approach, is presented to achieve the aim of search load improvement in the absence of speech quality degradation. As its name indicates, the top 1 pulse contribution in a global sorting, termed as the G1 pulse, is presumed to be one of the four optimal pulses, following which the rest of optimal pulses are located over the remaining three tracks through RCM.

Thus, it is an issue of our interest whether there exists a high correlation between the top 1 pulse contribution and the possibility that such pulse is indeed one of the optimal pulses. Hence, over entire tracks, a hit probability $p_h(n)$ in a global sorting is defined as

$$p_h(n) = \frac{\mathrm{NH}(n)}{\mathrm{TSF}}, \quad 1 \le n \le 40 \tag{6}$$

where NH(n) denotes the number of times that the nth pulse, in terms of the contribution priority, hits the optimal codevector and TSF the total number of testing subframes.

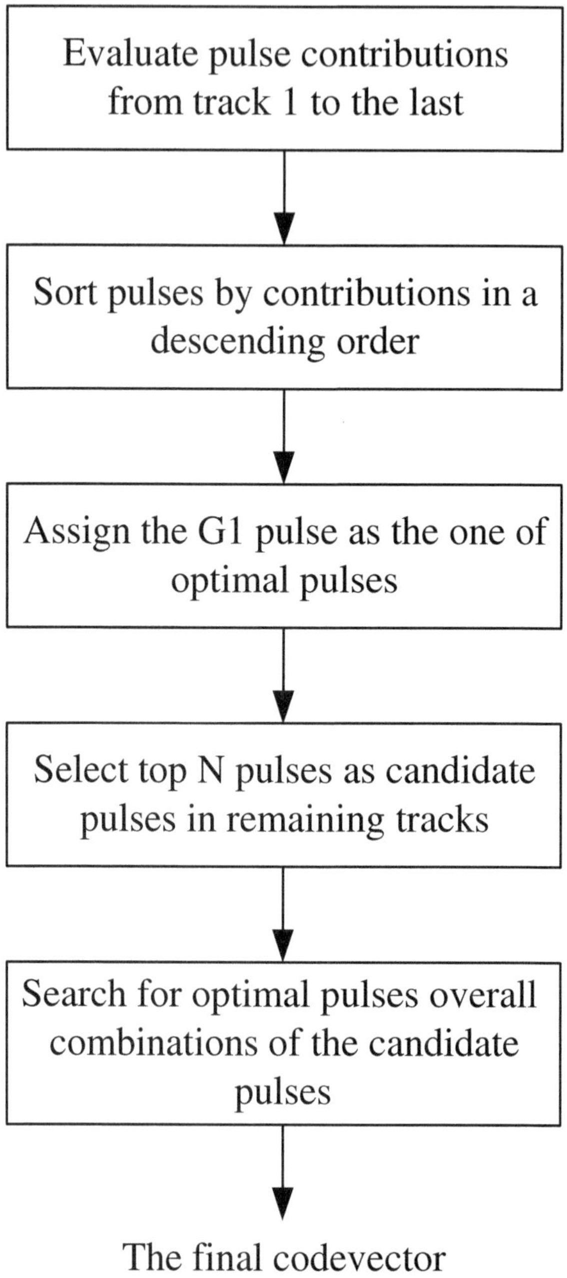

Figure 5 A system flowchart of the proposed search algorithm.

Table 3 Search accuracy comparison among various methods

Methods		Search accuracy for locating various numbers of intended pulses (%)				
		1 pulse	2 pulses	3 pulses	4 pulses (all right)	0 pulse (all wrong)
G.729A		98.3475	92.1456	80.9918	*68.6438*	1.6525
GPR	$R=1$	98.7032	90.7750	76.1053	*55.0718*	1.2968
	$R=2$	98.4946	91.8335	80.4779	*76.1053*	1.5054
	$R=3$	98.5246	92.1081	81.2227	*80.4779*	1.4754
	$R=4$	98.5283	92.1547	81.3393	*81.2227*	1.4717
IFPR		98.6810	92.4963	80.0048	*68.0824*	1.3190
RCM	$N=1$	99.3295	89.0873	55.8720	*17.3353*	0.6705
	$N=2$	98.3009	90.8486	73.6532	*50.3579*	1.6991
	$N=3$	98.8545	94.0394	83.5763	*70.7329*	1.1455
	$N=4$	99.2014	95.9617	89.2187	*81.8716*	0.7986
Proposed method	$N=1$	99.3295	89.0873	55.8720	*17.3353*	0.6705
	$N=2$	98.2054	89.8500	70.0755	*44.0814*	1.7946
	$N=3$	98.2060	91.2376	76.7365	*60.2498*	1.7940
	$N=4$	98.2369	92.1541	80.4684	*69.0717*	1.7631

Subsequently, a global sorting is conducted by pulse contribution over entire tracks. As tabulated in Table 2, it is seen in the G.729 algebraic codebook that there is a 0.8321 hit probability that the number 1 pulse is indeed one of the optimal pulses, while the hit probability drops dramatically from 0.8321 to 0.5857 in case the number 2 pulse acts as the optimal one. A graphic illustration of Table 2 is shown in Figure 4. The statistics on Table 2 is based on a speech database in Chinese language, containing 9,650 syllables out of 100 sentences for a duration over 41 min and 495,608 subframes, that is, TSF = 495,608.

The above analysis confirms that the top 1 pulse contribution in a global sorting has a highly hit probability that such pulse is indeed the one of optimal pulses. Thus, the Fixed-G1-RCM approach is presented in this study as an efficient way to further speed up the searching process. In this approach, the G1 pulse is presumed to be one of the four optimal pulses, following which the rest of optimal pulses are located over the remaining three tracks through RCM. In this context, the number of searches required is reduced to N^3 for $N \geq 2$. This proposal, as opposed to RCM, is developed in an attempt to considerably reduce the search load to N^3 from N^4 for identical value of N. Furthermore, since there is a high probability, up to 0.8321, that the top 1 pulse in the global sorting hits one of the optimal pulses, the speech quality can be well maintained on a condition that a search accuracy of approximately 50% is reached. Finally, this proposal is implemented as follows and is shown in Figure 5.

Step 1 Individual pulse contribution is evaluated by (5), and a sorting is made by pulse contribution within the associated track.

Step 2 The one with the global maximum pulse contribution, named as G1, is located out of all the top 1 pulses among all the tracks.

Step 3 G1 is presumed to be one of four optimal pulses, and then, the value of N is determined for the searching task conducted over the remaining three tracks through RCM.

Step 4 The searching task terminates the moment the combination of optimal pulses is acquired.

Furthermore, this proposal can be applied to an algebraic codebook search for other ACELP-based coders. Assuming that there are Nt tracks in a structured algebraic

Table 4 Search load comparison among various methods

Methods		Search load
G.729A		320
GPR	$R=1$	37
	$R=2$	64
	$R=3$	91
	$R=4$	118
IFPR		48
RCM	$N=1$	1
	$N=2$	16
	$N=3$	81
	$N=4$	256
Proposed method	$N=1$	1
	$N=2$	8
	$N=3$	27
	$N=4$	64

Table 5 Objective speech quality comparison among various methods

Methods		PESQ				MOS-LQO			
		Mean	STD	Max.	Min.	Mean	STD	Max.	Min.
G.729A		3.8126	0.0838	3.9640	3.5150	3.9502	0.1026	4.1230	3.5670
GPR	$R=1$	3.8145	0.0757	4.0850	3.5640	3.9521	0.0883	4.2350	3.6410
	$R=2$	3.8324	0.0777	4.0490	3.5420	3.9725	0.0906	4.2010	3.6110
	$R=3$	3.8376	0.0739	3.9830	3.5950	3.9786	0.0859	4.1370	3.6830
	$R=4$	3.8382	0.0749	4.0350	3.5560	3.9793	0.0867	4.1880	3.6310
IFPR		3.8276	0.0761	4.0830	3.5820	3.9672	0.0884	4.2330	3.6650
RCM	$N=1$	3.7083	0.0807	3.9050	3.3230	3.8250	0.1026	4.0550	3.3030
	$N=2$	3.8064	0.0819	4.0770	3.5190	3.9423	0.0971	4.2280	3.5800
	$N=3$	3.8256	0.0740	3.9880	3.5850	3.9651	0.0868	4.1420	3.6700
	$N=4$	3.8331	0.0735	3.9710	3.5570	3.9735	0.0859	4.1240	3.6320
Proposed method	$N=1$	3.7083	0.0807	3.9050	3.3230	3.8250	0.1026	4.0550	3.3030
	$N=2$	3.7937	0.0775	4.0220	3.5120	3.9279	0.0927	4.1760	3.5700
	$N=3$	3.8143	0.0766	4.0270	3.5640	3.9518	0.0906	4.1800	3.6410
	$N=4$	3.8312	0.0682	4.0320	3.6100	3.9717	0.0791	4.1860	3.7020

codebook, out of which Np pulses must be selected in each codebook search, then the respective numbers of searches required in RCM and this presented Fixed-G1-RCM searching approaches are given as

$$\text{Searches}(\text{RCM}, N) = \left(C_{Np}^{N}\right)^{Nt} = \left(\frac{N!}{Np!(N-Np)!}\right)^{Nt} \tag{7}$$

$$\text{Searches}(\text{Fixed-G1-RCM},\ N) = C_{Np-1}^{N-1}\left(C_{Np}^{N}\right)^{(Nt-1)} = \frac{(N-1)!}{(Np-1)!(N-Np)!}\left(\frac{N!}{Np!(N-Np)!}\right)^{(Nt-1)} \tag{8}$$

Taking a structured algebraic codebook in G.729 as an instance ($Nt=4$, $Np=1$), this proposal requires N^3 searches, while RCM requires as many as N^4 searches. In an arbitrary structured algebraic codebook, there is no doubt in this study, as opposed to RCM, that the G1 pulse with a higher probability of hitting one of the four optimal pulses gives rise to a less degree of speech quality degradation.

4 Experimental results

There are three experiments conducted in this work. The first is a search accuracy comparison among the full search and other search approaches. Subsequently, the second is a search load comparison among the preceding search approaches. The third is a series of objective and subjective speech quality testings among various approaches for comparison purposes. The test objects are those selected out of a speech database in Chinese language, containing 9,650 syllables out of 100 sentences for a duration over 41 min and 495,608 subframes.

For the brevity of the following discussion, the RCM approach with N candidate pulses is abbreviated as RCM-N, $1 \le N \le 8$. For instance, RCM-1 symbolizes the one with merely a candidate pulse extracted out of each track. Similarly, the GPR approach with the number R of repetitions is designated as GPR-R.

Table 6 Comparison of statistics on quantized MOS-LQO scores among various methods

MOS-LQO	G.729A	GPR-2	IFPR	Proposed method		
				$N=2$	$N=3$	$N=4$
3.50	0	0	0	0	0	0
3.55	0.01	0	0	0.01	0	0
3.60	0	0.01	0	0.01	0	0
3.65	0.01	0	0.02	0.01	0.01	0
3.70	0.02	0.01	0.01	0	0.04	0.01
3.75	0.04	0.02	0.01	0.03	0.02	0.03
3.80	0.06	0.02	0.04	0.06	0	0.02
3.85	0.04	0.05	0.03	0.10	0.09	0.03
3.90	0.08	0.10	0.11	0.19	0.14	0.14
3.95	0.25	0.25	0.26	0.26	0.22	0.23
4.00	0.28	0.24	0.29	0.26	0.28	0.31
4.05	0.17	0.23	0.19	0.05	0.17	0.19
4.10	0.04	0.06	0.01	0.01	0.02	0.02
4.15	0	0	0.02	0	0	0.01
4.20	0	0.01	0	0.01	0.01	0.01
4.25	0	0	0.01	0	0	0
4.30	0	0	0	0	0	0

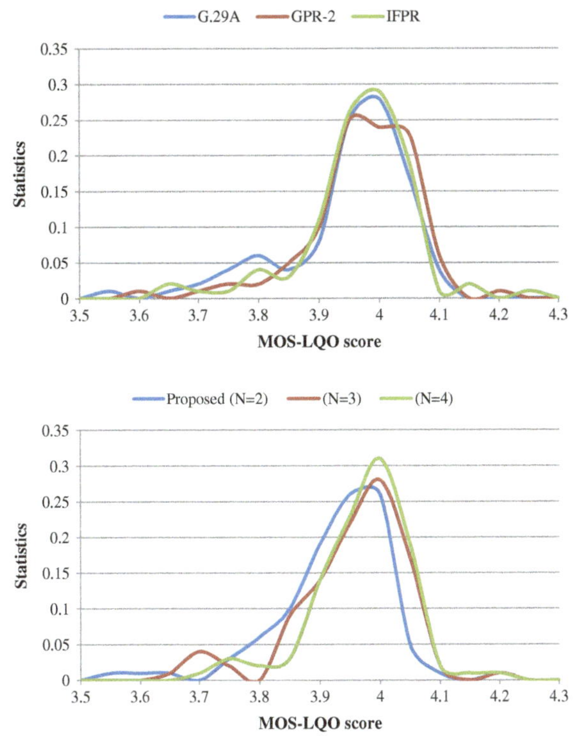

Figure 6 Histogram comparison of MOS-LQO scores among various approaches.

Tabulated in Table 3 is the search accuracy analysis among various approaches, that is, the hit probability of individual approach against the optimal pulse identified through a full search. During the search process, the best case is the one to successfully locate four intended pulses, the all right case, and the worst is to locate none, the all wrong case. As tabulated in Table 3, tracking the all right case as an instance, the accuracies made by G.729A, GPR-2, IFPR, and RCM-2 are 68.6438%, 76.1053%, 68.0824%, and 50.3579%, respectively, while that by the proposed method falls between 17.3353% ($N = 1$) and 69.0717% ($N = 4$).

Tabulated in Table 4 is the comparison of the search load, that is, the number of searches performed and those required in the evaluation of Q_k defined in (2). It is found that G.729A requires 320 searches, GPR-2 64, IFPR 48, RCM-2 16, and the proposed method a number somewhere between 1 ($N = 1$) and 64 ($N = 4$). Accordingly, a search complexity is as intended reduced.

Subsequently, the values of ITU-T P.862 perceptual evaluation of speech quality (PESQ) [15] and ITU-T P.862.1 mean opinion score and listening quality objective (MOS-LQO) [16] are evaluated for objective speech quality comparison among various approaches. Table 5 gives comparisons on PESQ and MOS-LQO, each including the mean, standard deviation (STD), maximum, and minimum values, respectively. In comparison with MOS-LQO, G.729A, all the approaches provide a comparable speech quality within a 1% deviation, except that RCM-1 exhibits a 3% drop.

Moreover, there are up to 100 MOS-LQO scores measured to make a histogram. The MOS-LQO scores are firstly quantized with a maximum quantization error of 0.025. For instance, a MOS-LQO score in the range of 3.475 to 3.525 is quantized as 3.50. Table 6 gives the statistics and normalization on quantized MOS-LQO scores for comparison purposes among G.729A, GPR-2, IFPR, and this proposal. Figure 6 is a graphic presentation of Table 6 for a clear view of the MOS-LQO score distribution.

On the other hand, a forced choice listening test is conducted for a subjective speech quality comparison among the proposed ($N = 2$), the GPR-2, and the IFPR search approaches. A set of ten arbitrary sentences was selected and processed by such three approaches in each test group. Each set of processed sentences was evaluated

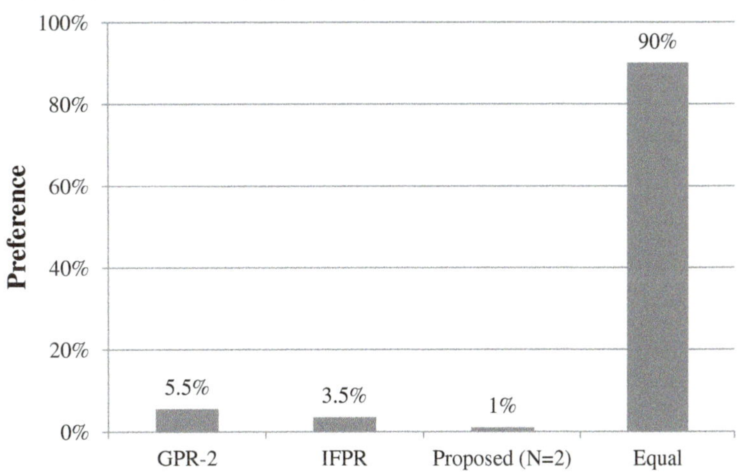

Figure 7 Subjective speech quality comparison among various approaches.

by 20 listeners. The preference score is evaluated for the three approaches by each single listener in each group. Presented in Figure 7 is a bar graph of subjective speech quality testing versus search strategy. Merely an extremely low percentage of listeners can tell a performance difference among these three search strategies, while as high as 90% of listeners cannot distinguish any differences for the test cases. This finding indicates that this proposal does provide a comparable subjective speech quality. Moreover, there is an evidence that little difference in MOS-LQO values reflects an indistinguishable subjective speech quality.

Furthermore, it is a point worthy of mention that with a marginal variation in MOS-LQO, a low level of search complexity signifies a superior system performance, e.g., the proposed method at $N = 2$ and $N = 3$. Particularly, it merely takes eight searches at $N = 2$, a figure tantamount to 12.5% of that required in GPR-2, 16.7% in IFPR, and 50% in RCM-2, but provides comparable speech quality.

5 Conclusions

An improved version of RCM approach is presented in this work as an efficient means to enhance the performance of search over an algebraic codebook when applied to a G.729A speech codec. It is experimentally demonstrated that this proposal requires as few as eight searches in the case of $N = 2$, that is, a search load tantamount to 2.5% of that in G.729A, 12.5% in GPR-2, 16.7% in IFPR, and 50% in RCM-2, but still provides a comparable speech quality. Thus, this proposal is validated as a superior candidate in the aspect of search performance. In addition, the proposed approach can be implemented to other ACELP-based speech coders.

Furthermore, this improved G.729A speech codec can be utilized to improve the VoIP performance on smart phone. As a consequence, the energy efficiency requirement is met for an extended operation time period due to computational load reduction.

Competing interests
The authors declare that they have no competing interests.

Acknowledgements
This research was financially supported by the National Science Council under Grant No. NSC 102-2221-E-167-027, Taiwan, Republic of China.

Author details
[1]Department of Electrical Engineering, National Taipei University of Technology, 1, Sec. 3, Chung-hsiao E. Rd., Taipei 10608, Taiwan. [2]Department of Electrical Engineering, National Chin-Yi University of Technology, 57, Sec. 2, Zhongshan Rd., Taiping Dist, Taichung 41170, Taiwan.

References
1. JP Adoul, P Mabilleau, M Delprat, S Morissette, Fast CELP coding based on algebraic codes, in *Proceedings of the IEEE International Conference on Acoustics, Speech, and Signal Processing* (Dallas, Texas, USA, 1987), pp. 1957–1960
2. C Laflamme, JP Adoul, R Salami, S Morissette, P Mabilleau, 16 kbps wideband speech coding technique based on algebraic CELP, in *Proceedings of the IEEE International Conference on Acoustics, Speech, and Signal Processing* (Toronto, Canada, 1991), pp. 13–16
3. R Salami, C Laflamme, JP Adoul, A Kataoka, S Hayashi, T Moriya, C Lamblin, D Massaloux, S Proust, P Kroon, Y Shoham, Design and description of CS-ACELP: a toll quality 8 kb/s speech coder. IEEE Trans. Speech Audio Process. **6**(2), 116–130 (1998)
4. SM Tsai, JF Yang, Efficient algebraic code-excited linear predictive codebook search. IEE Proc. Vis. Image Signal Process. **153**(6), 761–768 (2006)
5. R Salami, C Laflamme, B Bessette, JP Adoul, ITU-T G.729 Annex A: reduced complexity 8 kb/s CS-ACELP codec for digital simultaneous voice and data. IEEE Commun. Mag. **35**(9), 56–63 (1997)
6. FK Chen, JF Yang, YL Yan, Candidate scheme for fast ACELP search. IEE Proc. Vis. Image Signal Process. **149**(1), 10–16 (2002)
7. H Park, Y Choi, D Lee, Efficient codebook search method for ACELP speech codecs, in *Proceedings of the IEEE Speech Coding Workshop* (Tsukuba, Ibaraki, Japan, Tsukuba, Ibaraki, Japan, 2002), pp. 17–19
8. ED Lee, MS Lee, DY Kim, Global pulse replacement method for fixed codebook search of ACELP speech codec, in *Proceedings of the 2nd IASTED International Conference on Communications, Internet and Information Technology* (Scottsdale, Arizona, USA, 2003), pp. 372–375
9. B Geiser, P Jax, P Vary, H Taddei, S Schandl, M Gartner, C Guillaume, S Ragot, Bandwidth extension for hierarchical speech and audio coding in ITU-T Rec. G.729.1. IEEE Trans. Audio Speech Lang. Process. **15**(8), 2496–2509 (2007)
10. ED Lee, SH Yun, SI Lee, JM Ahn, Iteration-free pulse replacement method for algebraic codebook search. Electron. Lett. **43**(1), 59–60 (2007)
11. ED Lee, JM Ahn, Efficient fixed codebook search method for ACELP speech codecs. Lect. Notes Comput. Sci. **4413**, 178–187 (2007)
12. CY Yeh, YJ Su, Reduced candidate mechanism for an algebraic code-excited linear-prediction codebook search. IET Commun. **6**(17), 2864–2869 (2012)
13. FK Chen, GM Chen, BK Su, YR Tsai, Unified pulse-replacement search algorithms for algebra codebooks of speech coders. IET Signal Process. **4**(6), 658–665 (2010)
14. CY Yeh, CZ Zhuo, An efficient complexity reduction algorithm for G.729 speech codec. Comput. Math. Appl. **64**(5), 887–896 (2012)
15. ITU-T Recommendation P.862, *Perceptual evaluation of speech quality (PESQ): an objective method for end-to-end speech quality assessment of narrow-band telephone networks and speech codecs*, 2001
16. ITU-T Recommendation P.862.1, *Mapping Function for Transforming P.862 raw Result Scores to MOS-LQO*, 2003

7

A signal subspace approach to spatio-temporal prediction for multichannel speech enhancement

Adam Borowicz

Abstract

The spatio-temporal-prediction (STP) method for multichannel speech enhancement has recently been proposed. This approach makes it theoretically possible to attenuate the residual noise without distorting speech. In addition, the STP method depends only on the second-order statistics and can be implemented using a simple linear filtering framework. Unfortunately, some numerical problems can arise when estimating the filter matrix in transients. In such a case, the speech correlation matrix is usually rank deficient, so that no solution exists. In this paper, we propose to implement the spatio-temporal-prediction method using a signal subspace approach. This allows for nullifying the noise subspace and processing only the noisy signal in the signal-plus-noise subspace. As a result, we are able to not only regularize the solution in transients but also to achieve higher attenuation of the residual noise. The experimental results also show that the signal subspace approach distorts speech less than the conventional method.

Keywords: Signal subspace; Spatio-temporal prediction; Speech enhancement

Introduction

Speech enhancement is important for many applications including mobile communications, speech coding, speech recognition, and hearing aids. The traditional objective of multichannel speech enhancement is to recover the source speech signal from the outputs of an array of microphones. It is usually achieved by using the beamforming techniques [1-3]. The key idea of beamforming is to process signals of a microphone array, so as to extract the sounds that come from only one direction. In this way, it is possible to dereverberate speech, but the background noise can be reducted as well by avoiding noise directions. Unfortunately, in order to work reasonably well in a reverberant environment, these techniques usually require knowing the impulse responses of the acoustic room or their relative ratios. These parameters can be fixed, provided the geometry of the microphone array is known, or estimated adaptively [4], which in general is a difficult task, however.

Recently, the objective of multichannel speech enhancement has been reformulated, so that noise reduction can be achieved without dereverberating speech. In opposition to the beamforming techniques, the knowledge about the geometry of the microphone array is not required, and the optimal filter depends only on the second-order statistics of the noisy signal.

In [5], the authors presented the most common techniques of multichannel noise reduction based on linear filtering. In such solutions, the noise-free speech is estimated by a linear transformation of the observation vector. The simplest approach is to minimize the mean square error (MSE) between the noise-free and filtered speech signals at a given microphone, which leads to a multichannel version of the classical Wiener filter. In this case, some noise is reduced at the cost of the increased speech distortion, but we cannot explicitly control the trade-off between these quantities.

Speech estimation can also be considered as a constrained optimization problem, where the speech distortions are minimized subject to the residual noise power. This approach is used by the single-channel methods [6] and was implemented in a similar way using a signal subspace technique in [5]. Unlike the frequency

Correspondence: a.borowicz@pb.edu.pl
Department of Computer Graphics and Digital Media, Faculty of Computer Science, Bialystok University of Technology, Wiejska str. 45A, 15-351 Bialystok, Poland

domain methods, which are based on the discrete Fourier transform (DFT), the signal subspace approach decomposes the vector space of noisy signals into the speech-plus-noise subspace and noise-only subspace using the Karhunen-Loeve transform (KLT). Then, spectral weighting is performed only in the signal-plus-noise subspace. The components projected onto the noise-only subspace are simply nullified, which results in significantly better performance when compared to the conventional DFT-based methods, where the full-band (and thus erroneous) spectrum must be processed. Unfortunately, also in this case, it is impossible to reduce the residual noise without introducing speech distortions. Several single-channel approaches [7-9] that exploit the masking effects are known to make the speech distortion or the residual noise inaudible, but introducing psychoacoustics into multichannel speech enhancement is a challenging task. On the other hand, some hearing properties have been introduced in a beamforming technique [10], but the resulting improvement is not as great as in the single-channel case.

It seems that the major limitation of all these methods is that they use only temporal prediction. In fact, spatial correlations are implicitly embedded in the second-order statistics, or inter-channel correlation matrices, but are not explicitly used. Therefore, in [11,12], the authors proposed a novel technique based on the spatio-temporal prediction (STP). A DFT-based implementation of this technique has also been proposed [13,14], but in this case, the algorithm has been restricted to use only spatial prediction. It has been verified experimentally that the STP approach outperforms the classical beamforming techniques in terms of noise reduction [11]. In [5], it was proved analytically that by using the STP method, it is theoretically possible to reduce the residual noise without distorting the speech. However, a major drawback of the STP method is its numerical instability, as this approach assumes that speech correlation matrix is of full rank. Because this is not true for low power speech at transients, the solution must be regularized empirically in practice. Alternatively, under the uncertainty about the speech presence, the conditional estimators can be used [15]. Even if the speech correlation matrix is of full rank, the STP method requires many microphones to effectively reduce the residual noise.

In this paper, we propose a signal-subspace implementation of the STP method. By decomposing the signal vector space, we are able to limit processing to the signal-plus-noise subspace only. Thus, the numerical problems can be evaded in a more natural way. Since the noisy speech projected on the noise-only subspace can simply be nullified, the signal subspace approach allows for attenuating noise more, even for a small number of microphones. In addition, we have rederived the STP method using a notation slightly different from that in [5], in order to expose the possibility of denoising all microphone signals at once.

Signal model and linear filtering

Let us consider an array of N microphones with arbitrary geometry and a single speech source $s(k)$ located inside a reverberant enclosure. The observation signal at the nth microphone is given by:

$$y_n(k) = a_n(k) * s(k) + v_n(k) = x_n(k) + v_n(k), \quad (1)$$

where $*$ denotes convolution, a_n is the acoustic impulse response from the source to the nth microphone, and $x_n(k)$ and $v_n(k)$ are, respectively, the noise-free speech and the noise components received by the nth microphone. Such a mixing model is illustrated in Figure 1.

Usually data are processed in L-sample blocks. Thus, the signals can be represented using the vector-matrix notation as follows:

$$\mathbf{y}_n(k) = \left[y_n(k)\ y_n(k-1)\ \ldots\ y_n(k-L+1)\right]^T. \quad (2)$$

The estimate of the noise-free speech at the nth microphone can be obtained using a linear transformation of the observation vector:

$$\hat{\mathbf{x}}_n(k) = \mathbf{H}_n\mathbf{y}(k) = \mathbf{H}_n\left[\mathbf{x}(k) + \mathbf{v}(k)\right], \quad (3)$$

where:

$$\begin{aligned}\mathbf{y}(k) &= \left[\mathbf{y}_1^T(k)\ \mathbf{y}_2^T(k)\ \ldots\ \mathbf{y}_N^T(k)\right]^T,\\ \mathbf{x}(k) &= \left[\mathbf{x}_1^T(k)\ \mathbf{x}_2^T(k)\ \ldots\ \mathbf{x}_N^T(k)\right]^T,\\ \mathbf{v}(k) &= \left[\mathbf{v}_1^T(k)\ \mathbf{v}_2^T(k)\ \ldots\ \mathbf{v}_N^T(k)\right]^T.\end{aligned} \quad (4)$$

The vectors $\mathbf{x}_n(k)$ and $\mathbf{v}_n(k)$ denote the noise-free speech and the noise, respectively, and are defined similarly to Equation 2. $\mathbf{H}_n$ is a filtering matrix of size $L \times LN$. The estimation error is defined by:

$$\begin{aligned}\mathbf{e}(k) &= \hat{\mathbf{x}}_n(k) - \mathbf{x}_n(k)\\ &= \underbrace{(\mathbf{H}_n - \mathbf{U}_n)\mathbf{x}(k)}_{\mathbf{e}_x(k)} + \underbrace{\mathbf{H}_n\mathbf{v}(k)}_{\mathbf{e}_v(k)},\end{aligned} \quad (5)$$

where:

$$\mathbf{U}_n = \left[\mathbf{0}_{L\times(n-1)L}\ \mathbf{I}_L\ \mathbf{0}_{L\times(N-n)L}\right], \quad (6)$$

is a selection matrix of size $L \times LN$. The terms $\mathbf{e}_x(k)$ and $\mathbf{e}_v(k)$ denote the speech distortion and the residual noise, respectively.

For completeness, we also define the correlation matrix of an arbitrary vector $\mathbf{a}$ as:

$$\mathbf{R}_{\mathbf{aa}}(k) = E\left\{\mathbf{a}(k)\mathbf{a}^T(k)\right\}, \quad (7)$$

where $E\{.\}$ is the expectation operator. Assuming that the speech and noise are short-term stationary

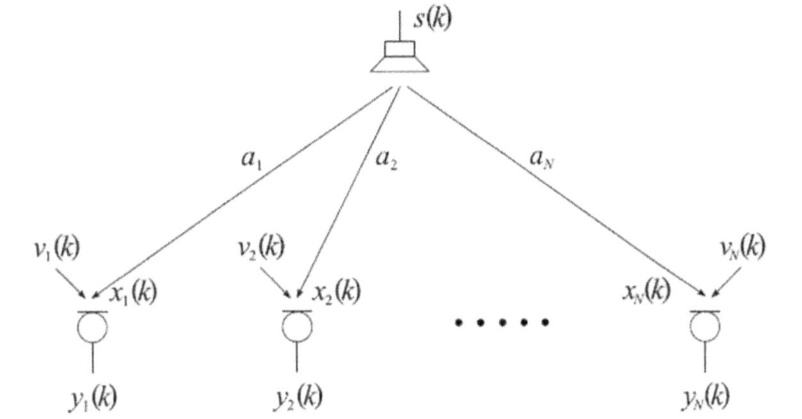

Figure 1 Multi-microphone signal model.

and uncorrelated processes, the correlation matrix of the noisy speech can be written as:

$$\mathbf{R_{yy}}(k) = \mathbf{R_{xx}}(k) + \mathbf{R_{vv}}(k). \quad (8)$$

Unless otherwise stated, all equations hold for any arbitrarily chosen point in time. Therefore, for the sake of brevity, the time index k is often omitted in the rest of this paper.

Spatio-temporal prediction

The STP method is based on the assumption that the microphone signals can be predicted not only in the time domain but also in the space domain [11]. In particular, the signal $\mathbf{x}_m(k)$ can be predicted from the signal $\mathbf{x}_n(k)$ using a linear filter matrix $\mathbf{W}_{n,m}$ such that:

$$\mathbf{x}_m(k) = \mathbf{W}_{n,m}^T \mathbf{x}_n(k), \quad m = 1, 2, 3, \ldots, N, \quad (9)$$

with $\mathbf{W}_{n,n} = \mathbf{I}_L$. The prediction matrices can be concatenated so as to form the $L \times NL$ matrix:

$$\mathbf{W}_n = \left[\mathbf{W}_{n,1}\ \mathbf{W}_{n,2}\ \ldots\ \mathbf{W}_{n,N}\right], \quad (10)$$

and:

$$\mathbf{x}(k) = \mathbf{W}_n^T \mathbf{x}_n(k). \quad (11)$$

By substituting Equation 11 into Equation 5 and assuming that $\mathbf{H}_n \mathbf{W}_n^T = \mathbf{I}_L$, we can deduce that the residual noise can be minimized without distorting speech. Thus, the constrained optimization problem is formulated as follows:

$$\min_{\mathbf{H}_n} \operatorname{tr} \left\{ E\left[\mathbf{e}_v(k)\mathbf{e}_v^T(k)\right]\right\} \text{ subject to } \mathbf{H}_n \mathbf{W}_n^T = \mathbf{I}_L. \quad (12)$$

The optimal filter matrix is found using the Lagrange multipliers method:

$$\mathbf{H}_n = \left(\mathbf{W}_n \mathbf{R_{vv}}^{-1} \mathbf{W}_n^T\right)^{-1} \mathbf{W}_n \mathbf{R_{vv}}^{-1}. \quad (13)$$

A solution exists if and only if $\mathbf{R_{vv}}$ is positive definite, and the matrix $\mathbf{W}_n$ is of rank L. As noise signals are usually stationary and have smooth spectra, $\mathbf{R_{vv}}$ has full rank and can be estimated using long-term averaging during speech pauses.

Unfortunately the prediction matrices $\mathbf{W}_{n,m}$ for $m \neq n$ are not known and have to be estimated. They can be found by solving the following minimization problem:

$$\min_{\mathbf{W}_{n,m}} E\left\{\left[\mathbf{x}_m(k) - \mathbf{W}_{n,m}^T \mathbf{x}_n(k)\right]^T \left[\mathbf{x}_m(k) - \mathbf{W}_{n,m}^T \mathbf{x}_n(k)\right]\right\}. \quad (14)$$

whose solution is given by:

$$\mathbf{W}_{n,m}^T = \mathbf{R}_{\mathbf{x}_m\mathbf{x}_n} \mathbf{R}_{\mathbf{x}_n\mathbf{x}_n}^{-1} \quad (15)$$

where $\mathbf{R}_{\mathbf{a}_i\mathbf{a}_j}$ stands for the (i, j)th $L \times L$ submatrix of the matrix $\mathbf{R_{aa}}$. The correlation matrices of the clean speech are unknown, and the vectors $\mathbf{x}_n(k)$ cannot be observed directly, but by using Equation 8 we can write:

$$\mathbf{R}_{\mathbf{x}_n\mathbf{x}_m} = \mathbf{R}_{\mathbf{y}_n\mathbf{y}_m} - \mathbf{R}_{\mathbf{v}_n\mathbf{v}_m}, \quad m = 1, 2, \ldots, N. \quad (16)$$

Thus, finally, we obtain the following expression for the prediction matrices:

$$\mathbf{W}_{n,m}^T = \left(\mathbf{R}_{\mathbf{y}_m\mathbf{y}_n} - \mathbf{R}_{\mathbf{v}_m\mathbf{v}_n}\right)\left(\mathbf{R}_{\mathbf{y}_n\mathbf{y}_n} - \mathbf{R}_{\mathbf{v}_n\mathbf{v}_n}\right)^{-1}. \quad (17)$$

In order to obtain a full rank matrix $\mathbf{W}_{n,m}$, the matrices $\mathbf{R}_{\mathbf{x}_m\mathbf{x}_n}$ and $\mathbf{R}_{\mathbf{x}_n\mathbf{x}_n}$ have to be positive definite. In [5], the authors suggest to estimate the filter matrix (Equation 13) only when the speech source is active, using a voice activity detector (VAD), but this generally does not prevent the matrix $\mathbf{W}_{n,m}$ from being rank deficient. Moreover, such a technique can introduce discontinuity effects at transients or/and increased residual noise during silence intervals. For low-power speech signals, the covariance matrix of the clean speech is usually positive semi-definite, or at least ill-conditioned, which means that in practice the STP method is numerically stable only for high signal-to-noise ratios (SNRs). The simplest solution is to add some white noise to the speech signal, so that the inverses in

Equation 13 and Equation 17 can be replaced with pseudoinverses and properly regularized [16]. However, all these approaches are rather empirical and need a careful adjustment. Thus, we need a more robust solution, which can be applied also to low power speech signals, especially at low SNRs.

Signal subspace approach

In the conventional STP method, data are processed in the vector space of the noisy speech. The key idea of the signal subspace approach is to decompose that vector space into the signal-plus-noise and noise-only subspaces and to process data only in the signal-plus-noise subspace, while the projection of the noisy signal onto the noise-only subspace is simply nullified. The dimensionality of the signal-plus-noise or, simply, signal subspace is closely related to the rank of the speech correlation matrix. Thus, by introducing the signal subspace approach to the STP method, we are able to not only increase the attenuation of the residual noise during silence intervals but also to avoid the ill-conditioning issues.

Let us rewrite Equation 13 more compactly. Please notice that the prediction matrix can be alternatively written as:

$$\mathbf{W}_n = \mathbf{R}^{-1}_{\mathbf{x}_n\mathbf{x}_n}\mathbf{U}_n\mathbf{R}_{\mathbf{xx}}, \tag{18}$$

and then, by substituting the above into Equation 13, we obtain:

$$\mathbf{H}_n = \mathbf{R}_{\mathbf{x}_n\mathbf{x}_n}\left(\mathbf{U}_n\mathbf{R}_{\mathbf{xx}}\mathbf{R}_{vv}^{-1}\mathbf{R}_{\mathbf{xx}}\mathbf{U}_n^T\right)^{-1}\mathbf{U}_n\mathbf{R}_{\mathbf{xx}}\mathbf{R}_{vv}^{-1}. \tag{19}$$

Since $\mathbf{R}_{\mathbf{vv}}$ is positive definite, the matrices $\mathbf{R}_{\mathbf{xx}}$ and $\mathbf{R}_{\mathbf{vv}}$ can be jointly diagonalized [17,18], i.e.:

$$\mathbf{R}_{\mathbf{vv}}^{-1/2}\mathbf{R}_{\mathbf{xx}}\mathbf{R}_{\mathbf{vv}}{}^{-1/2} = \mathbf{V}\Lambda\mathbf{V}^T, \tag{20}$$

where $\mathbf{V}$ denotes the orthogonal matrix of the eigenvectors, and $\Lambda = \operatorname{diag}\{\lambda_1, \ldots, \lambda_{NL}\}$ is the diagonal matrix of the corresponding eigenvalues. We also assume that the eigenvalues in Λ are arranged in descending order, i.e. $\lambda_i \geq \lambda_j$ for any $i < j$. The matrix $\mathbf{V}$ can also be interpreted as the KLT matrix of the whitened clean speech. Alternatively, it can be obtained using the eigendecomposition of the whitened noisy speech correlation matrix:

$$\mathbf{R}_{\mathbf{vv}}^{-1/2}\mathbf{R}_{\mathbf{yy}}\mathbf{R}_{\mathbf{vv}}{}^{-1/2} = \mathbf{V}\left(\Lambda + \mathbf{I}\right)\mathbf{V}^T. \tag{21}$$

As shown in [17], the vector space of the noisy speech can be decomposed using the square matrix:

$$\mathbf{B} = \mathbf{V}^T\mathbf{R}_{\mathbf{vv}}^{1/2} \tag{22}$$

which has full rank but is not necessarily orthogonal. Please notice that applying $\mathbf{B}^{-T}$ to the noisy signal is equivalent to whitening data before performing the subspace decomposition, so that the resulting coefficients are perfectly decorrelated in the transform domain, i.e.:

$$E\left[\tilde{\mathbf{y}}(k)\tilde{\mathbf{y}}^T(k)\right] = \Lambda + \mathbf{I}, \tag{23}$$

where $\tilde{\mathbf{y}}(k) = \mathbf{B}^{-T}\mathbf{y}(k)$. Thus, our correlation matrices can be expressed as follows:

$$\begin{aligned}\mathbf{R}_{\mathbf{yy}} &= \mathbf{B}^T(\Lambda + \mathbf{I})\mathbf{B}\\ \mathbf{R}_{\mathbf{xx}} &= \mathbf{B}^T\Lambda\mathbf{B}\\ \mathbf{R}_{\mathbf{vv}} &= \mathbf{B}^T\mathbf{B}\end{aligned} \tag{24}$$

Let $\mathbf{Q}_n = \Lambda\mathbf{B}\mathbf{U}_n^T$. Substituting the relations given in Equation 24 into Equation 19 results in the optimal filter matrix:

$$\mathbf{H}_n = \mathbf{U}_n\mathbf{B}^T\left[\mathbf{Q}_n\left(\mathbf{Q}_n^T\mathbf{Q}_n\right)^{-1}\mathbf{Q}_n^T\right]\mathbf{B}^{-T}. \tag{25}$$

Since $\mathbf{R}_{\mathbf{vv}}$ is positive definite, and $\mathbf{R}_{\mathbf{xx}}$ can be semi-positive definite, the dimension of the signal-plus-noise subspace is equal to the number of non-zero eigenvalues of the correlation matrix of the whitened clean speech. Assume that $NL = L_s + L_v$, where L_s and L_v denote the dimensions of the signal-plus-noise and noise-only subspaces, respectively. Thus, for $L_s < NL$, we can rewrite Equation 25 as follows:

$$\mathbf{H}_n = \mathbf{U}_n\mathbf{B}^T\begin{bmatrix}\Sigma_n & \mathbf{0}_{L_s\times L_v}\\ \mathbf{0}_{L_v\times L_s} & \mathbf{0}_{L_v\times L_v}\end{bmatrix}\mathbf{B}^{-T}, \tag{26}$$

where:

$$\Sigma_n = \mathbf{Q}_{n,1:L_s}\left[\mathbf{Q}_{n,1:L_s}^T\mathbf{Q}_{n,1:L_s}\right]^{-1}\mathbf{Q}_{n,1:L_s}^T \tag{27}$$

can be viewed as a reweighting matrix, with $\mathbf{Q}_{n,1:L_s}$ denoting sub-matrix of $\mathbf{Q}_n$ consisting rows from 1 to L_s. As can be seen the noisy signal is transformed using a non-orthogonal matrix $\mathbf{B}^{-T}$. The denoising is achieved by 'reweighting' the coefficients in the signal-plus-noise subspace using the matrix Σ_n and simply nullifying the noise-only subspace. In opposition to the conventional signal subspace approach, the reweighting matrix is not diagonal here but symmetric and idempotent.

Finally, the filtered signal is brought back to the time domain using the inverse transform $\mathbf{B}^T$.

In practice, L_s can be estimated as the number of the strictly positive eigenvalues, according to the following rule:

$$L_s \approx \underset{1\geq l\geq NL}{\arg\max}\ \{\lambda_l > \theta\}, \tag{28}$$

where the threshold θ is a some small positive number.

It can be noticed that $\mathbf{Q}_n^T\mathbf{Q}_n$ is invertible as long as $L_s \geq L$. However, even when this condition is not in force (which is fairly common at transients or during silence intervals), the inverse can be easily regularized. For example, if $L_s = L$, $\mathbf{Q}_{n,1:L}$ is a square matrix, and

$\Sigma_n = \mathbf{I}$, which means that the filter performs nullifying the noise subspace without cleaning the signal-plus-noise subspace, or that the residual noise can be effectively reduced without distorting the speech.

Therefore, in order to regularize the solution, the best we can do is to use the following rule:

$$\Sigma_n = \begin{cases} \mathbf{Q}_{n,1:L_s}\left[\mathbf{Q}_{n,1:L_s}^T \mathbf{Q}_{n,1:L_s}\right]^{-1}\mathbf{Q}_{n,1:L_s}^T, & L_s > L \\ \mathbf{I}_{L_s}, & \text{otherwise.} \end{cases} \tag{29}$$

Please also notice that if $N = 1$ and $L_s = L$, then the filter matrix is simply the identity matrix. For $N > 1$, it is possible to arrange matrices $\mathbf{H}_n$, $n = 1, 2, \ldots, N$ into the single filter matrix:

$$\mathbf{H}_P = \left[\mathbf{H}_1^T\ \mathbf{H}_2^T\ \ldots\ \mathbf{H}_N^T\right], \tag{30}$$

which can be used to estimate all noise-free microphone signals at once. Namely, the vector $\mathbf{x}(k)$ can be estimated as follows:

$$\mathbf{x}(k) \approx \hat{\mathbf{x}}(k) = \mathbf{H}_P \mathbf{y}(k). \tag{31}$$

The filter matrix $\mathbf{H}_P$ can also be written in a more convenient form:

$$\mathbf{H}_P = \left[\mathbf{U} \circ \left(\mathbf{B}^T \Lambda \mathbf{B}\right)\right]\left[\mathbf{U} \circ \left(\mathbf{B}^T \Lambda^2 \mathbf{B}\right)\right]^{-1} \mathbf{B}\Lambda\mathbf{B}^{-T}, \tag{32}$$

where:

$$\mathbf{U} = \mathbf{I}_N \otimes \mathbf{J}_{L\times L}, \tag{33}$$

and the operators $\circ$ and $\otimes$ stand for the Hadamard and the Kronecker products, respectively, and $\mathbf{J}_{L\times L}$ is the $L \times L$ matrix of ones.

The proposed approach can be verified analytically in terms of noise reduction and speech distortion. The noise reduction factor can be defined for any filter matrix $\mathbf{H}_n$ as follows:

$$\xi_{nr}(\mathbf{H}_n) = \frac{\mathrm{tr}\left\{E\left[\mathbf{U}_n \mathbf{v}\mathbf{v}^T \mathbf{U}_n^T\right]\right\}}{\mathrm{tr}\left\{E\left[\mathbf{H}_n \mathbf{v}\mathbf{v}^T \mathbf{H}_n^T\right]\right\}} = \frac{\mathrm{tr}\left\{\mathbf{U}_n \mathbf{R}_{\mathbf{vv}} \mathbf{U}_n^T\right\}}{\mathrm{tr}\left\{\mathbf{H}_n \mathbf{R}_{\mathbf{vv}} \mathbf{H}_n^T\right\}}. \tag{34}$$

It is expected that $\xi_{nr}(k) \geq 1$: the larger this factor, the lower residual noise. Usually, the noise is reduced at the cost of attenuating speech. Therefore, in order to quantify this attenuation, we define the speech reduction factor:

$$\xi_{sr}(\mathbf{H}_n) = \frac{\mathrm{tr}\left\{E\left[\mathbf{U}_n \mathbf{x}\mathbf{x}^T \mathbf{U}_n^T\right]\right\}}{\mathrm{tr}\left\{E\left[\mathbf{H}_n \mathbf{x}\mathbf{x}^T \mathbf{H}_n^T\right]\right\}} = \frac{\mathrm{tr}\left\{\mathbf{U}_n \mathbf{R}_{\mathbf{xx}} \mathbf{U}_n^T\right\}}{\mathrm{tr}\left\{\mathbf{H}_n \mathbf{R}_{\mathbf{xx}} \mathbf{H}_n^T\right\}} \tag{35}$$

and expect $\xi_{sr}(\mathbf{H}_n) \geq 1$. The output SNR of the filter $\mathbf{H}$ can be expressed in the following way:

$$\mathrm{SNR}(\mathbf{H}) = \frac{\mathrm{tr}\left\{\mathbf{H}_n \mathbf{R}_{\mathbf{xx}} \mathbf{H}_n^T\right\}}{\mathrm{tr}\left\{\mathbf{H}_n \mathbf{R}_{\mathbf{vv}} \mathbf{H}_n^T\right\}} = \mathrm{SNR}\frac{\xi_{nr}(\mathbf{H})}{\xi_{sr}(\mathbf{H})}, \tag{36}$$

where the SNR stands for the input SNR.

For $L_s \geq L$, the proposed approach is theoretically equivalent to the time-domain implementation of the STP method. In order to analyse performance of the proposed implementation for $L_s < L$, we consider the case of the white noise, for which $\mathbf{R}_{\mathbf{v}_n\mathbf{v}_n} = \sigma_{\mathbf{v}_n}\mathbf{I}$. Because the inverse $\left(\mathbf{Q}_n^T\mathbf{Q}_n\right)^{-1}$ does not exist for $L_s < L$, we use Equation 29. Then, by replacing Σ_n in Equation 26 with the identity matrix and by substituting it to Equation 34 and Equation 35, we obtain:

$$\xi_{nr}(\mathbf{H}_n) = \frac{L}{\sum_{i=1}^{L}\sum_{j=1}^{L_s} \mathbf{V}_{(n-1)L+i,j}^2} > 1 \tag{37}$$

$$\xi_{sr}(\mathbf{H}_n) = 1. \tag{38}$$

Since $\xi_{nr}(\mathbf{H}_n) > \xi_{sr}(\mathbf{H}_n)$, we always have $\mathrm{SNR}(\mathbf{H}) > \mathrm{SNR}$, or an improvement of the SNR.

Simulations

Although a full evaluation of the proposed approach, including listening tests, is out of the scope of this article, we have conducted some experiments using objective measurements. In this section, we compare the performances of the conventional time-domain implementation of the STP method and of the proposed approach based on the signal subspace.

Implementation

Both methods have been implemented in MATLAB. Instead of recalculating the filter from sample to sample, we collect the microphone recordings in overlapped buffers and process them frame-by-frame in a similar way as in [8] or [19]. Namely, we divide the microphone signals into frames of length N_f with 50% overlap. Each frame is partitioned into $M = N_f - L + 1$ shorter overlapping L-dimensional vectors. The sequence of these vectors is arranged into the trajectory matrix of size L-by-M. The trajectory matrices for all microphones are concatenated together so as to form the noisy speech matrix $\mathbf{Y}(k)$ of size LN-by-M so that:

$$\mathbf{Y}(k) = \left[\,\mathbf{y}(k)\ \mathbf{y}(k-1)\ \cdots\ \mathbf{y}(k-M+1)\,\right]. \tag{39}$$

As all required parameters are estimated, the effective filter matrix $\mathbf{H}_n$ is computed, and then all in-frame vectors are processed using the same matrix, i.e. $\hat{\mathbf{Y}}(k) = \mathbf{H}_n\mathbf{Y}(k)$. The enhanced vectors are obtained from the matrix $\hat{\mathbf{Y}}(k)$ using the diagonal averaging technique [19]. Finally, the frames are multiplied by the

Hanning window and synthesized using the overlap-add method.

The correlation matrix of the noisy speech can be estimated according to:

$$\mathbf{R_{yy}}(k) \approx \frac{1}{MN}\mathbf{Y}(k)\mathbf{Y}(k)^T, \tag{40}$$

being the outer product of the matrix $\mathbf{Y}(k)$. This estimate is the basis for computing both noise statistics and the KLT of the whitened signal (Equation 20). The matrix $\mathbf{R_{vv}}$ is estimated only during speech pauses as:

$$\mathbf{R_{vv}}(k) \approx \begin{cases} \alpha\mathbf{R_{vv}}(k-1) + (1-\alpha)\mathbf{R_{yy}}(k), & \text{if} \quad I(k) = 1 \\ \mathbf{R_{vv}}(k-1), & \text{otherwise} \end{cases} \tag{41}$$

where $0 < \alpha < 1$ is the forgetting factor, and $I(k)$ is the VAD output of the kth frame. In our simulations, the VAD was not implemented, and the speech pause/activity regions were marked manually.

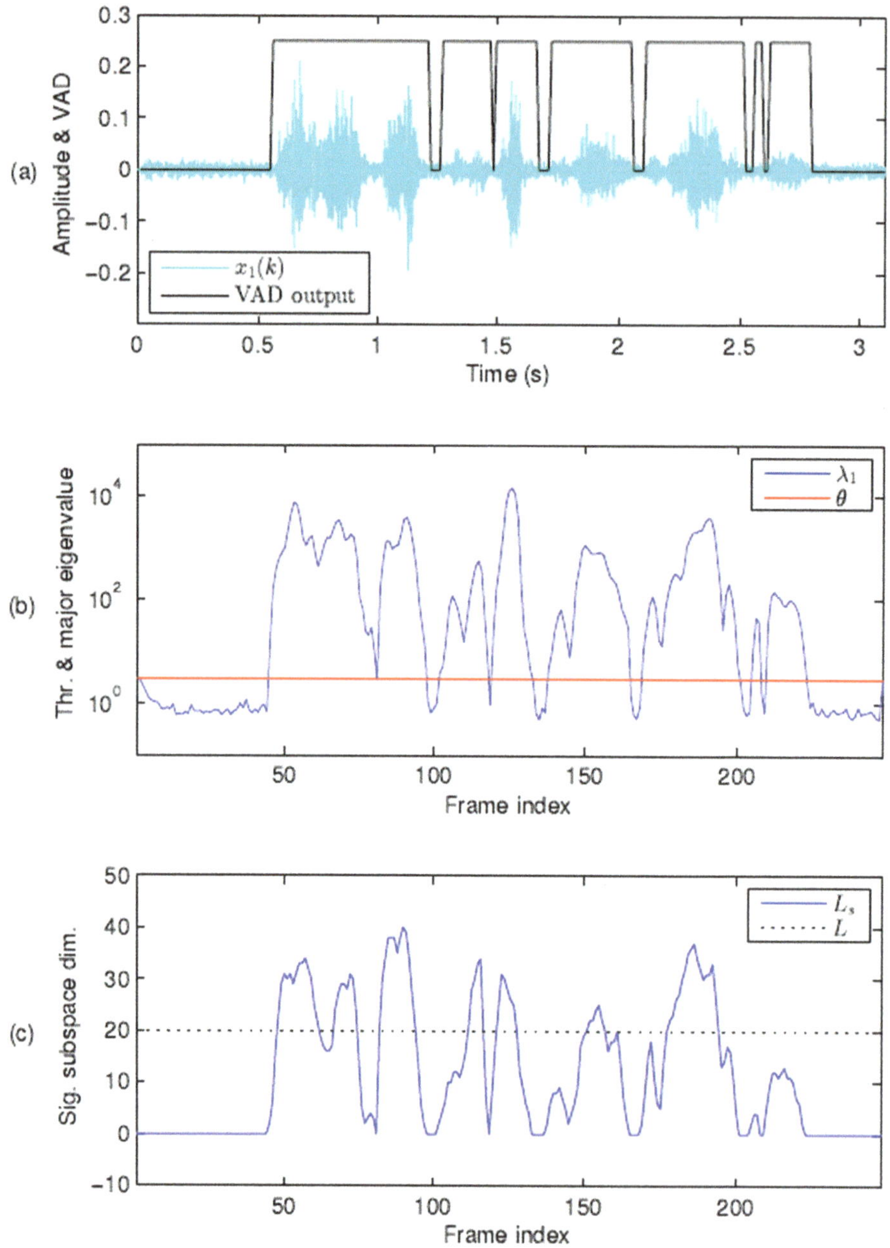

Figure 2 Estimation of the dimension of the signal-plus-noise subspace. (a) Example noisy speech signal at SNR = 10 dB. **(b)** The parameter θ and major eigenvalue of the whitened clean speech. **(c)** Estimate of the dimension of the signal-plus-noise subspace.

In most cases, the noise correlation matrix is positive definite, so that the computations of both whitening and unwhitening transformations ($\mathbf{R}_{\mathbf{vv}}^{-1/2}, \mathbf{R}_{\mathbf{vv}}^{1/2}$, respectively) should be numerically stable. The transformations can be calculated at once using the eigenstructures of the matrix $\mathbf{R}_{\mathbf{vv}} = \mathbf{V}_{\mathbf{v}}\Lambda_{\mathbf{v}}\mathbf{V}_{\mathbf{v}}{}^{T}$ in the following way:

$$\begin{aligned} \mathbf{R}_{\mathbf{vv}}^{-1/2} &= \mathbf{V}_{\mathbf{v}}\Lambda_{\mathbf{v}}^{-1/2}\mathbf{V}_{\mathbf{v}}{}^{T}, \\ \mathbf{R}_{\mathbf{vv}}^{1/2} &= \mathbf{V}_{\mathbf{v}}\Lambda_{\mathbf{v}}^{1/2}\mathbf{V}_{\mathbf{v}}{}^{T}, \end{aligned} \tag{42}$$

where $\mathbf{V}_{\mathbf{v}}$ denotes the orthogonal matrix of the eigenvectors, and $\Lambda_{\mathbf{v}}$ is the diagonal matrix of the corresponding eigenvalues.

In our experiments, we take $\alpha = 0.75$, $N_f = 400$, and $L = 20$. A proper choice of the value of the parameter θ seems to be crucial for the proposed implementation. In general, greater values of θ lead to cancellation of the residual noise, but a special care must be taken because low-power speech components can be also nullified. Therefore, the simplest solution is to fix this threshold, so that it is large enough to give $L_s = 0$ (or equivalently $\theta \gg \lambda_1$) during speech pauses. We found empirically that its value depends mainly on the bias of the estimator of the noise correlation matrix, i.e. on the forgetting factor α and the frame/window size N_f. In Figure 2c, we present the variability of the estimated dimension of the signal-plus-noise subspace for the parameter $\theta = 3$. Further experiments show that the optimal value of the parameter θ (in terms of speech distortion) does not depend on the input SNR. It can be observed that $L_s < L$ occurs fairly commonly, not only at transients, but also during speech activity.

In the case of the conventional implementation, all inverses in Equation 17 and Equation 13 were replaced with pseudoinverses. They were computed using singular value decomposition (SVD), and all singular values less than some tolerance were treated as zeros. In fact, that tolerance plays the same role as the parameter θ in the signal subspace approach. Thus, by setting it sufficiently large, it is possible to increase noise reduction. Unfortunately, the speech reduction factor is also increased. Additionally, we have found empirically that the optimal tolerance is SNR dependent. Therefore, during our simulations, all SVD-based pseudoinverses were computed using the default tolerance set by MATLAB.

Objective evaluation

The acoustic environment was simulated using the image method [20]. We assumed that the enclosure is rectangular with dimensions 6 × 5 × 2.8 (all dimensions and coordinates are in meters). A uniform linear array of eight microphones was placed along the x-axis, with spacing 0.1 and beginning from the first microphone at the position $(2.65, 4, 1)$. The locations of the microphones and the sound sources are shown in Figure 3. The source speech signal was sampled at 16 kHz. The signal was about 14-s-long and comprised of four short sentences uttered by male and female speakers (see Additional file 1). In order to represent general broadband signals the pink noise was chosen. The microphone signals were obtained by convolving the source speech signal with the generated

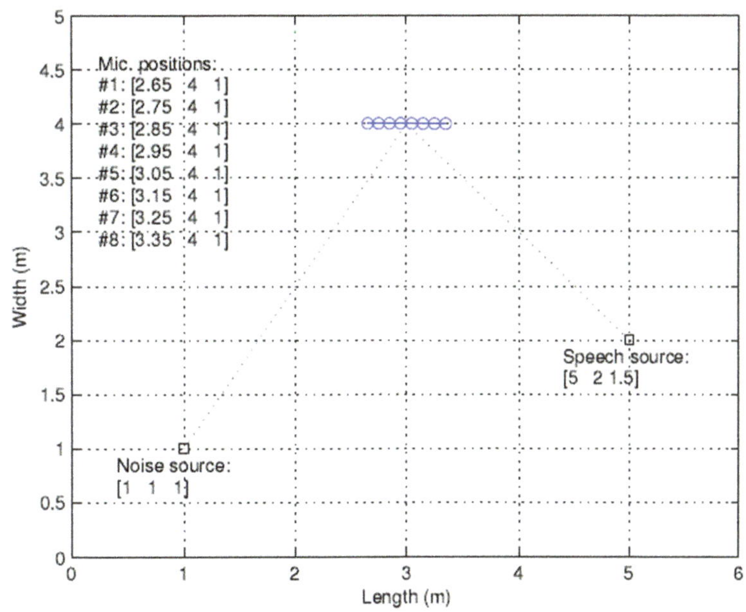

Figure 3 Floor plan of the simulated enclosure (all coordinates in meters).

impulse responses of a room, and by adding noise signals at SNRs ranged from −5 to 20 dB, in accordance with Equation 1. An example noisy speech sample is provided as the Additional file 2. In all experiments, we estimated the noise-free signal only at the first microphone, $n = 1$, which served as the reference microphone.

The SNR-based measures were used for evaluating the objective performance. The speech distortion measure (SD) was defined as the segmental signal-to-noise ratio, in which the noise was identified with the difference between the source signal and enhanced speech. The higher the value of this factor, the better the performance. The amount of reduced noise was measured using the noise attenuation (NA) factor defined as the mean ratio between the input noise power and output noise power.

Firstly, taking into consideration only on the first four microphones, we have evaluated the impact of the

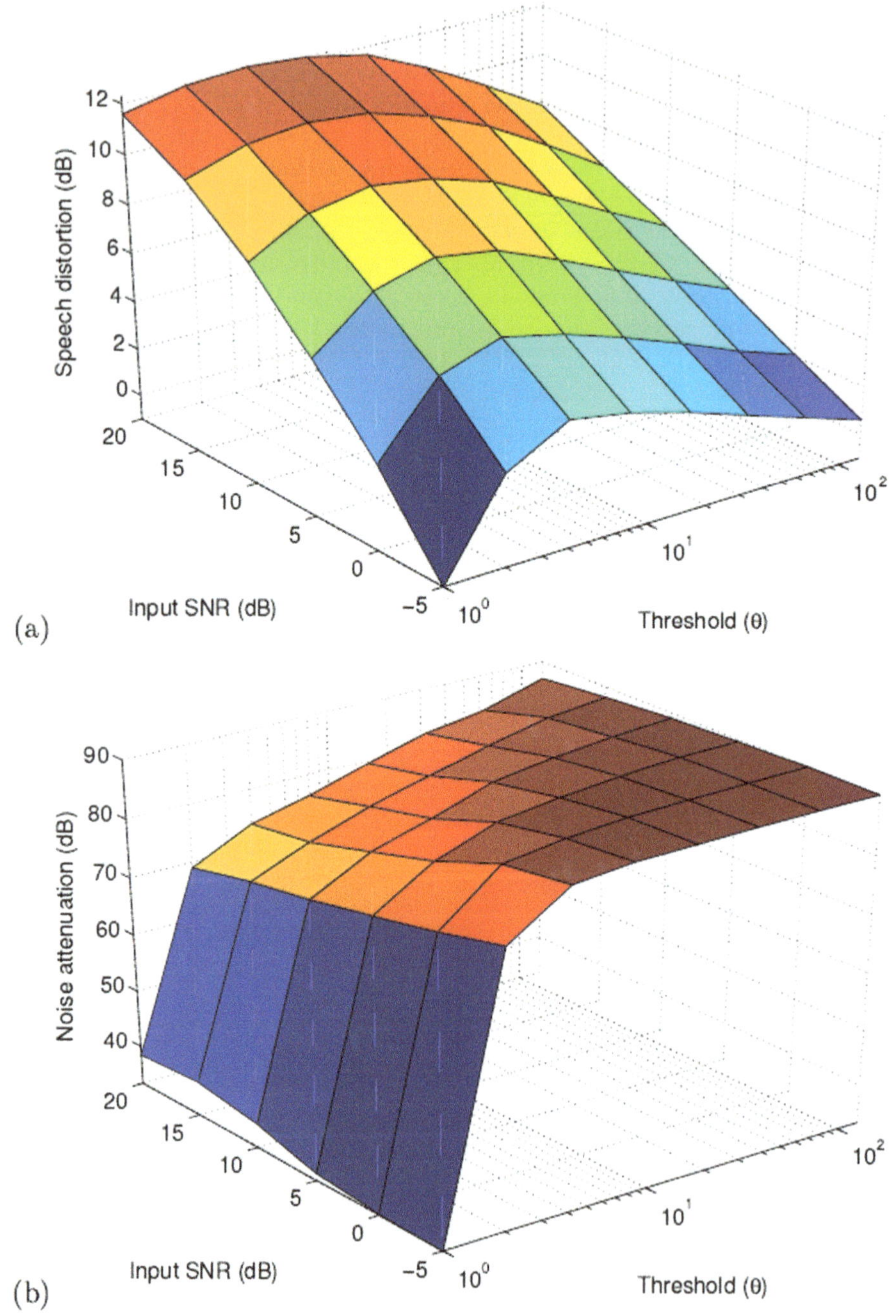

Figure 4 Adjustment of the parameter θ for $N = 4$. (a) Speech distortion measure (SD) and **(b)** noise attenuation factor (NA).

parameter θ on speech distortion and noise attenuation. The measured speech distortion, which is shown in Figure 4a, indicates rather weak influence of the parameter θ on the input SNR. The optimal value of θ is between 3 to 4 for all SNRs. On the other hand, the plot of the noise attenuation factor in Figure 4b, demonstrates that the higher the value of the θ, the higher noise attenuation.

The subsequent simulations were performed for $\theta = 3$ and $N = 2, 3, \ldots, 8$. For conciseness, we present in Figure 5 only the results of objective measurements of the systems with $N = 2, 4$, and eight microphones. Example recordings of the speech enhanced using conventional and proposed method are provided as Additional files 3 and 4, respectively.

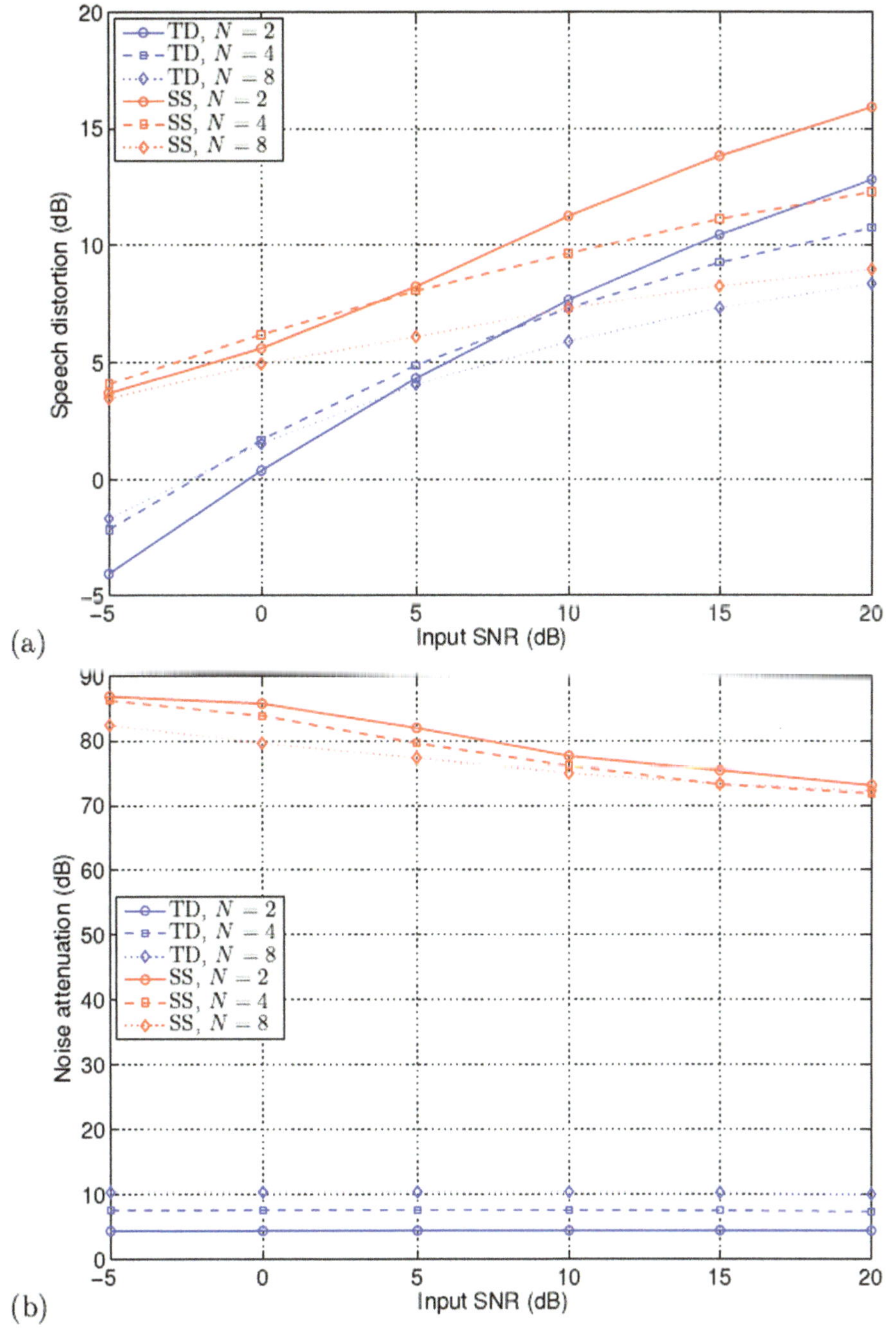

Figure 5 Objective measurement of the time-domain (TD) and signal subspace (SS) implementations. (a) Speech distortion and **(b)** noise attenuation factor.

It can easily be seen that the proposed method outperforms the conventional one, as it provides lower speech distortions and higher noise attenuation. Surprisingly, the speech distortion for the system with $N = 2$ microphones was lower than for the eight-microphone system, especially at high SNRs. A possible explanation of this phenomena is that for more microphones, the correlation matrix is larger, which makes the estimation less accurate. In practice, it makes sense to use more microphones only in the conventional time-domain method (in order to improve the noise attenuation). Figure 5a shows that the speech distortion can be also decreased but only at low SNRs.

Unlike the conventional method, the signal subspace approach does not require many microphones to work

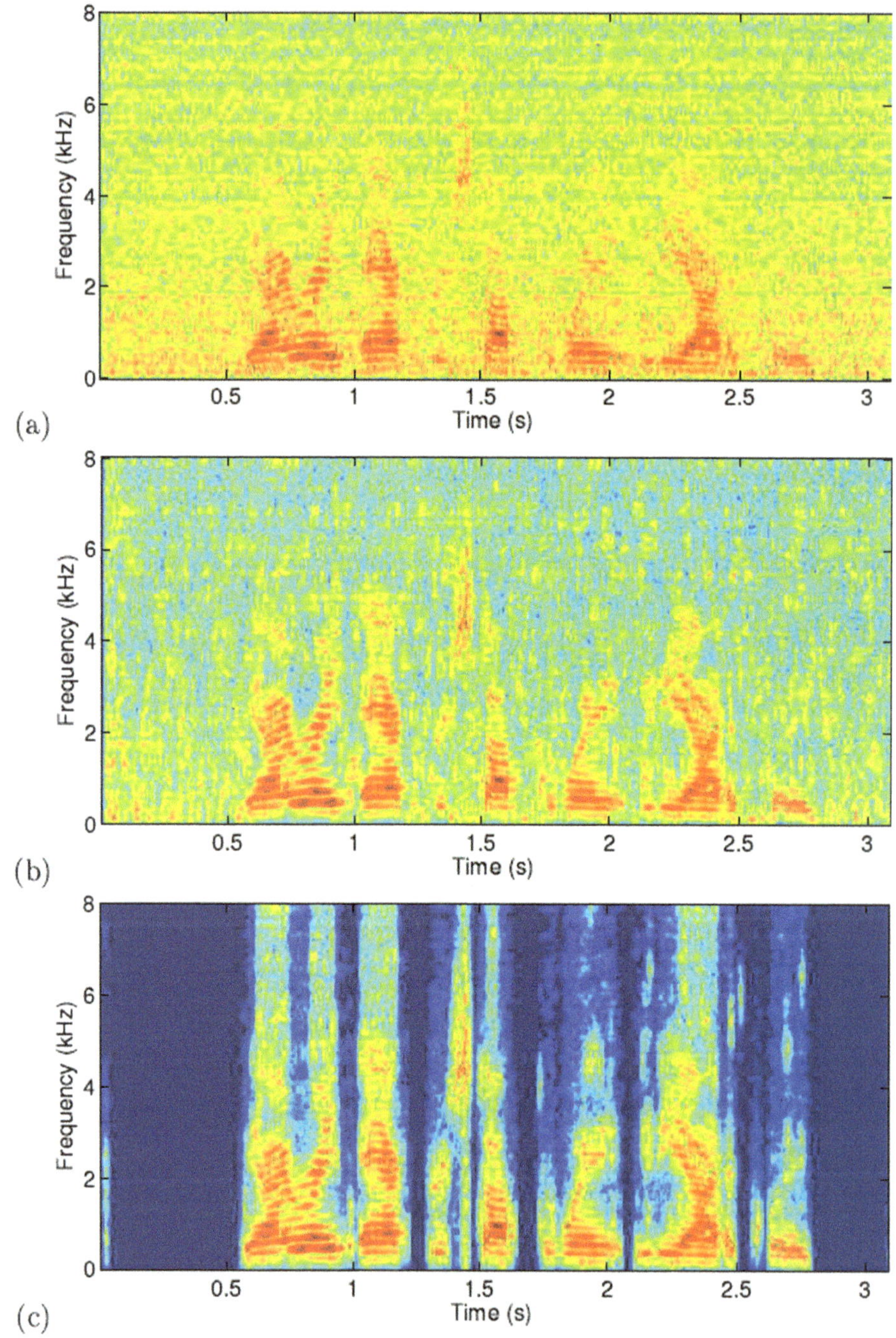

Figure 6 Speech spectrograms. (a) Noisy speech at microphone number 1 (input SNR = 10 dB). **(b)** Speech enhanced with time-domain STP method. **(c)** Speech enhanced with the signal subspace implementation of the STP method.

reasonably well. The proposed method removes the residual noise almost completely (NA = 70 to 90 dB) without introducing speech distortions or unnatural discontinuity effects at transients. This is not surprising, since the matrix Σ_n may contain only zeros during silence intervals, which is highly desirable in speech coding or automatic speech recognition (ASR) systems. On the other hand, complete cancellation of the noise is neither necessary nor desired in some applications, like mobile communication. In such cases, zero diagonal coefficients in Σ_n can be replaced with some small positive numbers.

The objective evaluation has been validated using spectrograms. Figure 6a shows the spectrogram of the noisy speech signal recorded at the first microphone, at SNR = 10 dB. The enhancement results for the conventional and proposed methods with $N = 4$ are presented below. Once again, we see that the proposed method offers incomparably higher noise attenuation during both speech pauses and voice activity periods. Unlike the time-domain implementation, the signal subspace approach does not generate musical tones (random peaks in the time-frequency plane). However, one should remember that this is an idealized situation, because the VAD has not been implemented, and speech/pause frames were marked manually. In practice, the VAD is difficult to implement, and its performance generally depends on the input SNR. Therefore, we expect some performance drop in real applications.

Conclusions

We have shown that the STP method can be implemented using a signal subspace approach. The conditions for uniqueness of a solution have been provided. We proposed Equation 29 as a simple rule that can be used when the speech correlation matrix is rank deficient. It has been verified analytically that the proposed approach can reduce noise without distorting the speech (as long as the parameter L_s is not less than the true rank of $\mathbf{R}_{\mathbf{yy}}$). In order to estimate the dimension of the speech-plus-noise subspace, we also used some sort of the thresholding technique. However, we have found empirically that, unlike in the conventional SVD-based regularization, a corresponding threshold (or the parameter θ) is not SNR dependent and can be adjusted to fixed value. The objective measurements show that the signal subspace approach outperforms the conventional one providing higher noise attenuation and lower speech distortion. We have also reported that the proposed implementation does not require as many microphones as its time-domain counterpart to work reasonably well.

Listening tests are usually difficult and time-consuming, thus they were not used to evaluate our approach.

In this article, we have introduced a novel notation that allows for estimating the speech signals at all microphones at once. This can potentially be useful if the system has to work as a preprocessor for a beamformer. Since the STP method relies only on the second-order statistics, it may find other applications in areas where multi-sensor data are processed, i.e. in the electroencephalography, as a means for enhancing EEG signals. These points have not been discussed here, but they are promising directions for future work.

Additional files

Additional file 1: Clean speech sample recorded at microphone number 1.

Additional file 2: Noisy speech sample recorded at microphone number 1 (at SegSNR = 0 dB).

Additional file 3: Speech signal enhanced using time-domain STP method.

Additional file 4: Speech signal enhanced using the proposed method.

Competing interests

The author declares that he has no competing interests.

Acknowledgements

This work was supported by the Polish National Science Centre under Decision No. DEC-2012/07/D/ST6/02454.

References

1. OL Frost, An algorithm for linearly constrained adaptive array processing. Proc. IEEE. **60**, pp. 926–935 (1972)
2. LJ Griffiths, CW Jim, An alternative approach to linearly constrained adaptive beamforming. IEEE Trans. Antennas Propag. **AP-30**(1), 27–34 (1982)
3. S Gannot, D Burshtein, E Weinstein, Signal enhancement using beamforming and nonstationarity with applications to speech. IEEE Trans. Signal Process. **49**(8), 1614–1626 (2001)
4. S Affes, Y Grenier, A signal subspace tracking algorithm for microphone array processing of speech. IEEE Trans. Speech Audio Process. **5**(5), 425–437 (1997)
5. Y Huang, J Benesty, J Chen, Analysis and comparison of multichannel noise reduction methods in a common framework. IEEE Trans. Audio, Speech, Lang. Process. **16**(5), 957–968 (2008)
6. Y Ephraim, HL Van Trees, A signal subspace approach for speech enhancement. IEEE Trans. Speech Audio Process. **3**(4), 251–266 (1995)
7. D Virette, P Scalart, C Lamblin, Analysis of background noise reduction techniques for robust speech coding. Proc. EUSIPCO. **3**, 297–300 (2002)
8. A Borowicz, A Petrovsky, Signal subspace approach for psychoacoustically motivated speech enhancement. Speech Comm. **53**(2), 210–219 (2011)
9. F Jabloun, B Champagne, Incorporating the human hearing properties in the signal subspace approach for speech enhancemnt. IEEE Trans. Speech Audio Process. **11**(6), 700–708 (2003)
10. A Borowicz, A Petrovsky, Incorporating auditory properties into generalised sidelobe canceller. Paper presented at the 2012 Proceedings of the 20th European Signal Processing Conference (EUSIPCO). Bucharest, Romania, 27–31 August 2012
11. J Chen, J Benesty, Y Huang, A minimum distortion noise reduction algorithm with multiple microphones. IEEE Trans. Audio, Speech, Lang. Process. **16**(3), 481–493 (2008)
12. J Benesty, J Chen, Y Huang, *Microphone Array Signal Processing*. (Springer, Berlin, Germany, 2008)
13. B Cornelis, M Moonen, J Wouters, Comparison of frequency domain noise reduction strategies based on multichannel wiener filtering and spatial prediction. Paper presented at the 2009 IEEE International Conference on Acoustics, Speech and Signal Processing. Taipei, 19–24 April 2009

14. J Benesty, J Chen, EAP Habets, *Speech Enhancement in the STFT Domain. SpringerBriefs in Electrical and Computer Engineering*. (Springer, Berlin, Germany, 2012)
15. EAP Habets, A distortionless subband beamformer for noise reduction in reverberant environments. Paper presented at the Proc. IWAENC, Tel Aviv. Israel, August 2010
16. PC Hansen, The truncated SVD as a method for regularization. BIT. **27**, 534–553 (1987)
17. Y Hu, PC Loizou, A generalized subspace approach for enhancing speech corrupted by colored noise. IEEE Trans. Speech Audio Process. **11**(4), 334–341 (2003)
18. H Lev-Ari, Y Ephraim, Extension of the signal subspace enhancement to colored noise. IEEE Signal Process. Lett. **10**(4), 104–106 (2003)
19. R Vetter, N Virag, P Renevey, JM Vesin, Single channel speech enhancement using principal component analysis and MDL subspace selection. Paper presented at the Proc. EUROSPEECH. Budapest, Hungary, 5–9 September 1999
20. JB Allen, DA Berkley, Image method for efficiently simulating small-room acoustics. J. Acoust. Soc. Am. **65**(4), 943 (1979)

Method for creating location-specific audio textures

Toni Heittola[1*], Annamaria Mesaros[1,2], Dani Korpi[1], Antti Eronen[3] and Tuomas Virtanen[1]

Abstract

An approach is proposed for creating location-specific audio textures for virtual location-exploration services. The presented approach creates audio textures by processing a small amount of audio recorded at a given location, providing a cost-effective way to produce a versatile audio signal that characterizes the location. The resulting texture is non-repetitive and conserves the location-specific characteristics of the audio scene, without the need of collecting large amount of audio from each location. The method consists of two stages: analysis and synthesis. In the analysis stage, the source audio recording is segmented into homogeneous segments. In the synthesis stage, the audio texture is created by randomly drawing segments from the source audio so that the consecutive segments will have timbral similarity near the segment boundaries. Results obtained in listening experiments show that there is no statistically significant difference in the audio quality or location-specificity of audio when the created audio textures are compared to excerpts of the original recordings. Therefore, the proposed audio textures could be utilized in virtual location-exploration services. Examples of source signals and audio textures created from them are available at www.cs.tut.fi/~heittolt/audiotexture.

1 Introduction

Virtual location-exploration services, such as Here Maps 3D [1], Google Street View [2], and Microsoft Streetside [3], provide ways to virtually explore different locations on the globe. In these services, a user is able to see a 360-degree view at various locations. The image material is typically collected using video cameras mounted on cars which drive around the streets of a city. Using an audio ambiance in conjunction with visual information would provide additional information and give the service a more 'real' feeling. In addition, audio ambiance would provide a rich source of information characterizing a location. A location could be, e.g., a busy street corner in New York or a quiet back alley in Tokyo. Users of such an exploration service may be interested in how crowded a certain location is at a certain time of the day, whether there is a lot of traffic noise on the adjacent street when choosing a hotel or in which parks around the potential hotels one can hear birds singing.

*Correspondence: toni.heittola@tut.fi
[1] Department of Signal Processing, Tampere University of Technology, P.O. Box 553, Tampere 33101, Finland
Full list of author information is available at the end of the article

Currently, all the available location-exploration services provide only visual description of the location, and no location-specific audio ambiances are used. One of the reasons may be the cost of collecting a comprehensive audio database: to record audio for such a service, one would need to stay at each location for a certain period of time in order to collect enough audio data for non-repeating audio ambiance. Current data collection methods involve driving by with a car while recording images and other type of information. Such method is not suitable for collecting audio material, as the sound produced by the car would interfere with the audio content. As crowd-sourced data is a feasible solution to data collection, it is attractive to be able to create location-specific audio ambiances in a cost-effective way, using a small amount of audio. In addition to the virtual location-exploration services, location-specific audio ambiances could be used in applications such as computer games to enhance the user experience.

Studies show that presenting visual and audio information together enhances perception in comparison to presenting only visual information [4,5]. Adding audio ambiance for the specific location, even having different ambiances depending on the time of the day, would enhance the user experience in the virtual location-exploration services. A few works that combine visual and audio representations for a location suggest using abstract or natural sounds for representing additional information [6,7] or using characteristic audio sequence for mapping environmental noise into city models [8]. However, none of these prior works use specific audio ambiances recorded at the location of interest.

According to listening tests, humans are able to discriminate everyday audio ambiances [9,10]. The listening test conducted in [9] showed a recognition performance of 69% with 24 everyday environments such as street, nature, restaurant, or cafeteria. In [10], same type of test was conducted with more descriptive audio categories. Some of the categories included in the test were representing different situations within the same environment, such as nature during daytime or during nighttime and street with traffic or with police car. The experiment reports a recognition performance of 82% with 15 categories.

A more detailed listening test, presented in [11], shows that humans are able to discriminate even specific locations based on the audio, such as different streets of the same urban area. This test was conducted as a preliminary study for developing location-based systems augmented with audio ambiances. The results indicate that audio ambiances from similar urban areas of the same city (e.g., different streets, different parks) differ enough to make them recognizable. Thus, location-specific audio ambiances should use audio recorded at the location of interest.

Our work proposes a method to create a new arbitrary long and representative audio ambiance for a location by using a small amount of audio recorded at the exact location. The method aims to retain the location-specific characteristics of the auditory scene while providing a new and versatile audio signal to represent the location. The steps involved in the method are the segmentation of the source audio signal into homogeneous segments, finding a new segment sequence with smooth timbral transitions, and creation of a new signal by concatenating segments in such a way that the segment boundaries do not create perceivable cuts. This new generated signal is an audio texture. The audio texture is constructed from the segments of the source audio signal by repeating them from time to time but never in their original order, resulting in a continuous, infinitely varying stream of audio. Despite the repeated segments, the audio texture should have a non-repetitive feel to the listener. In order to create location-specific audio texture, the source signal has to be recorded at the given location.

Previous studies for audio texture creation have been mostly concentrating on producing textures which are perceived as homogeneous in an interval of few seconds [12-15]. In these studies, textures are composed of small sound elements, which are occurring according to a specified higher level pattern. Typical target sounds for this type of audio textures are background sounds, e.g., rain, wind, and crowds. We use the term audio texture in our approach even though our goal is to synthesize a signal to represent the whole acoustic scene, not only background sounds. Our method aims to produce more heterogeneous audio textures than the ones typically targeted with audio texture synthesis methods, better suited for realistic auditory scenes. Therefore, longer sound elements and longer pattern intervals than in typical audio texture approaches are considered in our work. The reason for this is to keep homogeneous segments of the auditory scene intact (i.e., segments containing similar sound events) and produce a versatile but realistic representation of the source auditory scene.

Previous work in automatic generation of soundscapes used isolated sound event samples together with audio textures as components to create auditory scenes [16]. Our method avoids explicit usage of sound event samples when creating a synthesized representation of an auditory scene, as it is complicated to automatically find samples with unique content and acoustic properties that match a target scene.

The paper is organized as follows. Section 2 discusses related previous work, and Section 3 presents a detailed description of the audio texture creation method. Section 4 presents the details of a listening test created for the evaluation of the obtained audio textures, and Section 5 presents and discusses the results of the listening test. Finally, Section 6 presents conclusions and ideas for future work.

2 Related work

Previous work on synthesis of audio textures has concentrated on producing sounds with temporal homogeneity [12-14,16]. On the broader scope, the idea of audio textures is shared in many other domains, such as music textures [17] and musical mosaicing [18].

There are many application areas for audio texture synthesis in multimedia creation and games, and a wide range of different approaches have been proposed for it: noise filtering [19], physical modeling [13], wavelets [14], granular synthesis [12], corpus-based concatenative synthesis [16]. A comprehensive review of state of the art in audio texture synthesis approaches can be found in [15]. The method most related to our work, granular synthesis, uses small segments of an original signal, and some statistical model to synthesize the texture using those segments. The granular synthesis approach is extended in corpus-based concatenative synthesis with a large database containing segments from source audio which are organized according to audio descriptors (e.g., pitch and timbre) [20]. In the synthesis stage, the segments are retrieved from the database according to the specified descriptors and concatenated to form a texture.

Audio texture synthesis method for signals with simple structures is proposed in [12] for background sound creation and audio restoration. The main objective of the work is to produce a continuous and perceptually similar texture to the source signal. Segmentation of original audio is done using novelty detection method applied on the similarity matrix; sequencing of segments is based on transition probabilities.

Soundscape synthesis combining audio textural elements and sound object elements for virtual environments is proposed in [16]. The synthesis elements were retrieved from user-contributed unstructured audio corpus, categorized into these two layers, sound objects being the sounds meant to draw attention and textures being the sounds that form the ambiance. For obtaining elements for textures, segmentation is based on Bayesian information criterion with a minimum segment length separately chosen to three categories (natural, human, mechanical). The texture was created using concatenative synthesis, and sound objects were mixed at the final stage to form the soundscape. A subjective evaluation of the immersive properties of the synthesized soundscapes in virtual environments showed soundscapes to be acceptable in comparison to the original recording.

Music textures are a special case of audio textures, where the underlying musical structure is utilized in the segmentation process. The work in [17] describes various ways to create music textures, by using an onset detection method for selecting segments of audio. The segments between two onsets are the units to be rearranged. The work demonstrates recreating music pieces by selecting similar segments from other pieces and concatenating the segments. Applications include, e.g., creating audio textures, music remixing, and replacing corrupted segments in an encoded audio file with synthesized ones. Another example on creating new musical audio signals is the musical mosaicing presented in [18]. The input of this method is a sequence of higher level constraints, for example pitch. Based on this sequence, samples are retrieved from a large database and joined to create the mosaic.

The main difference between methods for creating music textures and methods which operate on more general signals like everyday audio ambiances is that music rearranging methods can utilize the metrical structure in selecting the segment positions. For example, [17] applies a similarity measure based on the metrical location of a segment. Everyday audio ambiances contain a lot of overlapping sound events, and usually, there is not a clear structure between events which could be utilized in the audio texture creation. In this case, the methods operate on more local constraints trying to ensure, for example, that there are no abrupt changes in the timbre of two adjacent segments.

The concatenation of different segments in creation of an audio texture can in fact be considered similar to the concatenative speech synthesis [21], where units representing phonemes are concatenated in a desired order. Unit selection rules are employed so that the best chain of candidate units is chosen from the database to obtain smooth and natural transitions from one to another. Unit selection is done based on acoustic parameters like the fundamental frequency (pitch), duration, position in the syllable, and neighboring phones, with as little as possible processing at the point of concatenation [21]. The concatenation of music segments in [17] does not deal with inconsistencies at segment boundaries. The boundaries are at onsets and this tends to mask the cuts.

There are multiple services available to collect geotagged sound samples, as part of acoustic ecology and soundscape evaluation. The Sound Around You [22] Project [23] collected audio data uploaded by participants using mobile devices. For each uploaded recording, the participants subjectively rated the soundscape in terms of, e.g, quality, pleasantness, and tranquility. The scope of the project was to study the perception of soundscapes and their link to human activities. A collaborative artistic project called the Urban Remix project also collected sounds recorded using mobile devices along with geotagged information [24]. These web-based services allow users to open a map view and listen to the sounds recorded at various locations. In Urban Remix, the user can also create new soundscapes by drawing a path on the map, which will create a mix of the sounds encountered along this path. Practically, these services offer the possibility of listening to the samples, usually of short duration, that other users recorded at specific locations.

Audio synthesis techniques would provide the possibility of creating longer and more versatile audio content to characterize a location. A method for generating location-specific audio is presented in [25] as an environmental sound synthesis for location-based sound search system. Their method uses spatial audio mixing based on geographical features and dynamic situations, requiring additional knowledge about the location (e.g., listener position, position of obstacles such as buildings). Our work has similar goal of creating location-specific audio, and using audio textures synthesis based on short original recording from a location avoids the collection of any additional location-specific information.

3 Audio texture creation

By an audio texture, we mean a new, unique, audio signal created based on a source audio signal acquired from a certain location. The texture should provide an audio representation of the location from which the source signal comes from by having the same generic acoustic properties and types of sound events that are characteristic to the location.

The proposed audio texture creation method consists of two stages: analysis of the source audio signal and synthesis of the audio texture. The main steps are illustrated in Figure 1. The analysis stage performs clustering and segmentation of the source audio in an unsupervised manner. The goal of the clustering analysis is to automatically find homogeneous audio segments from the source signal. Ideally, the segments would be representative of individual sound events from the auditory scene. In the synthesis stage, an audio texture is constructed by shuffling and concatenating these segments. The shuffling will be done in a way that takes into account the timbral continuity at the segment boundaries.

3.1 Similarity-based segmentation

The segmentation is performed in an unsupervised way by using the self-similarity of the source audio. The source audio is split into non-overlapping 4-s chunks. This means that the system is looking for homogeneous regions that are at least 4 s long. Mel-frequency cepstral coefficients (MFCCs) calculated in short frames are used to represent the rough spectral shape inside the chunks. The MFCCs have been successfully applied in speech and speaker recognition, various music signal classification tasks (artist, genre, song structure), environmental audio classification, audio event detection, and audio context classification [9,26]. For the proposed system, the 16 lower order static MFCCs and their delta coefficients are extracted in 20-ms frames with 50% overlapping. The feature vector also includes the zeroth order MFCC coefficient, which corresponds to the energy of the frame. This is done so that the segmentation process would react also to changes in signal energy.

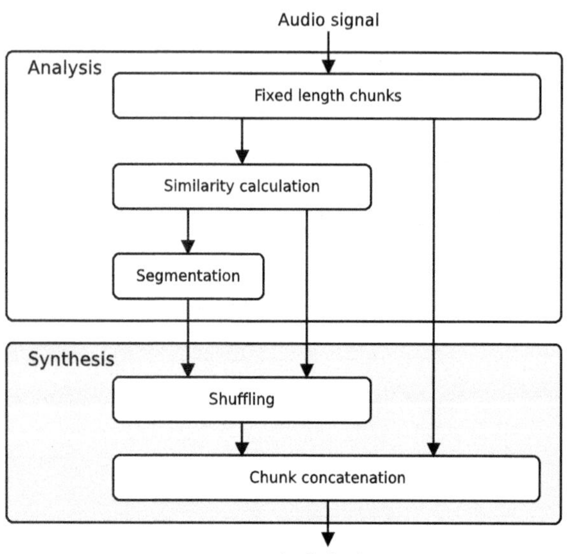

Figure 1 Main steps of the audio texture creation application.

Gaussian mixture models (GMMs) are trained for each chunk. Since the chunks have varying complexity, the agglomerative EM algorithm [27] is used to train GMM with the best amount of Gaussian components individually for each chunk. The used training algorithm is based on an information theoretic criterion for choosing the number of components. In this work, the maximum possible number of Gaussians is fixed to eight.

A distance matrix is formed by calculating the distance from each of the chunk-based GMMs to every other. The smaller the distance, the more similar the chunks are. Distances between two distributions, GMMs, are calculated using the empirical symmetric Kullback-Leibler divergence [28]. The distances between all the GMMs are collected into a matrix, as a representation related to the similarity between the 4-s chunks. Each row is a feature vector representative of the dissimilarity of one chunk to every other. The matrix reveals patterns in the source audio, and the acoustically homogeneous segments will appear as rectangular regions on the diagonal of the matrix [12,29].

Various methods exist for extracting the segment boundaries from a similarity matrix or a distance matrix. A common method, used for example in song structure segmentation, is to calculate a novelty curve, and its maxima are considered segment boundaries [12,29]. We will use a method based on clustering of the matrix rows [30]. The reasoning behind this is that a row of this matrix describes how close one chunk is to every other chunk, and this can be considered as a feature vector for that specific chunk.

A k-means algorithm is used to cluster the rows [30]. In this process, similar chunks of the source audio are clustered together. The number of clusters should ideally be related to the expected complexity of the auditory scene. Since we do not have reliable ways to estimate scene complexity, in this study, the number of clusters is fixed to 20. A chunk merging step assigns to the same segment the consecutive chunks (in time) that belong to the same cluster.

A simplified example of the distance matrix and clustering is shown in Figure 2. In this case, chunks 1 and 2 are similar because the calculated distance between them is small. They are both similar with chunk 4 since the calculated distances $d(1,4)$ and $d(2,4)$ are also small. During the clustering, these chunks will be assigned to the same cluster. At chunk merging, chunks 1 and 2 will be merged to form a segment of length 8 s and this will be one unit in the synthesis stage.

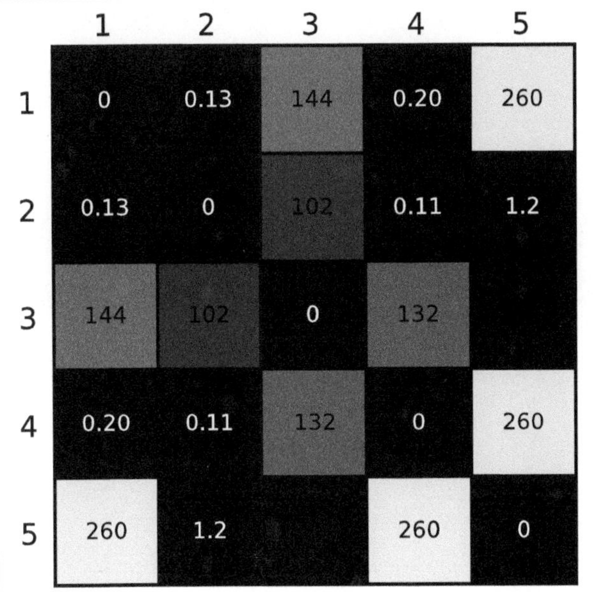

Figure 2 Exaggerated example of distance matrix between chunks.

Figure 3 presents a segmentation example. The upper panel shows the distance matrix, in which the darker color corresponds to smaller distance between chunks. The lower panel shows the outcome of the segmentation process. Each color in the lower panel corresponds to a specific cluster, and adjacent chunks belonging to the same cluster form a segment. The resulting segments are indicated also in the distance matrix by the rectangles on the diagonal. Their lengths are multiples of chunk length.

The outcome of these steps is a collection of segments of audio that are timbrally homogeneous, spanning all the source audio data. These segments will be used to create the texture by concatenating them as it will be explained in the following section.

3.2 Shuffling and concatenation

The synthesis stage of the system consists of a shuffling controlled by the similarity of the segments and their concatenation for obtaining the audio texture. As illustrated in Figure 1, the synthesis stage uses the segments obtained after the analysis stage and the similarity information that was also calculated during analysis.

A straightforward method to obtain a new signal based on segments is simply randomly shuffling them. However, this would most likely produce audible artifacts on the segment boundaries. In order to avoid the artifacts, we devise segment selection rules that will take into account timbre continuity at the boundaries of consecutive segments. We guide the segment selection using the similarity between the end chunk of the current segment and the start chunks of possible following segments. The synthesis quality difference between random segment shuffling and similarity-controlled segment shuffling was evaluated with a listening test presented in Section 4.

The first segment of the texture is selected randomly from the collection of all available segments. Distances between the last chunk of the selected segment and start chunks of all segments available are extracted from the similarity matrix calculated in the analysis stage. Candidate segments are ranked by the similarity (minimum distance), and the next segment is randomly selected from the top N candidates. The parameter N controls the change rate of auditory scene, and in this work, it is fixed to 10. The procedure is shown in Figure 4. This segment-order-randomization approach will generate a homogeneous and timbrally slowly changing auditory scene, where transitions between segments are indistinguishable.

The amount of audio data per location will most likely be small due to the cost of data collection. Thus, the system has to be able to utilize a low amount of source audio efficiently and have a sensible approach for reusing the segments while avoiding noticeable repetitions. Some constraints are set for the segment selection: at each time, it is not allowed that the next segment is a repetition of the current segment (avoids stuttering) or the segment which was originally following the current segment in the source signal (forces shuffling). A time constraint is also imposed so that re-usage of a segment is allowed only after a certain waiting period. We decided on waiting period of 20 s. This controls how repetitive the resulting audio texture will be. The diversity of audio texture will depend on the length and the complexity of source signal.

The chosen segments are concatenated one after another based on the order determined by the presented segment selection rules. The segments are linearly cross-faded within 2 s, outside of their boundaries, for smooth transitions. The cross-fading process is illustrated in Figure 5.

Example of source signals and synthesized audio textures are available at www.cs.tut.fi/~heittolt/audiotexture/.

4 Subjective evaluation

Three listening tests were conducted to study the proposed method for location-specific audio textures. The first listening test studied the benefits of using similarity-controlled segment shuffling over random segment shuffling (later denoted by experiment 1). The second test studied the relative difference in quality between a synthesized audio texture and an original audio sample (later denoted by experiment 2). First and second listening tests were conducted in a forced choice paired comparison setup. The third listening test studied the absolute quality of the synthesized audio textures and their ability for representing the auditory scene of various locations (later

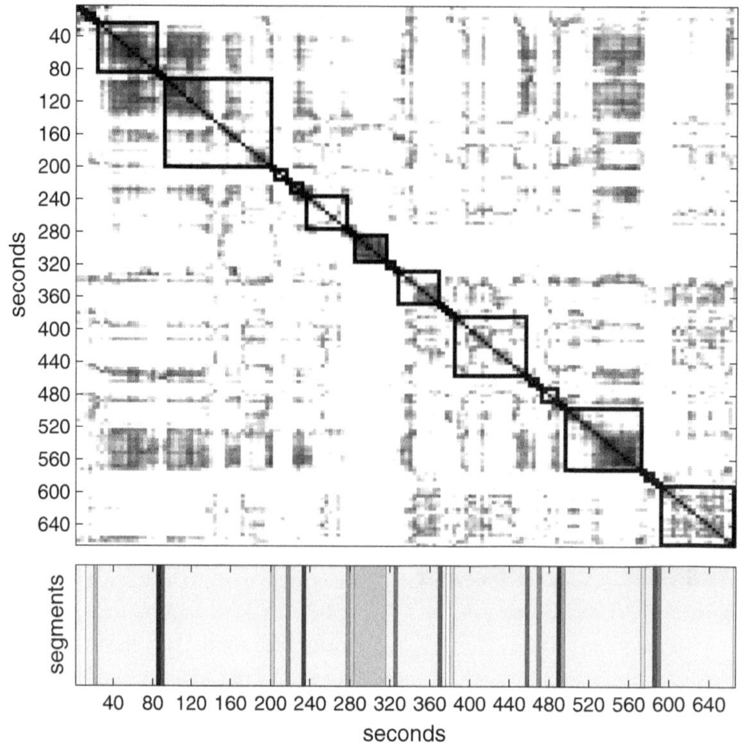

Figure 3 Example of a distance matrix (upper panel) and corresponding segmentation (lower panel). In the lower panel, the segments belonging to the same cluster are marked with the same color. The corresponding segmentation is marked on the distance matrix by the rectangles.

denoted by experiment 3). This test was conducted in an absolute rating test setup.

The tests had two goals. The first goal was to study the quality of the textures: can people notice a degradation in audio quality when they are played an automatically created texture of an environmental recording? The second goal was to study the representativeness of the automatically created textures: can an automatically created texture credibly represent a location, such as a street, even if it had some noticeable audio quality problems? These experiments studied the audio textures from two perspectives: on one hand, the audio textures were compared directly to reference samples in a forced choice test, and on the other hand, the audio textures were presented without a reference asking subjects to rate the representativeness for a certain location. The later setup is more close to the target application of virtual location-exploration services, where audio textures are used to represent certain locations without direct reference to real recordings.

In order to allow flexibility, the listening tests were carried out in a web browser via the Internet. This allowed subjects to do the test effortlessly when they had time for it. There was no requirement for the used equipments; however, the subjects were encouraged to listen to the audio samples with high-quality headphones. At the start of the experiment, they were given details about the experiments and put through a familiarization stage.

The experiment consisted of task pages presented sequentially to the test subject, each page containing one pair of samples or an individual sample depending on the test setup. The subjects controlled the playback of the audio sample with the playback button on the task page. It was emphasized to the subjects that it is important to listen through the whole audio samples before answering the questions. After answering the given question, the subject could proceed to the next task page by clicking a button. The order of the task pages was random for all subjects.

4.1 Audio samples

The audio data used in the experiments consists of recordings from different indoor and outdoor locations in the city of Tampere, Finland. The recordings were made using a binaural setup, with the recording person wearing the microphones in his ears during the recording. During the recording, some movement was permitted inside small radius of the location and the recording person was instructed to talk as little as possible. The recording equipment consists of a Soundman OKM II Klassik/studio A3 electret microphone (Berlin, Germany) and a Roland Edirol R-09 wave recorder (Hamamatsu, Japan). The samples were recorded at a sampling rate of

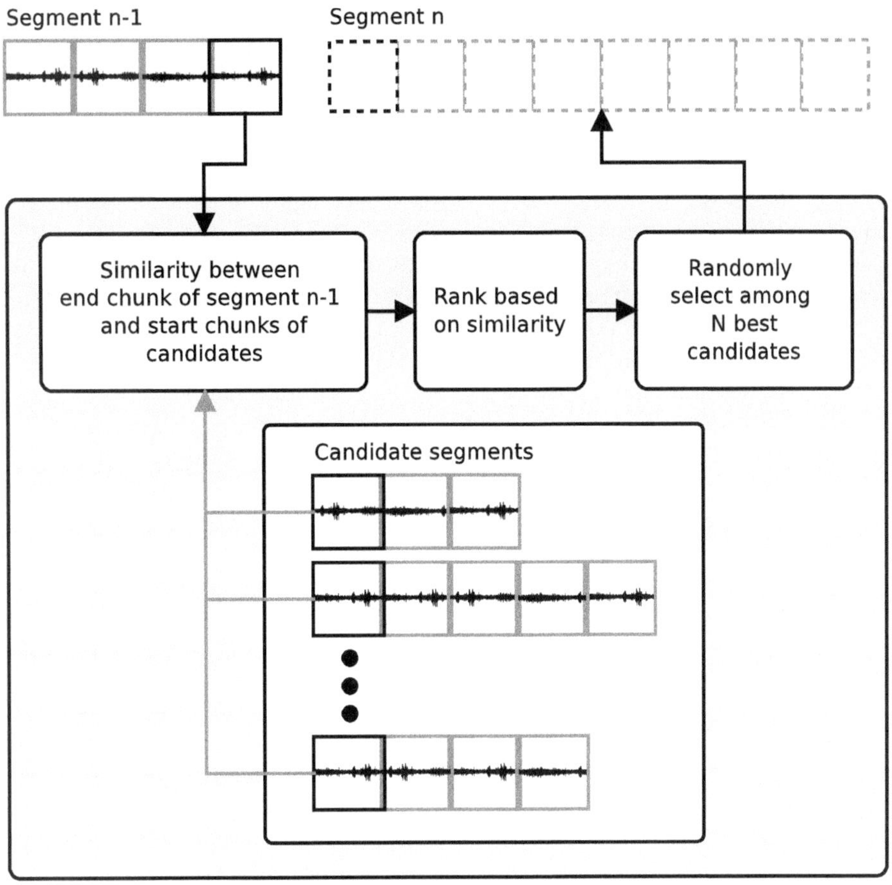

Figure 4 Similarity-controlled shuffling, finding candidate segments for the randomization.

44.1 kHz and 24-bit resolution. The stereo signals were averaged into single-channel signals for the experiments.

The amount of locations used in the listening tests was limited to four. This allowed to include sufficient amount of recordings per location to get statistically reliable results while keeping the length of the listening test per test subject reasonable. Selected locations were a pub, a restaurant, a street, and a track and field stadium during an athletics competition. These locations were selected considering that such public locations would be interesting for a virtual location-exploration service. From each of the selected locations, four audio recordings which durations ranged from 3 to 10 min were included in the listening tests, considering that this would be the typical length in crowdsourced material. The audio ambiance for the pub location contains mostly music and speech babble background, with occasional conversation in foreground. The restaurant location is a self-service type of student restaurant, located in a noisy hallway, ambiance composed mostly of background sounds like speech babble and sounds of cutlery and trays. The audio ambiance of the street location contains sound of cars passing by

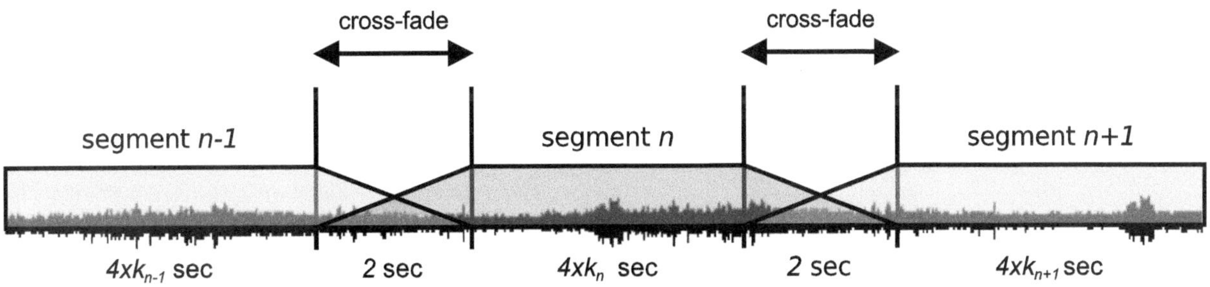

Figure 5 Synthesizing the audio from the segments; each segment comprises *k* chunks of length of 4 s.

and footstep sounds of pedestrians, with general distant traffic noise in the background. Track and field recordings were collected during the athletics competition. The audio ambiance of this location includes foreground events such as announcements and sound from the competitions on foreground and sounds of audience in the background.

4.2 Pair comparison tests

Experiments 1 and 2 used a forced choice setup, where test subjects were asked to select the more fluent and natural audio sample out of two presented samples. An example of a task page used in the tests can be seen in Figure 6. Test subjects were asked to select one sample out of a pair, based on the statement: 'Select the more fluent and natural audio sample. Audio samples are representing X location.' Before the listening test, the test subjects were guided to evaluate samples according to following cues: 'Is the order and continuity of the events in the sample natural? (fluent and natural sample), Or are there audible abrupt changes in the sound which break the fluent structure?'

The test setup for the experiments was chosen to force the test subjects to mark the preference even when there is no obvious difference between the samples. The similar setup of the two experiments allowed them to be implemented as a single test session. Pairs of samples for experiments 1 and 2 were presented in random order. This also prevents the subjects to adapt their choices during the test. The results of the two experiments were analyzed independently.

In experiment 1, a pair of samples included audio textures synthesized with two different methods, based on the same original recording. One of the samples was synthesized using the proposed similarity-controlled segment shuffling, and the other one was synthesized using a random segment shuffling approach. A 30-s excerpt was randomly selected from each audio texture. In order to ensure a fair comparison of the methods, only excerpts having at least three transitions between audio segments were selected. In experiment 2, a pair of samples included one audio texture synthesized using the proposed similarity-controlled segment shuffling and one real audio sample. A real audio sample with a length of 30 s was randomly selected from an original recording, and an audio texture was created based on the same recording. A 30-s excerpt was selected from the texture the same way as in experiment 1. The total length of test session (experiments 1 and 2) was approximately 40 min per subject, containing a total of 32 sample pairs, 16 pairs for each experiment. In total, 27 voluntary test subjects participated in the experiments 1 and 2. There were 22 male and 5 female test subjects aged between 20 and 40. People with audio research background and people without any special knowledge about audio signal processing were included in the listening tests. However, we consider these aspects of the test subjects as not having any influence on the results of the listening tests.

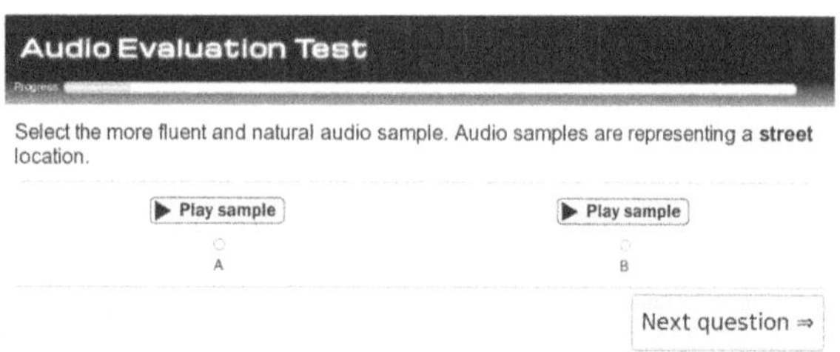

Figure 6 Caption of the experiments 1 and 2 as presented to the subjects.

4.3 Absolute rating test

The absolute quality and representativeness of the synthesized audio textures were studied in experiment 3. Real audio samples and synthesized audio textures produced by the proposed method were presented in random order to test subjects.

The listening test included four original recordings from each of the four selected locations. One audio texture was created based on each original audio recording, resulting in four audio textures available for each of the four locations. For each test subject, a 30-s excerpt was randomly extracted from each of the above audio signals. The length of the listening test was approximately 40 min per subject, containing a total of 32 excerpts (four textures and four original samples per each of the four locations). In order to ensure a fair rating, audio texture excerpts having at least three transitions between audio segments were selected, like in experiments 1 and 2.

The test consisted of several task pages, which were sequentially presented to the subject. An example of a task page can be seen in Figure 7, consisting of a single audio sample accompanied by the rating tickers for two statements:

- The audio quality of this audio sample is good.
- This audio sample sounds like it has been recorded at location X.

The ratings of the first statement provided information regarding the audio quality of the textures. The second statement simply asked how well the audio sample represents a certain type of location according to the subjects. This was used to reveal how realistic the audio textures were. The answers were given on a discrete 9-point scale, where 1 means disagree and 9 means agree.

In the familiarization stage of the listening test, an example of poor quality audio texture was presented. The example was created by random shuffle of the segments

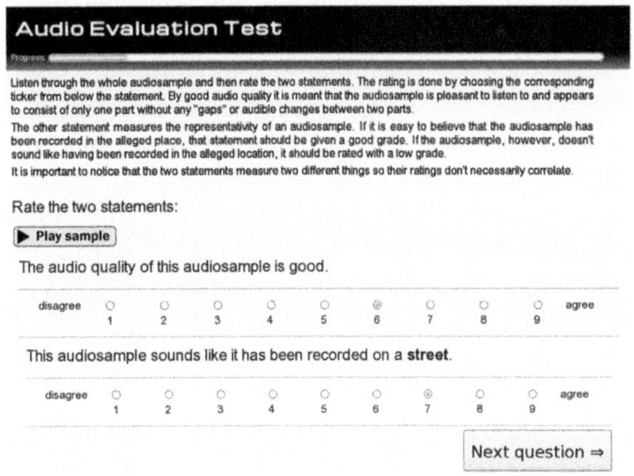

Figure 7 Caption of experiment 3 as presented to the subjects.

with no similarity-based segment selection or cross-fading while concatenating the segments. The segment transitions were easily audible in the example. The subject was informed that the example audio sample is such that it would receive the lowest ratings in terms of audio quality. A second sample was presented, this time a fragment of a real recording, and the subject was told that this example is such that it would receive the highest ratings in terms of audio quality.

The representativeness of an audio sample was explained textually: an audio sample represents a certain type of location well if it sounds as if it has been actually recorded there. It was also emphasized that an audio sample may have bad audio quality but still good representativeness for a certain location, so the two aspects of the evaluation should be kept separate.

In total, 21 voluntary test subjects participated in the listening experiment 3. Test subjects were different than in experiments 1 and 2, preventing test subjects from using prior knowledge of the signals. There were 13 male and 8 female test subjects aged between 17 and 40. As in experiments 1 and 2, both audio researchers and people without any special knowledge about audio signal processing were included in the listening test. However, we consider these aspects of the test subjects as not having any influence on the results of the listening tests.

5 Results and discussion

The results and statistical analysis of the implemented listening tests are presented in this section. Pair comparison tests, experiments 1 and 2, were implemented as one listening test session, but their results are analyzed separately.

5.1 Experiment 1

Experiment 1 studied the effect of segment shuffling method to the perceived fluency and naturalness of the synthesized audio texture. In the test, two samples were presented to test subjects: a reference sample synthesized by random segment shuffling and a sample synthesized by similarity-controlled segment shuffling. The expected outcome of this experiment was that the similarity-based segment shuffling is preferred because the transitions between segments should be concealed by similar audio content on either side of the cuts.

The results were analyzed using binomial test with a significance level of 5%, under the null hypothesis that the two options are equally likely to be selected. If the p-value is below the chosen significance level, this null hypothesis is rejected. This means that the choice is biased, so one option is more likely to be selected by the test subjects. The results are given in Table 1. There are 108 data points for each location, totaling 432 data points for the whole experiment. The 95% confidence interval and p-values resulted from the binomial test for each case are given in the table. The results indicate that the proposed method - the similarity-controlled segment shuffling-based synthesis - is able to outperform purely random segment shuffling in three out of four locations. The proposed method was preferred over the reference method in 61.1% of the overall cases, which is statistically significant.

There is significant difference in user preference for the track and field, street, and pub locations. In these locations, there are distinct foreground events - such as announcements, crowd sounds, cars passing by, conversation, which when segmented and randomly shuffled can result in very abrupt changes between adjacent segments. The restaurant location in the experiments is a self-service restaurant located in a noisy hallway with indistinguishable sounds of cutlery, trays, and speech babble forming seemingly static background ambiance, so it is understandable that timbre similarity does not offer much advantage. Based on the outcome of this test, we consider that for most locations, the similarity-controlled segment shuffling results in a more fluent and natural audio texture than a random shuffling of the segments.

Table 1 Average preference percentage and 95% confidence intervals for binomial test for experiment 1

Location	Reference preferred (%)	Proposed preferred (%)	p-value
Pub	39.8 ± 9.6	60.2 ± 9.6	0.0214
Restaurant	41.7 ± 9.6	58.3 ± 9.6	0.0507
Street	38.9 ± 9.5	61.1 ± 9.5	0.0132
Track and field	35.2 ± 9.4	64.8 ± 9.4	0.0013
Overall	38.9 ± 4.7	61.1 ± 4.7	0.0000

Reference method is random segment shuffling, and proposed method is similarity-controlled segment shuffling.

5.2 Experiment 2

Experiment 2 presented test subjects with a forced choice between an original audio sample and a synthesized one. Similar to experiment 1, the subjects were asked to choose the more fluent and natural sample. The results of the listening test are given in Table 2, together with p-values resulted from a binomial test with a significance level of 5%, under the null hypothesis that the two options are equally likely to be selected.

Overall test result for all cases shows that the null hypothesis must be rejected, and the test subjects are biased towards selecting the real recording. On smaller sets corresponding to individual locations, however, the user preference for the real recording is highlighted only for the pub location, while for restaurant, street, and track and field, the resulting p-values do not allow rejection of the null hypothesis. For the pub location, the presence of music might cause the difference in preferences as our method does not consider in any way musical structure, and while timbre might be similar between adjacent segments, music will certainly get fragmented. In other situations, it seems that the similarity-controlled shuffling is able to create sequences of segments that are realistic.

5.3 Experiment 3

In experiment 3, we presented the test subjects with an audio sample and an associated location. In this case, there was no given audio reference, therefore, no comparison to be made. The only thing the subjects had to take into account is how well the given audio characterizes the given location, from two points of view: audio quality and representativeness, scored on a scale from 1 to 9. This test refines the findings from experiment 2, by allowing subjects to rate audio quality and representativeness of a given audio for a given location based on personal experience from similar locations. The mean values of both measured aspects are listed in Table 3.

The ratings were analyzed using the statistical t-test with a 5% significance level. The null hypothesis for the t-test is that the two sets of ratings (audio sample or audio texture) have the same mean. Separate tests were performed for audio quality and representativeness. According to the results presented in Tables 4 and 5, p-values are higher than the chosen significance level. As consequence, the null hypothesis cannot be rejected.

Table 2 Average preference percentage and 95% confidence intervals for binomial test for experiment 2

Location	Reference preferred (%)	Proposed preferred (%)	p-value
Pub	63.0 ± 9.5	37.0 ± 9.5	0.0045
Restaurant	46.3 ± 9.8	53.7 ± 9.8	0.8067
Street	58.3 ± 9.6	41.7 ± 9.6	0.0507
Track and field	57.4 ± 9.7	42.6 ± 9.7	0.0743
Overall	56.2 ± 4.8	43.8 ± 4.8	0.0053

Reference method is original recording, and proposed method is similarity-controlled segment shuffling.

In this case, the result means that there is no statistically significant difference between the two sets. We cannot conclude that they are equal; however, we can perform an equivalence test to check if they are equivalent to a certain acceptable threshold. To test for equivalence, we perform two one-sided t-test (TOST) [31].

TOST has been designed specifically for bioequivalence testing of pharmaceutical products [32] and has been expanded into broader applications [33]. TOST has as null hypothesis that the difference of the two means is outside an interval. This interval is defined by the equivalence threshold θ, such that when the difference of means of the two sets is within $[-\theta, +\theta]$, the two sets are considered equivalent. The null hypotheses H_{01} and H_{02} considered in TOST are the following:

$$\begin{aligned} H_{01}: &\quad \mu_1 - \mu_2 > \quad \theta \\ H_{02}: &\quad \mu_1 - \mu_2 < \quad -\theta \end{aligned} \tag{1}$$

The choice of the equivalence threshold θ is well defined in bioequivalence testing, but for other disciplines, there are no guidelines for choosing it. The real recordings represent a control set, and in our analysis, we consider a threshold around the mean of scores given for the control set by the subjects. We performed two one-sided t-tests according to the two null hypotheses in Equation 1, for an equivalence threshold θ_{10} which is 10% of the mean of scores of the control set. The equivalence threshold is selected so that the equivalence interval at the middle of the grading scale equals to one unit.

The p-values resulted from this test, and the corresponding equivalence thresholds for each performed test are presented in Tables 4 and 5.

For a significance level of 5%, the overall set (sample size 336) passes the test both in audio quality and representativeness; as in both tests, the obtained p-value is smaller than the chosen significance level, allowing us to reject the null hypotheses. The tests performed for each separate location, however, do not allow us to reject the null hypotheses in majority of the cases. In this case, we cannot show a statistically significant equivalence at the given threshold. This can happen in experiments which have a small sample size (84 in this case), as the sample size, standard error, and equivalence threshold are related. In normal analysis situations, the equivalence threshold is chosen based on prior knowledge of the experiment and its intended application. When considering the audio quality for example, the size of the difference which can be considered to represent equivalence is not something intuitive - it is difficult to explain what an audio quality of 6.8 is and if it is equivalent to an audio quality of 7.1 or

Table 3 The mean values along with standard error of the ratings of audio quality and representativeness in experiment 3

Location	Audio quality		Representativeness	
	Audio texture	Audio sample	Audio texture	Audio sample
Pub	5.54 ± 0.24	5.93 ± 0.24	6.48 ± 0.24	6.57 ± 0.24
Restaurant	5.90 ± 0.22	5.83 ± 0.27	6.83 ± 0.21	6.86 ± 0.23
Street	5.73 ± 0.26	5.54 ± 0.25	6.87 ± 0.24	7.08 ± 0.23
Track and field	5.82 ± 0.27	6.12 ± 0.24	6.70 ± 0.27	6.57 ± 0.27
Average	5.75 ± 0.12	5.85 ± 0.13	6.72 ± 0.12	6.77 ± 0.12

not. For a complete analysis of our results, we determined at what threshold θ_c we would be able to conclude equivalence in each separate case, based on the critical t-value for a significance level of 5%. The obtained values are also presented in Tables 4 and 5.

5.4 Discussion

Based on the result of experiments 1 and 2, it seems that the similarity-controlled segment shuffling produces more fluent and natural audio texture than random shuffling; however, when given a choice, subjects selected the real recording over the audio texture. An equivalence between quality and representativeness grades for audio texture and audio samples can be concluded to certain degree.

For representativeness, the equivalence can be concluded for smaller threshold than for audio quality, as shown in experiment 3. This is understandable, since an audio texture is based on a real recording from the given location. Audio quality is of course affected by the perceived continuity of the auditory scene. In some cases, the sequentiality of events is broken by the shuffling, even if timbre continuity is considered: announcements in track and field and speech and music in pub. In such cases, for equivalence, we need to tolerate a larger threshold.

The restaurant location has overall the smallest equivalence threshold - we also noticed from experiments 1 and 2 that for this location, there was no obvious user preference for either option (random or similarity-based shuffling, method, or real recording). On the other hand, the track and field and the street locations have clear foreground sounds, and we can demonstrate equivalence only if we tolerate a one-unit threshold.

The problem of selecting an equivalence threshold beforehand for listening tests is difficult since there are no intuitive values when dealing with opinion scales. By using multiple types of experiments, we could however make some observations.

According to the results, the generated audio has comparable representativeness to real recording and a satisfactory quality. As a method for creating location-based audio ambiance, it is certainly a viable option, as it will avoid recording of large amounts of data or perceived repetitions if using smaller amount of audio. The audio texture has satisfactory properties for the proposed location, and users are likely to accept it as an audio ambiance for the location. The results showed that when facing a forced choice, in most cases, people prefer the real recording, but for the envisioned application, this is not the case. When a user is presented with audio ambiance of a location, he or she should not be bothered by artifacts or notice that it is artificially created audio, and our method was shown to fulfill these requirements.

6 Conclusions

We presented a method for audio texture creation for virtual location-exploration services. A short audio signal recorded at the given location is used to synthesize a

Table 4 The p-values obtained from performing location-wise t-tests and TOST with a 10% equivalence threshold θ_{10} for audio quality results from experiment 3

Location	t-test	TOST	θ_{10}	Equivalence with threshold
Pub	0.2359	0.5456	0.59	$\theta_c \leq 1.04$
Restaurant	0.8351	0.1372	0.58	$\theta_c \leq 0.75$
Street	0.5976	0.3152	0.55	$\theta_c \leq 0.90$
Track and field	0.4268	0.4014	0.61	$\theta_c \leq 1.04$
Overall	0.5425	0.0067	0.59	$\theta_c \leq 0.45$

According to the t-test, none of the cases has statistically significant difference. According to TOST, equivalence cannot be concluded for any location with θ_{10}.

Table 5 The p-values obtained from performing location-wise t-tests and TOST with a 10% equivalence threshold θ_{10} for representativeness results from experiment 3

Location	t-test	TOST	θ_{10}	Equivalence with threshold
Pub	0.7756	0.0939	0.66	$\theta_c \leq 0.75$
Restaurant	0.9385	0.0332	0.69	$\theta_c \leq 0.63$
Street	0.5161	0.1354	0.71	$\theta_c \leq 0.86$
Track and field	0.7325	0.1708	0.66	$\theta_c \leq 0.89$
Overall	0.7655	0.0002	0.68	$\theta_c \leq 0.38$

According to the t-test, none of the cases has statistically significant difference. According to TOST, equivalence can be concluded only for the restaurant location with θ_{10}.

new arbitrary-length signal, while conserving location-specific characteristics of the source audio. The proposed audio texture creation method consists of two main steps: analysis and synthesis. In the analysis stage, the source audio recording is segmented into homogeneous segments. In the synthesis stage, the segments are rearranged and concatenated to create the audio texture. The segment sequence is created by selecting next segment to the end of sequence randomly among timbrally most similar segments.

Three listening experiments were conducted to assess audio quality of the created audio textures. Forced choice experiments showed that the proposed method is better than random shuffling of segments, but user preference is biased toward the real recordings. With the absolute rating experiment, we showed that representativeness and audio quality of the created audio textures are comparable within certain threshold to real recordings. This means that automatically created audio textures could be used in a virtual location-exploration service in a credible way instead of original audio recordings. If the events in the source auditory scene are sequential, the resulting audio texture will most likely break the order of events and fail in being realistic. This should not be a problem for outdoor environments but can be annoying in sport events or music concerts where sound events might happen in a specific order.

There are several directions for future work. To increase the quality of the created audio textures for ambiances containing music, a music detector front-end could be used to detect sections containing music and perform the rearranging of the audio segments by following music-specific rules such as rhythm and structural continuity in addition to general timbral similarity. In addition, estimation of the auditory scene complexity could help in determining the number of clusters which should be used in the modeling of segments.

In addition, methods for creating audio textures from more than one recording should be studied. Nowadays, it is easy to obtain crowdsourced data; this would be an excellent source of audio content for such a system. A larger set of source recordings would allow creating more versatile content compared to rearranging segments from a single recording. However, this would also need paying attention to possible acoustical mismatches between recordings, e.g., recorded at different time of the day.

Competing interests

The authors declare that they have no competing interests.

Acknowledgements

This work was financially supported by the Academy of Finland under the grant 265024 (Heittola and Virtanen). This work was partially supported from the grant 251170 (Mesaros) Finnish Centre of Excellence Program (2012-2017).

Author details

[1]Department of Signal Processing, Tampere University of Technology, P.O. Box 553, Tampere 33101, Finland. [2]Department of Signal Processing and Acoustics, Aalto University, P.O. Box 13000, Aalto 00076, Finland. [3]Nokia Research Center, Visiokatu 3, Tampere 33720, Finland.

References

1. Nokia, Here Maps 3D (2013). http://here.com/3D, Accessed 15 November 2013
2. Google, Google Maps with Street View (2013). http://www.google.com/streetview, Accessed 15 November 2013
3. Microsoft, Microsoft Streetside (2013). http://www.microsoft.com/maps/streetside.aspx, Accessed 15 November 2013
4. J Vroomen, Bd Gelder, Sound enhances visual perception: cross-modal effects of auditory organization on vision. J. Exp. Psychol.: Human Perception and Performance. **26**(5), 1583–1590 (2000)
5. F Frassinetti, N Bolognini, E Ladavas, Enhancement of visual perception by crossmodal visuo-auditory interaction. Exp. Brain Res. **147**(3), 332–343 (2002)
6. JB Krygier, Visualization in modern cartography, in *Sound and Geographic Visualization* (Pergamon Press, London, UK, 1994), pp. 149–166
7. R MacVeigh, RD Jacobson, Increasing the dimensionality of a geographic information system (GIS) using auditory display, in *Proceedings of the 13th International Conference on Auditory Display* (McGill University, Montreal, Canada, 2007), pp. 530–535
8. J Schiewe, AL Kornfeld, Framework and potential implementations of urban sound cartography, in *12th AGILE International Conference on Geographic Information Science* (AGILE, Hannover, Germany, 2009)
9. A Eronen, V Peltonen, J Tuomi, A Klapuri, S Fagerlund, T Sorsa, G Lorho, J Huopaniemi, Audio-based context recognition. IEEE Trans. Audio, Speech, and Language Process. **14**, 321–329 (2006)
10. S Chu, S Narayanan, CC Kuo, Environmental sound recognition with time-frequency audio features. IEEE Trans. Audio, Speech, and Language Process. **17**(6), 1142–1158 (2009)
11. D Korpi, T Heittola, T Partala, A Eronen, A Mesaros, T Virtanen, On the human ability to discriminate audio ambiances from similar locations of an urban environment. Personal and Ubiquitous Comput. **17**(4), 761–769 (2013)
12. L Lu, L Wenyin, HJ Zhang, Audio textures: theory and applications. IEEE Trans. Speech and Audio Process. **12**(2), 156–167 (2004)
13. D Menzies, Physically motivated environmental sound synthesis for virtual worlds. EURASIP J. Audio, Speech, and Music Process. **2010**, 1–11 (2010)
14. S Kersten, P Purwins, Sound texture synthesis with hidden Markov tree models in the wavelet domain, in *Sound and Music Computing Conference* (SMC, Barcelona, Spain, 2010)
15. D Schwarz, ed. by G Peeters, State of the art in sound texture synthesis, in *Proceedings of the 14th International Conference on Digital Audio Effects (DaFx-11)* (IRCAM–Centre Pompidou, IRCAM, Paris, France, 2011), pp. 221–231
16. N Finney, J Janer, Soundscape generation for virtual environments using community-provided audio databases, in *W3C Workshop: Augmented Reality on the Web* (Barcelona, Spain, 2010)
17. T Jehan, Creating music by listening. PhD thesis, Massachusetts Institute of Technology, Cambridge, MA, USA, 2005
18. A Zils, F Pachet, Musical mosaicing, in *Proceedings of the COST G-6 Conference on Digital Audio Effects (DaFx-01), Volume 2* (University of Limerick, Limerick, Ireland, 2001), pp. 39–44
19. McJ Dermott, A Oxenham, E Simoncelli, Sound texture synthesis via filter statistics, in *IEEE Workshop on Applications of Signal Processing to Audio and Acoustics* (IEEE Signal Processing Society, New Paltz, NY, USA, 2009), pp. 297–300
20. D Schwarz, Corpus-based concatenative synthesis. IEEE Signal Process. Mag. **24**(2), 92–104 (2007)
21. AJ Hunt, AW Black, Unit selection in a concatenative speech synthesis system using a large speech database, in *IEEE International Conference on Acoustics, Speech, and Signal Processing* (IEEE Signal Processing Society, Atlanta, GA, 1996), pp. 373–376

22. Sound Around You (2013). http://www.soundaroundyou.com, Accessed 15 November 2013
23. C Mydlarz, I Drumm, T Cox, Application of novel techniques for the investigation of human relationships with soundscapes, in *INTER-NOISE and NOISE-CON Congress and Conference Proceedings, Volume 2011* (Institute of Noise Control Engineering, Osaka, Japan, 2011), pp. 738–744
24. J Freeman, DiC Salvo, M Nitsche, S Garrett, Soundscape composition and field recording as a platform for collaborative creativity. Organised Sound. **16**(3), 272–281 (2011)
25. S Innami, H Kasai, Super-realistic environmental sound synthesizer for location-based sound search system. IEEE Trans. Consumer Electron. **57**(4), 1891–1898 (2011)
26. T Heittola, A Mesaros, A Eronen, T Virtanen, Context-dependent sound event detection. EURASIP J. Audio, Speech, and Music Process. **2013**(1-13), 1 (2013)
27. M Figueiredo, J Leitão, A Jain, ed. by E Hancock, M Pelillo, Berlin on fitting mixture models, in *Energy Minimization Methods in Computer Vision and Pattern Recognition, Volume 1654 of Lecture Notes in Computer Science* (Springer Berlin Heidelberg, Germany, 1999), pp. 54–69
28. T Virtanen, M Helen, Probabilistic model based similarity measures for audio query-by-example, in *IEEE Workshop on Applications of Signal Processing to Audio and Acoustics* (IEEE Signal Processing Society, New Paltz, NY, USA, 2007), pp. 82–85
29. J Foote, Automatic audio segmentation using a measure of audio novelty, in *IEEE International Conference on Multimedia and Expo, Volume 1* (IEEE Computer Society, New York, NY, USA, 2000)
30. A Jacobs, ed. by G Antoniou, G Potamias, C Spyropoulos, and D Plexousakis, Using self-similarity matrices for structure mining on news video, in *Advances in Artificial Intelligence, Volume 3955 of Lecture Notes in Computer Science* (Springer Berlin Heidelberg, Berlin, Germany, 2006), pp. 87–94
31. DJ Schuirmann, A comparison of the two one-sided tests procedure and the power approach for assessing the equivalence of average bioavailability. J. Pharmacokinetics and Biopharmaceutics. **15**(6), 657–680 (1987)
32. GB Limentani, MC Ringo, F Ye, ML Bergquist, EO McSorley, Beyond the *t*-test: statistical equivalence testing. Analytical Chem. **77**(11), 221–226 (2005)
33. BL Stegner, AG Bostrom, TK Greenfield, Equivalence testing for use in psychosocial and services research: an introduction with examples. Evaluation and Program Plann. **19**(3), 193–198 (1996)

9

Deep neural network-based bottleneck feature and denoising autoencoder-based dereverberation for distant-talking speaker identification

Zhaofeng Zhang[1], Longbiao Wang[1*], Atsuhiko Kai[2], Takanori Yamada[2], Weifeng Li[3] and Masahiro Iwahashi[1]

Abstract

Deep neural network (DNN)-based approaches have been shown to be effective in many automatic speech recognition systems. However, few works have focused on DNNs for distant-talking speaker recognition. In this study, a bottleneck feature derived from a DNN and a cepstral domain denoising autoencoder (DAE)-based dereverberation are presented for distant-talking speaker identification, and a combination of these two approaches is proposed. For the DNN-based bottleneck feature, we noted that DNNs can transform the reverberant speech feature to a new feature space with greater discriminative classification ability for distant-talking speaker recognition. Conversely, cepstral domain DAE-based dereverberation tries to suppress the reverberation by mapping the cepstrum of reverberant speech to that of clean speech with the expectation of improving the performance of distant-talking speaker recognition. Since the DNN-based discriminant bottleneck feature and DAE-based dereverberation have a strong complementary nature, the combination of these two methods is expected to be very effective for distant-talking speaker identification. A speaker identification experiment was performed on a distant-talking speech set, with reverberant environments differing from the training environments. In suppressing late reverberation, our method outperformed some state-of-the-art dereverberation approaches such as the multichannel least mean squares (MCLMS). Compared with the MCLMS, we obtained a reduction in relative error rates of 21.4% for the bottleneck feature and 47.0% for the autoencoder feature. Moreover, the combination of likelihoods of the DNN-based bottleneck feature and DAE-based dereverberation further improved the performance.

Keywords: Speaker recognition; Bottleneck features; Denoising autoencoder; Deep neural network; Reverberant speech

1 Introduction

Although speaker recognition has been researched for many years, most applications still require a microphone located near the speaker. However, many applications would benefit from speaker recognition through distant-talking speech capture, where the speaker is able to speak at some distance from the microphones. While in this task, even in quiet conditions, the microphone records not only the direct sound of the specific speaker but also reverberation signals. A reverberation signal is created when a sound or signal is reflected, causing a large number of reflections to build up and then decay as the sound is absorbed by the surfaces of objects in the space, which could include walls, furniture, people, and air.

Owing to the effects of reverberation, the accuracy of distant-talking speaker identification is significantly reduced. According to [1], approaches for dealing with reverberation can be classified as front-end- or back-end-based approaches. Approaches of the former type attempt to reduce the effect of reverberation from the observed speech signal [2-5], while the latter methods attempt to modify the acoustic model and/or decoder to suit a reverberant environment [6,7]. In this paper, we focus

*Correspondence: wang@vos.nagaokaut.ac.jp
[1] Nagaoka University of Technology, 1603-1 Kamitomioka-cho, Nagaoka, Niigata, 940-2188, Japan
Full list of author information is available at the end of the article

on front-end-based approaches for distant-talking speaker identification.

Many front-end-based techniques have been proposed for robust automatic speech recognition (ASR) and speaker recognition in distant-talking environments [2,4,5,8-18].

Cepstral mean normalization (CMN) [19-22] is considered the most general approach for dereverberation. However, the length of an impulse response in a distant-talking environment is usually much longer than the size of the analysis window in short-term spectral analysis. Therefore, CMN cannot compensate for late reverberation. Several studies have focused on mitigating this problem [4,5,13,17,23].

Beamforming [8,24], which is a simple and robust means of spatial filtering, can be used to suppress any signal from noise or the direction of reflection; therefore, it is effective for dereverberation [13,25]. Recently, a two-stage beamforming approach [26] was presented for dereverberation and noise reduction. The first stage comprises a delay-and-sum beamformer that generates a reference signal containing a spatially filtered version of the desired speech and the interference. The second stage uses the filtered microphone signals and the noisy reference signal to estimate the desired speech. However, good performance cannot be achieved, particularly when the reverberation is very strong.

In [27,28], a method based on mean subtraction using a long-term spectral analysis window was proposed. The results showed that while subtracting the mean of the log magnitude spectrum improved ASR performance, the improvement was not sufficient, especially in the presence of significant late reverberation. A reverberation compensation method for speaker recognition using spectral subtraction [29], in which late reverberation is treated as additive noise, was proposed in [4], while a method based on multistep linear prediction (MSLP) was proposed in [5,17] for both single and multiple microphones. This method first estimates late reverberation using long-term MSLP and then suppresses this with the subsequent spectral subtraction. Wang et al. proposed a distant-talking speech recognition method based on generalized spectral subtraction (SS) [30] employing the multichannel least mean squares (MCLMS) algorithm [13,31,32]. The authors further extended their method to distant-talking speaker recognition and proposed an efficient computational method for combining the likelihoods of dereverberant speech using multiple compensation parameter sets [23]. The drawback of the above approaches is that the estimation of late reverberation is not very accurate, and thus, adequate improvement cannot be achieved.

To construct a more robust representation of each cepstral feature distribution, a feature warping method was proposed [4,33]. Such methods warp the distribution of a cepstral feature stream to a standardized distribution over a specified time interval. In addition, a feature transformation approach was presented for robust distant-talking speaker recognition [34]. The transformation is applied to distorted features before mapping them to a normal distribution and aims to decorrelate the feature vectors making them more amendable to the diagonal covariance Gaussian mixture model (GMM).

Neural network-based approaches have been proposed for feature mapping and dereverberation for speech/speaker recognition [35,36] because of their flexible representations. Bottleneck features extracted by a multilayer perceptron (MLP) can be used for nonlinear feature transformation and dimensionality reduction [35]. The MLP is trained by a backpropagation algorithm from random initial parameters. Then, the bottleneck features are extracted by dimensionality reduction of several frames of cepstral coefficients. The combination of bottleneck features and cepstral coefficients is better than the conventional mel-frequency cepstral coefficients (MFCCs). However, deep networks of MLPs with many hidden layers have a high computational cost and cannot learn in layers further away from the top layer. Nugraha et al. proposed a neural network-based method to map a reverberant feature in a log-melspectral domain to its corresponding anechoic feature [36]. The results show that cascading neural network-based dereverberation significantly improves speaker recognition compared with other dereverberation approaches. Many studies have shown that cepstral features such as MFCCs are very efficient for speaker recognition; however, extending this method directly to cepstral domain dereverberation is very difficult.

Recently, deep neural network (DNN)-based approaches have been successful in many speech and image processing fields [37-40]. Deep belief networks, which employ an unsupervised pre-training method using a restricted Boltzmann machine (RBM) [39,41], have also been proposed to train better initial values of deep networks [37]. DNNs with pre-training achieve better performance than, for example, conventional MLPs without pre-training on ASR [39,40] and large vocabulary business search tasks [38]. Denoising autoencoders (DAEs) have been shown to be effective in many noise reduction applications because higher level representations and increased flexibility of the feature mapping function can be learned [42,43]. Ishii et al. applied a DAE to spectral domain dereverberation [44] and found that the word accuracy of large vocabulary continuous speech recognition improved from 61.4% to 65.2% for the JNAS (speech corpus for large vocabulary continuous speech recognition research) database [45]. However, the suppressed spectral domain feature needs to be converted to a cepstral domain feature, and the subsequent performance improvement is not sufficient.

Few studies have focused on a DNN-based approach for distant-talking speaker recognition. By removing reverberation, we can expect to improve the speech/speaker recognition performance. However, very little research has focused on the differences between speech and speaker recognition in a distant-talking environment. For speech recognition, it is necessary to maximize the inter-phoneme variation while minimizing the intra-phoneme variation in the feature space. For speaker recognition, on the other hand, the focus is on speaker variation instead of phoneme variation. These characteristics mean some methods that are effective in speech recognition may not be as effective for speaker recognition, especially in a hands-free environment [46]. Therefore, the effect of DNN-based feature mapping and dereverberation on distant-talking speaker recognition is still unknown.

In our preliminary experiment, we found that DNN-based cepstral domain feature mapping is efficient for distant-talking speaker recognition [47]. In this paper, we present DNN-based bottleneck feature mapping, DAE-based cepstral domain dereverberation, and a combination of the two for distant-talking speaker recognition. For the DNN-based bottleneck feature (BF-DNN), we noted that DNNs can transform the reverberant speech feature to a new feature space with greater discriminative classification ability for distant-talking speaker recognition. In addition, by using multiple contexts (frames) for input data, the bottleneck features can reduce the influence of reverberation over several frames.

For neural network-based dereverberation, previous studies have shown that the spectral domain feature is efficient for the ASR task [44]. Noting that many speaker recognition systems adopt a cepstral domain feature as the direct input, it is meaningful to discover the performance of the cepstral domain DAE-based dereverberation method. Cepstral domain DAE-based dereverberation transforms the cepstrum of reverberant speech to that of clean speech. Moreover, the dimensions of the spectral domain-based features are greater than those of the cepstral domain-based ones. This introduces greater difficulties in learning a DAE with a deep architecture. Thus, it is expected that DAE-based cepstral domain dereverberation would be more efficient than DAE-based spectral domain dereverberation for speaker identification under distant-talking environments.

The DNN-based bottleneck feature is a method for extracting discriminant features while DAE-based dereverberation is a method for suppressing reverberation. Thus, they have a strong complementary nature, and a combination of the two methods should be very efficient in distant-talking speaker identification. Therefore, the likelihood of the bottleneck features extracted from the DNN and that of cepstral domain DAE-based dereverberation are combined linearly. A block diagram of the complete system is shown in Figure 1. In the training stage, DAE and BF-DNN models for feature transformation and speaker models with transformed features are trained. In the test stage, first, MFCCs extracted from the reverberant speech are input to the DAE and BF-DNN models for feature transformation. Then, the transformed features and speaker models are used to calculate the likelihood of each speaker. Finally, the likelihoods of DAE-based and BF-DNN-based features are combined and the target speaker is determined.

We also analyzed the optimal neural network architecture and parameters of the DNN-based bottleneck feature and DAE-based dereverberation for distant-talking speaker identification.

The remainder of this paper is organized as follows: Section 2 presents some basic theory for constructing and training DNNs, while an outline of the DNN-based bottleneck feature and DAE-based dereverberation method is given in Section 3. Section 4 discusses the development and evaluation of an experiment for distant-talking speaker recognition in reverberant environments. Finally, Section 5 summarizes the paper.

2 Overview of restricted Boltzmann machine

In speech recognition, DNN has been successfully used for modeling the posterior probability of state. In this work, for non-linear feature transformation, we used DNN, which can suppress the reverberation and transform the original feature to a discriminative feature for reverberant speech. A basic training strategy involved multiple phases. First, pre-training of the DNN was accomplished by training an unsupervised RBM and stacking them in a deep belief network (DBN). Second, optimization with back-propagating, referred to as fine-tuning, discriminatively trains the DNN using supervised signals. Meanwhile, in the pre-training phase of the DAE task, the encoder network was also trained layer by layer as a stack on RBM. In this section, we briefly introduced the RBM [39,41].

2.1 Restricted Boltzmann machine

The RBM is a bipartite graph as shown in Figure 2.

It has both visible and hidden layers in which visible units representing observations are connected to hidden units that learn to represent features using weighted connections. An RBM is restricted in that there are no visible-visible or hidden-hidden connections. Different types of RBMs are used for binary and real-valued input. Bernoulli-Bernoulli RBMs are used to convert binary stochastic variables to binary stochastic variables, while Gaussian-Bernoulli RBMs are used to convert

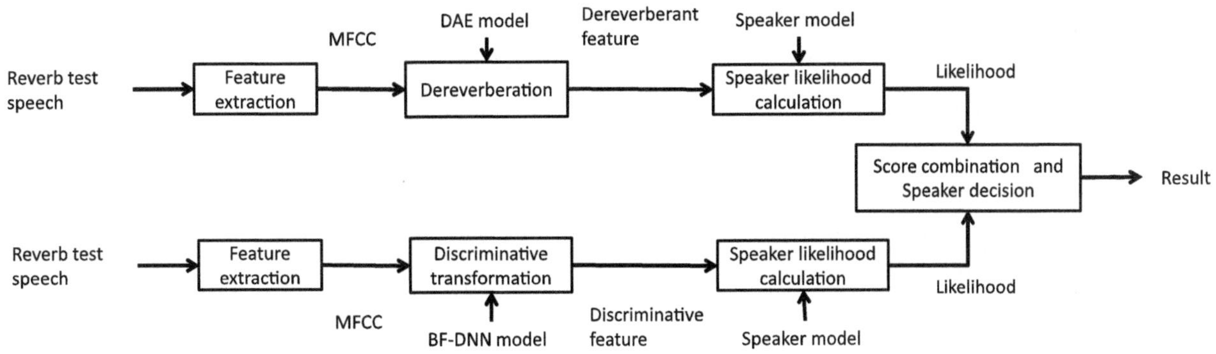

Figure 1 Block diagram of the proposed speaker recognition system.

real-valued stochastic variables to binary stochastic variables.

In a Bernoulli-Bernoulli RBM, the weights on the connections and the biases of the individual units define a probability distribution over the joint states of the visible and hidden units via an energy function. The energy of the joint configuration is given by

$$E(\mathbf{v},\mathbf{h}|\theta) = -\sum_{i=1}^{\mathcal{V}}\sum_{j=1}^{\mathcal{H}} w_{ij}v_ih_j - \sum_{i=1}^{\mathcal{V}} a_iv_i - \sum_{j=1}^{\mathcal{H}} b_jh_j, \quad (1)$$

where $\theta = (\mathbf{w}, \mathbf{a}, \mathbf{b})$ and w_{ij} represents the symmetric interaction term between visible unit i and hidden unit j with a_i and b_j their respective bias terms. $\mathcal{V}$ and $\mathcal{H}$ denote the numbers of visible and hidden units, respectively.

The maximum likelihood estimation of an RBM is to maximize the log likelihood $\log p(\mathbf{v}|\theta)$ of parameter θ. Therefore, the weight update equation is given by

$$\Delta w_{ij} = \epsilon(\langle v_ih_j\rangle_{\text{data}} - \langle v_ih_j\rangle_{\text{model}}), \quad (2)$$

where ϵ is the learning rate, $\langle\cdot\rangle_{\text{data}}$ is the expectation that v_i and h_j are on together in the training set, while $\langle\cdot\rangle_{\text{model}}$ is the same expectation calculated from the model. Because computing $\langle v_ih_j\rangle$ is expensive, we use a contrastive divergence approximation to compute the gradient. It is possible to compute $\langle v_ih_j\rangle$ by applying Gibbs sampling.

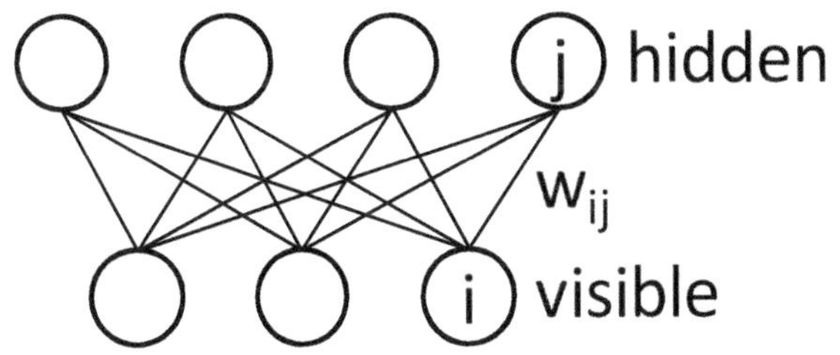

Figure 2 Graphical representation of an RBM.

2.2 DNN structure and training

DBNs are configured hierarchically by connecting pre-trained RBMs. The top layer of a DBN is a softmax layer, with the softmax operation given as

$$p(l|\mathbf{h}) = \frac{\exp(b_l + \sum_i h_iw_{il})}{\sum_m \exp(b_m + \sum_i h_iw_{im})}, \quad (3)$$

where b_l is the bias of the label and w_{il} is the weight of hidden unit i in the top layer to label l.

After configuring the DBN using RBMs, it is discriminatively trained using the backpropagation algorithm [48] to maximize the log probability of the class labels. In general, after discriminative training, a DBN is called a DNN.

In particular, we used the algorithm from [37] to train a DNN. In the pre-training phase, we first initialized the RBMs with random values. We then subdivided all training datasets into mini-batches, with 128 data vectors for unsupervised pre-training. Each hidden layer was pre-trained for 50 passes. The weight was updated after each mini-batch. For the DNN training phase, also referred to as the fine-tuning phase, we used the method of the conjugate gradient algorithm. We repeated the fine-tuning for 100 epochs updating the entire training set. The learning rate for the weights was 0.03 and for biases was 0.1.

3 DNN-based bottleneck feature and DAE-based dereverberation

3.1 Bottleneck features extracted from a DNN

Bottleneck features were generated from an MLP [35] in which one of the internal layers has a small number of hidden units relative to the size of the other layers. The multilayer network to obtain the bottleneck features is shown in Figure 3. In this example, the number of hidden layers (including the bottleneck layer) is set to 5. The number of hidden units in the innermost layer is smaller than that in the other layers. We call this the bottleneck layer.

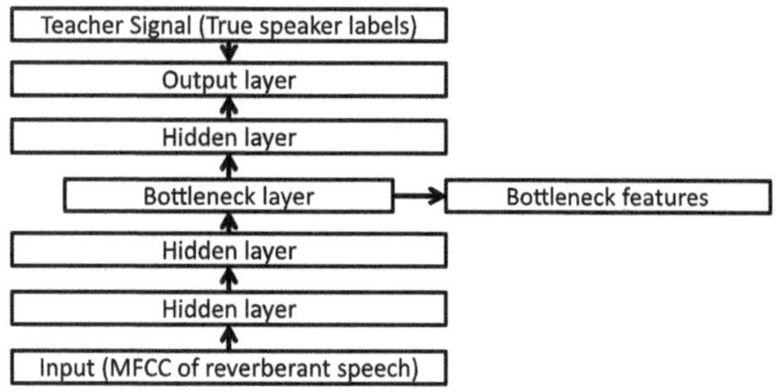

Figure 3 Flowchart of bottleneck feature extraction. In the training stage for a bottleneck DNN, the input of the DNN was a frame, or multi frames, of reverberant speech. The teacher signal was the true speaker labels. For a trained DNN, giving a frame or multiframes of reverberant speech, we took the bottleneck layer as the output feature.

In our work, both MLPs without pre-training and DNNs with pre-training were used as multilayer networks. In the pre-training step, we trained each layer of the RBM to construct a DBN using the common DBN training. With the pre-training step, the DBN achieved better initial values of the neural network. This structured bottleneck layer could be treated as a nonlinear mapping of input features. In addition, it was possible to enhance the identification ability of bottleneck features by discriminative training, which was expected to mitigate the influence of reverberation on speaker identification.

We used the speaker labels as the teacher signal. DNN's can be trained by backpropagating derivatives of a cost function that measures the cross entropy between the target outputs and the actual outputs produced for each training case.

The initial value of the MLP was generated randomly in the range −0.5 to 0.5, while the initial value of the DBN was determined by unsupervised pre-training. After initialization, supervised discriminative training was performed for both the MLP without pre-training and DBN with pre-training. Finally, the bottleneck features extracted from the bottleneck layer of the DNN were used to train the speaker model.

3.2 Denoising autoencoder for cepstral domain dereverberation

An autoencoder is a type of artificial neural network whose output is a reconstruction of the input and which is often used for dimensionality reduction.

The autoencoder training phase aims to find a value for the parameter vector, which minimizes the value between the input and teacher signals. This minimization is usually carried out by minimizing the cross entropy using conjugate gradients. Because it was difficult to directly optimize weights in a deep autoencoder with many layers, an initialization step called pre-training was conducted.

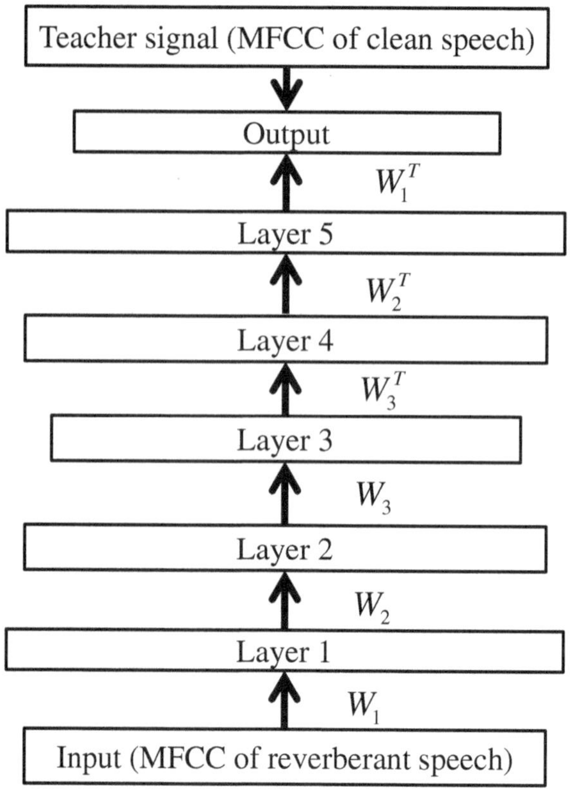

Figure 4 Topology of denoising autoencoder for cepstral domain dereverberation. In the training phase of the DAE, the cepstrum of reverberant speech was set as the input, and the cepstrum of the corresponding clean speech was set as the teacher signal. After training, the DAE network was able to generate the output of reverberation suppressed features.

DAEs share the same structure as autoencoders, but the input data are a noisy version of the output data. Autoencoders use feature mapping to convert noisy input data into clean output and, thus, have been used for noise removal in the field of image processing [42,49]. Ishii et al. applied a DAE to spectral domain dereverberation [44]. However, the suppressed spectral domain feature needs to be converted to a cepstral domain feature, and this improvement in performance was inadequate. In this

Table 1 Dataset descriptions

Data type	Usage	Data set
Training data	To train the DNN and GMM	100 (speakers) × 5 (utterances) × 3 (environments)
Development data	To determine the settings of the DNN (layers, batches, etc.) in the experimental step	Same as above
Test data	To test the speakers in this dataset in the evaluation step	100 (speakers) × 20 (utterances) × 5 (environments)

Table 2 Details of recording conditions for impulse response measurement

Array number	Room	Array type	RT60 (s)
	(a) CENSREC-4 database for training		
1	Japanese style room	Linear	0.40
2	Japanese style bath	Linear	0.60
3	Elevator hall	Linear	0.75
	(b) RWCP database for testing		
4	Echo room (cylinder)	Circle	0.38
5	Tatami-floored room (S)	Circle	0.47
6	Tatami-floored room (L)	Circle	0.60
7	Conference room	Circle	0.78
8	Echo room (panel)	Linear	1.30

RT60 (second), reverberation time in room. S, small. L, large.

paper, we applied a DAE for cepstral domain dereverberation because there were many speaker recognition systems that adopted cepstral domain features as their direct input. It is meaningful to evaluate the performance with cepstral domain-based DAE features of speaker recognition. Given a pair of speech samples, that is, clean speech and the corresponding reverberant speech, the DAE learns the nonlinear conversion function that converts reverberant speech features into clean speech. In general, reverberation is dependent on both the current and several previous observation frames. In addition to the vector of the current frame, vectors of past frames were concatenated to form input.

For cepstral feature X_i of the observed reverberant speech of the $i-th$ frame, cepstral features of $N-1$ frames before the current frame are concatenated with those of the current frame to form a cepstral vector of N frames. Output O_i of the nonlinear transformer based on the DAE is given by

$$O_i = f_L(\ldots f_l(\ldots f_2(f_1(X_i, X_{i-1}, \ldots, X_{i-N}))), \quad (4)$$

where f_l is the nonlinear transformation function in layer l, and N is the number of frames to be used as input features.

The topology of the cepstral domain DAE for dereverberation is shown in Figure 4. In this example, the number of hidden layers was set to five. In Figure 4, $W_i (i = 1, 2, 3)$ denotes the weighting of the different layers and W_i^T shows the transposition of W_i [a]. That is, W_1, W_2, and W_3 were the encoder matrices and W_1^T, W_2^T, and W_3^T were the decoder matrices, respectively. To train a DAE, we used DBNs [50] for pre-training because they can obtain accurate initial values of the deep-layer neural networks. To obtain a pre-trained RBM, we trained the second hidden layer using a Bernoulli-Bernoulli RBM and the third hidden layer using a Gaussian-Bernoulli RBM. DBNs are hierarchically configured by connecting these pre-trained RBMs. Here W_1, W_2, and W_3 are learned automatically, while W_1^T, W_2^T, and W_3^T are generated from W_1, W_2, and W_3, respectively.

After pre-training, a backpropagation algorithm was applied to adjust the parameters of autoencoder. Backpropagation algorithm modified the weights of autoencoder to reduce the cross entropy error between the teacher signal and the output value when a pair of signals is given (an input signal and an ideal teacher signal pairs.). In this paper, the input signal is the cepstral feature of reverberant speech and the ideal teacher signal is the cepstral feature of clean speech. The conjugate gradient algorithm was used to adjust the relative weightings of the units to minimize the cross entropy error for each training case [37].

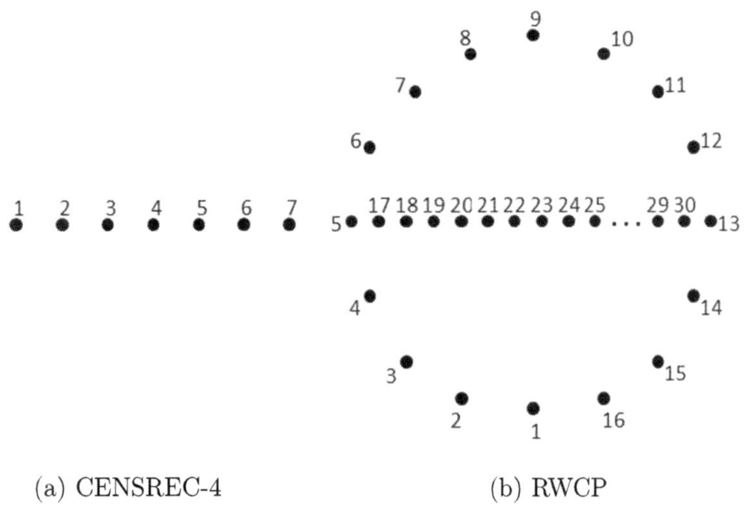

Figure 5 Graphic illustration of microphone array. **(a)** CENSREC-4 **(b)** RWCP.

Table 3 Channel numbers corresponding to Figure 5 used for dereverberation

	Linear array	Circular array
CENSREC-4	1, 3, 5, 7	—
RWCP	17, 21, 25, 29	1, 5, 9, 13

3.3 Combination of DNN-based bottleneck feature and DAE-based dereverberation

We used a GMM as our speaker model owing to its convenience and effectiveness in conventional speaker recognition. In this paper, our methods were combined by GMM likelihood. The likelihood of a DNN-based bottleneck feature-based GMM likelihood was linearly coupled with that of the DAE-based one to produce a new score L^n_{comb} given by

$$L^n_{\mathrm{comb}} = (1-\alpha)L^n_{\mathrm{BF}} + \alpha L^n_{\mathrm{DAE}}, n = 1, 2, \ldots, N, \tag{5}$$

where L^n_{BF} and L^n_{DAE} are the likelihoods produced by the n-th bottleneck feature-based model and DAE-based model, respectively. N was the number of speakers registered and α denoted the weighting coefficients. The speaker with the maximum likelihood was selected as the target speaker.

4 Experiments

Our proposed method was evaluated on both simulated and actual data. Settings for the simulated data and speaker identification experiment are discussed in Section 4.1, while experimental results are presented in Sections 4.2.1 to 4.2.3. Section 4.2.1 describes the development experiment, while Section 4.2.2 evaluates our proposed method on simulated data. Section 4.2.3 investigates the effect of different training data. Regarding the experiment on actual data, details of the training data (comprising artificially created reverberant speech), the actual evaluation data, and evaluation experiment are described in Section 4.2.4.

Table 4 Reverberation methods

	Reverberation methods	
	Conventional methods	
1	CMN	MFCC with CMN
2	MCLMS-SS	Multichannel least mean squares with spectral subtraction
3	MSLP-SS	Multistep linear prediction with spectral subtraction
4	BF-MLP	Bottleneck feature extracted from multilayer perceptron
	DNN-based feature transformation methods	
5	BF-DNN	Bottleneck feature extracted from deep neural network
6	DAE	Denoising autoencoder-based cepstral-domain dereverberation
7	DAE + BF-DNN	Combination of DAE and BF-DNN

Table 5 Conditions for speaker recognition

	Values
Sampling frequency	16 kHz
Frame length	25 ms
Frame shift	10 ms
Feature space	25 dimensions with CMN (12 MFCCs + Δ + Δpower)
Acoustic model	GMMs with 128 diagonal covariance matrices

4.1 Experimental setup

We used clean speech convoluted with various impulse responses to generate simulated data for the dereverberation experiment. For the simulated data, eight multichannel impulse responses were selected from the Real World Computing Partnership (RWCP) sound scene database [51] and the CENSREC-4 database [52]. These were convoluted with clean speech to create artificial reverberant speech. A large-scale database, the Japanese Newspaper Article Sentence (JNAS) [45] corpus, was used as the source for clean speech. Table 1 describes the development, training, and test datasets. Since the training and development datasets are the same, we refer to both as the training dataset. Utterances from 100 speakers (50 male and 50 female) were used for development and to train parameters for the DAE, BF-DNN, and GMMs. For each speaker, we used three types of artificial impulses (CENSREC-4) convoluted into five different sentences unless there was a special expression. Thus, in total, 1,500 sentences (15 sentences per speaker $\times$ 100 speakers) were used to train the DAE, BF-DNN, and GMMs. Each speaker provided 20 utterances for the test data. The average duration of training and test utterances was about 3.9 and 5.6 s, respectively.

Table 2 lists the impulse responses for the training and test sets. The impulse responses were collected by microphone arrays, as illustrated in Figure 5. The channel numbers corresponding to Figure 5 used for

Table 6 Initial parameters of DNN

	Values
Number of layers	5
Number of units in each layer	1,024
Context size	9

Table 7 Learning parameters of DNN

	Value
Batch size	128
Learning rate of Bernoulli-Bernoulli RBM	0.02
Learning rate of Gaussian-Bernoulli RBM	0.002
Weight decay	0.0002
Number of iterations in pre-training	50
number of iterations in fine-tuning	100

dereverberation are shown in Table 3. For the RWCP database, a four-channel circular or linear microphone array was taken from a circular + linear microphone array (30 channels). The circular array had a diameter of 30 cm. The microphones in the linear microphone array were located at 2.83-cm intervals. Impulse responses were measured at several positions 2 m from the microphone array. For the CENSREC-4 database, four-channel microphones were taken from a linear microphone array (seven channels in total), with the microphones located at 2.125-cm intervals. Impulse responses were measured at several positions 0.5 m from the microphone array.

In this study, we compared seven dereverberation methods, briefly described in Table 4.

For each method, we performed delay-and-sum beamforming. For *method 1*, only CMN with beamforming was used to reduce the reverberation (denoted as 'CMN'). For comparison, MCLMS-SS- [32] and MSLP-SS [17]-based dereverberation was performed in *method 2* and *method 3*, respectively. The MCLMS-SS and MSLP-SS methods both treated late reverberation as additional noise and used the spectral subtraction method to suppress it. We also performed bottleneck feature extraction without pre-training, denoted as 'BF-MLP' (*method 4*) [35]. *Method 5* (denoted as 'BF-DNN'),*method 6* (denoted as 'DAE'), and *method 7* (denoted as 'DAE + BF-DNN') represent methods introduced in this paper. For all methods, dereverberant speaker models were trained using artificial reverberant speech with three types of CENSREC-4 impulse responses (see Table 2a) and suppressed by the corresponding dereverberant method. The features of dereverberant speech were used to train the dereverberant speaker models.

Table 8 Recognition rates of DNN with varying numbers of units in each layer for training data (%)

Method	Number of units in each layer	CENSREC-4 database			Ave.
		0.40	0.60	0.75	
BF-DNN	512	84.90	78.95	83.15	82.33
BF-DNN	1,024	88.40	83.90	88.10	86.80
BF-DNN	2048	87.45	83.25	87.05	85.92

Table 9 Recognition rates of BF-DNN with and without pre-training for training data (%)

Method	Pre-training	CENSREC-4 database			Ave.
		0.40	0.60	0.75	
BF-DNN	With	88.40	83.90	88.10	86.80
BF-MLP	Without	84.85	78.15	83.50	82.17

Table 5 lists the conditions for speaker identification. We used 25-dimensional MFCCs and GMMs [53,54] with 128 mixtures. The MFCC features were normalized with the mean of the entire training data. GMMs were trained using three kinds of reverberant speech corresponding to three kinds of impulse responses. The conditions for the MCLMS and MSLP-based methods were the same as those in [55] and [5], respectively. In the MCLMS and MSLP methods, the spectral floor parameter was set to 0.15, while the noise overestimation factor and exponent parameter were set to 0.5.

Bottleneck and DAE features for distant-talking speaker identification were extracted from the MFCC features. Since the details of the parameters of a DNN are determined by the training data, this is discussed in the next section.

4.2 Experimental results

4.2.1 *Results of simulated development experiment*

Since the number of DNN layers and units in each layer need to be set before DNN learning, we used development experiments to determine the optimal parameter settings for this approach. The initial numbers of DNN structures were set according to Table 6, while the parameters were set according to Table 7, based on the settings in [39]. Five utterances taken from each of 50 male and 50 female speakers were used as training data (Table 1).

The structure of a DNN is determined by: 1) the units in each layer; 2) presence or absence of pre-training; and 3) the number of layers. These parameters for the bottleneck feature DNN and DAE were determined empirically.

Table 10 Recognition rates of BF-DNN with a varying number of layers for training data (%)

Method	Number of layers	CENSREC-4 database			Ave.
		0.40	0.60	0.75	
BF-DNN	3	56.60	48.65	55.45	53.57
BF-DNN	5	88.40	83.90	88.10	86.80
BF-DNN	7	90.40	86.40	89.20	88.67
BF-DNN	9	91.80	88.35	91.15	90.43

Table 11 Recognition rates of DAE with different sized contexts for the training data (%)

Method	Context size	CENSREC-4 database			Ave.
		0.40	0.60	0.75	
DAE	9 (c1+p8)	94.30	90.85	93.80	92.98
DAE	9 (c1+p4+l4)	93.40	89.60	93.15	92.05
DAE	9 (c1+l8)	93.50	89.65	92.75	91.07

Determining the DNN for bottleneck features (BF-DNN)

First, we determined the units in each layer. Table 8 shows the speaker recognition rates for the bottleneck feature DNN (denoted as BF-DNN in the table) with different unit settings in each layer. Initially, we set the number of layers to five. In theory, more units in each layer achieve better performance in recognition tasks. Conversely, too large a number of units may lead to over-learning, which causes diminished performance. Moreover, in the bottleneck layer, we need to compress the units to 25 dimensions. Thus, we chose 1,024 as the optimal number of units for a BF-DNN in the evaluation experiment.

Next, we investigated whether unsupervised pre-training is necessary for a BF-DNN. With pre-training, the BF-DNN achieves better performance (Table 9). The reason for this is that the multiple layers of the neural network present a much better starting point for a discriminative phase and converge faster [37]. We refer to this phase as 'fine turning'. Thus, pre-training was applied in the evaluation step.

Third, we considered how many layers would be appropriate for the BF-DNN. Table 10 shows the effect of the number of layers in the BF-DNN. We can see that more layers achieve better performance. Because the value of the teacher signal is significantly different from that of the input signal, the system needs more layers to transform the MFCC of reverberant speech to a teacher signal (true speaker label). Moreover, it has been shown that with fewer layers, the recognition performance of a system is extremely poor. We used nine layers for the BF-DNN in the evaluation experiment because more layers would have increased the time needed to train the BF-DNN, while resulting in only a relatively modest improvement in performance.

Table 12 Recognition rates of DAE with varying numbers of layers for training data (%)

Method	Number of layers	CENSREC-4 database			Ave.
		0.40	0.60	0.75	
DAE	1	95.35	90.65	95.20	93.73
DAE	3	95.40	91.50	94.70	93.87
DAE	5	94.30	90.85	93.80	92.98
DAE	7	89.65	84.40	87.60	87.22
DAE	9	87.85	82.75	87.50	86.03

Table 13 Optimal parameters for BF-DNN

	Values
Number of layers	9
Number of units in each layer	1,024
Context size	9 (p4 + c1 + l4)

Determining the DNN for the denoising autoencoder

In DAE learning, because of the duration of the reverberation, we cannot fully represent reverberation in a single frame. Thus, in the input layer, not only the current frame but also its neighboring frames are needed. Here, we need to determine another parameter that controls the input vector size, namely, context-size. We compared three kinds of contexts with a context size of nine: 1) left context (the current frame (c1 for short) + the previous eight frames (p8)); 2) left and right contexts (p4 + c1 + next 4 frames (l4)); and 3) right context (c1 + p8) (see Table 11). The results show that the best performance was obtained with a context size of nine (c1+ p8) in Table 11 because in a reverberant environment, the current frame could be affected by the previous frames. Thus, we used the setting of c1+ p8 in the subsequent step.

Units in the DAE refer to a setting in a BF-DNN. Pre-training is needed in the DAE for the same reason as in the BF-DNN.

The number of layers in the DAE must also be pre-determined. Table 12 shows the effect of the number of layers on the DAE. Contrary to the BF-DNN, fewer layers result in better performance. This can be explained by the complex structure of DNNs: too many layers cause an increase in the transformation magnitude of fine-tuning convergence, with the output being overlearned. Contrary to the BF-DNN, the values of the input MFCC of reverberant speech and teacher signal MFCC of clean speech are similar. An appropriate number of layers is sufficient for this task. Based on the experimental results, we used three layers for the DAE in the evaluation experiment.

Table 14 Optimal parameters for denoising autoencoder

	Values
Number of layers	3
Number of units in each layer	1,024
Context size	9 (c1+p8)

Table 15 Distant-talking speaker identification rates for evaluation data (%)

Method	RT60 of test data (s) (RWCP data)					Ave.
	0.38	0.47	0.60	0.78	1.30	
(a) Conventional methods						
CMN	79.70	76.05	75.55	74.40	75.75	76.29
MCLMS-SS	82.25	79.70	78.75	78.05	81.30	80.01
MSLP-SS	82.85	78.60	78.50	78.00	75.70	78.73
BF-MLP	72.35	69.30	64.05	64.90	63.25	66.70
(b) DNN-based feature transformation methods						
BF-DNN	87.90	84.95	82.45	84.00	82.15	84.29
DAE	92.10	89.70	87.60	89.45	88.10	89.39
DAE + BF-DNN	94.20	92.20	90.65	91.95	90.70	91.94

Determining the parameters for the combination of the two systems

Because the BF-DNN system tries to find discriminative features while the DAE system aims to create a transformation that can transform reverberant features into clean features, these two systems are complementary in nature and both perform well. Thus, we considered that a combination of likelihoods of these two systems could achieve better performance using Equation 5. To determine the linear combination parameter α, we varied α from 0.1 to 0.9 in steps of 0.1 and computed the recognition rate in each step. The maximum recognition rate was obtained for $\alpha = 0.4$. Thus, this value was used in the evaluation experiment.

4.2.2 Experimental results of simulated evaluation data

The optional parameters determined in the development step are summarized in Tables 13 and 14. Table 15 compares the results of distant-talking speaker identification using conventional and the proposed methods. The results show that both BF-DNN and DAE performed better than the conventional methods in all five different reverberant environments. The speaker recognition rates based on CMN for distant-talking conditions are very low because late reverberation cannot be suppressed. Conventional late reverberation suppression methods such as MCLMS-SS and MSLP-SS do not achieve sufficient improvement either. Both of the DNN-based feature mapping methods outperform the conventional dereverberation methods. Our nonlinear transformation-based approaches have a more flexible representation ability, which is more suited to distant-talking speech with a complex distribution.

The reason for the improvement in BF-DNN learning is illustrated directly in Figure 6. For the BF-DNN, we performed linear discriminant analysis to reduce the dimensions of the utterances of 20 speakers from 25 to 2 and showed them in two directions of coordinating axes. The distribution of speaker's features is clearly distinguished here. The BF-DNN changes the features to a space that is easily distinguished. We also applied the proposed method with and without pre-training (bottleneck feature MLP in Table 15). The pre-trained method achieved better performance. The unsupervised pre-training step enhanced the distinguishing characteristics (in our experiment, these are the dereverberation characteristics) to obtain good initial parameters for the neural network. The supervised training step then leads these distinguishing characteristics in the right direction, hence the need for a pre-training step.

As for the DAE, the improvement in recognition rate can be explained by the fact that the DAE always learns

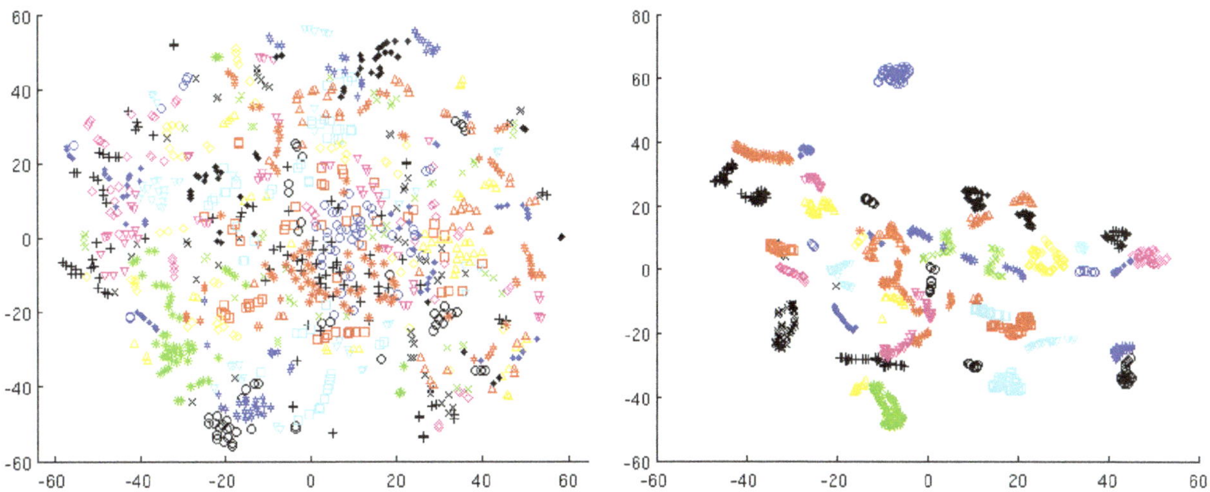

Figure 6 Feature space before and after the BF-DNN. We used PCA to reduce the speaker features to two dimensions. The lift graph shows the locations of the different speakers using different colors in a two-dimensional plane. The boundary of each speaker is not clear. The right graph shows the features after the BF-DNN in two dimensions. Here, the boundaries are clearer than those on the left graph.

Table 16 Three different training sets used to train the DNN and GMM

Training set number	DAE and BF-DNN training data	GMM training data
1	5 (utterances) × 100 (speakers) × 3 (environments)	5 (utterances) × 100 (speakers) × 3 (environments)
2	Same as training set 1	10 (utterances) × 100 (speakers) × 3 (environments)
3	10 (utterances) × 100 (speakers) × 3 (environments)	Same as training set 2

a vector field toward the higher probability regions and minimizes the variational lower bound on a generative model [42].

The DAE retains speaker characteristics and suppresses the reverberation by nonlinear feature mapping. The BF-DNN classifies speaker characteristics in the right direction. Therefore, BF-DNN-based discriminant features and DAE-based dereverberation have a strong complementary nature. A linear combination of the likelihoods of the DAE and BF-DNN was also evaluated. We used Equation 5 to obtain the combination. The weights of the DAE and BF-DNN likelihoods were 0.6 and 0.4, respectively, corresponding to the settings in the development experiment. The combination method performed better than all the individual methods. The average reduction in relative error rate was 66.0%, 59.7%, and 62.1% for CMN, MSLP-SS, and MCLMS-SS methods.

4.2.3 *Investigation of the effect of varying sizes of training data*

We also investigated how the result changes with a varying amount of training data. In this experiment, we doubled the training data for each speaker and compared the recognition results with the different training sets. Details of the training sets used in this section are given in Table 16. For all training sets, the test set was the same (20 sentences per speaker × 5 environments).

Variations in the results are shown in Figure 7. Recognition performance improves with more training data. Using twice as much GMM training data and retaining the same training data to train the DNN-based feature transformation model (experiment 2), the relative error rates of all methods were reduced by more than 40%. When doubling the size of the training data for both GMMs and DNNs, the recognition results of DNN-based feature transformation approaches are further improved. The DNN-based method outperformed both MCLMS and MSLP-based dereverberation under all conditions.

4.2.4 *Experimental results of actual environmental data*

We also used reverberant speech from an actual environment in our experiment. The recording setting was the same as in our previous work [55]. The speech was collected in a meeting room with dimensions 7.7 m × 3.3m × 2.5m ($D \times W \times H$). The utterances were collected from 20 male speakers. Each speaker uttered nine training phrases, which were recorded by an adjacent microphone. To train GMMs, the clean speech recorded by the adjacent microphone in the actual reverberant environment was convoluted with three types of impulse responses from the CENSREC-4 database to create artificial reverberant

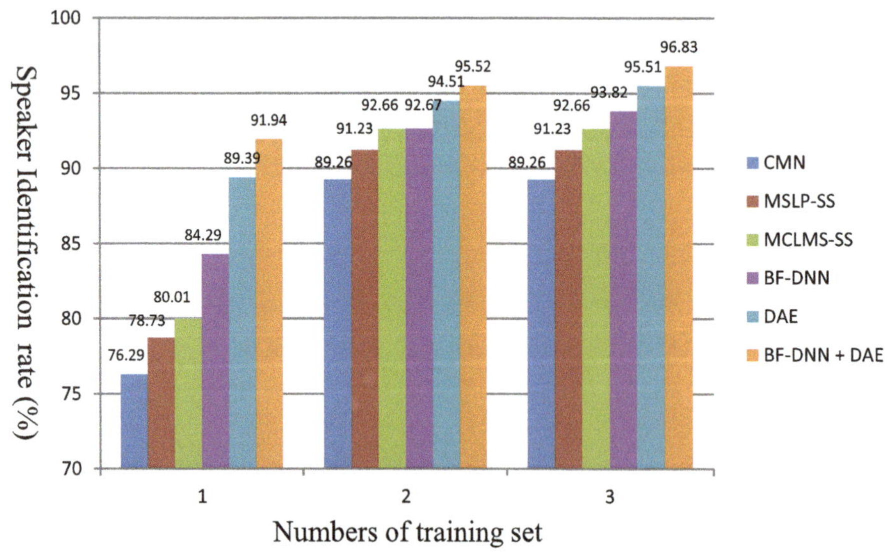

Figure 7 Average speaker identification rates using different training sets.

Table 17 Speaker identification rates in actual environment (%)

Method	Recognition rate
CMN	72.5
MCLMS-SS	87.8
MSLP-SS	83.8
BF-DNN	88.0
DAE	91.0
DAE + BF-DNN	92.5

speech. For the detailed conditions, please refer to [55]. Thus, 540 sentences (9 sentences per speaker × 3 environments × 20 speakers) were used to train the GMMs. To avoid overlearning, the 540 sentences plus the training data (1,500 sentences) shown in Table 1 were used to train BF-DNN and DAE. We retained the same neural network settings used in the experiment with simulated data (see Tables 13 and 14). For the test data, 400 utterances (20 sentences per speaker × 20 speakers) recorded by a distant four-channel microphone array were used.

The results are shown in Table 17. For the real data, a similar tendency to that found in the simulated data was observed. DAE and BF-DNN outperformed CMN, MSLP, and MCLMS. By combining the likelihoods of the DAE and BF-DNN-based features, a further improvement was achieved.

5 Conclusions

In this paper, we presented two robust distant-talking speaker identification methods based on DNNs and using bottleneck and DAE features, respectively. Bottleneck and DAE features extracted from the DNN were used to train a GMM for speaker identification. These methods achieved recognition rates of 84.29% (bottleneck DNN) and 89.39% (DAE) compared with 78.73% for the conventional MSLP and 80.01% for MCLMS in an artificial reverberant environment. Results comparing an MLP without pre-training with a DNN with pre-training show that pre-training is effective for distant-talking speaker identification. Moreover, speaker recognition performance is further improved by combining the likelihoods of the bottleneck and DAE features.

In an actual reverberant environment, BF-DNN- and DAE-based approaches also worked better than MSLP-SS and MCLMS-SS methods. The combination of DAE- and BF-DNN-based methods outperformed other methods.

Recently, Weninger et al. proposed a method for combining spectral subtraction with reverberation time estimation-based dereverberation and DAE [56]. They used reverberant and dereverberant speech to train the deep recurrent denoising autoencoder. As the DAE was trained with prior knowledge of dereverberant speech, it could learn the relationship between clean, reverberant, and dereverberant speech. This provides good motivation for our future work, that is, combining DAE with MCLMS-based dereverberation.

Endnote

[a] W_i and Wi_1^T correspond to f_L in Equation 4.

Competing interests
The authors declare that they have no competing interests.

Acknowledgements
This work was partially supported by a research grant from the Kayamori Foundation of Informational Science Advancement.

Author details
[1]Nagaoka University of Technology, 1603-1 Kamitomioka-cho, Nagaoka, Niigata, 940-2188, Japan. [2]Shizuoka University, 3-5-1 Johoku, Naka-ku, Hamamatsu, Shizuoka 432-8561, Japan. [3]Tsinghua University, 1 Qinghuayuan, Beijing 100084, China.

References

1. T Yoshioka, A Sehr, M Delcroix, K Kinoshita, R Maas, T Nakatani, W Kellermann, Making machines understand us in reverberant rooms: robustness against reverberation for automatic speech recognition. IEEE Signal Process. Mag. **29**(6), 114–126 (2012)
2. M Wu, D Wang, A two-stage algorithm for one-microphone reverberant speech enhancement. IEEE Trans. ASLP. **14**(3), 774–784 (2006)
3. M Delcroix, T Hikichi, M Miyoshi, Precise dereverberation using multi-channel linear prediction. IEEE Trans. ASLP. **15**(2), 430–440 (2007)
4. Q Jin, T Schultz, A Waibel, Far-field speaker recognition. IEEE Trans. ASLP. **15**(7), 2023–2032 (2007)
5. K Kinoshita, M Delcroix, T Nakatani, M Miyoshi, Suppression of late reverberation effect on speech signal using long-term multiple-step linear prediction. IEEE Trans. Audio Speech Lang. Process. **17**(4), 534–545 (2009)
6. H Hirsch, H Finster, A new approach for the adaptation of HMMs to reverberation and background noise. Speech Commun. **50**(3), 244–263 (2008)
7. A Sehr, R Maas, W Kellermann, Reverberation model-based decoding in the logmelspec domain for robust distant-talking speech recognition. IEEE Trans. ASLP. **18**(7), 1676–1691 (2010)
8. BD Van Veen, KM Buckley, Beamforming: a versatile approach to spatial filtering. IEEE assp mag. **5**(2), 4–24 (1988)
9. EA Habets, in *Proc. IEEE ICASSP*. Multi-channel speech dereverberation based on a statistical model of late reverberation, vol. 4, (2005), pp. 173–176
10. S Gannot, M Moonen, Subspace methods for multimicrophone speech dereverberation.EURASIP. J Appl. Signal Process. **2003**(1), 1074–1090 (2003)
11. S Subramaniam, AP Petropulu, C Wendt, Cepstrum-based deconvolution for speech dereverberation. IEEE Trans. Speech Audio Process. **4**(5), 392–396 (1996)
12. H Maganti, M Matassoni, in *Proceedings of INTERSPEECH-2010*. An auditory based modulation spectral feature for reverberant speech recognition, (2010), pp. 570–573
13. L Wang, K Odani, A Kai, Dereverberation and denoising based on generalized spectral subtraction by multi-channel LMS algorithm using a small-scale microphone array.Eurasip. J. Adv. Signal Process. **2012**(12), 1–11 (2012)
14. L Wang, N Kitaoka, S Nakagawa. Eurasip J. Appl. Signal Process. **2006**(95491), 1–11 (2006)
15. Q Jin, Y Pan, T Schultz, Far-field speaker recognition. Proc. of ICASSP-2006. **1**, 937–940 (2006)

16. SO Sadjadi, JHL Hasnen, Hilbert envelope based features for robust speaker identification under reverberant mismatched conditions. Proc. IEEE ICASSP, 5448–5451 (2011)
17. K Kinoshita, M Delcroix, T Nakatani, M Miyoshi, in *Proceedings of IEEE ICASSP 2006*. Spectral subtraction steered by multi-step forward linear prediction for single channel speech dereverberation, (2006), pp. 817–820
18. TB Hughes, HS Kim, JH DiBiase, HF Silverman, Performance of an HMM speech recognizer using a real-time tracking microphone array as input. IEEE Trans. Speech, and Audio Process. **7**(3), 346–349 (1999)
19. S Furui, Cepstral analysis technique for automatic speaker verification. IEEE Trans. Acoust. Speech Signal Process. **29**(2), 254–272 (1981)
20. F Liu, R Stern, X Huang, A Acero, in *Proc. ARPA Speech and Nat. Language Workshop*. Efficient cepstral normalization for robust speech recognition, (1993), pp. 69–74
21. L Wang, N Kitaoka, S Nakagawa, Robust distant speaker recognition based on position dependent cepstral mean normalization. Proc. of Interspeech. **2005**, 1977–1980 (2005)
22. L Wang, N Kitaoka, S Nakagawa, in *Proc. of ICASSP*. Robust distant speech recognition by combining position-dependent CMN with conventional CMN, (2007), pp. 817–820
23. L Wang, Z Zhang, A Kai, Hands-free speaker identification based on spectral subtraction using a multi-channel least mean square approach.Proc. ICASSP. **2013**, 7224–7228 (2013)
24. TB Hughes, HS Kim, JH DiBiase, HF Silverman, Performance of an HMM speech recognizer using a real-time tracking microphone array as input. IEEE Trans. Speech Audio Process. **7**(3), 490–497 (1997)
25. ML Seltzer, B Raj, RM Stern, Likelihood-maximizing beamforming for robust hands-free speech recognition. IEEE Trans. Speech Audio Process. **12**(5), 489–498 (2004)
26. EA Habets, J Benesty, in *Hands-free Speech Commun. Microphone Arrays (HSCMA) 2011 Joint Workshop on. IEEE*. Joint dereverberation and noise reduction using a two-stage beamforming approach, (2011), pp. 191–195
27. D Gelbart, N Morgan, *Evaluating long-term spectral subtraction for reverberant ASR*. (Madonna di Campiglio, Italy, 2001)
28. D Gelbart, N Morgan Double the trouble: handling noise and reverberation in far-field automatic speech recognition. INTERSPEECH, 968–971 (2002)
29. S Boll, Suppression of acoustic noise in speech using spectral subtraction. IEEE Trans. Acous. Speech Signal Process. **27**(2), 113–120 (1979)
30. BL Sim, YC Tong, JS Chang, CT Tan, A parametric formulation of the generalized spectral subtraction method. IEEE Trans. Speech Audio Process. **6**(4), 328–337 (1998)
31. Y Huang, J Benesty, J Chen, Optimal step size of the adaptive multi-channel LMS algorithm for blind SIMO identification. IEEE Signal Process. Lett. **12**(3), 173–175 (2005)
32. L Wang, N Kitaoka, S Nakagawa, Distant-talking speech recognition based on spectral subtraction by multi-channel LMS algorithm. IEICE TransInf. Syst., 659–667 (2011)
33. J Pelecanos, S Sridharan, in *Proc. of Speaker Odyssey 2001 conference*. Feature warping for robust speaker verification, (2011)
34. D Zhu, B Ma, H Li, Q Huo, in *Proc. of ICASSP 2007*. A generalized feature transformation approach for channel robust speaker verification, (2007)
35. Y Konig, L Heck, M Weintraub, K Sonmez, in *Proc. of RLA2C, ESCA workshop on Speaker Recognition and its Commercial and Forensic Applications*. Nonlinear discriminant feature extraction for robust text-independent speaker recognition, (1998), pp. 72–75
36. A Nugraha, K Yamamoto, S Nakagawa, Single-channel dereverberation by feature mapping using cascade neural networks for robust distant speaker identification and speech recognition. Eurasip. J Adv. Signal Process. **2014**(13), 1–31 (2014)
37. G Hinton, R Salakhutdinov, Reducing the dimensionality of data with neural networks. Science. **313**(5786), 504–507 (2006)
38. D Yu, ML Seltzer, in *Proc. of Interspeech*. Improved bottleneck features using pretrained deep neural networks, (2011), pp. 237–240
39. G Hinton, DYu L Deng, GE Dahl, A Mohamed, N Jaitly, A Senior, V Vanhoucke, P Nguyen, TN Sainath, B Kingsbury, Deep neural networks for acoustic modeling in speech recognition: the shared views of four research groups. IEEE Signal Process. Mag. **29**(6), 82–97 (2012)
40. A Mohamed, GE Dahl, G Hinton, Acoustic modeling using deep belief networks. IEEE Trans. Audio Speech Language Process. **20**, 12–22 (2012)
41. G Hinton, A practical guide to training restricted Boltzmann machines machine learning group, 2010-003 (2010)
42. P Vincent, H Larochelle, I Lajoie, Y Bengio, PA Manzagol, Stacked denoising autoencoders: learning useful representations in a deep network with a local denoising criterion. J. Mach. Learn. Res. **11**, 3371–3408 (2010)
43. X Lu, Y Tsao, S Matsuda, C Hori, Speech enhancement based on deep denoising autoencoder. Proc. Interspeech, 436–440 (2013)
44. T Ishii, H Komiyama, T Shinozaki, Y Horiuchi, S Kuroiwa, in *Proc. Interspeech*. Reverberant speech recognition based on denoising autoencoder, (2013), pp. 3512–3516
45. K Itou, M Yamamoto, K Takeda, T Kakezawa, T Matsuoka, T Kobayashi, K Shikano, S Itahashi, JNAS: Japanese speech corpus for large vocabulary continuous speech recognition research. J. Acoust. Soc. Jpn. (E). **20**(3), 199–206 (1999)
46. L Wang, N Kitaoka, S Nakagawa, Robust distant speaker recognition based on position-dependent CMN by combining speaker-specific GMM with speaker-adapted HMM. Speech Commun. **49**(6), 501–513 (2007)
47. T Yamada, L Wang, A Kai, Improvement of distant-talking speaker identification using bottleneck features of DNN. Proc. Interspeech, 3661–3664 (2013)
48. D Rumelhart, G Hinton, R Williams. Learning representations by back-propagating error.Nature. **323**(6088), 533–536 (1986)
49. P Vincent, H Larochelle, Y Bengio, PA Manzagol, in *Proceedings of the 25th International Conference on Machine learning*. Extracting and composing robust features with denoising autoencoders (ACM, 2008), pp. 1096–1103
50. GE Hinton, S Osindero, YW Teh. A fast learning algorithm for deep belief nets.Neural comput. **18**(7), 1527–1554 (2006)
51. S Nakamura, K Hiyane, F Asano, T Nishiura, T Yamada, in *Proc. of LREC*. Acoustical sound database in real environments for sound scene understanding and hands-free speech recognition, (2000), pp. 965–968
52. T Nishiura, M Nakayama, Y Denda, N Kitaoka, K Yamamoto, T Yamada, S Tsuge, C Miyajima, M Fujimoto, T Takiguchi, in *Proc. of INTERSPEECH*. Evaluation framework for distant-talking speech recognition under reverberant environments, (2008), pp. 968–971
53. DA Reynolds, Speaker identification and verification using Gaussian mixture speaker models. Speech Commun. **17**(1–2), 91–108 (1995)
54. DA Reynolds, TF Quatieri, R Dunn, Speaker verification using adapted Gaussian mixture models. Dig. Signal Process. **10**(1–3), 19–41 (2000)
55. Z Zhang, L Wang, A Kai, Distant-talking speaker identification by generalized spectral subtraction-based dereverberation and its efficient computation. Eurasip J. Audio Speech Music Process. **2014**(15), 1–12 (2014)
56. F Weninger, S Watanabe, Y Tachioka, B Schuller, in *Proc. of IEEE International conference Acoustics, Speech and Signal Processing (ICASSP)*. Deep recurrent de-noising auto-encoder and blind de-reverberation for reverberated speech recognition, (2014), pp. 4623–4627

10

Expanded three-channel mid/side coding for three-dimensional multichannel audio systems

Shi Dong, Ruimin Hu*, Xiaochen Wang, Yuhong Yang and Weiping Tu

Abstract

Three-dimensional (3D) audio technologies are booming with the success of 3D video technology. The surge in audio channels makes its huge data unacceptable for transmitting bandwidth and storage media, and the signal compression algorithm for 3D audio systems becomes an important task. This paper investigates the conventional mid/side (M/S) coding method and discusses the signal correlation property of three-dimensional multichannel systems. Then based on the channel triple, a three-channel dependent M/S coding (3D-M/S) method is proposed to reduce interchannel redundancy and corresponding transform matrices are presented. Furthermore, a framework is proposed to enable 3D-M/S compress any number of audio channels. Finally, the masking threshold of the perceptual audio core codec is modified, which guarantees the final coding noise to meet the perceptual threshold constraint of the original channel signals. Objective and subjective tests with panning signals indicate an increase in coding efficiency compared to Independent channel coding and a moderate complexity increase compared to a PCA method.

Introduction

Recently, 3D audio has attracted more attention and developed fast following the booming market of 3D movie. Many 3D audio technologies are now introduced into audio-involved applications to replace the surround sound system to provide superior sound localization and an immersive feeling. Wave field synthesis (WFS), Ambisonics and vector-based amplitude panning (VBAP) are the three most well-developed technologies. WFS generally follows Huygens principle to reconstruct the original sound field [1]. Research institutions such as IDMT of Fraunhofer and IRCAM in France have an intensive study in WFS, and attempt to bring WFS into theater and live transmission of concert. Ambisonics utilizes spherical harmonic functions to recording sound field and driving loudspeakers, its loudspeakers have rigorous configuration and give a good sound field reconstruction in the center [2]. VBAP follows the tangent law in a three-dimensional space using three adjacent loudspeakers to form a sound vector. For its simplicity, VBAP is the most common algorithm in 3D signal panning [3]. A 3D system like 22.2 multichannel system proposed by NHK in Japan utilizes VBAP to generate 3D sound image [4]. The 22.2 multichannel system is also included in the developing MPEG-H standard for rendering 3D audio scene.

There is a clear trend that 3D audio technology will become mature gradually and replace stereo and surround sound [5]. However, a main and common feature of 3D audio technologies is the great number of sound channels. For instance, WFS system always contains dozens and even hundreds of audio channels. The 22.2 system has three layers and 24 audio channels. Although the Ambisonics system can have flexible order and channel number, it usually uses dozens of channels because fewer channels will cause quality deterioration. Comparing with a two-channel stereo and a 5.1 surround sound, the increasing of audio channel causes a dramatical 3D audio data increase. A report from Fraunhofer shows 37 Mbps is needed for live transmission of WFS [6]. For the 22.2 multichannel system, uncompressed data also reaches 28 Mbps [7]. Currently, storage media and transmission bandwidth can hardly afford those huge data size. So the compression of 3D multichannel audio signals becomes an important subject.

The well-known Spatial Audio Coding (SAC) models the signals as virtual sound sources in the frequency domain, extracts the interchannel level difference (ICLD)

*Correspondence: hurm1964@gmail.com
National Engineering Research Center for Multimedia Software, Computer School, Wuhan University, Wuhan, China

and interchannel time difference (ICTD) and interchannel coherence (IC) to represent the direction and width of virtual sound source and downmixes the multichannels to reduce redundancy [8-11]. The idea of using downmixed sources with spatial parameters was later developed into Spatial Audio Object Coding (SAOC) for efficiently coding the multiple input spatial audio objects with interactive and personalized rendering ability [12]. Recently, some other investigations have been published to increase the compression efficiency for multichannel 3D audio signals. In 2007, Goodwin and Jot proposed a PCA-based multichannel compression framework for parametric coding [13], which can enhance specific audio scenarios and provide robust spatial audio coding. In 2008, Cheng et al. proposed the Spatially Squeezed Surround Audio Coding (S^3AC) for parametrically compressing the Ambisonics signal [14]. In 2009, Hellerud used an inter-channel prediction-based coding method to remove the redundancy between Ambisonics channels [15], which has low algorithm delay but high computational complexity. Tzagkarakis used a sinusoidal model and linear prediction to parameterize the separate spot microphone channels, then downmixed the residual signals. This coding scheme is more suitable for multichannel signals with weak correlation, and such scenarios require Independent channel decoding [16]. In 2010, Pinto et al. utilized a space/time-frequency transform to decompose the WFS signals into plane waves and evanescent waves. By discarding the evanescent waves and perceptually coding the plane wave signals, coding gain is obtained. Coding efficiency increases along with the number of audio channels, because the transform decomposition accuracy depends on the spatial resolution which is the number of WFS channels [17,18]. In 2013, Cheng further proposed a Spatial Localization Quantization Point (SLQP) codec using localization cues to compress the 3D audio signals [19,20]. Since SLQP extracts the spatial cues and downmixes the channels, it achieved high compression ratio for SLQP signals and other 3D audio systems.

In order to increase the coding efficiency at high bitrates, some non-parametric coding schemes were developed. Yang proposed a scalable multichannel codec, using the Karhunen-Loeve Transform (KLT) to remove the interchannel redundancy to realize scalable multichannel audio coding [21]. Mid/side (M/S) coding was introduced by J.D. Johnston [22] and adopted by many audio codec such as MPEG2-Layer III and MPEG4-AAC. In 2003, Liu et al. proposed a bit allocation method for M/S coding based on allocation entropy, which increases the objective quality by allocating more bits to high energy channel in M/S coding [23]. In 2008, Derrien et al. proposed an error model for M/S coding. The error model enables tuning of the quantizer used for channels M and S at the encoder with respect to the distortion of L and R at the decoder side, which increased the coding efficiency of M/S without much complexity [24]. Since M/S coding works as the simplest interchannel prediction, Krueger generalized it using linear prediction instead of M/S transformation and residual signal instead of difference signal [25]. In 2012, Schafer further developed Krueger's method, the multichannel case, which has low algorithmic delay [26]. Recently, M/S coding was combined with parametric stereo coding at low bitrates in the MPEG-USAC standard [27] by predicting the residual channel using spatial cue-based parameters, which aimed to bridge the stereo quality gap between low bitrates and high bitrates [28]. M/S coding also works alone at high bitrates utilizing a novel complex prediction to achieve better performance [29].

The above model-based codec and parametric codec can offer a considerable compression ratio. However, those methods need to know the direction of the real audio source to do objective-oriented coding, or estimated a virtual source direction to do downmixing and parametric coding. In practice, such as live recording, it is very difficult to obtain the real audio source direction. Downmixing and parametric coding will cause interchannel interference such as 'tone leakage' artifacts when channel signals differ greatly [30]. Furthermore, the computational complexity of an audio codec should be acceptable while maintaining enough coding efficiency, and parametric coding can only achieve a performance gain at low bitrates. This paper focuses on the situation that only the multichannel signals of audio sources are recorded, instead of their directions. And we consider high-quality/high-bitrate application and focus on the non-parametric coding method. Section 'M/S coding in 3D space' describes the conventional M/S coding process and presents a three-channel Dependent M/S coding (3D-M/S) method. The main idea is to expand M/S coding to three-dimensional audio by designing a new transform matrix, which remove the redundancy of three channels in 3D space rather than just two channels in the horizontal plane. Section '3D-M/S psychoacoustic model' discusses the psychoacoustic model for transformed 3D-M/S signals. Section 'Framework for general channel configuration' specifies a new framework enables 3D-M/S to be applied to a more general channel configuration. Section 'Experiment' gives a comparison of 3D-M/S coding with PCA coding and Independent channel coding to justify the performance of compression ratio and computational complexity. Section 'Conclusion' summarizes and concludes this paper.

M/S coding in 3D space

Conventional M/S coding

M/S coding is based on the fact that most stereo channels are strongly correlated. By simply transforming the stereo channel pair into the M/S domain, core codec encodes a summation channel and a difference channel instead of the original channels. The difference channel has much lower energy than the original channel, so more frequency bins can be quantized to similar and smaller quantized values, which leads to the entropy of the resulting quantised time-frequency samples is lower and hence lossless coding using Huffman coding achieves a higher compression rate. To illustrate how M/S coding works, a generalized sine stereo model is used. Here, a stereo pair is denoted as a vector $\mathbf{V_0} = (C_L, C_R)$ where

$$\begin{aligned} C_L &= S\sin\theta \\ C_R &= S\cos\theta \end{aligned} \tag{1}$$

S is the virtual audio source, θ is the stereo panning angle and $\theta \in \left[0, \frac{\pi}{2}\right]$. The M/S coding can be denoted as two transform matrices $\mathbf{M_0}$ and $\mathbf{M_1}$, the summation vector of $\mathbf{M_1}$ is denoted as $\mathbf{V_1} = \left(\frac{\sqrt{2}}{2}, \frac{\sqrt{2}}{2}\right)$

$$\mathbf{M_0} = \begin{bmatrix} 1 & 0 \\ 0 & 1 \end{bmatrix} \mathbf{M_1} = \begin{bmatrix} \frac{\sqrt{2}}{2} & \frac{\sqrt{2}}{2} \\ \frac{\sqrt{2}}{2} & -\frac{\sqrt{2}}{2} \end{bmatrix} \tag{2}$$

And the mid/side channels are obtained by multiplying input signals with $\mathbf{M_1}$ as $(C_M, C_S) = \mathbf{M_1}\mathbf{V_0}$, where

$$\begin{aligned} C_M &= \frac{\sqrt{2}}{2}C_L + \frac{\sqrt{2}}{2}C_R = S\sin\left(\theta + \frac{\pi}{2}\right) \\ C_S &= \frac{\sqrt{2}}{2}C_L - \frac{\sqrt{2}}{2}C_R = S\sin\left(\theta - \frac{\pi}{2}\right) \end{aligned} \tag{3}$$

In practice, subband energy or masking threshold will be used instead of C_L, C_R in $\mathbf{V_0}$. Only when two channels is sufficiently correlated, for example energy difference is less than a threshold Thr = 2 dB as shown in (4), will the M/S mode be used to avoid being frequently transformed and recalculating the masking threshold [22].

$$\left| 10\log_{10}\left(\frac{C_L}{C_R}\right)^2 \right| \leq \text{Thr} \tag{4}$$

To discuss the switching condition more conveniently, the M/S switching condition will be expressed using the distance between vectors $\mathbf{V_0}$ and $\mathbf{V_1}$. Given $\frac{C_L}{C_R} = \tan\theta$, (4) can be denoted as

$$\frac{1}{\sqrt{1+10^{\frac{\text{Thr}}{10}}}} \leq \cos\theta \leq \frac{1}{\sqrt{1+10^{-\frac{\text{Thr}}{10}}}} \tag{5}$$

The left side of Figure 1 shows the M/S switching condition (5) in vector space, and Figure 2 illustrates a stereo signal and corresponding transformed sum and difference signals. Equation (3) and the two figures denote that when input signal vector $\mathbf{V_0}$ is close to $\mathbf{V_1}$, where $\cos\theta \approx \frac{1}{\sqrt{2}}$ and $\theta \approx \frac{\pi}{4}$, the switching condition (4) will be satisfied. The difference signal has less amplitude than the original signals, and M/S coding will be used. Since θ is the angle of $\mathbf{V_0}$, switching condition (4) can be represented by the inner product between signal vector $\mathbf{V_0}$ and summation vector $\mathbf{V_1}$

$$\frac{\langle \mathbf{V_0}, \mathbf{V_1} \rangle}{|\mathbf{V_0}|\,|\mathbf{V_1}|} \geq \text{Thr}_v \tag{6}$$

This is an equivalent expression to the energy condition. It indicates that only when the input signal vector is close enough to the summation vector of a M/S transform matrix, this matrix will be used. This idea will be helpful when later discussing the 3D-M/S coding where more than one transform matrix exists. Here, Thr_v is

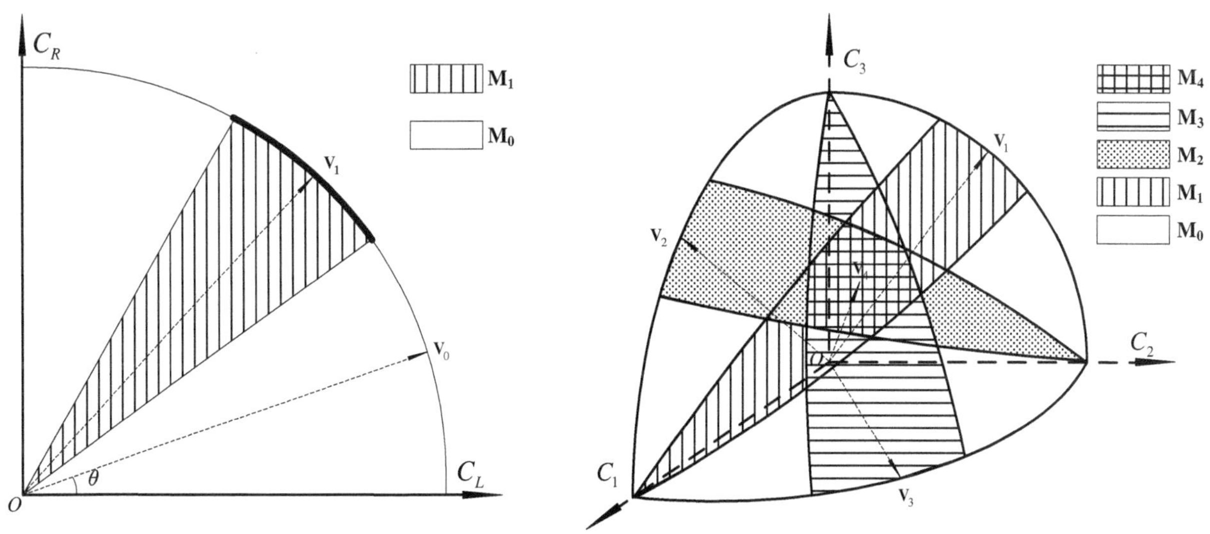

Figure 1 Switching condition of M/S and 3D-M/S method. Conventional M/S switching condition in vector space (left). 3D-M/S swiching condition in vector space (right).

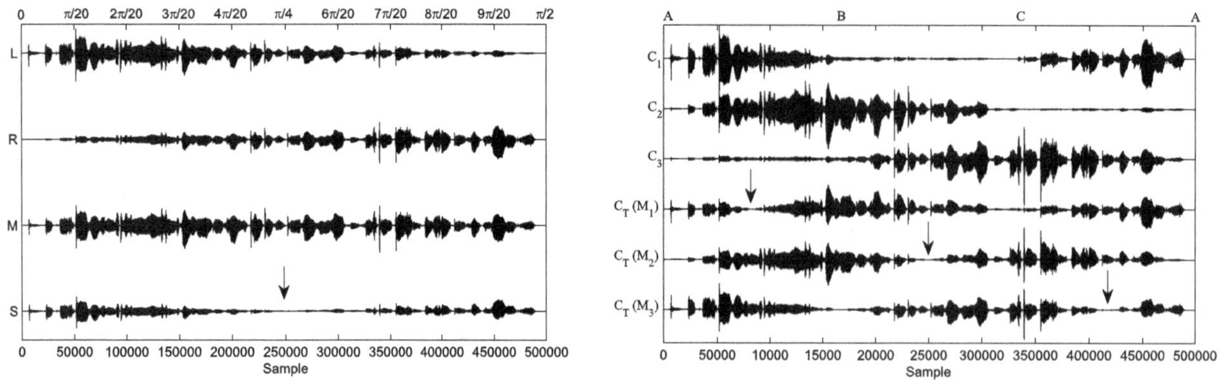

Figure 2 Two-dimensional stereo and three-dimensional VBAP signals and their transformed signals. The left and right channels of a stereo signal and its summation (M) and difference (S) signals (left). The sound source moves from the left loudspeaker to the right by the panning angle θ. When sound source is close to one loudspeaker and θ tends to be 0 and $\frac{\pi}{2}$, the original signal has less dynamic range. When sound source is close to the center and θ tends to be $\frac{\pi}{4}$, the transformed signal has less dynamic range. The three channels of a VBAP signal and its 3D-M/S signals (right). The sound source moves from A to B, B to C and C to A shown in Figure 3. When sound source is close to the center of two channels, transformed signal C_T has less dynamic range with corresponding matrix $\mathbf{M_1}$, $\mathbf{M_2}$, and $\mathbf{M_3}$.

the corresponding switching threshold of Thr in vectorial distance and

$$\mathrm{Thr}_v = \frac{\sqrt{2}}{2}\left(\frac{1}{\sqrt{1+10^{\frac{\mathrm{Thr}}{10}}}} + \frac{1}{\sqrt{1+10^{-\frac{\mathrm{Thr}}{10}}}}\right) \quad (7)$$

M/S coding in three-dimensional space

For diffuse audio sources like ambient sound, the multichannel signals are weakly correlated. For directional audio sources, the signals are highly correlated which make it possible to reduce the interchannel redundancy. To maintain the stability of the sound image, in a stereo and surround audio system, the virtual source is always being panned or recorded by the two most adjacent channels. So two adjacent channels always have the maximum similarity, and M/S coding and parametric coding are performed based on a two-channel unit. But for Ambisonics, VBAP and 22.2 multichannel systems, sound channels are spherically configured in 3D space as shown in Figure 3. In a VBAP system, three adjacent channels form a directional sound image and have the maximum correlation. In other

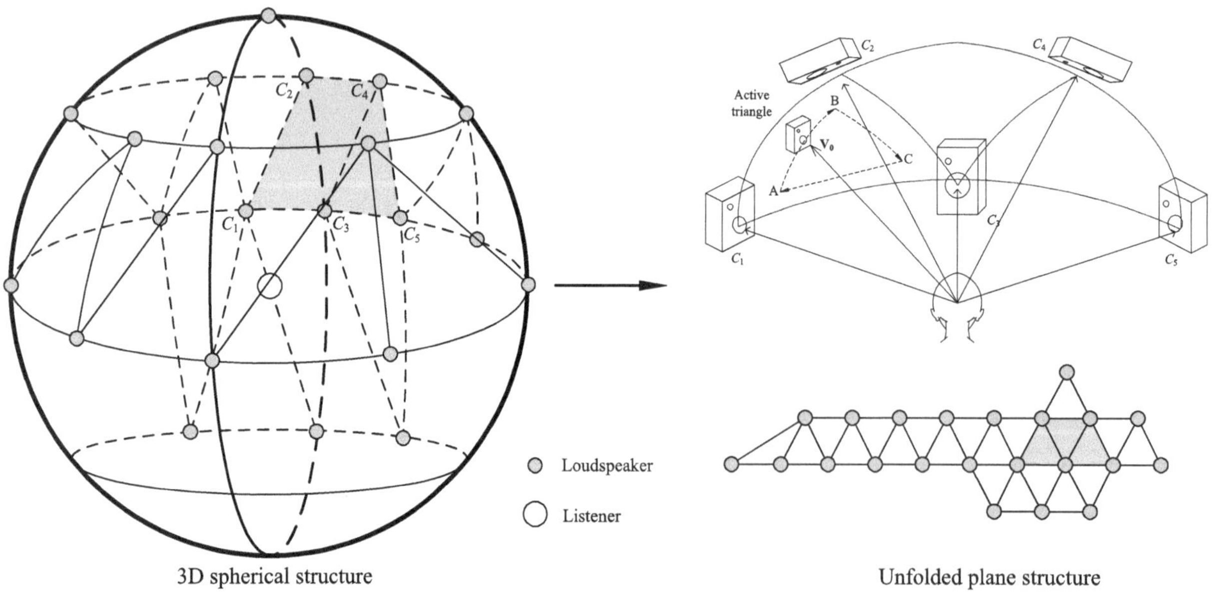

Figure 3 The 22.2 multichannel 3D audio system. A loudspeaker configuration of the 22.2 multichannel system on the sphere surface in 3D space and its plane triangle structure.

3D systems such as Ambisonics, three channels cover a basic area of 3D space and adjacent three channels also have the most similar signals [31]. If the multichannel signals are still grouped and compressed on the channel pair basis, like Ando et al. recent work for coding 22.2 multichannel signals with the AAC codec [32], there will be more than one adjacent channel for each channel. It will require extra computation to select the channel pair but may still lead to a possible to mismatch. For example, in Figure 3, channel C_1, C_2 and C_3 formed a virtual source. If C_1 and C_2 were grouped for parametric or M/S coding, C_3 would be grouped with another less correlated channels, which decreases the coding performance. And if we dynamically group the signals, we have to use complex correlation analysis algorithm to analyze its six adjacent channels. Because the channel pair grouping is based on frequency subbands, it will not only increase the codec complexity dramatically but will also be unable to reduce the overall redundancy that exist in more than two channels. In brief, the conventional channel pair unit should be redesigned for 3D audio systems.

Considering the basic unit for any 3D surface is a triangle, a spherical multichannel configuration can be easily unfolded to a plane triangle structure as shown in Figure 3. The above 3D systems also utilize three or more channels to produce 3D audio effects, which leads to interchannel signal redundancy existing in or more than three channels. Hence, channel triples should be the basic unit to remove interchannel redundancy rather than channel pair in conventional coding schemes. More specifically, in a VBAP system, the input signal $\mathbf{V}_0 = (C_1, C_2, C_3)$ is calculated following the tangent model in 3D space as

$$\begin{aligned} C_1 &= S\sin\theta\cos\varphi \\ C_2 &= S\sin\theta\sin\varphi \\ C_3 &= S\cos\theta \end{aligned} \tag{8}$$

where $\theta, \varphi \in \left[0, \frac{\pi}{2}\right]$, which determine the gain factor of the three channels.

There are infinite possible situations, where a virtual source can be located. But for the VBAP model, those possibilities can be reduced to three basic situations. First, the virtual source is located near the position of one channel. This situation corresponds to the source panned mainly using one channel, or one channel forms a virtual source with another two channels which are out of current three loudspeakers. This situation is similar to stereo audio with only one active channel, so no transform is performed and $\mathbf{M}_0$ will be used. Second, the virtual source is located between two channels. This situation corresponds to the source panned mainly using two channels, or two channels form a virtual source with one other channel out of the current three channels. This situation is similar to conventional stereo audio, and M/S coding can be applied. However, the M/S transform matrix must be modified to adapt to the three channels condition which is expressed in Equation 9. Third, the source is panned using all the three channels. This is a new situation that stereo audio never contains. To remove the interchannel redundancy, a new transform matrix $\mathbf{M}_4$ is designed following the rule of conventional M/S coding. The first vector is the summation of three channels, and the rest vectors are orthogonal with the first vector. To guarantee the conservation of energy after transformation, unit vectors are used. This matrix realizes the sum-difference processing for 3D channel, and guarantees that when three channel signals are nearly the same, two channels primarily contain the difference signal.

$$\begin{aligned}
\mathbf{M}_0 &= \begin{bmatrix} 1 & 0 & 0 \\ 0 & 1 & 0 \\ 0 & 0 & 1 \end{bmatrix} \quad
\mathbf{M}_1 = \begin{bmatrix} 1 & 0 & 0 \\ 0 & \frac{\sqrt{2}}{2} & \frac{\sqrt{2}}{2} \\ 0 & \frac{\sqrt{2}}{2} & -\frac{\sqrt{2}}{2} \end{bmatrix} \\
\mathbf{M}_2 &= \begin{bmatrix} \frac{\sqrt{2}}{2} & 0 & \frac{\sqrt{2}}{2} \\ 0 & 1 & 0 \\ \frac{\sqrt{2}}{2} & 0 & -\frac{\sqrt{2}}{2} \end{bmatrix}
\mathbf{M}_3 = \begin{bmatrix} \frac{\sqrt{2}}{2} & \frac{\sqrt{2}}{2} & 0 \\ \frac{\sqrt{2}}{2} & -\frac{\sqrt{2}}{2} & 0 \\ 0 & 0 & 1 \end{bmatrix} \\
\mathbf{M}_4 &= \begin{bmatrix} \frac{\sqrt{3}}{3} & \frac{\sqrt{3}}{3} & \frac{\sqrt{3}}{3} \\ \frac{\sqrt{2}}{2} & -\frac{\sqrt{2}}{2} & 0 \\ \frac{\sqrt{6}}{6} & \frac{\sqrt{6}}{6} & -\frac{\sqrt{6}}{3} \end{bmatrix}
\end{aligned} \tag{9}$$

An example is shown in Figure 2. It can be observed that when the source is close to the center of two or three channels, a corresponding matrix can produce difference signals with lower dynamic range compared to the original channel signals. Under a certain masking threshold, far less bits are required for quantizing the difference signals which brings the coding gain.

Then we discuss the switching condition for 3D-M/S as shown on right side of Figure 1, and the transformed mid/second/third channels are shown in Table 1. Firstly, if all three channel signals are almost the same which means the input vector is close to spherical triangle center $\mathbf{V}_4$ ($\cot\theta \approx \frac{\sqrt{2}}{2}$ and $\varphi \approx \frac{\pi}{4}$), matrix $\mathbf{M}_4$ will be chosen to give two difference channels. Secondly, if there are only two channels satisfying the conventional M/S switching condition and the projection of input vector is close to $\mathbf{V}_1$ ($\cot\theta \approx \sin\varphi$) or $\mathbf{V}_2$ ($\cot\theta \approx \cos\varphi$) or $\mathbf{V}_3$ ($\varphi \approx \frac{\pi}{4}$), 3D-M/S will select the matrix having the nearest distance from the input vector. The distance is measured by vector distance following expression (6). Compared with conventional switching condition, it can be seen that conventional M/S coding works on the two-dimensional space and has two switching areas. 3D-M/S switching condition is an expansion of M/S coding, where its input vector works in three-dimensional space and has five switching

Table 1 Transformed channel signals with five matrices

	C_M	C_S	C_T
M_0	C_1	C_2	C_3
M_1	C_1	$\frac{\sqrt{2}}{2}S\sin\theta(\cot\theta+\sin\varphi)$	$\frac{\sqrt{2}}{2}S\sin\theta(\cot\theta-\sin\varphi)$
M_2	$\frac{\sqrt{2}}{2}S\sin\theta(\cot\theta+\cos\varphi)$	C_2	$\frac{\sqrt{2}}{2}S\sin\theta(\cot\theta-\cos\varphi)$
M_3	$S\sin\theta\sin(\varphi+\frac{\pi}{4})$	$S\sin\theta\sin(\varphi-\frac{\pi}{4})$	C_3
M_4	$\frac{\sqrt{6}}{3}S\sin\theta(\cot\theta+\sin(\varphi+\frac{\pi}{4}))$	$S\sin\theta\sin(\varphi-\frac{\pi}{4})$	$\frac{\sqrt{6}}{3}S\sin\theta(\frac{\sqrt{2}}{2}\sin(\varphi+\frac{\pi}{4})-\cot\theta)$

The expression of transformed signals using five matrices with parameters θ, φ.

areas. Following the vector distance switching condition, the switching rule of 3D-M/S can be denoted as

$$\text{mode} = \begin{cases} M_4, \text{ if } & \frac{\langle \mathbf{V_0}, \mathbf{V_4} \rangle}{|\mathbf{V_0}||\mathbf{V_4}|} \geq \text{Thr}_v \\ M_i, \text{ else if } & \frac{\langle \mathbf{V_{0i}}, \mathbf{V_i} \rangle}{|\mathbf{V_{0i}}||\mathbf{V_i}|} \geq \text{Thr}_v \\ \quad \text{and} & \frac{\langle \mathbf{V_{0i}}, \mathbf{V_i} \rangle}{|\mathbf{V_{0i}}||\mathbf{V_i}|} \geq \frac{\langle \mathbf{V_{0j}}, \mathbf{V_j} \rangle}{|\mathbf{V_{0j}}||\mathbf{V_j}|}, \ \forall j \neq i \\ M_0, \text{ else} \end{cases} \tag{10}$$

where $i, j \in \{1, 2, 3\}$, $\mathbf{V_{01}} = (0, C_2, YC_3)$, $\mathbf{V_{02}} = (C_1, 0, C_3)$, $\mathbf{V_{03}} = (C_1, C_2, 0)$ are the two channel projections of input vector $\mathbf{V_0}$. $\mathbf{V_1} = \left(0, \frac{\sqrt{2}}{2}, \frac{\sqrt{2}}{2}\right)$, $\mathbf{V_2} = \left(\frac{\sqrt{2}}{2}, 0, \frac{\sqrt{2}}{2}\right)$, $\mathbf{V_3} = \left(\frac{\sqrt{2}}{2}, \frac{\sqrt{2}}{2}, 0\right)$, $\mathbf{V_4} = \left(\frac{\sqrt{3}}{3}, \frac{\sqrt{3}}{3}, \frac{\sqrt{3}}{3}\right)$ are the summation vectors of each transform matrix.

3D-M/S psychoacoustic model

The perceptual threshold is the maximum amount of noise for an audio signal that will not degrade its subjective quality. 3D-M/S coding transforms the original signals into summation and difference signals and then sends these to the core codec. The perceptual thresholds calculated by the core codec are for the transformed signals, which can not guarantee the noise in original signals is imperceptible. So after transformation, the perceptual threshold calculation in the core encoder needs a redesign. An example is the conventional M/S stereo coding, where the masking threshold model for the main channel and side channel is revised to avoid perceptible noise in reconstructed signals [23]. To derive the masking threshold for 3D-M/S signals, we reference the deducing method in [23] and expand it to channel triple case. Consider the transformed signals $(C_M, C_S, C_T) = \mathbf{M_4}\mathbf{V_0}^T$ as

$$\begin{aligned} C_M &= \frac{C_1+C_2+C_3}{\sqrt{3}} \\ C_S &= \frac{C_1-C_2}{\sqrt{2}} \\ C_T &= \frac{C_1+C_2-2C_3}{\sqrt{6}} \end{aligned} \tag{11}$$

where C_M is the first signal after transformation and C_S and C_T are the second and third signals, respectively. After core codec quantization, independent noise is introduced into the three signals which is denoted as N_M, N_S and N_T. So at the decoder side:

$$\begin{aligned} C_M + N_M &= \frac{\hat{C}_1+\hat{C}_2+\hat{C}_3}{\sqrt{3}} \\ C_S + N_S &= \frac{\hat{C}_1-\hat{C}_2}{\sqrt{2}} \\ C_T + N_T &= \frac{\hat{C}_1+\hat{C}_2-2\hat{C}_3}{\sqrt{6}} \end{aligned} \tag{12}$$

where $\hat{C}_1$, $\hat{C}_2$ and $\hat{C}_3$ are the reconstructed signals. Compared with the original signals shown in (11), the noise energy for the original signals can be obtained

$$\begin{aligned} \sigma^2_{C_1-\hat{C}_1} &= \tfrac{1}{3}N_M^2 + \tfrac{1}{2}N_S^2 + \tfrac{1}{6}N_T^2 \leq T_{C_1} \\ \sigma^2_{C_2-\hat{C}_2} &= \tfrac{1}{3}N_M^2 + \tfrac{1}{2}N_S^2 + \tfrac{1}{6}N_T^2 \leq T_{C_2} \\ \sigma^2_{C_3-\hat{C}_3} &= \tfrac{1}{3}N_M^2 + \tfrac{2}{3}N_T^2 \leq T_{C_3} \end{aligned} \tag{13}$$

where $\sigma^2_{C_1-\hat{C}_1}$, $\sigma^2_{C_2-\hat{C}_2}$ and $\sigma^2_{C_3-\hat{C}_3}$ are the expected noise energy for original signals. T_{C_1}, T_{C_2} and T_{C_3} are the masking thresholds calculated for the original signals. The sufficient conditions that guarantee the expected noise energy will not exceed masking threshold are $N_M^2, N_S^2, N_T^2 \leq \min(T_{C_1}, T_{C_2}, T_{C_3})$. This means the noise energy for any transformed signal must be less than the minimum threshold of their original signals. So the masking thresholds for the transformed signals can be derived from the masking threshold of the original signals as

$$T_{C_M} = T_{C_S} = T_{C_T} \leq \min(T_{C_1}, T_{C_2}, T_{C_3}) \tag{14}$$

The same results can be deduced for other matrices.

Framework for general channel configuration

3D-M/S only works for three channels, actually all 3D audio systems contain dozens of channels. A general channel configuration is shown in Figure 4, where channel C corresponds to a loudspeaker. Here, a framework is proposed based on 3D-M/S coding as shown in Figure 5. Because all spatially placed loudspeakers can be decomposed into basic triangle units, this structure will enable 3D-M/S coding to work for arbitrary channel configurations. The framework processes the audio channels triangle by triangle until all channels are coded. C_M is the summation channel and C_S and C_T are the second and third channels, respectively. Every 3D-M/S unit shares two channels with the previous unit and only one new

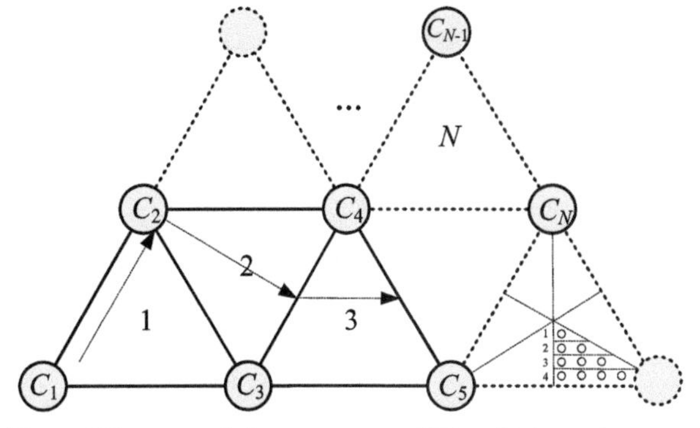

Figure 4 The general plane structure of 3D multichannel systems. A general plane structure for 3D multichannel audio systems and experimental sound source setting.

channel is added in. So, it only needs to compress the channel which contains the signal of the new channel. For all matrices M_0, M_1, M_2, M_3 and M_4, C_T is the third channel after 3D-M/S transform. Because every unit outputs only one that contains a new input channel, the whole coding framework keeps the number of channels exactly the same as original input signals. And because the output channel contains either the difference signal or original signal, coding gain can be obtained. The original signals can be obtained by multiplying 3D-M/S inverse transform matrix subband by subband at the decoder side. This framework is also suitable for other methods. For example, replacing the 3D-M/S with PCA, the codec can achieve better interchannel redundancy removing performance.

Experiment

The experiment used five channels (C_1, C_2, C_3, C_4, C_5) in spherical 22.2 multichannel configuration as shown in Figure 4. Considering that PCA is the best decorrelation transform theoretically and Independent channel coding is widely used for 22.2 multichannel compression, the experiment compared the proposed 3D-M/S method with PCA and Independent channel coding in bitrate, complexity and objective quality. Three MPEG test sequences (es01 voice signal, sc03 symphony music signal, si02 castanets transient signal, mono 48-kHz sampling) were used as the moving virtual sources following the VBAP rule, four sequences (si03, si01, sc01, es02) were used as the discrete fixed-position virtual sources. The virtual sources and respective azimuth and altitude panning angle are generated on a per-frame basis. Here, only point virtual sources were used to test the best performance of three methods, as subband signals can be regarded as point sources in subband coding when bandwidths are small enough. Signals with decorrelated elements are beyond the scope of VBAP model and will decrease the coding performance, for its difference signals retains high energy which depends on the correlation and the energy of the decorrelated elements. Uncorrelated signals with independent audio content is tested in the end.

Three basic virtual sound movements were used to cover some basic possible virtual source locations. The three movements in a triangle are movement 1 from point to point (virtual source es01), movement 2 from point to edge (virtual source sc03) and movement 3 from edge to edge (virtual source si02). Three movements are as shown

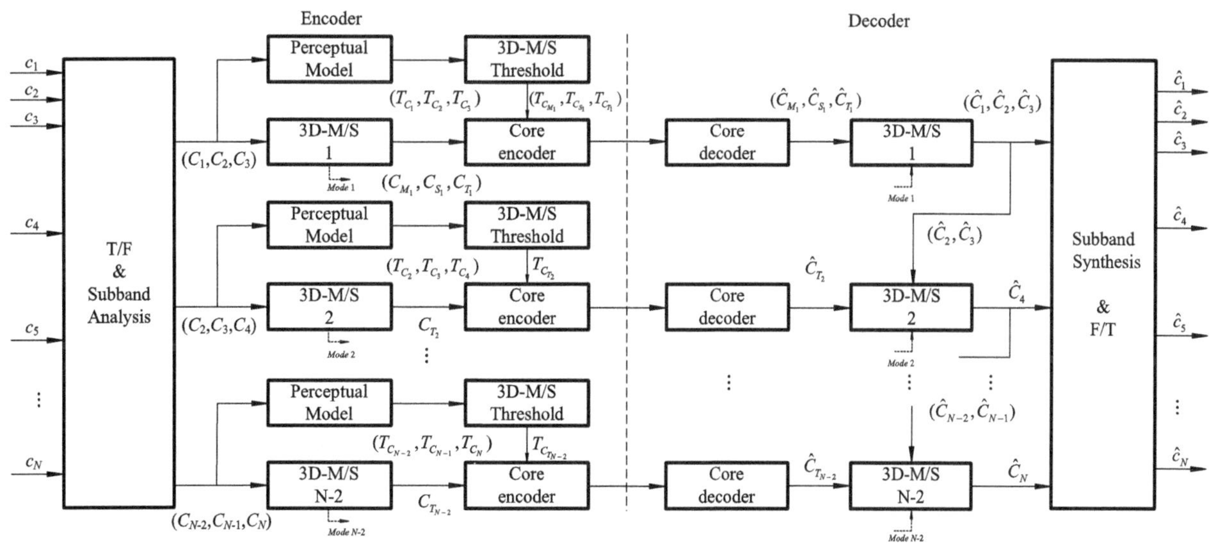

Figure 5 Three-channel unit-based coding framework. The framework using 3D-M/S coding for *N* channels.

Table 2 Discrete virtual source setting and objective results

	θ	φ				Average
		$\frac{7\pi}{32}$	$\frac{5\pi}{32}$	$\frac{3\pi}{32}$	$\frac{\pi}{32}$	
Ind	$\frac{11\pi}{36}$	20.76	–	–	–	20.76
3D-M/S	$\frac{11\pi}{36}$	21.45	–	–	–	21.45
PCA	$\frac{11\pi}{36}$	21.95	–	–	–	21.95
Ind	$\frac{13\pi}{36}$	15.37	21.14	–	–	18.26
3D-M/S	$\frac{13\pi}{36}$	16.07	21.88	–	–	18.98
PCA	$\frac{13\pi}{36}$	16.54	22.74	–	–	19.64
Ind	$\frac{15\pi}{36}$	23.48	23.28	22.08	–	22.95
3D-M/S	$\frac{15\pi}{36}$	23.77	22.93	22.54	–	23.08
PCA	$\frac{15\pi}{36}$	23.86	23.78	23.49	–	23.71
Ind	$\frac{17\pi}{36}$	21.82	19.06	18.93	19.36	19.79
3D-M/S	$\frac{17\pi}{36}$	21.88	18.52	19.02	19.51	19.73
PCA	$\frac{17\pi}{36}$	21.92	21.36	19.18	20.57	20.76

SNR (dB) of three methods for 10 discrete virtual source positions at about 64 kbps.

in Figure 4. Different sources were used for different panning patterns, because the experiments were designed to compare three methods on the same condition and change the condition simultaneously to see how three methods work on different kinds of sources and locations, instead of comparing the performance of one method on different virtual source locations with the same source. Considering the symmetry in the triangle, 10 discrete virtual source positions in $\frac{1}{6}$ triangle as shown in Table 2 and Figure 4 are used which equally divided the $\frac{1}{6}$ triangle: si03 was panned to $\left(\theta = \frac{11\pi}{36}, \varphi = \frac{7\pi}{32}\right)$; si01 was divided into two sequences and panned to $\left(\theta = \frac{13\pi}{36}, \varphi = \frac{7\pi}{32}\right)$, $\left(\theta = \frac{13\pi}{36}, \varphi = \frac{5\pi}{32}\right)$; sc01 was divided into three sequences and panned to $\left(\theta = \frac{15\pi}{36}, \varphi = \frac{7\pi}{32}\right)$, $\left(\theta = \frac{15\pi}{36}, \varphi = \frac{5\pi}{32}\right)$ and $\left(\theta = \frac{15\pi}{36}, \varphi = \frac{3\pi}{32}\right)$; es02 was divided into four sequences and panned to $\left(\theta = \frac{17\pi}{36}, \varphi = \frac{7\pi}{32}\right)$, $\left(\theta = \frac{17\pi}{36}, \varphi = \frac{5\pi}{32}\right)$, $\left(\theta = \frac{17\pi}{36}, \varphi = \frac{3\pi}{32}\right)$ and $\left(\theta = \frac{17\pi}{36}, \varphi = \frac{\pi}{32}\right)$.

The 3D-M/S and PCA was used in each subband in the frequency domain. The three encoders were realized based on FAAC-1.28, and decoders were based on FAAD2-2.7. AAC-LC was used as the core codec and only the long window was enabled for simplification. To avoid the influence of dynamic bandwidth setting of the FAAC, the experiment fixed the bandwidth at 12 kHz with 35 subbands.

Independent channel coding: Audio signals were sent into the core codec and compressed directly.

3D-M/S: The vector was calculated using the subband energy of three channels from AAC psychoacoustic module with no extra energy computation. Then 3D-M/S matrix switching was performed and 3 bits were used per mode parameter. The transformed signals were sent into the core codec, and the masking threshold was modified accordingly.

PCA: The eigenvectors were calculated for each subband. Subband signals were transformed using eigenvector matrix and then sent into core codec. The covariance matrix was quantized and transmitted to the decoder following a previous KLT-based multichannel audio coding scheme [21], with 4 bits per non-redundant element.

Objective evaluation

Complexity was measured by the running time of each method on PC (CPU: Intel Core2 Duo P8600 2.53GHz, RAM: 8GB). The main application scenario of the proposed method is medium and high bitrate, and the sound pressure at the center listening point is the most

Table 3 Bitrate setup and overall SNR

	Quality	Bitrate/channel (kbps)								Average
	(SNR)	C_1	C_2	C_3	C_4	C_5	P_1	P_2	P_3	
Ind	16.08	61.6	60.8	60.3	58.7	58.7	–	–	–	61.2
3D-M/S	18.21		229.5		21.7	57.9	4.9	4.9	4.9	64.7
PCA	18.87		133.2		25.2	42.8	39.3	39.3	39.3	63.8

Overall SNR (dB) and bitrate of three methods. For PCA and 3D-M/S, C_1, C_2 and C_3 share a total bitrate, and C_4 and C_5 are the C_T channels. P_1, P_2 and P_3 are the parameter bitrates for triangles 1, 2 and 3.

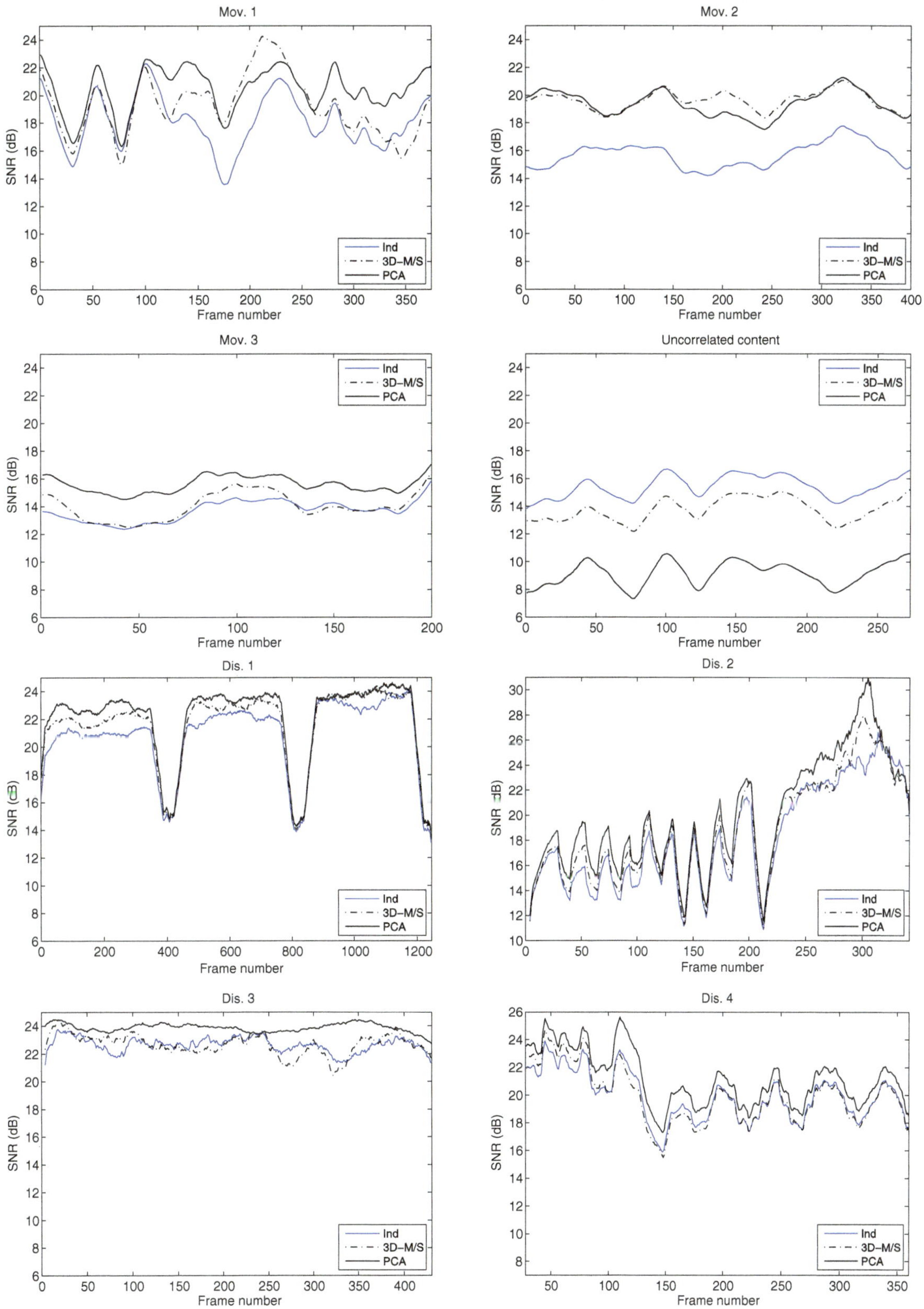

Figure 6 Objective evaluation results. Average SNR (12 kHz bandwidth) per frame for Independent channel coding, 3D-M/S and PCA methods in movements 1, 2 and 3, uncorrelated content and discrete positions 1, 2, 3 and 4.

important in multichannel audio and the VBAP panning method. Here we compared the sound pressure SNR (original and reconstructed signal sound pressure in the center listening point) to measure the objective quality following the sound pressure equation in [33]

$$p(\omega) = G\sum_{j=1}^{5} \frac{e^{-ikr_j}}{r_j} C_j(\omega), \tag{15}$$

where $p(\omega)$ is the sound pressure, G is a proportionality coefficient, $k = \frac{\omega}{c}$ is the wave number and c is the wave speed. r is the distance from the loudspeaker to the listening point, and in the spherical 22.2 multichannel, all loudspeakers have the same distance. And the SNR is calculated by

$$\mathrm{SNR} = \frac{\int_0^{2\pi} |p(\omega)|^2 d\omega}{\int_0^{2\pi} |p(\omega) - \hat{p}(\omega)|^2 d\omega} \tag{16}$$

Firstly, the bitrates of all three methods have been set to be nearly identical to compare the performance shown in Table 3. The bitrates were adopted by modifying the masking threshold in core codec and iterating two nested loops until rate constrain satisfied. Figure 6 shows the SNR curves for each method, where each SNR curve was smoothed to omit the details. The smoothing method is a typical average filtering over 20 frames length. An overall SNR was presented in Table 3 to give an overview by averaging the SNR for all frames. The three SNR curves have a downtrend because transient signal and symphony signal have a wider bandwidth than the voice signal, which require more bits to achieve the similar SNR. When the virtual source came close to the middle of the two channels (between C_1, C_2 around the 200th frame in Mov. 1, between C_3, C_4 for all frames in Mov. 2, between C_3, C_4 around the start frames, between C_3, C_5 around the 100th frame and between C_4, C_5 around the end frames in Mov. 3), 3D-M/S gets a higher SNR than the Independent channel coding and close to PCA. Moveover, around the 200th frame in Mov. 1 and Mov. 2, where all two and three channels are nearly the same, $\mathbf{M_3}$ and $\mathbf{M_4}$ can remove redundancy to the largest extent and outperform the PCA method. This is because some transformed subband signals came below the masking threshold and more bits were reserved for summation channel. The same results can be seen in the discrete virtual sources in Table 2, where $\varphi = \frac{7\pi}{32}$ $(\varphi \approx \frac{\pi}{4})$ and $(\theta = \frac{11\pi}{36}, \varphi = \frac{7\pi}{32})$ $(\theta = \frac{13\pi}{36}, \varphi = \frac{5\pi}{32})$ $(\theta = \frac{15\pi}{36}, \varphi = \frac{3\pi}{32})$ $(\theta = \frac{17\pi}{36}, \varphi = \frac{\pi}{32})$ $(\cot\theta \approx \sin\varphi)$. But when the virtual source located beyond the middle of two channels, such as $(\theta = \frac{15\pi}{36}, \varphi = \frac{5\pi}{32})$, M/S coding cannot bring coding gain. In conclusion, if the input signals are located in one of the five switching areas, the coding gain can be obtained by transforming them into summation and difference signals.

Secondly, the PCA parameter bitrate of 39.3 kbps/channel is considerably higher than 3D-M/S. If the three channels have little correlation (e.g. channels with different contents or ambient sound), the transformed signals will not save any bits and cause the decrease of coding efficiency. To test the three methods under such condition, the virtual sources of three different signals were fixed at three channels and coded all at 64 kbps. The experimental result is shown in Figure 6. We can see Independent channel coding achieves the best performance in this case; meanwhile, 3D-M/S degrades about 1 dB and PCA degrades nearly 7 dB. It is because, for PCA requirement, too many bits are used for parameters which now cannot bring any coding gain. But for 3D-M/S, parameter bits for modes are only 4.9 kbps/channel. It will not reduce the coding efficiency much for medium and high bitrate conditions, which is the main application scenario of M/S coding. Although the high bitrate for PCA can be alleviated by reducing the refresh rate of PCA parameters, but it will decrease the coding performance on VBAP signals at the same time.

Finally, Table 4 demonstrates the complexity of each method. Due to the computation for covariance matrix and eigenvector matrix, PCA increases about 30% complexity compared with the original AAC codec. For 3D-M/S, the matrix switching and signal transformation increased by about 11% complexity.

Subjective evaluation

Eight subjects who are actively working in the domain of audio compression participated in the subjective test based on ITU MUSHRA [34] method using a 3.5-kHz low-pass filtered original channels as anchor, and the test was carried out in a quite room with five-spherical channel configuration as shown in Figure 7. Subjects were required to evaluate the sound quality and sound orientation separately, and can draw the perceived sound position and movement to help their rating.

The MUSHRA test results are shown in Figure 8. For the sound quality, the subjective result generally matches

Table 4 Time complexity

	Complexity (s)		
	Encoder	Decoder	Ratio
Ind	2.382	0.223	100.0%
3D-M/S	2.604	0.306	111.7%
PCA	2.977	0.416	130.2%

Time complexity of three methods, the length of the three sequences are 27 s.

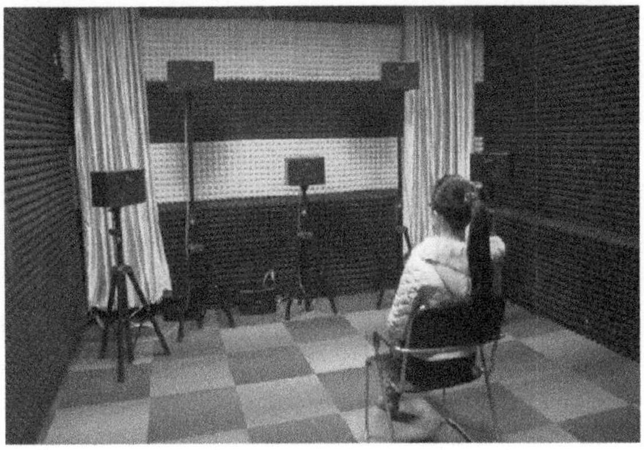

Figure 7 Subjective test setup. Subjective test environment and configuration.

the SNR result where PCA and 3D-M/S got a subjective quality improvement compared with the Independent channel coding. For the sound orientation, subjects reported that the voice and castanet signals were easier to locate. Although the 3.5-kHz anchor signal was poor in sound quality test, some subjects indicated its virtual source position was not so bad in the movement than the discrete virtual source. This may be because it is easier to distinguish the direction change for fixed sound source than the moving sound source. For 3D-M/S, subjects felt the virtual source had a sharper position in the center of two loudspeakers compared with Independent channel coding. On the whole, the subjective improvement in sound orientation exists but is not as obvious as the sound quality.

From the above results on three point sources and uncorrelated signals, it can be observed that both PCA and 3D-M/S method get about 13% SNR improvement

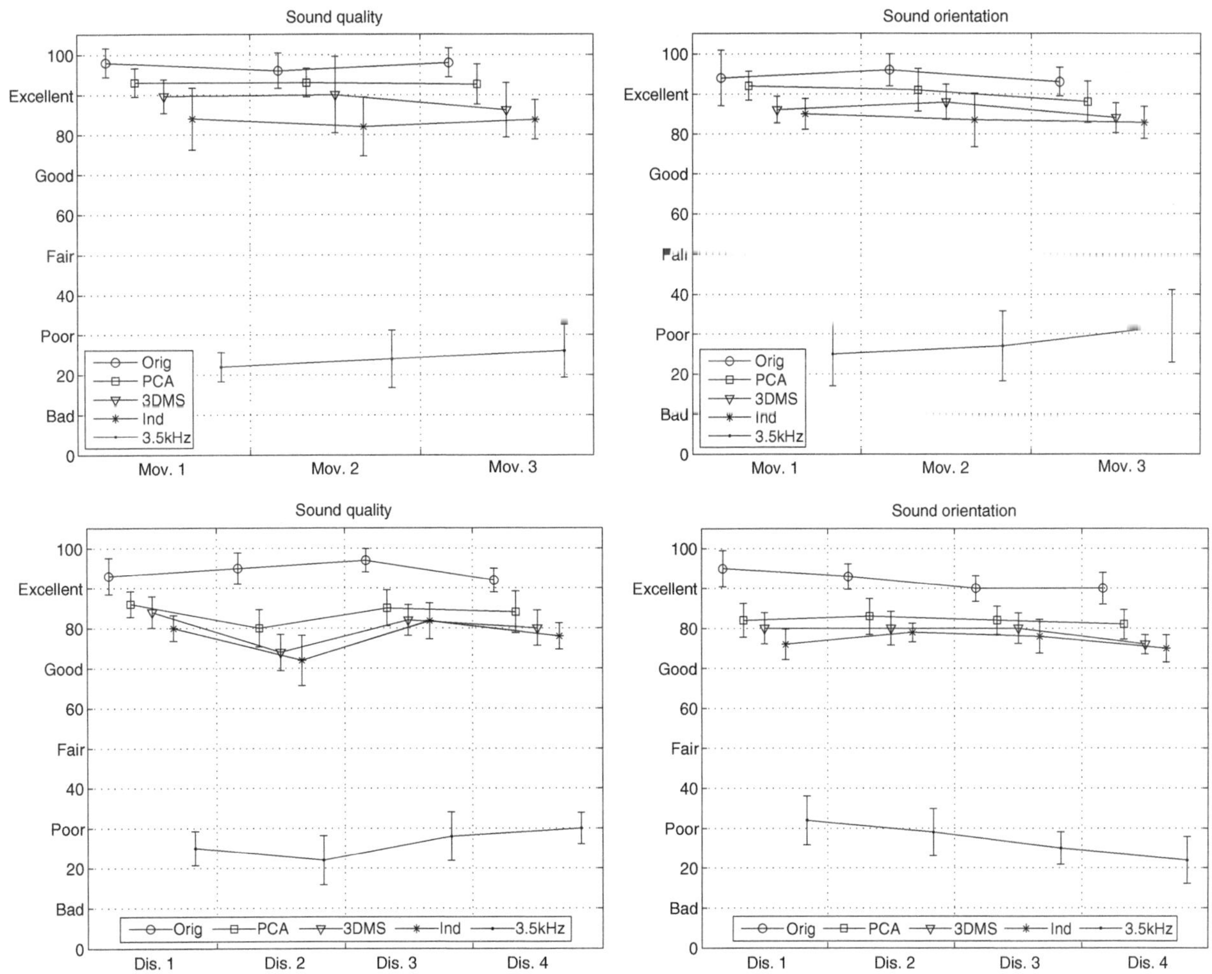

Figure 8 Subjective evaluation results. MUSHRA test results on sound quality and orientation for Independent channel coding, 3D-M/S and PCA methods in movements 1, 2 and 3 and discrete positions 1, 2, 3 and 4 with 95% confidence interval and the mean.

for each channel. But the complexity of 3D-M/S is much lower than PCA to achieve similar performance. It can be explained that the fixed matrix transform can be regarded as some special vectors in PCA. The special vectors are chosen based on the assumption that channel signals are either quite similar or quite different. This assumption may not be always true for the diversity of subband signals, but it makes a good compromise between coding efficiency and complexity.

Conclusion

This paper proposed a 3D-M/S coding method, which inherits the low complexity of conventional M/S coding. Moreover, 3D-M/S performs the sum and difference coding triple by triple, rather than couple by couple of the conventional method. This structure is more suitable for a 3D multichannel audio configuration, because adjacent three channels form a triangle and will have the maximum redundancy in spatial configured 3D audio channels. Besides, it is also convenient to unfold 3D audio multichannel structure into plane triangles. Combining the proposed framework, 3D-M/S and PCA methods can be applied to more than three channels. An experiment on VBAP signals indicates the performance of proposed method with relatively low complexity, comparing to the PCA and independent channel coding. Considering the development of 3D audio technology and its requirement for compression efficiency, a low complexity 3D audio codec will be promising and preferable for practical application.

Competing interests

The authors declare that they have no competing interests.

Acknowledgements

This work was supported by the National Natural Science Foundation of China (nos. 61231015, 61102127, 61201340, 61201169) and Natural Science Foundation of Hubei (nos. 2011CDB451, 2012FFB04205).

References

1. AJ Berkhout, D de Vries, P Vogel, Acoustic control by wave field synthesis. J. Acoust. Soc. Am. **93**(5), 2764–2778 (1993)
2. MA Gerzon, Ambisonics in multichannel broadcasting and video. J. Audio Eng. Soc. **33**(11), 859–871 (1985)
3. J Cooperstock, Multimodal telepresence systems. IEEE Signal Process. Mag. **28**, 77–86 (2011)
4. A Staff, Multichannel audio systems and techniques. J. Audio Eng. Soc. **53**(4), 329–335 (2005)
5. F Rumsey, Cinema sound for the 3-D era. J. Audio Eng. Soc. **61**(5), 340–344 (2013)
6. J Nettingsmeier, Birds on the wire - WFS live transmission project report. Tech. rep., Fraunhofer 2008
7. S Sakaida, K Iguchi, N Nakajima, Y Nishida, A Ichigaya, E Nakasu, M Kurozumi, S Gohshi, The super hi-vision codec, in *IEEE International Conference on Image Processing, 2007. ICIP 2007, Volume 1*, (2007), pp. I-21–I-24
8. F Baumgarte, C Faller, Binaural cue coding-part I: psychoacoustic fundamentals and design principles. IEEE Trans. Speech Audio Process. **11**(6), 509–519 (2003)
9. C Faller, F Baumgarte, Binaural cue coding-part II: schemes and applications. IEEE Trans. Speech Audio Process. **11**(6), 520–531 (2003)
10. W Oomen, E Schuijers, B Brinker den, J Breebaart, Advances in parametric coding for high-quality audio, in *Audio Engineering Society Convention 114*, (2003)
11. J Breebaart, Par van de S, A Kohlrausch, E Schuijers, Parametric coding of stereo audio. EURASIP J. Adv. Sig. Pr. **2005**(9), 561917 (2005)
12. J Herre, S Disch, New concepts in parametric coding of spatial audio: from SAC to SAOC, in *2007 IEEE International Conference on Multimedia and Expo*, (2007), pp. 1894–1897
13. M Goodwin, J Jot, Primary-ambient signal decomposition and vector-based localization for spatial audio coding and enhancement, in *IEEE International Conference on Acoustics, Speech and Signal Processing, 2007. ICASSP 2007, Volume 1*, (2007), pp. I-9–I-12
14. B Cheng, C Ritz, I Burnett, A spatial squeezing approach to ambisonic audio compression, in *IEEE International Conference on Acoustics, Speech and Signal Processing, 2008. ICASSP 2008*, (2008), pp. 369–372
15. E Hellerud, A Solvang, U Svensson, Spatial redundancy in Higher Order Ambisonics and its use for lowdelay lossless compression, in *IEEE International Conference on Acoustics, Speech and Signal Processing, 2009. ICASSP 2009* (2009), pp. 269–272
16. C Tzagkarakis, A Mouchtaris, P Tsakalides, A multichannel sinusoidal model applied to spot microphone signals for immersive audio. IEEE Trans. Audio Speech Lang. Process. **17**(8), 1483–1497 (2009)
17. F Pinto, M Vetterli, Wave field coding in the spacetime frequency domain, in *IEEE International Conference on Acoustics, Speech and Signal Processing, 2008. ICASSP 2008*, (2008), pp. 365–368
18. F Pinto, M Vetterli, space-time-frequency processing of acoustic wave fields: theory, algorithms, and applications. IEEE Trans. Signal Process. **58**(9), 4608–4620 (2010)
19. B Cheng, Spatial squeezing techniques for low bit-rate multichannel audio coding. PhD thesis. University of Wollongong 2011
20. B Cheng, C Ritz, I Burnett, X Zheng, A general compression approach to multi-channel three-dimensional audio. IEEE Trans. Audio Speech Lang. Process. **21**(8), 1676–1688 (2013)
21. D Yang, H Ai, C Kyriakakis, CC Kuo, High-fidelity multichannel audio coding with Karhunen-Loeve transform. IEEE Trans. Speech Audio Process. **11**(4), 365–380 (2003)
22. J Johnston, A Ferreira, Sum-difference stereo transform coding, in *1992 IEEE International Conference on Acoustics, Speech, and Signal Processing, 1992. ICASSP-92, Volume 2* (1992), pp. 569–572
23. CM Liu, WC Lee, YH Hsiao, M/S coding based on allocation entropy, in *Proceedings of the 6th International Conference on Digital Audio Effects (DAFx-03)*, (2003)
24. O Derrien, G Richard, A new model-based algorithm for optimizing the MPEG-AAC in MS-Stereo. IEEE Trans. Audio Speech Lang. Process. **16**(8), 1373–1382 (2008)
25. H Krueger, P Vary, A new approach for low-delay joint-stereo coding, in *2008 ITG Conference on Voice Communication (SprachKommunikation)*, (2008), pp. 1–4
26. M Schafer, P Vary, Hierarchical multi-channel audio coding based on time-domain linear prediction, in *2012 Proceedings of the 20th European Signal Processing Conference (EUSIPCO)*, (2012), pp. 2148–2152
27. M Neuendorf, M Multrus, N Rettelbach, G Fuchs, J Robilliard, J Lecomte, S Wilde, S Bayer, S Disch, C Helmrich, R Lefebvre, P Gournay, B Bessette, J Lapierre, K Kjorling, H Purnhagen, L Villemoes, W Oomen, E Schuijers, K Kikuiri, T Chinen, T Norimatsu, CK Seng, E Oh, M Kim, S Quackenbush, B Grill, MPEG unified speech and audio coding-the ISO/MPEG standard for high-efficiency audio coding of all content types, in *Audio Engineering Society Convention 132* (Audio Engineering Society, 2012)
28. M Multrus, M Neuendorf, J Lecomte, G Fuchs, S Bayer, J Robilliard, F Nagel, S Wilde, D Fischer, J Hilpert, N Rettelbach, C Helmrich, S Disch, R Geiger, B Grill, ed. by A Heuberger, G Elst, and R Hanke, MPEG unified speech and audio coding - bridging the gap, in *Microelectronic Systems* (Springer Berlin Heidelberg Berlin, Heidelberg, 2011), pp. 351–362
29. C Helmrich, P Carlsson, S Disch, B Edler, J Hilpert, M Neusinger, H Purnhagen, RettelbachN, J Robilliard, L Villemoes, Efficient transform coding of two-channel audio signals by means of complex-valued stereo prediction, in *2011 IEEE International Conference on Acoustics, Speech and Signal Processing (ICASSP)*, (2011), pp. 497–500
30. CM Liu, HW Hsu, WC Lee, Compression artifacts in perceptual audio coding. IEEE Trans. Audio Speech Lang. Process. **16**(4), 681–695 (2008)

31. F Zotter, M Frank, All-round ambisonic panning and decoding. J. Audio Eng. Soc. **60**(10), 807–820 (2012)
32. A Ando, T Sugimoto, K Irie, Coding of 22.2 multichannel audio signal by MPEG-AAC, in *IEICE Tech. Rep., Volume 113 of EA2013-46*, (2013), pp. 75–80
33. A Ando, Conversion of multichannel sound signal maintaining physical properties of sound in reproduced sound field. IEEE Trans. Audio Speech Lang. Process. **19**(6), 1467–1475 (2011)
34. ITU-T, *Method for the subjective assessment of intermediate sound quality (MUSHRA)*, (2001)

11

Source ambiguity resolution of overlapped sounds in a multi-microphone room environment

Rupayan Chakraborty*, Climent Nadeu and Taras Butko

Abstract

When several acoustic sources are simultaneously active in a meeting room scenario, and both the position of the sources and the identity of the time-overlapped sound classes have been estimated, the problem of assigning each source position to one of the sound classes still remains. This problem is found in the real-time system implemented in our smart-room, where it is assumed that up to two acoustic events may overlap in time and the source positions are relatively well separated in space. The position assignment system proposed in this work is based on fusion of model-based log-likelihood ratios obtained after carrying out several different partial source separations in parallel. To perform the separation, frequency-invariant null-steering beamformers, which can work with a small number of microphones, are used. The experimental results using all the six microphone arrays deployed in the room show a high assignment rate in our particular scenario.

Keywords: Source position assignment; Null steering beamforming; Acoustic event detection; Simultaneous sounds; Fuzzy integral-based fusion

1. Introduction

Sound is a rich source of information. For that reason, machine audition [1] plays an important role in many applications. In particular, for meeting room scenarios, knowledge of the identity of possibly simultaneous sounds that take place in a room at a given time and their position in space is relevant to automatically describe social and human activities [2-5], to increase the robustness of speech processing systems operating in the room [6], to assist video conferencing [7], etc.

Acoustic event detection (AED) systems try to determine the identity of an occurring sound and the time interval when it is produced [2-4]. Acoustic source localization (ASL) systems estimate its position in space [5,8-11]. Both tasks become much more challenging when there exists sound simultaneity, i.e., several sounds overlapping in time and in a given room. For example, after the CLEAR'07 international evaluations [12], where AED was carried out with meeting room seminars, it became clear that time overlapping of acoustic events (AEs) was a major source of detection errors [13].

In the concrete scenario used for the experiments, a typical meeting room acoustic scene is considered, where only a person is speaking at a given time and other non-speech sounds may happen simultaneously with the speaker's voice. Therefore, we have to deal with the problem of detecting and localizing an acoustic event that may be temporally overlapped with speech. The detection of overlapping events may be tackled with different approaches, either at the signal level, at the model level, or at the decision level. In [13-15], a model-based approach was adopted for detection of events in that meeting room scenario with two sources, one of which is always speech and the other one is an acoustic event from a list of 11 predefined events. Thus, besides the mono-event acoustic models, additional acoustic models were considered for each AE overlapped with speech, so the number of models was doubled (22 in that case). That approach is used in the current real-time system implemented in the smart-room of Universitat Politècnica de Catalunya (UPC), which includes both AED and ASL [14].

In that model-based approach, a permutation problem exists. In fact, the AED system gives the hypothesized identities of the overlapped sounds, but does not associate each of them to one of the available source positions

* Correspondence: rupayan.chakraborty@upc.edu
TALP Research Center, Department of Signal Theory and Communications, Universitat Politècnica de Catalunya (UPC), Barcelona 08034, Spain

that are provided by the ASL system. The same problem may be encountered by using other AED approaches, for instance, if a blind source separation technique is used prior to the detection of each of the separated events. To solve that source ambiguity problem, a position assignment (PA) system that performs a one-to-one correspondence between the set of source positions and the set of class labels is presented in this paper.

The proposed PA system, a preliminary version of which was presented in [15,16], consists of three stages: beamforming-based signal separation, model-based likelihood calculations, and fusion of log-likelihood ratios over the set of beamformers. Frequency-invariant null-steering beamformers are designed for the small microphone arrays existing in the smart-room. In addition, the likelihoods coming from both a speech model and an acoustic event model are combined in order to improve the system accuracy. The work presented in this paper is an extension and improvement of the work reported in [15,16]. On the contrary, to that previous published work, in the PA system reported here, all the six microphone arrays available in the room are employed in the experiments. For taking the decision, the scores obtained from each array are combined using either a product of likelihood ratios or a fuzzy integral-based fusion technique [17-19]. Experiments are carried out for the scenario described above and with a set of signals collected in the smart-room. The observed PA accuracy is larger than 95%.

The acoustic scenario is presented in Section 2 together with the signal database used for experimentation. The position assignment system is described in Section 3. Experiments are reported in Section 4, along with some practical issues about the system implementation and the used metrics. Conclusions are presented in Section 5.

2. Acoustic scenario and database

Figure 1 shows the smart-room of the UPC, with the position of its six T-shaped four-microphone arrays on the walls. The linear arrays of three microphones are used in the experiments. The total number of considered acoustic event classes is 12, including speech, as shown in Table 1. In the working scenario, it is assumed that speech is always produced at one side of the room (either left or right), and the other AEs are produced at the other side.

For the offline design of the system, whose real-time version is later implemented in the room, a database is needed. It was recorded using a spatial distribution of the AE sources and the speech source, as depicted in Figure 1. The position of the speaker was rather stable, but the other AEs were produced within broad areas of the room layout. There are eight recorded sessions (S01 to S08) of isolated AEs, where six different persons had participated and performed each AE several times. Note that, though in a real meeting room scenario the speaker may be placed at either the left or the right side of the room, in the database its position is fixed. This will not constrain the usefulness of the results, because the system will not make use of that knowledge.

As in [14], we have used for training, development, and testing up to eight sessions of audio data with isolated acoustic events. Each session was recorded with all the six T-shaped microphone arrays (24 microphones). The overlapped signals of the database were generated adding those AE signals recorded in the room with a speech signal, also recorded in the room for a single speaker from all the 24 microphones. To do that, for each AE instance, a segment with the same length was extracted from the speech signal starting from a random position and added to the AE signal. The mean power of speech was made equivalent to the mean power of the overlapping AE. That addition of signals produces an increment of the background noise level, since it is included twice in the overlapped signals; however, going from isolated to overlapped signals, the SNR reduction is slight: from 18.7 to 17.5 dB. The average duration of the events is 500 ms, and the reverberation time of the room is around 450 ms. Signals were recorded at 44.1 kHz sampling frequency and further converted to 16 kHz.

3. Source position assignment

The block diagram of the whole system that performs position assignment from the outputs of the acoustic event detection and localization systems is depicted in Figure 2. The model-based AED system outputs either one or two AE hypothesis. On the other hand, in the online implementation at the UPC's smart-room, the ASL system provides either one or two source positions. Hence, there are four different possibilities for mapping the one/two detected events into the one/two detected positions. As can easily be noticed, there exists an ambiguity in three out of those four possibilities. This work is focused on the most general case, where two events are detected, i.e., E (one of the 11 possible AEs) and 'sp,' and also two source positions: P_1 and P_2. Hence, the position assignment (PA) block actually is a binary classifier that assigns E to either P_1 or P_2.

If the problem of assigning the two events to the two positions is solved, the other two cases with ambiguity can also be solved using the same approach. In this section, the aim is to design a system that can be deployed in real time in the room to resolve that ambiguity in the correspondence between detected AEs and acoustic source positions.

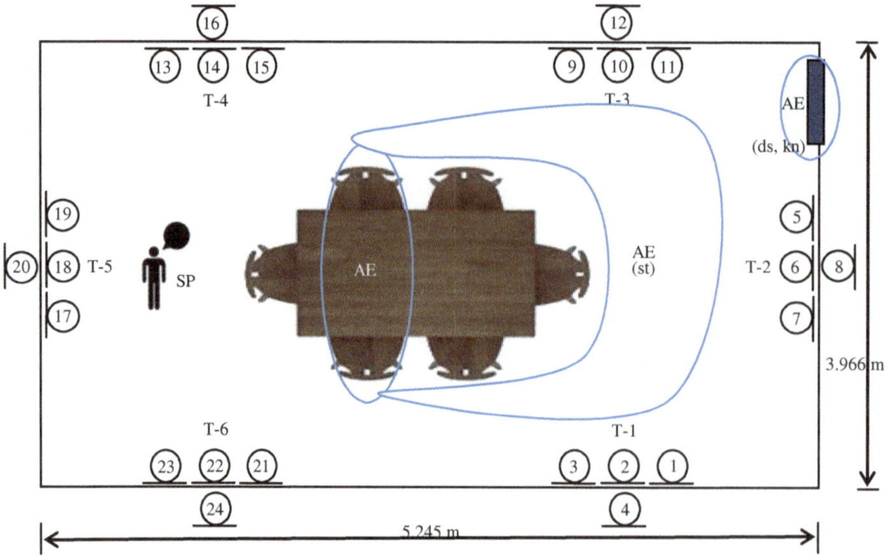

Figure 1 Smart-room (4 m × 5.3 m × 4 m) layout, with the positions of microphone arrays (T-*i*), acoustic events (AE), and speaker (SP).

3.1. The AED and ASL subsystems

The two-source AED system included in Figure 2, which was developed in previous work [13], employs a model-based approach with one microphone. All accepted sound combinations are modeled, i.e., the AED system has a model for each class whether it is an isolated event or a combination of events. This approach does not require a prior separation of the two overlapped signals, but requires a number of models that may be too large. In our particular meeting room scenario, however, the approach is feasible because 11 AEs are considered, which may be overlapped only with one class, speech, so only 22 models are required [14,15]. The ASL system, also developed in previous works [14], is based on the steered response power with phase transform (SRP-PHAT) algorithm, which uses 24 microphones available in the room.

Table 1 Acoustic classes and their number of occurrences

Label	Event type	Number of occurrences
[ap]	Applause	88
[cl]	Spoon/cup jingle	96
[cm]	Chair moving	242
[co]	Cough	90
[ds]	Door open/slam	256
[kj]	Key jingle	82
[kn]	Door knock	79
[kt]	Keyboard typing	89
[pr]	Phone ring	101
[pw]	Paper work	91
[st]	Step	205

3.2. The position assignment system

The scheme of the PA system, which is shown in Figure 3 for one array, has at its front-end two null-steering beamformers (NSBs), which work in parallel. The main beam of each NSB is steered towards the desired source and a null is placed in the direction of the interfering source, so each NSB will nullify a different source signal. Thus, the contribution of one of the simultaneous sounds to the beamformer output is expected to be lower than its contribution to the beamformer input. Indeed, beamforming is based on the prior knowledge of the direction of the desired source and the interferent source, which can be provided by an ASL system. Thus, each NSB requires two inputs: (1) the multi-microphone signal and (2) the position coordinates or direction of arrival (DOA) of the sources.

Each of the beamformers is followed by feature extraction (FE) and likelihood computation (LC). In this work, hidden Markov model and Gaussian mixture model (HMM-GMM) are employed, for both acoustic events and speech. Given the AE class E, the model for E and the model for speech (sp) are needed for the likelihood computations. Finally, a decision block makes the assignment based on the computed log-likelihoods.

3.3. Null-steering beamforming

Null-steering beamforming is one of the earliest, but potentially very useful beamforming techniques. It belongs to a class of very popular and widely used beamforming techniques called multiple side lobe cancellers (MSC) [20-22]. NSB adapts the sensor array pattern by steering the main beam towards the desired source and placing nulls in the direction of the interfering sources [23]. The

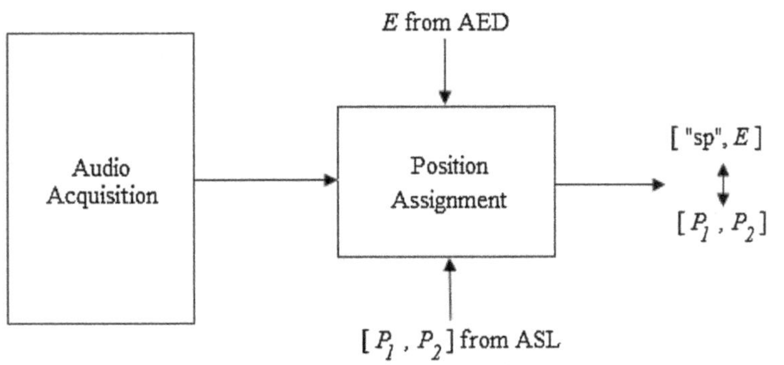

Figure 2 Block diagram of the whole system.

solution for the weight matrix in this type of beamformer is achieved by setting to unity the desired response at the direction of the target sound and setting it to zeros at the direction of the interfering sources. In our particular scenario, only one interfering source has to be nulled, so only two microphone signals are required to get a solution for the weight matrix. However, as in each T-shaped array of the room, there are three linearly spaced microphones; all three of them will be used in the experiments.

Given the broadband characteristics of the audio signals, to determine the beamformer coefficients, a technique called frequency-invariant beamforming (FIB) is used [24-27]. The method, proposed in [25] and [26], uses a numerical approach to construct an optimal frequency-invariant response for an arbitrary array configuration with a very small number of microphones, and it is capable of nulling several interfering sources simultaneously. As depicted in Figure 4, the FIB method first decouples the spatial selectivity from the frequency selectivity by replacing the set of real sensors by a set of virtual ones, which are frequency invariant. Then, the same array coefficients can be used for all frequencies.

An alternative frequency-dependent approach was explored for a single array in our previous works [15,16], using either a time domain implementation or a frequency domain implementation. However, in spite of carrying out a careful frequency tuning, the obtained PA accuracy was just slightly higher than the one from the FIB-based system for the time domain implementation [16]. On the other hand, the alternative FIB technique does not require frequency tuning, and thus, it is less dependent on the concrete scenario. For those reasons, FIB was chosen in the work reported here.

3.4. Single-array classification stage

As shown in Figure 3, the classification stage of the PA system with a single array consists of feature extraction, followed by log-likelihood calculation and a binary decision block. Features are extracted from the audio signals with a frame length of 30 ms and a frame shift of 20 ms. As features, frequency-filtered log filter-bank energies (FF-LFBEs), which were developed for speech recognition, are used [28]. These features are uncorrelated, similarly to the conventional MFCC features, but unlike the latter, the FF-LFBE features keep the frequency

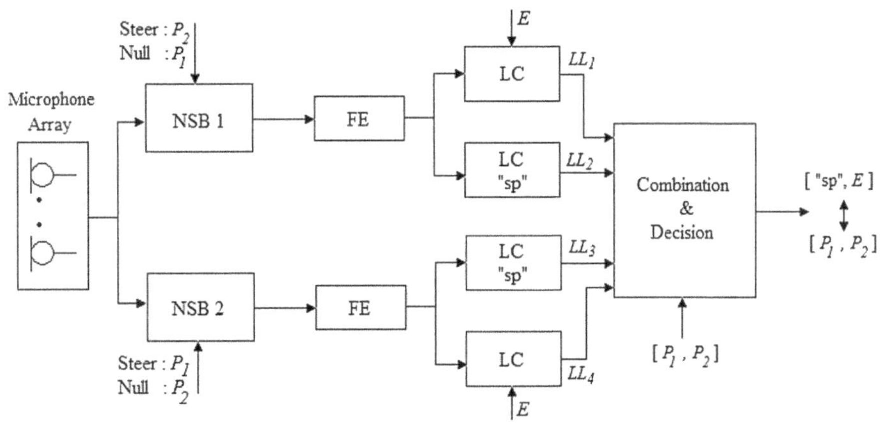

Figure 3 Position assignment system for a single array.

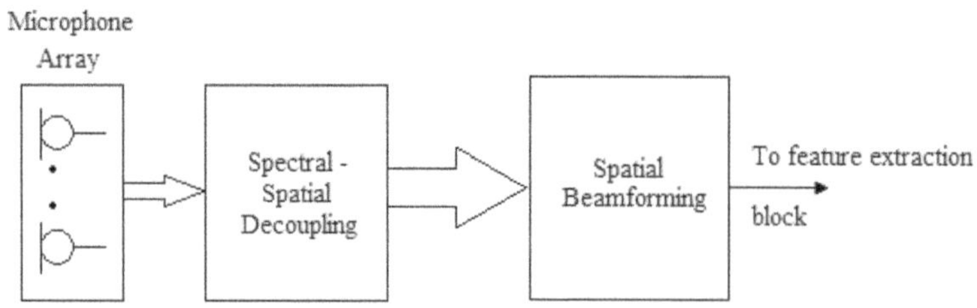

Figure 4 Frequency-invariant beamforming.

localization property of the LFBEs. In the experiments, a 32-dimension feature vector (16 FF-LFBEs and their first temporal derivatives) is used. As shown in the scheme of Figure 3, there is a set of two likelihood calculators for each parallel channel, one of them to calculate the model based log-likelihood for the AE label (E), provided by the AED system, and the other to calculate it for speech. As in [14], here also hidden Markov models (HMMs) are employed, where Gaussian mixture models (GMMs) are used to calculate the emission probabilities [29]. Thirty-two Gaussian components with diagonal covariance matrices are used per model. There is one left-to-right HMM with three emitting states for each AE and speech. Eleven HMMs are trained with isolated events using the Baum-Welch algorithm [29]. The HTK toolkit is used for training and testing this HMM-GMM system [30].

To make the mapping between source positions and event identities, the decision block uses the four log-likelihood scores computed from the HMM-GMM models. Those four scores, which are indicated in Figure 3 as LL_i, $i = 1, 2, 3, 4$, are grouped in two log-likelihood ratios LLR_1 and LLR_2, one for each beamformer path, and the following single-array score S is computed:

$$S = LLR_1 + LLR_2 = (LL_1 - LL_2) + (LL_3 - LL_4) \quad (1)$$

If S is positive, the AE E is associated to the position P_2, and if S is negative, it is associated to P_1. Let us illustrate it with a particular case. Assume that P_1 truly corresponds to speech and P_2 to the acoustic event E. When using the AE model, it is expected to get comparatively higher log-likelihood from the output of NSB1 (LL_1) than from the output of NSB2 (LL_4). For the clean speech model, it is expected to get comparatively higher log-likelihood from the output of NSB2 (LL_3) than from the output of NSB1 (LL_2). If that is the case, the decision is taken that speech is at P_1 and E is at P_2, which is the correct decision. Note that with this type of combination, the decision block gives equal importance to all the four likelihood calculator outputs.

In order to get the most from the available information, in the current scheme, unlike in [15], the classification stage includes a speech model besides the AE model. Indeed, the system could also work with only either the speech model-based classifier or the AE model-based classifier. To study the contribution of each one of the models, all those options have been tested, and the results are reported in Section 4. For the decision, if only either the AE or the speech-based classifier is used, just either LL_1-LL_4 or LL_3-LL_2, respectively, is needed.

3.5. Multi-array fusion

As it is mentioned earlier, all the six three-microphone linear arrays deployed in the room are used in the position assignment system. For taking the assignment decision, the six sets of scores LLR_1 and LLR_2, computed as indicated in Equation 1, are combined either with a uniformly weighted average [31] of the 12 values or by fuzzy integral-based fusion [19]. In the following, the latter technique is presented.

3.5.1. FI-based optimized fusion

The scores at the output of the classification stage can be linearly combined by using an optimal fusion approach that assigns an individual weight to each of them. However, a more sophisticated weighting technique that considers all subsets of information sources: the fuzzy integral (FI) approach, is considered in this work [19].

Let us denote by h_i, $i = 1, 2, \ldots, N$, the set of output scores (LLR_1 and LLR_2) of the $N/2$ single-array systems. Assuming that the sequence h_i, $i = 1, 2, \ldots, N$, is ordered in such a way that $h_1 \leq \ldots \leq h_N$, the Choquet fuzzy integral can be computed as

$$M_{FI}(\mu, h) = \sum_{i=1}^{N} [\mu(i, \ldots, N) - \mu(i+1, \ldots, N)] h_i \quad (2)$$

where $\mu(N+1) = 0$. The value $\mu(Q)$ can be viewed as a weight related to a subset Q of the set Z of N information sources. It is called the fuzzy measure, and if Q and T are subsets of Z, it has to meet the following conditions:

Boundary: $\mu(\phi) = 0, \mu(Z) = 1$

Monotonicity: $Q \subseteq T \Rightarrow \mu(Q) \leq \mu(T)$

In this work, a supervised gradient-based training algorithm for learning the fuzzy measures from the training data with cross-validation is used [18,19].

4. Experimental work

The PA experiments are done under the assumption that there is always an AE overlapped with speech. It is assumed that the identity of the AE event is known, to avoid the propagation of the AED errors to the PA system. Additionally, it is assumed that the approximate position in the room of the AE source and the speaker are known. Thus, the PA system only has to make a binary decision (the AE is from position P_1 or position P_2), which will be either correct or incorrect.

To design and evaluate the performance of the system, the position assignment rate (PAR) metric for a given AE class is defined as the quotient between the number of correct decisions and the total number of occurrences of that class in the testing database. Then, the PAR will be averaged over the classes to have the final evaluation measure. For reference, a second metric is also considered, called Diff_LL, which is the value of the S score from Equation 1 provided that the assignment is correct (LL_1-LL_4 for the AE-based system, or LL_3-LL_2 for the speech-based one, or S when both the AE model and the speech model are used). Actually, that score can be considered as an estimate of the degree of source separation carried out by the beamformers when a correct assignment is made.

In the PA system from Figure 3, there exist two FIB-based NSBs at the front end. The design of the beam formers for each particular AE sample requires the DOA angles corresponding to the target and the null, i.e., the DOAs from the source positions P_1 and P_2. Two different options regarding the approximate positions of the acoustic events from which the DOAs are extracted have been considered. First, the same approximate DOA for the whole set of acoustic events for each array has been considered. It is obtained as a DOA average over the AE source positions, which are known from visual inspection during recording. In this case, a beam pattern with a broader main lobe (as shown in the right side of Figure 5) to approximately encompass all the positions of the acoustic events has been designed. And, in the second option, the position of the event estimated by an ASL system based on the SRP-PHAT technique has been used. Therefore, in that case, the beam steers to the direction of the specific event position. It is worth to mention here that the AE source positions are estimated using a one-source ASL system instead of a two-source one, in order to avoid more propagation of errors from the ASL system to the PA system. Regarding the speech source position, the speaker's position specified during recording has been used for all experiments.

4.1. Results and discussion

To assess the performance of the PA system depicted in Figure 3, several experiments have been conducted. The testing results are obtained with all the eight recording sessions (S01 to S08), using a leave-one-out criterion, and averaging over all the testing dataset. In all the FI-based fusions, a 5-fold cross-validation on the training data to stop the training process and avoid over-fitting is used. To check the performance of the PA system when either only the AE model or only the speech model is used, the experiments for the array T6 is performed, using visually inspected positions for AEs and a broad beam. The results are shown in Table 2. It is worth mentioning that the AE source positions and the speech source position are physically rather well separated from the viewpoint of the array T6.

It can be observed, from the results in Table 2, that the combination of the two models with the S score, which averages the scores LLR_1 and LLR_2, improves the

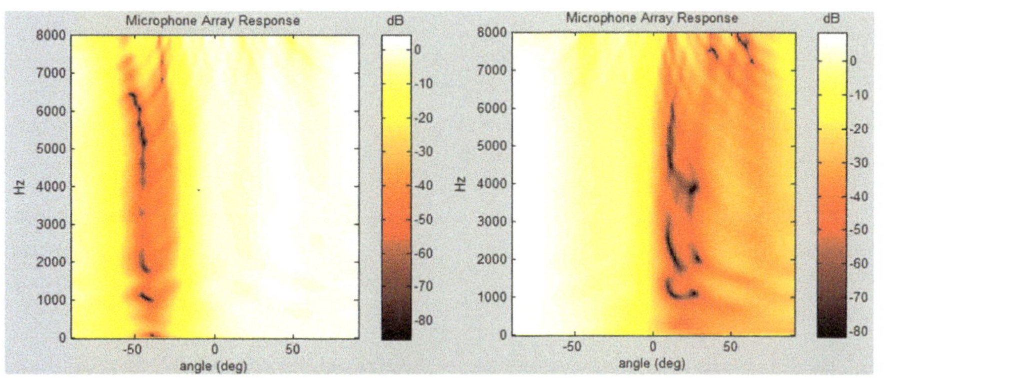

Figure 5 Beam pattern of the FIB. Left: null towards speech; right: null towards the AEs in the case of a wide beam encompassing the DOAs of all the AEs.

Table 2 PA rate and Diff_LL for the PA system with the T6 array alone

	AE model	Speech model	Average of LLR scores	FI-based fusion
PAR (%)	86.2	81	87.1	91.2
Diff_LL	1.44	1.07	2.53	2.88

performance of the system with respect to the use of only one type of model. The improvement is much more noticeable using the FI-based fusion of the two scores. Notice also that the AE model-based system works much better than the speech model-based one. In fact, the former uses a more specific model, because the speech model is obtained from the whole set of speech sounds. In that table, the Diff_LL score is also shown. Notice that, in general, it is well correlated with the PAR one. However, there is a large difference between the values of Diff_LL for the AE-based case and the LL combination case, in contrast with the very small difference there is in terms of PAR. It means that the use of both models allows achieving a much stronger confidence on the PA decision when it is correct.

Table 3 shows the PAR scores when all six microphone arrays are employed, either alone or in combination. The results are given for the two types of DOA settings mentioned above. Also, two types of intra-array combinations are considered, as in Table 2 (second column of Table 3): the average of LLR scores given by Equation 1 and the FI-based fusion. The testing results are obtained with all the eight recording sessions (S01 to S08), using a leave-one-out criterion, averaging over all the testing dataset, and tabulated with the population standard deviations.

In the first columns of Table 3, we have the PAR (in %) with the standard deviations when each one of the arrays is used alone. Notice that the PAR scores of the upper half of the array numbers (T4, T5, and T6) are higher than the ones of the lower half. It could be expected, since for those arrays the angle between the acoustic event positions and the speech source position is larger than that for the other arrays (T1, T2, and T3).

Note from the two last columns in Table 3 that the accuracy obtained from either the average of LLR scores or the FI-based fusion of the whole set of arrays is higher than the accuracy obtained from any of the single arrays. Comparing both types of fusion, the FI one shows a noticeable better performance for both DOA setting cases, arriving to a PA error of only 4.3%.

The use of intra-array FI-based fusion improves the PAR scores with respect to using a uniformly weighted average of LLR scores, especially for the upper half arrays. Therefore, though by employing the FI approach there is the cost of having to learn the fuzzy measures from data, it may be a good choice when the quality of the signal separation is not too low, like it presumably happens with the upper half arrays

Regarding the type of DOA setting, the ASL-estimated AE position-based system works always better than the one that uses an average DOA based on visual inspection (i.e., a broad-beam nulling angle). That could be expected, since the beam pattern is specific of each event occurrence, whereas the broad beam encompasses all the angles of the AE source positions. While the latter design simplifies the overall system, as it does not require a precise source position and may avoid an additional external ASL block, it is specific of the given scenario, so it has to be redesigned when the scenario changes.

5. Conclusions

An attempt is made in this paper to resolve the source identification ambiguity that appears when an acoustic event, which overlaps with speech, is detected. A position assignment system has been proposed and tested. It firstly consists of a set of frequency-invariant null-steering beamformers that carry out different partial signal separations for each microphone array. The beamformers are followed by model-based likelihood calculations, using both the acoustic event model and the speech model, to obtain two likelihood ratios, whose combination gives a final score per array. Using the fuzzy integral for that intra-array combination and also for the fusion of the six array scores, the best assignment error is obtained, which is smaller than 5%. It is worth noticing that, though the position assignment system has been developed for the problem encountered in the current scenario, its scheme can be extended to more than two sources and to different types of sound overlap combinations. Future work will be devoted to that.

Table 3 PA performance (%) with standard deviation for each single array and for the two combinations

DOA setting	Intra-array combination	T1	T2	T3	T4	T5	T6	Inter-array combination	
								Average of LLR scores	FI-based fusion
Broad-beam nulling angle	Average of LLR scores	83.1 ± 1.9	77.3 ± 2.6	81.3 ± 1.9	88.9 ± 2	88.2 ± 1.8	87.1 ± 1.8	89.8 ± 2	93.5 ± 1.9
	FI-based fusion	83.5 ± 1.9	77.1 ± 2.5	82.3 ± 1.8	92.8 ± 1.4	92.7 ± 1.5	91.2 ± 1.6	-	93.6 ± 1.7
ASL-estimated AE positions	Average of LLR scores	88.2 ± 1.8	85.6 ± 2.3	89.8 ± 1.7	91.2 ± 1.6	92.1 ± 1.7	91 ± 1.9	93.6 ± 1.8	95.4 ± 1.7
	FI-based fusion	88.3 ± 1.7	84.9 ± 2.1	90.2 ± 1.6	92.7 ± 1.3	93 ± 1.4	92.2 ± 1.4	-	95.7 ± 1.5

Competing interests

The authors declare that they have no competing interests.

Acknowledgements

This work has been supported by the Spanish project SARAI (TEC2010-21040-C02-01).

References

1. W Wang, *Machine Audition: Principles, Algorithms and Systems* (IGI Global, 2010)
2. A Waibel, R Stiefelhagen, *Computers in Human Interaction Loop* (Springer, New York, 2009)
3. A Temko, C Nadeu, D Macho, R Malkin, C Zieger, M Omologo, Acoustic event detection and classification, in *Computers in the Human Interaction Loop*, ed. by A Waibel, R Stiefelhagen (Springer, New York, 2009), pp. 61–73
4. X Zhuang, X Zhou, MA Hasegawa-Johnson, TS Huang, Real-world acoustic event detection. Pattern Recogn. Lett. **31**, 1543–1551 (2010)
5. M Omologo, P Svaizer, Acoustic event localization using crosspower-spectrum phase based technique, in *ICASSP* (Adelaide, 1994)
6. T Nishiura, S Nakamura, Study of environmental sound source identification based on hidden Markov model for robust speech recognition. J. Acoust. Soc. Am. **114**(4), 2399 (2003)
7. H Wang, P Chu, Voice source localization for automatic camera pointing system in videoconferencing, in *ICASSP* (Munich, 1997), pp. 187–190
8. J DiBiase, HF Silverman, M Brandstein, Microphone arrays. Robust localization in reverberant rooms, in *Microphone Arrays: Signal Processing Techniques and Applications*, ed. by M Brandstein, D Ward (Springer, New York, 2001)
9. D Wang, GJ Brown, in *Computational Auditory Scene Analysis: Principles, Algorithms, and Applications*, ed. by D Wang, GJ Brown (Wiley-IEEE, 2006)
10. J Dmochowski, J Benesty, Steered beamforming approaches for acoustic source localization, in *Speech Processing in Modern Communication*, ed. by I Cohen, J Benesty, S Gannot, vol. 12 (Springer, Berlin, 2010), pp. 307–337
11. J Velasco, D Pizarro, J Macias-Guarasa, Source localization with acoustic sensor arrays using generative model based fitting with sparse constraints. Sensors **12**(10), 13781–13812 (2012)
12. CLEAR, *Classifications of Events, Activities and Relationships: Evaluation and Workshop* (Baltimore, 2007)
13. A Temko, C Nadeu, Acoustic event detection in meeting-room environments. Pattern Recogn. Lett. **30/14**, 1281–1288 (2009)
14. T Butko, F Gonzalez Pla, C Segura, C Nadeu, J Hernando, Two-source acoustic event detection and localization: online implementation in a smart-room, in *Proc. EUSIPCO* (Barcelona, 2011)
15. R Chakraborty, C Nadeu, T Butko, Detection and positioning of overlapped sounds in a room environment, in *Proc. Interspeech* (Portland, 2012)
16. R Chakraborty, C Nadeu, T Butko, Binary position assignment of two known simultaneous acoustic sources, in *Proc. IberSPEECH* (Madrid, 2012)
17. M Grabisch, Fuzzy integral in multi-criteria decision-making. Fuzzy Set Syst **69**(3), 279–298 (1995)
18. S Chang, S Greenberg, Syllable-proximity evaluation in automatic speech recognition using fuzzy measures and a fuzzy integral, in *Proc. IEEE Fuzzy Systems Conference* (St. Louis, 2003), pp. 828–833
19. A Temko, D Macho, C Nadeu, Fuzzy integral based information fusion for classification of highly confusable non-speech sounds. Pattern Recogn. **41**(5), 1831–1840 (2008)
20. BD Van Veen, KM Buckley, Beamforming: a versatile approach to spatial filtering. ASSP Mag IEEE **5**(2), 4–24 (1988)
21. SP Applebaum, Adaptive arrays. IEEE Trans Antenna Propag **24**, 585–595 (1976)
22. AS Feng, DL Jones, Localization-based grouping, in *Computational Auditory Scene Analysis: Principles, Algorithm, and Applications*, ed. by D Wang, GJ Brown (IEEE/Wiley-Interscience, 2006)
23. O Hoshuyama, A Sugiyama, Robust adaptive beamforming, in *Microphone Arrays: Signal Processing Techniques and Applications*, ed. by M Brandstein, D Ward (Springer, New York, 2001)
24. DB Ward, RA Kennedy, RC Williamson, Theory and design of broadband sensor arrays with frequency invariant far-field beam patterns. J. Acoust. Soc. Am. **97**, 1023–1034 (1995)
25. LC Parra, Least squares frequency invariant beamforming, in *Proc. Workshops on Applications of Signal Processing on Acoustics and Audio* (IEEE, New York, 2005)
26. LC Parra, Steerable frequency-invariant beamforming for arbitrary arrays. J. Acoust. Soc. Am. **119**(6), 3839–3847 (2006)
27. Y Zhao, W Liu, RJ Langley, Design of frequency invariant beamformers in subbands, in *Proc. IEEE/SP 15th Workshop on Statistical Signal Processing* (Cardiff, 2009)
28. C Nadeu, D Macho, J Hernando, Frequency and time filtering of filter-bank energies for robust HMM speech recognition. Speech Comm. **34**, 93–114 (2001)
29. L Rabiner, B Juang, *Fundamentals of Speech Recognition* (Prentice Hall, 1993)
30. S Young, G Evermann, D Kershaw, G Moore, J Odell, D Ollason, D Povey, V Valtchev, P Woodland, *The HTK Book (for HTK Version 3.2)* (Cambridge University, 2002)
31. LI Kuncheva, *Combining Pattern Classifiers: Methods and Algorithms* (Wiley-Interscience, New Jersey, 2004)

12

Speech enhancement with an acoustic vector sensor: an effective adaptive beamforming and post-filtering approach

Yue Xian Zou[1*], Peng Wang[1], Yong Qing Wang[1], Christian H Ritz[2] and Jiangtao Xi[2]

Abstract

Speech enhancement has an increasing demand in mobile communications and faces a great challenge in a real ambient noisy environment. This paper develops an effective spatial-frequency domain speech enhancement method with a single acoustic vector sensor (AVS) in conjunction with minimum variance distortionless response (MVDR) spatial filtering and Wiener post-filtering (WPF) techniques. In remote speech applications, the MVDR spatial filtering is effective in suppressing the strong spatial interferences and the Wiener post-filtering is considered as a popular and powerful estimator to further suppress the residual noise if the power spectral density (PSD) of target speech can be estimated properly. With the favorable directional response of the AVS together with the trigonometric relations of the steering vectors, the closed-form estimation of the signal PSDs is derived and the frequency response of the optimal Wiener post-filter is determined accordingly. Extensive computer simulations and a real experiment in an anechoic chamber condition have been carried out to evaluate the performance of the proposed algorithm. Simulation results show that the proposed method offers good ability to suppress the spatial interference while maintaining comparable log spectral deviation and perceptual evaluation of speech quality performance compared with the conventional methods with several objective measures. Moreover, a single AVS solution is particularly attractive for hands-free speech applications due to its compact size.

Keywords: Speech enhancement; Acoustic vector sensor; Beamforming; Post-filtering; Power spectral density estimation

1 Introduction

As the presence of background noise significantly deteriorates the quality and intelligibility of speech, enhancement of speech signals has been an important and challenging problem and various methods have been proposed in the literature to tackle this problem. Spectral subtraction, Wiener filtering, and their variations [1] are commonly used for suppressing additive noise, but they are not able to effectively suppress spatial interference. In order to eliminate spatial interferences, beamforming techniques applied to microphone array recordings can be employed [2-9]. Among these, the minimum variance distortionless response (MVDR) beamformer known as the Capon beamformer and their equivalent generalized sidelobe cancellers (GSC) work successfully in remote speech enhancement applications [2]. However, the performance of MVDR-type methods is proportional to the number of array sensors used, thus limiting their application. Moreover, the MVDR beamformer is not effective at suppressing additive noise, leaving residual noise in its output. As a result, the well-known Wiener post-filtering solution normally can be employed to further reduce the residual noise from the output of the beamformer [7]. Recently, speech enhancement using the acoustic vector sensor (AVS) array has received research attention due to the merit of spatial co-location of microphones and signal time alignment [5,10-12]. Compared with the traditional microphone array, the compact structure (occupying a volume of approximately 1 cm^3) makes the AVS much more attractive in portable speech enhancement applications. Research showed that the AVS array beamformer with the MVDR method [5,10] successfully suppresses spatial interferences but fails to effectively suppress background noise. The integrated MVDR and Wiener post-filtering method

* Correspondence: zouyx@pkusz.edu.cn
[1]ADSPLAB/ELIP, School of Electronic Computer Engineering, Peking University, Shenzhen 518055, China
Full list of author information is available at the end of the article

using AVS array [12] offers good performance in terms of suppression of spatial interferences and background additive noise, but it requires more than two AVS units as well as the good voice activity detection (VAD) technique.

In this paper, we focus on developing a speech enhancement solution capable of effectively suppressing spatial interferences and additive noise at a less computational cost using only one AVS unit. More specifically, by exploring the unique spatial co-location property (the signal arrives at the sensors at the same time) and the trigonometric relations of the steering vectors of the AVS, a single AVS-based speech enhancement system is proposed. The norm-constrained MVDR method is employed to form the spatial filter, while the optimal Wiener post-filter is designed by using a novel closed-form power spectral density (PSD) estimation method. The proposed solution does not depend on the VAD technique (for noise estimation) and hence has advantages of small size, less computation cost, and the ability to suppress both spatial interferences and background noise.

The paper is organized as follows. The data model of an AVS and the frequency domain MVDR (FMV) with a single AVS are presented in Section 2. The detailed derivation of the closed-form estimation of the signal PSDs for an optimal Wiener post-filtering (WPF) using the AVS structure is given in Section 3. The proposed norm-constrained FMV-effective Wiener post-filtering (NCFMV-EWPF) algorithm for speech enhancement is presented in Section 4. Simulation results are presented in Section 5. Section 6 concludes our work.

2 Problem formulation

2.1 Data model for an AVS unit

An AVS unit generally consists of four co-located constituent sensors, including one omnidirectional sensor (denoted as the o-sensor) and three orthogonally oriented directional sensors depicted as the u-sensor, v-sensor, and w-sensor, respectively. As an example, Figure 1 shows a data capture system with an AVS unit. In this paper, focusing on deriving the algorithm and making the derivation clear, let us assume that there is one target speech $s(t)$ at $(\theta_s, \phi_s) = (90°, \phi_s)$ and one interference signal $s_i(t)$ at $(\theta_i, \phi_i) = (90°, \phi_i)$ impinging on this AVS unit, where $\phi_s, \phi_i \in [0°, 360°)$ are the azimuth angles. Since $s(t)$ and $s_i(t)$ arrive in the horizontal plane, we only need the u-sensor, v-sensor, and o-sensor to capture signals from the AVS unit. The angle difference between $s(t)$ and $s_i(t)$ is defined as

$$\Delta\phi = \phi_s - \phi_i \tag{1}$$

The corresponding steering vectors are given by

$$\boldsymbol{v}(\phi_s) = [u_s, v_s, 1]^T = [\cos\phi_s, \sin\phi_s, 1]^T \tag{2}$$

$$\boldsymbol{v}(\phi_i) = [u_i, v_i, 1]^T = [\cos\phi_i, \sin\phi_i, 1]^T \tag{3}$$

where $[.]^T$ denotes the vector/matrix transposition.

In the cases that room reverberation does not exist, the received data of the AVS can be modeled as [13]

$$\boldsymbol{x}_{\text{avs}}(t) = \boldsymbol{v}(\phi_s)s(t) + \boldsymbol{v}(\phi_i)s_i(t) + \boldsymbol{n}_{\text{avs}}(t) \tag{4}$$

where $\mathbf{n}_{\text{avs}}(t)$ is assumed as the additive white Gaussian noise at the AVS unit. Specifically, we have the following definitions:

$$\boldsymbol{x}_{\text{avs}}(t) = [x_u(t), x_v(t), x_o(t)]^T \tag{5}$$

$$\boldsymbol{n}_{\text{avs}}(t) = [n_u(t), n_v(t), n_o(t)]^T \tag{6}$$

where $x_u(t)$, $x_v(t)$, and $x_o(t)$ are the received data of the u-, v-, and o-sensor, respectively, and $n_u(t)$, $n_v(t)$, and $n_o(t)$ are the captured noise at the u-, v-, and o-sensor, respectively. The task of speech enhancement with an AVS is to estimate $s(t)$ from $\boldsymbol{x}_{avs}(t)$.

In this study, without loss of generality, we follow the commonly used assumptions [4]: (1) $s(t)$ and $s_i(t)$ are mutually uncorrelated; (2) $n_u(t)$, $n_v(t)$, and $n_o(t)$ are mutually uncorrelated.

2.2 FMV with a single AVS

The MVDR beamformer is considered as one of the most popular beamforming methods for suppressing spatial interferences in remote speech applications. In this subsection, we present the formulation of the frequency domain MVDR beamformer (FMV) with two sensors (u-sensor and v-sensor) of the AVS unit. From (2) to (4), the data received by the u sensor and the v-sensor can be modeled as [14]

$$\boldsymbol{x}(t) = [x_u(t), x_v(t)]^T = \boldsymbol{a}(\phi_s)s(t) + \boldsymbol{a}(\phi_i)s_i(t) + \boldsymbol{n}(t) \tag{7}$$

where

$$\boldsymbol{a}(\phi_s) = [u_s, v_s]^T = [\cos\phi_s, \sin\phi_s]^T \tag{8}$$

and

$$\boldsymbol{a}(\phi_i) = [u_i, v_i]^T = [\cos\phi_i, \sin\phi_i]^T \tag{9}$$

The frequency domain formulation of the data model of (7) is given by

$$\boldsymbol{X}(f) = \boldsymbol{a}(\phi_s)S(f) + \boldsymbol{a}(\phi_i)S_i(f) + \boldsymbol{N}(f) \tag{10}$$

where $\boldsymbol{X}(f) = [X_u(f), X_v(f)]^T$ and $\boldsymbol{N}(f) = [N_u(f), N_v(f)]^T$. The beamforming is then performed by applying a complex

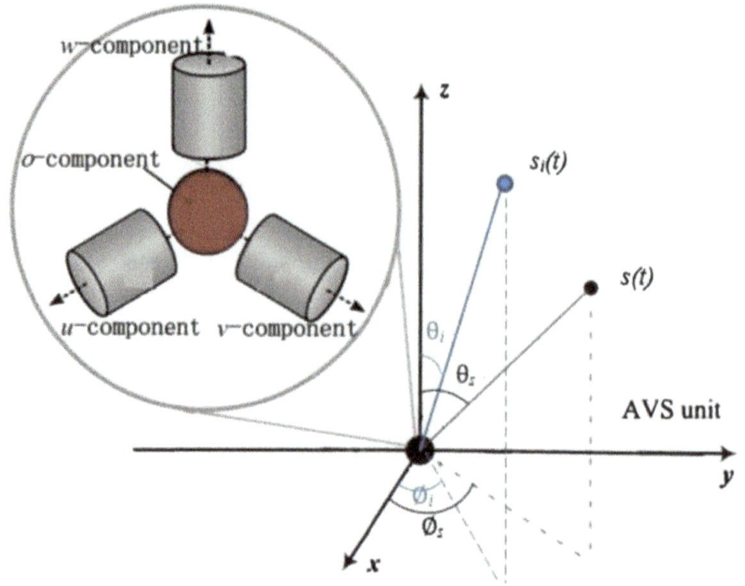

Figure 1 Illustration of one AVS data capturing system [13,14].

weight to the captured signals, and the output of the FMV can be denoted as

$$Y(f) = \boldsymbol{w}^H(f)\boldsymbol{X}(f) = \boldsymbol{w}^H(f)\boldsymbol{a}(\phi_s)S(f) + \boldsymbol{w}^H(f)(\boldsymbol{a}(\phi_i)S_i(f) + \boldsymbol{N}(f)) \tag{11}$$

where $(.)^H$ denotes the Hermitian transposition. $\boldsymbol{w}^H(f) = [w_u(f), w_v(f)]$ is the weight vector of the FMV. Let us define

$$g(\phi_s, f) = \boldsymbol{w}^H(f)\boldsymbol{a}(\phi_s) \tag{12}$$

$$g(\phi_i, f) = \boldsymbol{w}^H(f)\boldsymbol{a}(\phi_i) \tag{13}$$

Obviously, $g(\phi_s, f)$ and $g(\phi_i, f)$ can be viewed as the spatial response gains of the FMV to the target spatial signal $S(f)$ and the spatial interference signal $S_i(f)$, respectively. Substituting (12) and (13) into (11), we can get

$$Y(f) = \boldsymbol{w}^H(f)\boldsymbol{X}(f) = g(\phi_s, f)S(f) + g(\phi_i, f)S_i(f) + \boldsymbol{w}^H(f)\boldsymbol{N}(f) \tag{14}$$

The basic idea of designing the optimal FMV is to maintain $g(\phi_s, f) = 1$ for $S(f)$ while minimizing the output signal power ($P_{YY} = E[Y(f)Y^*(f)]$) of the FMV to suppress other undesired sources. Hence, the optimal weight vector of the FMV can be obtained by solving the constrained optimization cost function [2]:

$$\boldsymbol{w}_{\text{FMV}}(f) = \underset{w}{\arg\min}\ P_{YY} \quad \text{subject to } g(\phi_s, f) = 1, \text{and } P_{YY} = \boldsymbol{w}^H(f)\boldsymbol{R}_{\boldsymbol{x}}(f)\boldsymbol{w}(f) \tag{15}$$

where $\mathbf{R}_x(f) = E[\boldsymbol{X}(f)\boldsymbol{X}^H(f)]$ is the autocorrelation matrix of the received data of the FMV. The optimal solution of (15) is given as [2]

$$\boldsymbol{w}_{\text{FMV}}(f) = \frac{\boldsymbol{R}_{\boldsymbol{x}}^{-1}(f)\boldsymbol{a}(\phi_s)}{\boldsymbol{a}^T(\phi_s)\boldsymbol{R}_{\boldsymbol{x}}^{-1}(f)\boldsymbol{a}(\phi_s)} \tag{16}$$

Equation 16 is the standard form of the FMV. It is clear that when $\boldsymbol{a}(\phi_s)$ is fixed (speech target is static), $\boldsymbol{w}_{\text{FMV}}(f)$ depends on the estimate of $\boldsymbol{R}_{\boldsymbol{x}}^{-1}(f)$. There are several methods that have been proposed to estimate $\boldsymbol{R}_x(f)$ [1], and the diagonal loading technique is one of the robust algorithms aiming at avoiding the non-singularity in (16), which leads to a norm-constrained FMV (NCFMV) as shown in (17) [3]:

$$\boldsymbol{w}_{NC}(f) = \frac{(\boldsymbol{R}_{\boldsymbol{x}}(f) + \gamma\boldsymbol{I})^{-1}\boldsymbol{a}(\phi_s)}{\boldsymbol{a}^T(\phi_s)(\boldsymbol{R}_{\boldsymbol{x}}(f) + \gamma\boldsymbol{I})^{-1}\boldsymbol{a}(\phi_s) + \sigma} \tag{17}$$

where $\boldsymbol{I}$ is an identity matrix, γ is the positive loading factor, and σ is a small positive number to avoid the denominator becoming zero. It is expected that the NCFMV will greatly suppress the spatial unwanted signals. Obviously, the output of the NCFMV can be derived as follows with (17), (12), (13), and some simple manipulations:

$$Y(f) = \boldsymbol{w}_{NC}^H(f)\boldsymbol{X}(f) = S(f) + \boldsymbol{w}_{NC}^H(f)\boldsymbol{a}(\phi_i)S_i(f) + \boldsymbol{w}_{NC}^H(f)\boldsymbol{N}_{un}(f) = S(f) + g(\phi_i, f)S_i(f) + \boldsymbol{w}_{NC}^H(f)\boldsymbol{N}_{un}(f) \tag{18}$$

2.3 The estimation of the power spectral density

As discussed above, the NCFMV is only effective in suppressing the spatial interferences. In this section, a new solution has been proposed by incorporating the well-known Wiener post-filter (WPF) to further suppress the residual noise in beamformer output $Y(f)$ in (18).

Basically, according to the formulation of the Wiener filter in the frequency domain, to estimate $S(f)$ from $Y(f)$, the frequency response of the Wiener filter is given by [6,8]

$$W_{pf}(f) = \psi_{YS}(f)/\psi_{YY}(f) \approx \psi_{SS}(f)/\psi_{YY}(f) \tag{19}$$

where $\psi_{YS}(f)$ is the cross-power spectrum density (CSD) of $S(f)$ and $Y(f)$ and $\psi_{YY}(f)$ is the power spectral density (PSD) of $Y(f)$. Generally, $Y(f)$ are considered uncorrelated to interferences, and we can approximately get the second equation in (19) via (18). From (19), it is clear that a good estimate of $\psi_{SS}(f)$ and $\psi_{YY}(f)$ from $X(f)$ and $Y(f)$ are very crucial to the performance of the WPF. There are some PSD estimation algorithms that have been proposed under different spatial-frequency joint estimation schemes. For single-channel application as an example, the voice activity detection (VAD) method is usually applied to get the noise and speech segments, and then the spectrum subtraction algorithm can be used to remove noise components before estimating $\psi_{SS}(f)$. Moreover, for microphone array post-filtering schemes, $\psi_{SS}(f)$ can be estimated from the available multichannel signals, which are assumed to be within an incoherent noise environment [6].

Motivated by the unique properties of the AVS, where multichannel signals are available (u-, v, and o-sensor signals) and there exists a trigonometric relationship between the steering vectors $\boldsymbol{a}(\phi_s)$ and $\boldsymbol{a}(\psi_i)$ of the AVS, in this paper, we will derive a closed-form solution to estimate $\psi_{SS}(f)$ and $\psi_{YY}(f)$ to form an optimal WPF. The system diagram proposed is shown in Figure 2.

3 The formulation of the Wiener post-filter

3.1 Derivation of the estimate of CSD and PSD

For presentation clarity, let us define the notation of the cross-power spectral density (CSD) between $\alpha(f)$ and $\beta(f)$ as

$$\psi_{\alpha\beta}(f) = E[\alpha(f)\beta^*(f)] \tag{20}$$

From (10), we have

$$X_u(f) = \cos(\phi_s)S(f) + \cos(\phi_i)S_i(f) + N_u(f) \tag{21}$$

$$X_v(f) = \sin(\phi_s)S(f) + \sin(\phi_i)S_i(f) + N_v(f) \tag{22}$$

For presentation simplicity, the frequency index f will be dropped in the following derivation. Ideally, the additive noises of u-, v-, and o-sensors have the same power, and then we have

$$\psi_{NN} = E\left[N_u N_u^*\right] = E\left[N_v N_v^*\right] = E\left[N_o N_o^*\right] \tag{23}$$

It is noted that the assumption of equal power for all channels used in (23) is not true for the real signals recorded by the AVS unit, but these can be calibrated in practice [15]. With (18), (21), (22), and (23), the CSD and PSD of signals can be derived following the definition given in (20):

$$\begin{aligned}\psi_{uu} &= E\left[X_u X_u^*\right] = \cos^2(\phi_s)E[SS^*] + \cos^2(\phi_i)E\left[S_i S_i^*\right] \\ &+E\left[N_u N_u^*\right] = \cos^2(\phi_s)\psi_{SS} + \cos^2(\phi_i)\psi_{S_iS_i} + \psi_{NN}\end{aligned} \tag{24}$$

$$\begin{aligned}\psi_{vv} &= E\left[X_v X_v^*\right] = \sin^2(\phi_s)E[SS^*] + \sin^2(\phi_i)E\left[S_i S_i^*\right] \\ &+E\left[N_v N_v^*\right] = \sin^2(\phi_s)\psi_{SS} + \cos^2(\phi_i)\psi_{S_iS_i} + \psi_{NN}\end{aligned} \tag{25}$$

$$\psi_{oo} = E\left[X_o X_o^*\right] = \psi_{SS} + \psi_{S_iS_i} + \psi_{NN} \tag{26}$$

$$\psi_{YY} = E[YY^*] = \psi_{SS} + g^2(\phi_i)\psi_{S_iS_i} + ||\boldsymbol{w}_{NC}||^2\psi_{NN} \tag{27}$$

$$\psi_{u+v} = \psi_{uu} + \psi_{vv} = \psi_{SS} + \psi_{S_iS_i} + 2\psi_{NN} \tag{28}$$

$$\psi_{Yo} = E\left[YX_o^*\right] = \psi_{SS} + g^*(\phi_i)\psi_{S_iS_i} \tag{29}$$

$$\psi_{oY} = E[X_o Y^*] = \psi_{SS} + g(\phi_i)\psi_{S_iS_i} \tag{30}$$

From (24) to (30), it is clear that there are seven equations with four unknown variables ψ_{NN}, $g(\phi_i)$, $\psi_{S_iS_i}$, and $\psi_{S_iS_i}$. Hence, using (28) and (26), the PSD of noise can be derived as

$$\psi_{NN} = \psi_{u+v} - \psi_{oo} \tag{31}$$

Similarly, the gain response of the NCFMV on the interference S_i can be given by

$$g(\phi_i) = (\psi_{oY} - \psi_{YY} + ||\boldsymbol{w}_{NC}||^2\psi_{NN})/(\psi_{oo} - \psi_{Yo} - \psi_{NN}) \tag{32}$$

Moreover, the PSD of the interference S_i and the target speech S can be derived, respectively, as follows:

$$\psi_{S_iS_i} = (\psi_{oo} - \psi_{oY} - \psi_{NN})/(1 - g(\phi_i)) \tag{33}$$

$$\psi_{SS} = \psi_{oY} - g(\phi_i)\psi_{S_iS_i} \tag{34}$$

3.2 The proposed EWPF method and some discussions

Till now, we have mathematically derived the closed-form expressions of the ψ_{SS} in (34), ψ_{YY} in (27), and W_{pf} in (19). Since Y, X_u, X_v, and X_o can be measured, the estimates of ψ_{SS} and ψ_{YY} can be determined accordingly. Hence, (33), (34), (27), and (19) describe the basic form of

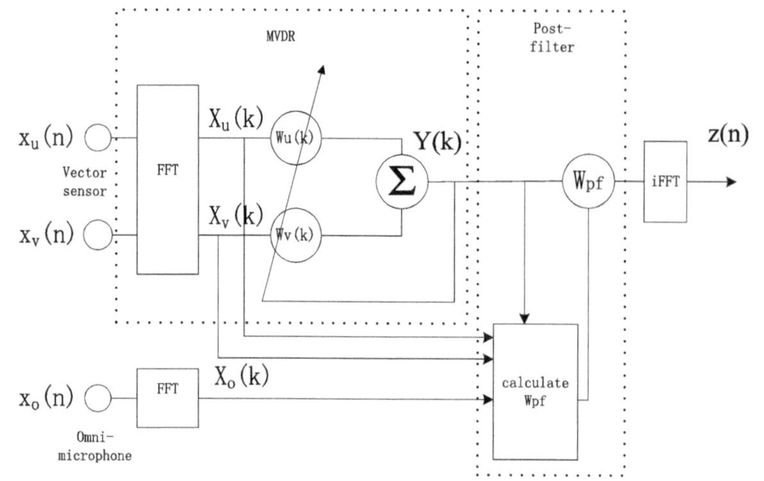

Figure 2 Our proposed speech enhancement system using single AVS.

our proposed effective Wiener post-filtering algorithm for further enhancing the speech with an AVS (here, we term it as EWPF for short). In the following context, we will have some discussions on our proposed EWPF method.

Firstly, to implement the EWPF, the cross-power spectral density $\psi_{\alpha\beta}(f)$ needs to be estimated. It is well known that the recursive update formula is a popular approach:

$$\hat{\psi}_{\alpha\beta}(f,l) = \lambda\hat{\psi}_{\alpha\beta}(f,l-1) + (1-\lambda)\alpha(f,l)\beta^*(f,l) \quad (35)$$

where l is the frame index and $\lambda \in (0, 1]$ is the forgetting factor.

Secondly, it is noted that when $\Delta\phi$ defined in (1) is close to or equal to 0, the denominator in (32) goes to 0. To avoid this situation, one small positive factor σ_r should be added to the denominator of (32) and we get

$$\hat{g}(\phi_i) = (\hat{\psi}_{oY}-\hat{\psi}_{YY} + \|\mathbf{w}_{NC}\|^2\hat{\psi}_{NN})/(\hat{\psi}_{oo}-\hat{\psi}_{Yo}-\hat{\psi}_{NN} + \sigma_r) \quad (36)$$

Thirdly, analyzing the properties of $g(\phi_i)$, we observe the following: (1) If the target source $s(t)$ is considered as short-time spatially stationary (approximately true for speech applications), $\mathbf{w}_{NC}$ in (17) can be updated every L_u frames for reducing computational complexity. Therefore, from the definition of (13), the gain $g(\phi_i)$ will remain unchanged within L_u frames. However, $\hat{\psi}_{\alpha\beta}(f,l)$ is estimated frame by frame via (35); therefore, a more accurate estimation of $g(\phi_i)$ can be achieved by averaging over L_u frames. (2) From (36), it is clear that the small denominator will lead to a large variation of $g(\phi_i)$, reflecting incorrect estimates since the NCFMV is designed to suppress rather than to amplify the interference. Hence, it is reasonable to apply a clipping function $f_c(x, b)$ (see (43)) to remove the outliers in the estimate of $\hat{g}(\phi_i)$.

4 The proposed NCFMV-EWPF algorithm

Similar to the existing remote speech enhancement applications, our proposed algorithm is implemented in the frequency domain by segmenting the received signal into frames and then the short-time Fourier transforms (STFTs) are applied. Specifically, to determine $\mathbf{w}_{NC}$ in (17), the estimate of the $\mathbf{R}_x$ is given by [10]

$$\hat{\mathbf{R}}_x(k) = \frac{1}{F}\begin{bmatrix} C_dA(k) & B(k) \\ C(k) & C_dD(k) \end{bmatrix} \quad (37)$$

where k is the frequency bin index and $k = 1,2,\ldots, K$. C_d is a constant slightly greater than the one that helps avoid matrix singularity. F is the frame number used for estimating $\mathbf{R}_x(k)$, and in our study it is set as $F = 2L_u$. Let us define $X_u(k,l)$ and $X_v(k,l)$ as the kth component of the spectrum of the lth frame of $x_u(n)$ and $x_v(n)$, respectively, and we have

$$A(k) = \sum_{i=0}^{F-1} X_u^*(k,l-i)X_u(k,l-i) \quad (38)$$

$$B(k) = \sum_{i=0}^{F-1} X_u^*(k,l-i)X_v(k,l-i) \quad (39)$$

$$C(k) = \sum_{i=0}^{F-1} X_v^*(k,l-i)X_u(k,l-i) \quad (40)$$

$$D(k) = \sum_{i=0}^{F-1} X_v^*(k,l-i)X_v(k,l-i) \quad (41)$$

From (37) to (41), we can see the autocorrelation matrix $\mathbf{R}_x(k)$ is estimated by using the F most recent fast

Fourier transforms (FFTs). Therefore, the robust estimation of $W_{pf}(k)$ in (19) asks for the robust estimation of $g(\phi_i,k)$. According to the discussions in Section 3.2, we adopt the following estimation:

$$\hat{g}(\phi_i,k) = \frac{1}{L_u}\sum_{l=L_1}^{L_2} f_c\left(\frac{\hat{\psi}_{oY}(k,l)-\hat{\psi}_{YY}(k,l)+\|\mathbf{w}_{NC}(k,l)\|^2\hat{\psi}_{NN}(k,l)}{\hat{\psi}_{oo}(k,l)-\hat{\psi}_{Yo}(k,l)-\hat{\psi}_{NN}(k,l)+\sigma_r}, b\right) \quad (42)$$

where $L_1 = \text{fix}((l-1)/L_u)L_u + 1$, $L_2 = \text{fix}((l-1)/L_u)L_u + L_u$, fix(.) is the floor operation, b is a predefined threshold, and $f_c(x,b)$ is the clipping function and defined as

$$f_c(x,b) = x \text{ when } 0 < x \le b \text{ else } f_c(x,b) = 0 \quad (43)$$

For presentation completeness, the proposed NCFMV-EWPF algorithm is summarized in Algorithm 1.

5 Simulation study

The performance evaluation of our proposed NCFMV-EWPF algorithm has been carried out in this section. The commonly used performance measurement metrics have been adopted, which include the following:

1. Output signal to interference plus noise ratio (SINR) defined as [7]

$$\text{SINR} = 10\log\left(\|z_s(t)\|^2/\|x_o(t)-z_s(t)\|^2\right) \quad (44)$$

where $z_s(t)$ is the enhanced speech of the system and $x_o(t)$ is the received signal of the o-sensor. Moreover, a segmental output SINR is calculated on a frame-by-frame basis and then averaged over the total number frames to get more accurate prediction of perceptual speech quality [7].

2. Log spectral deviation (LSD), which is used to measure the speech distortion and defined as [16]

$$\text{LSD} = \left\|\ln\left(\psi_{ss}(f)/\psi_{zz}(f)\right)\right\| \quad (45)$$

Algorithm 1 The pseudo-code of the proposed NCFMV-EWPF algorithm

Step 1: Signal segmentation: Hamming window with 50% overlapped

Step 2: Start $l = 1{:}F$:

2.1 Compute STFTs: $X_o(k,l)$, $X_v(k,l)$, $X_v(k,l)$

2.2 If mod(l, $F/2$):

Calculate the estimate of $\hat{\boldsymbol{R}}_x(k)$ by (37)

Compute $\boldsymbol{w}_{NC}(k)$ by (17)

End if

2.3 Compute the beamformer output: $Y(k,l) = \boldsymbol{w}_{NC}^H(k)\boldsymbol{X}(k,l)$

2.4 Using (35), compute $\hat{\psi}_{YY}(k,l) = \lambda\hat{\psi}_{YY}(k,l-1) + (1-\lambda)Y(k,l)Y^*(k,l)$

2.5 Similarly, compute $\hat{\psi}_{oY}(k,l)$, $\hat{\psi}_{Yo}(k,l)$, $\hat{\psi}_{oo}(k,l)$, $\hat{\psi}_{uu}(k,l)$, $\hat{\psi}_{vv}(k,l)$

2.6 Compute $\hat{\psi}_{u+v}(k,l)$ by (28) and $\hat{\psi}_{NN}(k,l)$ by (31)

2.7 If mod(l, L_u):

Calculate $\hat{g}(\phi_i,k)$ by (42)

End if

2.8 For $l = l - L_u + 1{:}l$:

Calculate $\hat{\psi}_{SS}(k,l)$ by (34), Compute $\hat{W}_{pf}(k,l)$ by (19)

End for l

2.9 Compute the WPF filter output in frequency domain

$Z(k,l) = \hat{W}_{pf}(k,l)Y(k,l)$

End l %start

Step 3: Compute the output in time domain:

Using the inverse FFT and performing an overlap, add the frames to generate the time domain output.

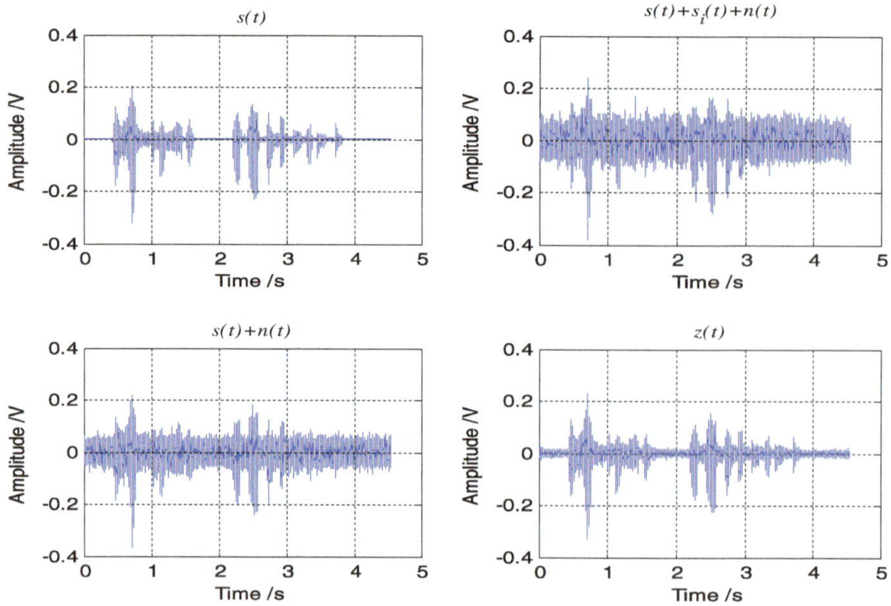

Figure 3 Waveforms. $s(t)$ is the target speech, $s_i(t)$ is the interference speech, $n(t)$ is the additive noise, and $z(t)$ is the enhanced speech signal.

where $\psi_{ss}(f)$ is the PSD of the target speech and $\psi_{zz}(f)$ is the PSD of the enhanced speech. It is clear that the smaller LSD indicates the less speech distortion. Similar to the calculation of SINR, the segmental LSD is computed.

3. Perceptual evaluation of speech quality (PESQ) [17]: To evaluate the performance of the speech enhancement algorithms, ITU-PESQ software [17] is utilized.

In addition, we also compared the performance of the Zelinski post-filter (ZPF) [4], NCFMV [5], and NCFMV-ZPF [6] algorithms under the same conditions to our proposed algorithm. The setup of the single AVS unit is shown in Figure 1.

In computer simulation studies, for each trial, a male speech lasting about 5 s acts as the target speech $s(t)$ and babble speech taken from the Noisex-92 database [18] acts as the interference speech $s_i(t)$. One set of the typical waveforms used in our simulation studies is shown in Figure 3.

5.1 Experiments on simulated data

5.1.1 Experiment 1: the SINR performance under different noise conditions

In this experiment, we have carried out nine trials (numbered as trial 1 to trial 9) to evaluate the performance of the algorithms under different spatial and additive noise conditions [9]. The experimental settings are as follows: The sampling rate is set to be 16 kHz and a 512-point FFT is used. The target speaker is located at (90°, 45°) and the interference speaker is set at (90°, 0°). For the proposed NCFMV-EWPF algorithm, parameters are set as $\lambda = 0.6$, $\sigma_r = 10^{-3}$, $L_u = 4$, $\gamma = \sigma = 10^{-5}$,

Table 1 SINR-out for different algorithms (dB)

Algorithm	ZPF [4]	NCFMV [5]	NCFMV-ZPF [6]	NCFMV-EWPF	SINR-input (dB)
Trial 1 ($\mathbf{n}_{avs}(t) = 0$ and $s_i(t) \neq 0$)	2.7	12.6	14.8	*26.2*	0
Trial 2 ($\mathbf{n}_{avs}(t) = 0$ and $s_i(t) \neq 0$)	7.8	12.8	16.4	*34.0*	5
Trial 3 ($\mathbf{n}_{avs}(t) = 0$ and $s_i(t) \neq 0$)	13.1	13.4	18.3	*28.3*	10
Trial 4 ($\mathbf{n}_{avs}(t) \neq 0$ and $s_i(t) = 0$)	8.1	2.0	7.8	*8.3*	0
Trial 5 ($\mathbf{n}_{avs}(t) \neq 0$ and $s_i(t) = 0$)	*13.5*	6.5	*13.5*	13.2	5
Trial 6 ($\mathbf{n}_{avs}(t) \neq 0$ and $s_i(t) = 0$)	*17.6*	9.1	*17.0*	16.5	10
Trial 7 ($\mathbf{n}_{avs}(t) \neq 0$ and $s_i(t) \neq 0$)	3.1	8.1	11.9	*14.2*	0
Trial 8 ($\mathbf{n}_{avs}(t) \neq 0$ and $s_i(t) \neq 0$)	8.3	10.3	14.7	*18.0*	5
Trial 9 ($\mathbf{n}_{avs}(t) \neq 0$ and $s_i(t) \neq 0$)	13.6	12.4	18.9	*21.2*	10

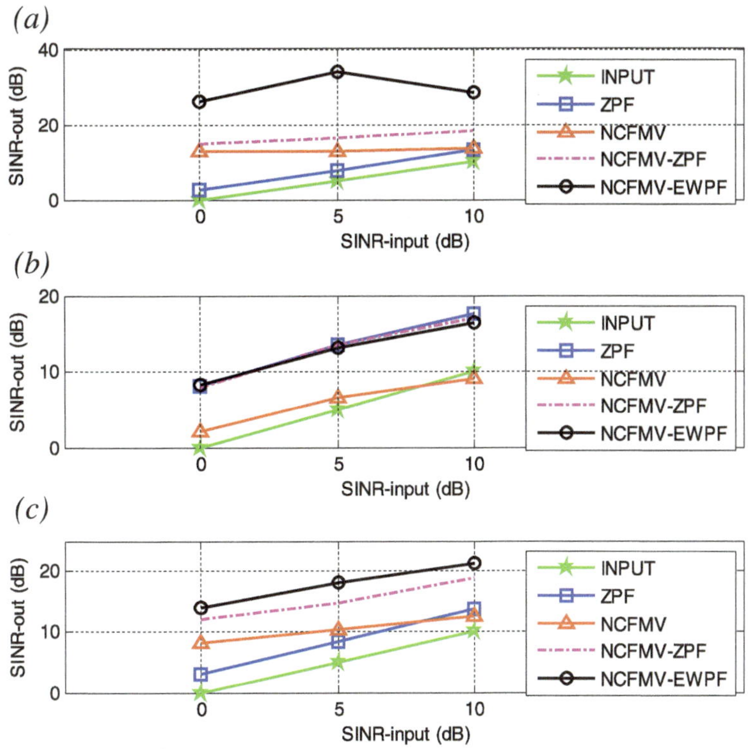

Figure 4 The performance of SINR-out versus SINR-input. (a) $n_{avs}(t) = 0$ and $s_i(t) \neq 0$. **(b)** $n_{avs}(t) \neq 0$ and $s_i(t) = 0$. **(c)** $n_{avs}(t) \neq 0$ and $s_i(t) \neq 0$.

$C_d = 1.1$, and $b = 6$, which produced the best experimental results under this specific setup. For comparison algorithms, the parameter settings are set as the same as those in the relevant papers. The experimental results are listed in Table 1.

As shown in Table 1, the best performance for different conditions is addressed in italics. The proposed NCFMV-EWPF algorithm outperforms other algorithms in terms of SINR-out in trials 1 to 4 and trials 7 to 9, and gives comparable performance in trial 5 and inferior performance in trial 6. It is noted that, in trials 4 to 6, there is no spatial interference considered (i.e., $s_i(t) = 0$). The performance for trial 5 indicates that the proposed NCFMV-EWPF is not as effective as the ZPF in suppressing the additive noise with higher SNR (SNR > 10 dB) when spatial interference is not present. Therefore, these experimental results demonstrate the superior capability of the proposed NCFMV-EWPF in suppressing the spatial and adverse additive interferences. For visualization purposes, the results in Table 1 have also been plotted in Figure 4, where the x-axis represents the SINR of the signal captured by the AVS (termed as SINR-input) and the y-axis represents the SINR of the enhanced speech (termed as SINR-out).

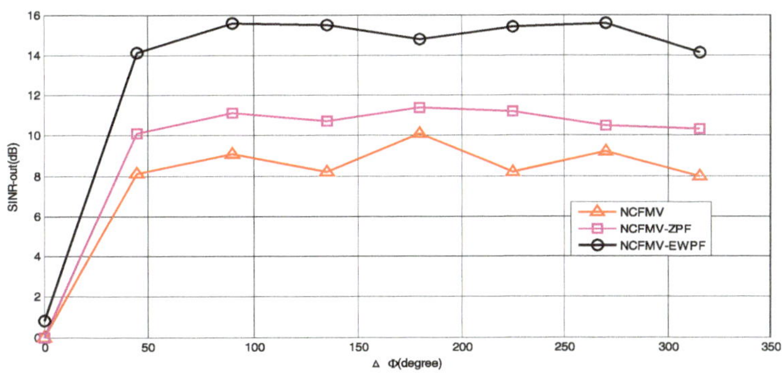

Figure 5 The SINR-out versus Δφ (NCFMV [5]).

Table 2 Performance comparison

	LSD	PESQ	SINR (dB)
INPUT	2.64	2.04	0.02
ZPF [4]	1.90	2.24	3.09
NCFMV [5]	2.55	2.29	8.12
NCFMV-ZPF [6]	*1.84*	*2.52*	11.94
NCFMV-EWPF (our proposed)	*1.84*	2.50	*13.85*

Results from the best performing methods are italicized.

5.1.2 Experiment 2: the impact of the angle between the target and interference speakers

This experiment evaluates the impact of the angle between the target and interference speakers ($\Delta\phi = \phi_s - \phi_i$) on the performance of the NCFMV-EWPF algorithm. The results of the SINR-out versus $\Delta\phi$ are shown in Figure 5, where the same experimental settings as those used for trial 7 in experiment 1 were adopted except the target speech location ϕ_s varied from (90°,0°) to (90°,360°) with 45° increments. From Figure 5, it is clear to see that when $\Delta\phi \rightarrow 0°$ (the target speaker moves closer to the interference speaker), for both algorithms, the SINR-out drops significantly and almost goes to 0. This means the speech enhancement is very much limited under this condition. However, when $\Delta\phi > 0°$, the SINR-out gradually increases. It is quite encouraging to see that the SINR-out of our proposed NCFMV-EWPF algorithm is superior to that of the NCFMV algorithm for all angles. Moreover, the SINR-out of our proposed NCFMV-EWPF algorithm maintains about 15 dB when $\Delta\phi \geq 45°$.

5.1.3 Experiment 3: SINR, LSD, and PESQ performance

In this experiment, we adopted three performance metrics (SINR, LSD, and PESQ) to evaluate the performance of the algorithms. The same experimental settings of those used in experiment 1 were employed, where the SINR-input is set as 0 dB, the target speaker is located at (90°, 45°), and the interference speaker is at (90°,0°) ($\Delta\phi = 45°$). The experimental results are given in Table 2. It can be seen that the overall performance of our proposed NCFMV-EWPF algorithm is superior to that of other comparison algorithms. The LSD and PESQ performance of the NCFMV-EWPF algorithm is comparable to that of the NCFMV-ZPF [6] algorithm. It is encouraging to see that the proposed NCFMV-EWPF algorithm is able to effectively suppress the interference and additive noise while maintaining good speech quality and less distortion.

5.2 Experiments on recorded data in an anechoic chamber

5.2.1 Experiment 4: the SINR-out performance with different speakers

In this experiment, we conducted the speech enhancement by using the recorded data from Ritz's lab [19]. The experimental setup is shown in Figure 6. The AVS has been built by two Knowles NR-3158 pressure gradient sensors (u-sensor and v-sensor) and one Knowles EK-3132 sensor (o-sensor) (Knowles Electronics Inc., Itasca, IL, USA). Recordings were made of 10 different speech sentences from the IEEE speech corpus [20] in an anechoic chamber and background noise only from computer servers and air conditioning. The anechoic chamber is similar to the noise field: $\boldsymbol{n}_{\text{avs}}(t) = 0$ and $s_i(t) \neq 0$. The sampling rate was 48 kHz and then down-sampled to 16 kHz for speech enhancement. The speakers were placed in front of the AVS at a distance of 1 m. Target speech was located at a fixed position (90°, 45°), while interference speech was located at (90°, 90°). Ten trials were carried out using the 10 different target speeches.

The experimental results are shown in Figure 7. The x-axis represents the number of trials, and the y-axis

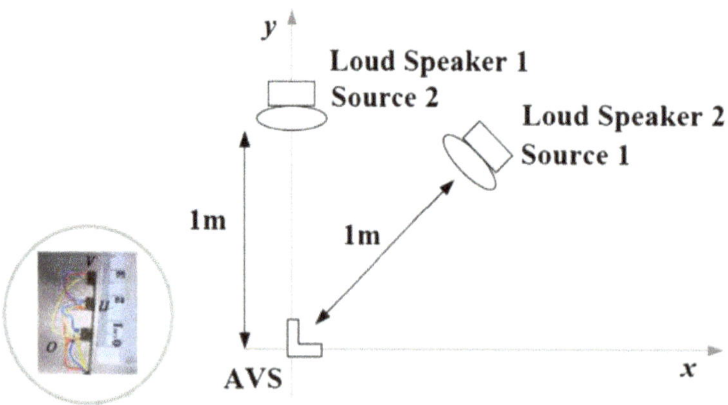

Figure 6 The illustration of the experimental setup in an anechoic chamber in Ritz's lab [19]. The leftmost image is the AVS used, $s(t)$ is the target speech (source 1) at 45°, and $s_i(t)$ is the interference speech (source 2) at 90°.

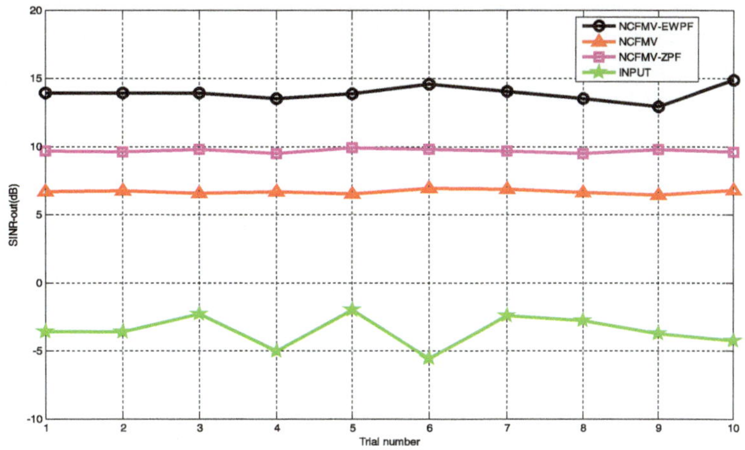

Figure 7 The SINR-out performance versus trial number (NCFMV [5]).

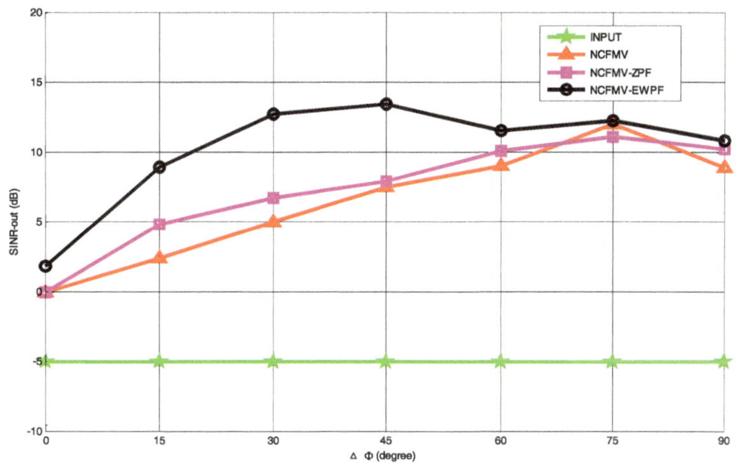

Figure 8 The SINR-out versus $\Delta\varphi$ (NCFMV [5]).

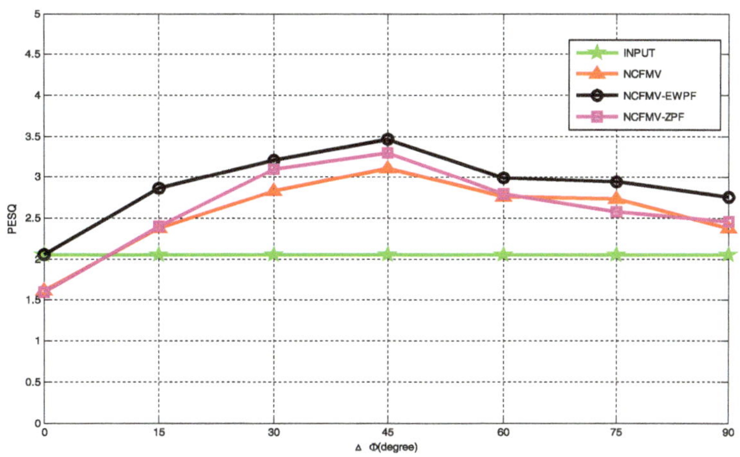

Figure 9 The PESQ performance versus $\Delta\varphi$ (NCFMV [5]).

represents the SINR of the enhanced speech (in dB). It is clear to see that the proposed NCFMV-EWPF algorithm provides superior SINR-out performance for all trails when the SINR-input of the recorded data is at about −5 dB. The experimental results with the real recorded data further validate the effectiveness of the proposed NCFMV-EWPF in suppressing the strong competing speech.

5.2.2 Experiment 5: the impact of the angle between the target and interference speakers

Similar to experiment 2, this experiment evaluates the impact of the angle between the target and interference speakers ($\Delta\phi = |\phi_s - \phi_i|$) on the performance of the NCFMV-EWPF algorithm. The results of the SINR-out versus $\Delta\phi$ are shown in Figure 8, where the experimental setup is the same as that of experiment 4 except that the angle of the target speaker (ϕ_s) varies from (90°,90°) to (90°,0°) with 15° decrement.

From Figure 8, it is clear to see that the performance of the proposed NCFMV-EWPF algorithm is superior to that of the NCFMV algorithm for all $\Delta\phi$ values. Compared to the results shown in Figure 5 using the simulated data, similar conclusions can be drawn for the proposed NCFMV-EWPF algorithm. More specifically, with the recorded data, when $\Delta\phi > 15°$, the proposed NCFMV-EWPF algorithm can effectively enhance the target speech.

5.2.3 Experiment 6: PESQ performance versus $\Delta\varphi$

In this experiment, we only adopted one performance metrics (PESQ) to evaluate the performance of the algorithms. The same experimental settings as those used in experiment 5 were employed, where the angle of the interference speaker (ϕ_i) was fixed at (90°,90°) and the angle of the target speaker (ϕ_s) varied from (90°,90°) to (90°,0°) with 15° decrement. The experimental results are given in Figure 9. It can be seen that the overall performance of PESQ for our proposed NCFMV-EWPF algorithm is superior to that of the comparison algorithm for all angle differences. This experiment also demonstrates the ability of the proposed NCFMV-EWPF algorithm in effectively suppressing the interference and additive noise while maintaining good speech quality and less distortion when $\Delta\phi > 15°$.

6 Conclusions

In this paper, a novel speech enhancement algorithm named as NCFMV-EWPF has been derived with a single AVS unit by an efficient closed-form estimation of the power spectral densities of signals. The results of computer simulation show that the proposed NCFMV-EWPF algorithm outperforms the existing ZPF, NCFMV, and NCFMV-ZPF algorithms, in terms of suppressing the competing speaker and noise field. The results of real experiments show that compared with the NCFMV algorithms, the proposed NCFMV-EWPF algorithm can effectively suppress the competing speech and additive noise while maintaining good speech quality and less distortion. In addition, it is noted that the NCFMV-EWPF algorithm does not require the VAD technique, which not only reduces the computational complexity but also provides more robust performance in a noisy environment, such as the higher output SINR, less speech distortion, and better speech intelligibility. It is expected that this novel approach developed in this paper is a suitable solution for implementation within hands-free speech recording systems.

Competing interests

The authors declare that they have no competing interests.

Acknowledgements

This work is partially supported by the National Natural Science Foundation of China (No. 61271309) and the Shenzhen Science & Technology Fundamental Research Program (No. JCY201110006). It was also partially supported by the Australian Research Council Grant DP1094053.

Author details

[1]ADSPLAB/ELIP, School of Electronic Computer Engineering, Peking University, Shenzhen 518055, China. [2]School of Electrical, Computer, and Telecommunications Engineering, University of Wollongong, Wollongong 2522, Australia.

References

1. S Boll, Suppression of acoustic noise in speech using spectral subtraction. IEEE Trans. Acoust. Speech Signal Process **27**(2), 113–120 (1979)
2. LJ Griffiths, CW Jim, An alternative approach to linearly constrained adaptive beamforming. IEEE Trans. Antennas Propag. **30**(1), 27–34 (1982)
3. YX Zou, SC Chan, B Wan, J Zhao, *Recursive robust variable loading MVDR beamforming in impulsive noise environment*, vol. 1–4 (Paper presented at the IEEE ASIA Pacific conference on circuits and system, Macao, 2008), pp. 988–991
4. R Zelinski, *A microphone array with adaptive post-filtering for noise reduction in reverberant rooms* (Paper presented at the IEEE international conference on acoustics, speech, and signal processing (ICASSP), New York, 1988)
5. ME Lockwood, DL Jones, Beamformer performance with acoustic vector sensors in air. J. Acoust. Soc. Am. **119**, 608–619 (2006)
6. IA McCowan, H Bourlard, Microphone array post-filter based on noise field coherence. IEEE Trans. Speech Audio Process **11**(6), 709–716 (2003)
7. J Benesty, MM Sondhi, Y Huang, *Springer Handbook of Speech Processing* (Springer, Berlin-Heidelberg, 2008)
8. SV Vaseghi, *Advanced Digital Signal Processing and Noise Reduction*, 2nd edn. (John Wiley & Sons ltd, Chichester, 2000)
9. J Bitzer, KU Simmer, KD Kammeyer, *Multichannel noise reduction algorithms and theoretical limits* (Paper presented at EURASIP European signal processing conference (EUSIPCO), Rhodes, 1998)
10. ME Lockwood, DL Jones, RC Bilger, CR Lansing, WDO Brien, BC Wheeler, AS Feng, Performance of time-and frequency-domain binaural beamformers based on recorded signals from real rooms. J. Acoust. Soc. Am. **115**, 379 (2004)
11. M Shujau, CH Ritz, IS Burnett, *Speech enhancement via separation of sources from co-located microphone recordings* (Paper presented at IEEE international conference on acoustics, speech and signal processing (ICASSAP), Dallas, 2010)
12. PKT Wu, C Jin, A Kan, *A multi-microphone speech enhancement algorithm tested using acoustic vector sensor* (Paper presented at the 12th international workshop on acoustic echo and noise control, Tel-Aviv-Jaffa, 2010)
13. B Li, YX Zou, *Improved DOA estimation with acoustic vector sensor arrays using spatial sparsity and subarray manifold* (Paper presented at IEEE

international conference on acoustics, speech and signal processing (ICASSP), Kyoto, 2012)
14. W Shi, YX Zou, B Li, CH Ritz, M Shujau, J Xi, *Multisource DOA estimation based on time-frequency sparsity and joint inter-sensor data ratio with single acoustic vector sensor* (Paper presented at IEEE international conference on acoustics, speech and signal processing (ICASSP), Vancouver, 2013)
15. M Shujau, *In air acoustic vector sensors for capturing and processing of speech signals, Dissertation* (University of Wollongong, , 2011)
16. R Gray, A Buzo, JA Gray, Y Matsuyama, Distortion measures for speech processing. IEEE Trans. Acoust. Speech Signal Process **28**(4), 367–376 (1980)
17. ITU-T, *Recommendation P.862 - Perceptual Evaluation of Speech Quality (PESQ): An Objective Method for End-to-End Speech Quality Assessment of Narrow-Band Telephone Networks and Speech Codecs* (International Telecommunication Union - Telecommunication Standardization Sector, Geneva, 2001)
18. NOISEX-92: http://www.speech.cs.cmu.edu/comp.speech/Section1/Data/noisex.html
19. CH Ritz, IS Burnett, *Separation of speech sources using an acoustic vector sensor* (Paper presented at IEEE international workshop on multimedia signal processing, Hanzhou, 2011)
20. IEEE Subcommittee, IEEE recommended practice for speech quality measurements. IEEE Trans. Audio Electro-acoustics **AU-17**(3), 225–246 (1969)

A modified Wiener filtering method combined with wavelet thresholding multitaper spectrum for speech enhancement

Yanna Ma* and Akinori Nishihara

Abstract

This paper proposes a new speech enhancement (SE) algorithm utilizing constraints to the Wiener gain function which is capable of working at 10 dB and lower signal-to-noise ratios (SNRs). The wavelet thresholded multitaper spectrum was taken as the clean spectrum for the constraints. The proposed algorithm was evaluated under eight types of noises and seven SNR levels in NOIZEUS database and was predicted by the composite measures and the SNR_{LOSS} measure to improve subjective quality and speech intelligibility in various noisy environments. Comparisons with two other algorithms (KLT and wavelet thresholding (WT)) demonstrate that in terms of signal distortion, overall quality, and the SNR_{LOSS} measure, our proposed constrained SE algorithm outperforms the KLT and WT schemes for most conditions considered.

Keywords: Speech enhancement; Multitaper spectrum; Wavelet thresholding; Constrained Wiener filtering

1 Introduction

The objective of speech enhancement (SE, also called noise reduction) algorithms is to improve one or more perceptual aspects of the noisy speech by decreasing the background noise without affecting the intelligibility of the speech [1]. Research on SE can be traced back to 40 years ago with two patents by Schroeder [2], where an analog implementation of the spectral magnitude subtraction method was described. Since then, the problem of enhancing speech degraded by uncorrelated additive noise, when only the noisy speech is available, has become an area of active research [3]. Researchers and engineers have approached this challenging problem by exploiting different properties of speech and noise signals to achieve better performance [4].

SE techniques have a broad range of applications, from hearing aids to mobile communication, voice-controlled systems, multiparty teleconferencing, and automatic speech recognition (ASR) systems [4]. The algorithms can be summarized into four classes: spectral subtractive [5-8], sub-space [9,10], statistical model-based [11-13], and Wiener-type [3,14-16] algorithms.

Much progress has been made in the development of SE algorithms capable of improving speech quality [17,18] which was evaluated mainly by the objective performance criteria such as signal-to-noise ratio (SNR) [19]. However, SE algorithm that improves speech quality may not perform well in real-world listening situations where background noise level and characteristics are constantly changing [20]. The first intelligibility study done by Lim [21] in the late 1970s found no intelligibility improvement with the spectral subtraction algorithm for speech corrupted in white noise at −5 to 5 dB SNR. Thirty years later, a study conducted by Hu and Loizou [1] found that none of the examined eight different algorithms improved speech intelligibility relative to unprocessed (corrupted) speech. Moreover, according to [1], the algorithms with the highest overall speech quality may not perform the best in terms of speech intelligibility (e.g., logMMSE [12]). And the algorithm which performs the worst in terms of overall quality may perform well in terms of preserving speech intelligibility (e.g., KLT [9]). To our knowledge, very few speech enhancement algorithms [22-25] claimed to improve speech intelligibility by subjective tests for either normal-hearing listeners

*Correspondence: mayanna@nh.cradle.titech.ac.jp
Department of Communications and Integrated Systems, Tokyo Institute of Technology, Meguro-ku, Tokyo 152-8552, Japan

or hearing-impaired listeners. Hence, we focused in this paper on improving performance on speech intelligibility of the SE algorithm.

From [19], we know that the perceptual effects of attenuation and amplification distortion on speech intelligibility are not equal. Amplification distortion in excess of 6.02 dB (region III) bears the most detrimental effect on speech intelligibility, while the attenuation distortion (region I) was found to yield the least effect on intelligibility. Region I+II constraints are the most robust in terms of yielding consistently large benefits in intelligibility independent of the SE algorithm used. However, in order to divide those three regions [19], the estimated magnitude spectrum needs to be compared with the clean spectrum which we usually do not have in real circumstances.

In this paper, we explored the multitaper spectrum which was shown in [26] to have good bias and variance properties. The spectral estimate was further refined by wavelet thresholding the log multitaper spectrum in [16]. The refined spectrum was proposed in this paper to be used as an alternative of the clean spectrum. Then, the region I+II constraints were imposed and incorporated in the derivation of the gain function of the Wiener algorithm based on *a priori* SNR [3]. We have experimentally evaluated its performance under a variety of noise types and SNR conditions.

The structure of the rest of this paper is organized as follows. Section 2 provides the background information on wavelet thresholding the multitaper spectrum, and Section 3 presents the proposed approach which imposes constraints on the Wiener filtering gain function. Section 4 contains the speech and noise database and metrics used in the evaluation. The simulation results are given in Section 5. Finally, a conclusion of this work and the discussion are given in Section 6.

2 Wavelet thresholding the multitaper spectrum

In real-world scenarios, the background noise level and characteristics are constantly changing [20]. Better estimation of the spectrum is required to alleviate the distortion caused by SE algorithms. For speech enhancement, the most frequently used power spectrum estimator is direct spectrum estimation based on Hann windowing. However, windowing reduces only the bias not the variance of the spectral estimate [27]. The multitaper spectrum estimator [26], on the other hand, can reduce this variance by computing a small number (L) of direct spectrum estimators (eigenspectra) each with a different taper (window) and then averaging the L spectral estimates. The underlying philosophy is similar to Welch's method of modified periodogram [27].

The multitaper spectrum estimator is given by

$$\widehat{S}^{mt}(\omega) = \frac{1}{L}\sum_{k=0}^{L-1}\widehat{S}_k^{mt}(\omega) \tag{1}$$

with

$$\widehat{S}_k^{mt}(\omega) = \left|\sum_{m=0}^{N-1} a_k(m)x(m)e^{-j\omega m}\right|^2, \tag{2}$$

where N is the data length, and a_k is the kth sine taper used for the spectral estimate $\widehat{S}_k^{mt}(\cdot)$, which is proposed by Riedel and Sidorenko [28] and defined by

$$a_k(m) = \sqrt{\frac{2}{N+1}}sin\frac{\pi k(m+1)}{N+1}, m = 0,\cdots,N-1. \tag{3}$$

The sine tapers were proved in [28] to produce smaller local bias than the Slepian tapers, with roughly the same spectral concentration.

The multitaper estimated spectrum can be further refined by wavelet thresholding techniques [29-31]. Improved periodogram estimates were proposed in [29], and improved multitaper spectrum estimates were proposed in [30,31]. The underlying idea behind those techniques is to represent the log periodogram as 'signal' plus the 'noise', where the signal is the true spectrum and the noise is the estimation error [32]. It was shown in [33] that if the eigenspectra defined in Equation 2 are assumed to be uncorrelated, the ratio of the estimated multitaper spectrum $\widehat{S}^{mt}(\omega)$ and the true power spectrum $S(\omega)$ conforms to a chi-square distribution with $2L$ degrees of freedom, i.e.,

$$\upsilon(\omega) = \frac{\hat{S}^{mt}(\omega)}{S(\omega)} \sim \frac{\chi^2_{2L}}{2L},\quad 0 < \omega < \pi. \tag{4}$$

Taking the log of both sides, we get

$$\log\hat{S}^{mt}(\omega) = \log S(\omega) + \log\upsilon(\omega). \tag{5}$$

From Equation 5, we know that the log of the multitaper spectrum can be represented as the sum of the true log spectrum plus a $\log\chi^2$ distributed noise term. It follows from Bartlett and Kendall [34] that the distribution of $\log\upsilon(\omega)$ is with mean $\phi(L) - \log(L)$ and variance $\phi'(L)$, where $\phi(\cdot)$ and $\phi'(\cdot)$ denote, respectively, the digamma and trigamma functions. For $L \geq 5$, the distribution of $\log\upsilon(\omega)$ will be close to a normal distribution [35]. Hence, provided L is at least 5, the random variable $\eta(\omega)$

$$\eta(\omega) = \log\upsilon(\omega) - \phi(L) + \log(L) \tag{6}$$

will be approximately Gaussian with zero mean and variance $\sigma_\eta^2 = \phi'(L)$. If $Z(\omega)$ is defined as

$$Z(\omega) = \log\hat{S}^{mt}(\omega) - \phi(L) + \log(L), \tag{7}$$

then we have

$$Z(\omega) = \log S(\omega) + \eta(\omega), \tag{8}$$

i.e., the log multitaper power spectrum plus a known constant ($\log(L) - \phi(L)$) can be written as the true log power spectrum plus approximately Gaussian noise $\eta(\omega)$ with zero mean and known variance σ_η^2 [30].

The model in Equation 8 is well suited for wavelet denoising techniques [36-39] for eliminating the noise $\eta(\omega)$ and obtaining a better estimate of the log spectrum. The idea behind refining the multitaper spectrum by wavelet thresholding can be summarized into four steps [16].

- Obtain the multitaper spectrum using Equations 1 to 3 and calculate $Z(w)$ using Equation 7.
- Apply a standard periodic discrete wavelet transform (DWT) out to level q_0 to $Z(w)$ to get the empirical DWT coefficients $z_{j,k}$ at each level j, where q_0 is specified in advance [40].
- Apply a thresholding procedure to $z_{j,k}$.
- The inverse DWT is applied to the thresholded wavelet coefficients to obtain the refined log spectrum.

3 Speech enhancement based on constrained Wiener filtering algorithm

Among the numerous techniques that were developed, the Wiener filter can be considered as one of the most fundamental SE approaches, which has been delineated in different forms and adopted in various applications [4]. The Wiener gain function is the least aggressive, in terms of suppression, providing small attenuation even at extremely low SNR levels.

A block diagram of the proposed SE algorithm is shown in Figure 1. The initial four frames are assumed to be noise only. The algorithm can be described as follows. The input noisy speech signal is decomposed into frames of 20-ms length with an overlap of 10 ms by the Hann window. Each segment was transformed using a 160-point discrete Fourier transform (DFT). The spectrum of the segmented noisy and noise signal are estimated by the multitaper method and then further refined by wavelet thresholding technique. The estimated 'clean' spectrum was gotten from the refined multitaper estimated noisy and noise spectrum. On the other hand, the noise-corrupted sentences were enhanced by the Wiener algorithm based on *a priori* SNR estimation [3]. The region I+II constraints were then imposed on the enhanced spectrum. Finally, the inverse fast Fourier transform (FFT) was applied to obtain the enhanced speech signal.

The implementation details of the proposed method can be described in the following four steps. For each speech frame,

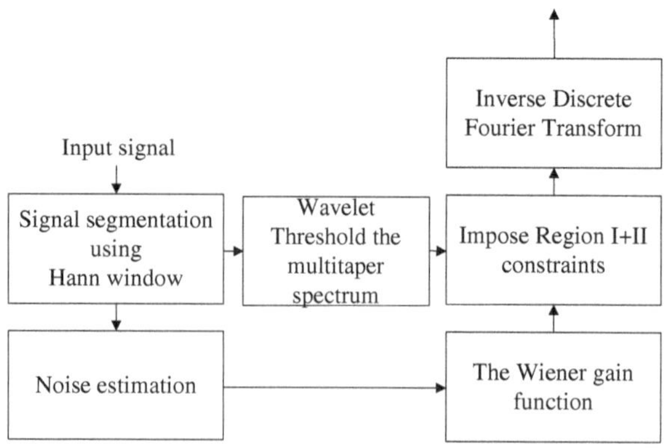

Figure 1 Block diagram of the proposed speech enhancement algorithm.

- compute the multitaper power spectrum $\widehat{S}_{\mathbf{y}}^{mt}$ of the noisy speech $\mathbf{y}$ using Equation 1 and estimate the multitaper power spectrum $\widehat{S}_{\mathbf{x}}^{mt}$ of the clean speech signal by $\widehat{S}_{\mathbf{x}}^{mt} = \widehat{S}_{\mathbf{y}}^{mt} - \widehat{S}_{\mathbf{n}}^{mt}$, where $\widehat{S}_{\mathbf{n}}^{mt}$ is the multitaper power spectrum of the noise. $\widehat{S}_{\mathbf{n}}^{mt}$ can be obtained using noise samples collected during speech absent frames. Here, L is set to 16. Any negative elements of $\widehat{S}_{\mathbf{x}}^{mt}$ are floored as follows:

$$\widehat{S}_{\mathbf{x}}^{mt} = \begin{cases} \widehat{S}_{\mathbf{y}}^{mt} - \widehat{S}_{\mathbf{n}}^{mt}, & \text{if } \widehat{S}_{\mathbf{y}}^{mt} > \widehat{S}_{\mathbf{n}}^{mt} \\ \beta \widehat{S}_{\mathbf{n}}^{mt}, & \text{if } \widehat{S}_{\mathbf{y}}^{mt} \leq \widehat{S}_{\mathbf{n}}^{mt}, \end{cases} \tag{9}$$

where β is the spectral floor set to $\beta = 0.002$.

- compute $Z(\omega) = log\widehat{S}_{\mathbf{y}}^{mt}(\omega) - \phi(L) + log(L)$ and then apply the DWT of $Z(\omega)$ out to level q_0 to obtain the empirical DWT coefficients $z_{j,k}$ for each level j, where q_0 is specified to be 5 [40]. Threshold the wavelet coefficients $z_{j,k}$ and apply the inverse DWT to the thresholded wavelet coefficients to obtain the refined log spectrum, $log\widehat{S}_{\mathbf{y}}^{\omega mt}(\omega)$, of the noisy singal. Repeat the above procedure to obtain the refined log spectrum, $log\widehat{S}_{\mathbf{n}}^{\omega mt}(\omega)$, of the noise signal. The estimated power spectrum $\widehat{S}_{\mathbf{x}}^{\omega mt}(\omega)$ of the clean speech signal can be estimated using

$$\widehat{S}_{\mathbf{x}}^{\omega mt}(\omega) = \widehat{S}_{\mathbf{y}}^{\omega mt}(\omega) - \widehat{S}_{\mathbf{n}}^{\omega mt}(\omega) \tag{10}$$

- let $Y(\omega, t)$ denote the magnitude of the noisy spectrum at time frame t and frequency bin ω estimated by the method in [41]. Then, the estimate of the signal spectrum magnitude is obtained by multiplying $Y(\omega, t)$ with a gain function $G(\omega, t)$ as $\hat{X}(\omega, t) = G(\omega, t) \cdot Y(\omega, t)$. The Wiener gain function is based on the a *priori* SNR and is given by

$$G(\omega, t) = \sqrt{\frac{\mathrm{SNR}_{\mathrm{prio}}(\omega, t)}{1 + \mathrm{SNR}_{\mathrm{prio}}(\omega, t)}}, \tag{11}$$

where SNR_{prio} is the *a priori* SNR estimated using the decision-directed approach [3,19] as follows:

$$SNR_{prio} = \alpha \cdot \frac{X_M^2(\omega, t-1)}{\hat{P}_D^2(\omega, t-1)} + (1-\alpha) \cdot \max\left[\frac{Y^2(\omega,t)}{\hat{P}_D^2(\omega,t)} - 1,\ 0\right], \quad (12)$$

where $\hat{P}_D^2(\omega, t)$ is the estimate of the power spectral density of background noise, and α is a smoothing constant (typically set to $\alpha = 0.98$).

- to maximize speech intelligibility, the final enhanced spectrum, $X_M(\omega, t)$, can be obtained by utilizing the region I+II constraints to the enhanced spectrum $\hat{X}(\omega, t)$ as follows:

$$X_M(\omega, t) = \begin{cases} \hat{X}(\omega, t), & \text{if } \hat{X}(\omega, t) < 2\widehat{S}_{\mathbf{x}}^{wmt}(\omega) \\ 0 & \text{else} \end{cases} \quad (13)$$

Finally, the enhanced speech signal can be obtained by apply the inverse FFT of $X_M(\omega, t)$.

The above estimator was applied to 20-ms duration frames of the noisy signal with 50% overlap between frames. The enhanced speech signal was combined using the overlap and add method.

4 Evaluation setup

The proposed SE algorithm was tested using a speech database that was corrupted by eight different real-world noises at different SNRs. The system was evaluated using both the composite evaluation measures proposed in [42] and the SNR_{LOSS} measure proposed in [43].

4.1 Database description

For the evaluation of SE algorithms, NOIZEUS [44] is preferred since it is a noisy speech corpus recorded by [18] to facilitate comparison of SE algorithms among different research groups [20]. The noisy database contains thirty IEEE sentences [45] which were recorded in a sound-proof booth using Tucker Davis Technologies (TDT; Alachua, FL, USA) recording equipment. The sentences were produced by three male and three female speakers (five sentences/speaker). The IEEE database was used as it contains phonetically balanced sentences with relatively low word-context predictability. The 30 sentences were selected from the IEEE database so as to include all phonemes in the American English language. The sentences were originally sampled at 25 kHz and downsampled to 8 kHz.

To simulate the receiving frequency characteristics of the telephone handsets, the intermediate reference system (IRS) filter used in ITU-T P.862 [46] for evaluation of the perceptual evaluation of speech quality (PESQ) measures was independently applied to the clean and noise signal [17]. Then, noise segment of the same length as the speech signal was randomly cut out of the noise recordings, appropriately scaled to reach the desired SNR levels (−8, −5, −2, 0, 5, 10, and 15 dB) and finally added to the filtered clean speech signal. Noise signals were taken from the AURORA database [47] and included the following recordings from different places: train, babble (crowd of people), car, exhibition hall, restaurant, street, airport, and train station. Therefore, in total, there are 1,680 (30 sentences × 8 noises × 7 SNRs) noisy speech segments in the test set.

4.2 Performance evaluation

The performance of an SE algorithm can be evaluated both subjectively and objectively. In general, subjective listening test is the most accurate and preferable method for evaluating speech quality and intelligibility. However, it is time consuming and cost expensive. Recently, many researchers have placed much effort on developing objective measures that would predict subjective quality and intelligibility with high correlation [42,43,48,49] with subjective listening test. Among them, the composite objective measures [42] were proved to have high correlation with subjective ratings and, at the same time, capture different characteristics of the distortions present in the enhanced signals [35], while the SNR_{LOSS} measure [43] was found appropriate in predicting speech intelligibility in fluctuating noisy conditions by yielding a high correlation for predicting sentence recognition. Therefore, the composite objective measures and the SNR_{LOSS} measure were adopted to predict the performance of the proposed SE algorithm on subjective quality and speech intelligibility, respectively.

4.2.1 *The composite measures to predict subjective speech quality*

The composite objective measures are obtained by linearly combining existing objective measures that highly correlate with subjective ratings. The objective measures include segmental SNR (*segSNR*) [18], weighted-slope spectral (*WSS*) [50], *PESQ* [51], and log likelihood ration (*LLR*) [18].

The three new composite measures obtained from multiple linear regression analysis are given below:

- C_{sig}: A five-point scale of signal distortion (SIG) formed by linearly combining the *LLR*, *PESQ*, and *WSS* measures (Table 1).
- C_{bak}: A five-point scale of noise intrusiveness (BAK) formed by linearly combining the *segSNR*, *PESQ*, and *WSS* measures (Table 1).
- C_{ovl}: The mean opinion score of overall quality (OVRL) formed by linearly combining the *PESQ*, *LLR*, and *WSS* measures.

Table 1 Scale of signal distortion, background intrusiveness and overall quality

Scale of signal distortion	Scale of background intrusiveness	Scale of overall quality
5- very natural, no degradation	5- not noticeable	5- excellent
4- fairly natural, little degradation	4- somewhat noticeable	4- good
3- somewhat natural, somewhat degraded	3- noticeable but not intrusive	3- fair
2- fairly unnatural, fairly degraded	2- fairly conspicuous, somewhat intrusive	2- poor
1- very unnatural, very degraded	1- very conspicuous, very intrusive	1- bad

The three new composite measures obtained from multiple linear regression analysis are given below:

$$C_{\mathrm{sig}} = 3.093 - 1.029 \cdot \mathrm{LLR} + 0.603 \cdot \mathrm{PESQ} - 0.009 \cdot \mathrm{WSS} \tag{14}$$

$$C_{\mathrm{bak}} = 1.634 + 0.478 \cdot \mathrm{PESQ} - 0.007 \cdot \mathrm{WSS} + 0.063 \cdot \mathrm{segSNR} \tag{15}$$

$$C_{\mathrm{ovl}} = 1.594 + 0.805 \cdot \mathrm{PESQ} - 0.512 \cdot \mathrm{LLR} - 0.007 \cdot \mathrm{WSS} \tag{16}$$

The correlation coefficients between the three composite measures and real subjective measures are given in Table 2 [42]. All three parameters should be maximized in order to get the best performance.

4.2.2 The SNR_{LOSS} measure to predict speech intelligibility

The SNR loss in band j and frame m is defined as follows [43]:

$$SL(j,m) = \mathrm{SNR}_X(j,m) - \mathrm{SNR}_{\hat{X}}(j,m), \tag{17}$$

where $\mathrm{SNR}_X(j,m)$ is the input SNR in band j, $\mathrm{SNR}_{\hat{X}}(j,m)$ is the SNR of the enhanced signal in the jth frequency band at the mth frame.

Assuming the SNR range is restricted to $[-\mathrm{SNR}_{\mathrm{Lim}}, \mathrm{SNR}_{\mathrm{Lim}}]$ dB ($\mathrm{SNR}_{\mathrm{Lim}} = 3$ in this paper), the $SL(j,m)$ term is then limited as follows:

$$\hat{SL}(j,m) = \min(\max(SL(j,m), -\mathrm{SNR}_{\mathrm{Lim}}), \mathrm{SNR}_{\mathrm{Lim}}) \tag{18}$$

and subsequently mapped to the range of [0, 1] using the following equation:

Table 2 Correlation coefficients between the composite measures and subjective measure

	C_{sig}	C_{bak}	C_{ovl}
SIG	0.7		
BAK		0.58	
OVRL			0.73

$$\mathrm{SNR}_{\mathrm{LOSS}}(j,m) = \begin{cases} -\dfrac{C_-}{\mathrm{SNR}_{\mathrm{Lim}}}\hat{SL}(j,m), & \text{if } \hat{SL}(j,m) < 0 \\ \dfrac{C_+}{\mathrm{SNR}_{\mathrm{Lim}}}\hat{SL}(j,m), & \text{if } \hat{SL}(j,m) \geq 0 \end{cases} \tag{19}$$

where C_- and C_+ are parameters (fixed to be 1 in this paper) controlling the slopes of the mapping function which was defined in the range of [0, 1]; therefore, the frame $\mathrm{SNR}_{\mathrm{LOSS}}$ is normalized to the range of $0 \leq \mathrm{SNR}_{\mathrm{LOSS}}(j,m) \leq 1$. The average $\mathrm{SNR}_{\mathrm{LOSS}}$ is finally computed by averaging $\mathrm{SNR}_{\mathrm{LOSS}}(j,m)$ over all frames in the signal as follows:

$$\overline{\mathrm{SNR}_{\mathrm{LOSS}}} = \frac{1}{M}\sum_{m=0}^{M-1} f\mathrm{SNR}_{\mathrm{LOSS}}(m), \tag{20}$$

where M is the total number of data segments in the signal, and $f\,\mathrm{SNR}_{\mathrm{LOSS}}(m)$ is the average (across bands) SNR loss computed as follows:

$$f\mathrm{SNR}_{\mathrm{LOSS}}(m) = \frac{\sum_{j=1}^{K} W(j) \cdot \mathrm{SNR}_{\mathrm{LOSS}}(j,m)}{\sum_{j=1}^{K} W(j)}, \tag{21}$$

where $W(j)$ is the weight (i.e., band importance function [52]) placed on the jth frequency band and was taken from Table B.1 in the ANSI standard [52].

The implementation of the $\mathrm{SNR}_{\mathrm{LOSS}}$ measure was supplied in the website of the authors in [43]. The smaller the value of the $\mathrm{SNR}_{\mathrm{LOSS}}$ measure is, the better performance of the SE algorithm is achieved.

5 Simulation results

The evaluation of the subjective quality and intelligibility of the speech enhanced by our proposed SE algorithm are reported in this section. Three other SE schemes, namely, wavelet thresholding (WT) [16], KLT [9], and Wiener algorithm with clean signal present (Wiener_Clean) [19], were also evaluated in order to gain a comparative analysis of the proposed SE algorithm. The KLT algorithm was proved in [1] and [22] by subjective tests to perform well in terms of preserving speech intelligibility for normal hearing listeners and improving speech intelligibility significantly for cochlear implant users in regard to recognition of sentences corrupted by stationary noises, respectively. The Wiener_Clean algorithm was taken as the ground truth in this paper because there is clean signal used in the algorithm. The unprocessed noisy signal (UP) was also evaluated by the SNR_{LOSS} measure for comparison purposes. The implementations of these three schemes were taken from the implementations in [18].

5.1 Performance of predicting subjective quality

5.1.1 *Performance average over all eight kinds of noise*

In Figure 2, the proposed algorithm is compared with WT and KLT algorithms in terms of the composite measures averaging over all eight noises for seven SNRs. The four objective measures (*LLR*, *segSNR*, *WSS*, and *PESQ*) that composed the composite measures were also given in the first row for reference. The Wiener_Clean algorithm, as the ground truth, performed the best for all four objective evaluation measures. According to [42], the *LLR* measure performed the best in terms of predicting signal distortion, while the *PESQ* measure gave the best prediction for both noise intrusiveness and overall speech quality. From the first row of Figure 2, we can notice that our proposed algorithm gives better performance than both WT and KLT in terms of the *LLR* measure for all seven SNRs tested. Moreover, when SNR is smaller than 5 dB, our proposed algorithm also performed better than both WT and KLT for the *PESQ* measure.

The second row of Figure 2 shows the composite measures, which include C_{sig}, C_{bak}, and C_{ovl}, estimated by the combination of all those four objective measures expressed in the first row. In terms of both signal distortion C_{sig} and overall quality C_{ovl}, our proposed method performs the best when SNR is less than 10 dB. Specifically speaking, for overall quality measure C_{ovl}, the proposed algorithm improved 10.94%, 18.94%, 21.63%, 23.66%, and 6.67% for $-8, -5, -2, 0$, and 5 dB, respectively, when compared with the KLT method. In general, the proposed algorithm achieved 13.88% and 6.40% improvement for C_{sig} and C_{ovl}, respectively, when average over all seven tested SNR levels. However, for C_{bak}, the WT and KLT algorithms give similar and better results than our proposed one when SNR is no smaller than 0 dB. The improvement was 0.98%, 6.98%, 11.11%, and 16.55% for 0, 5, 10, and 15 dB, respectively. In average, the WT and KLT methods were 5.14% better than our proposed algorithm in terms of background intrusiveness C_{bak}.

5.1.2 *Performance average over seven SNRs*

Figure 3 shows the three different composite measures averaged over seven SNRs for eight kinds of noise computed for WT, KLT, Wiener_Clean, and proposed SE algorithms. The Wiener_Clean algorithm still works as the ground truth here. From Figure 3, it is clear that in terms of C_{sig}, the KLT works much better than WT. Hence, the proposed algorithm is compared with only the KLT method here. We observe that on average, the proposed algorithm is better than the KLT method in terms of C_{sig} for train (9.19%), babble (15.93%), car (14.51%), exhibition hall (7.74%), restaurant (16.74%), street (13.42%), airport (16.64%), and train station (16.23%) noises. The number in the bracket indicates the C_{sig} by which our proposed

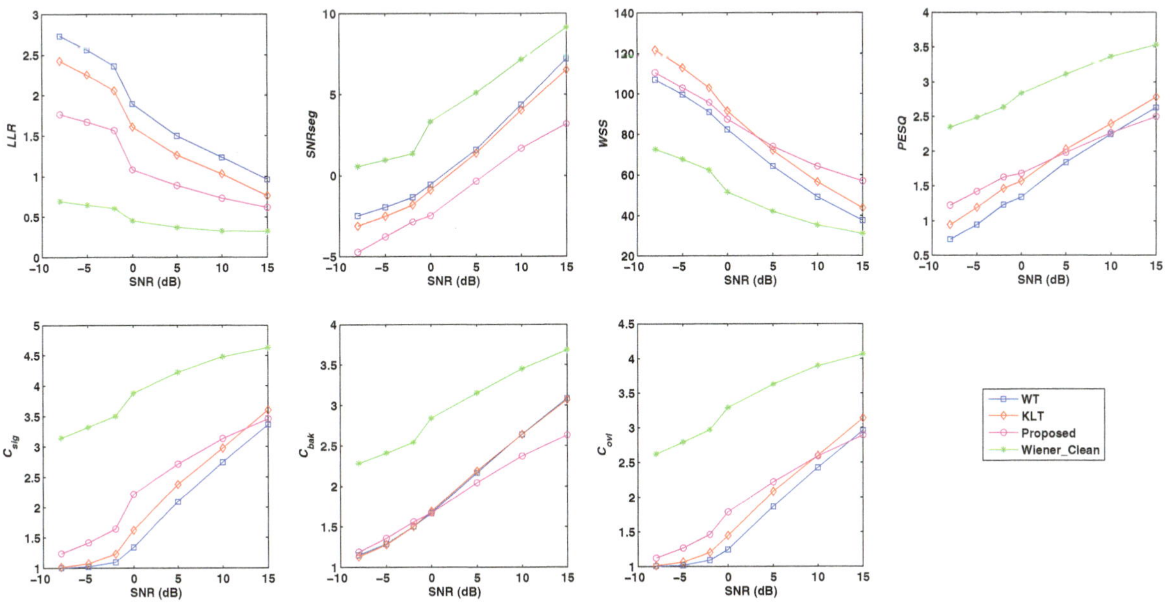

Figure 2 Averaged over seven SNRs. The composite measure comparisons for four SE schemes (WT, KLT, proposed, and Wiener_Clean) averaged over seven SNRs ($-8, -5, -2, 0, 5, 10$, and 15 db) for eight kinds of noise.

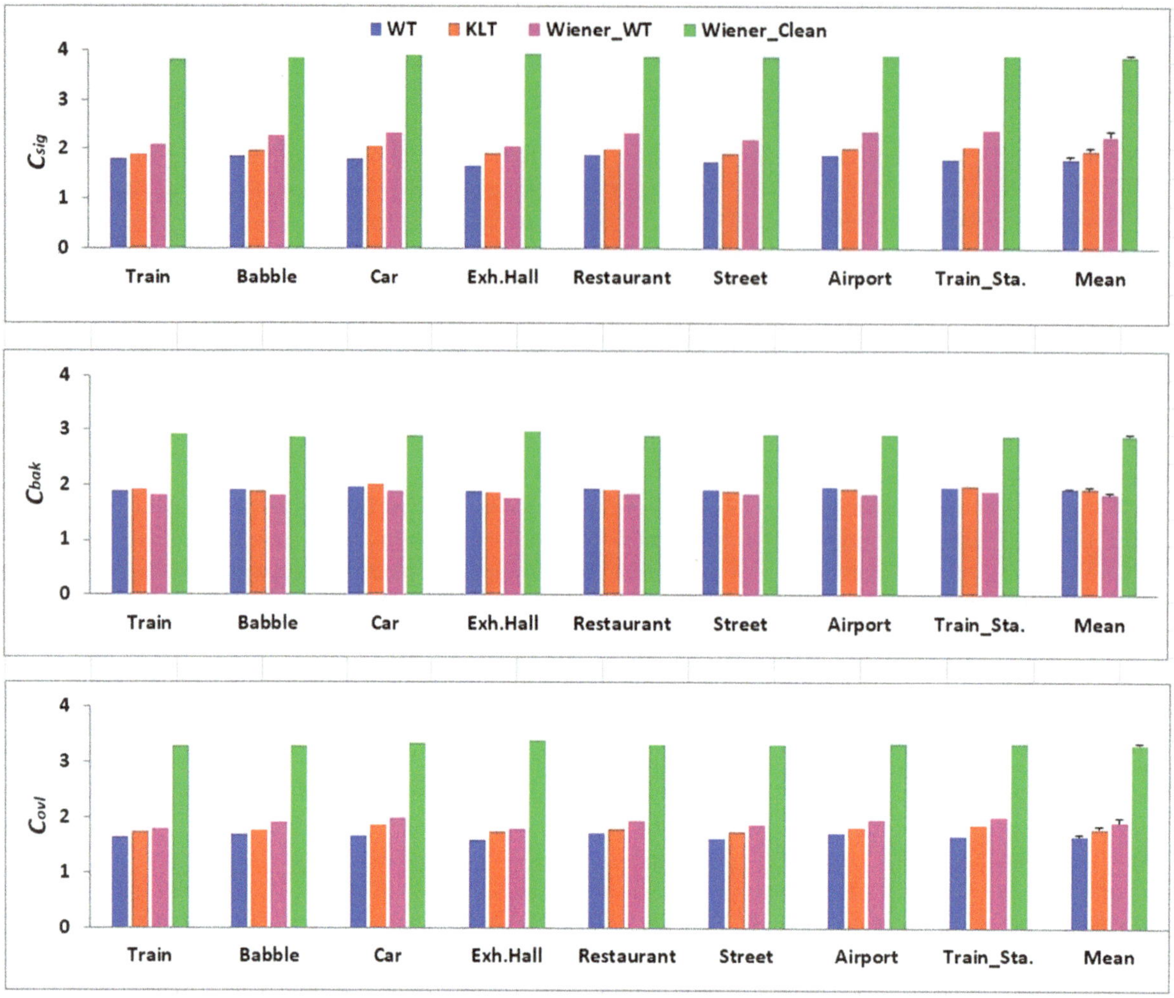

Figure 3 Averaged over eight kinds of noise. The composite measure comparisons for four SE schemes (WT, KLT, proposed, and Wiener_Clean) averaged over eight kinds of noise for seven SNRs (−8, −5, −2, 0, 5, 10, and 15 dB).

algorithm is better than the KLT method. The mean C_{sig} over all eight noise types of our proposed SE algorithm is 13.88% better than that of the KLT method. Furthermore, the proposed SE algorithms outperforms the KLT in terms of C_{ovl} by an average of 6.40% over all eight kinds of noise that were considered. However, in terms of background intrusiveness C_{bak}, the KLT algorithm gives an average of 5.14% better results than our proposed algorithm.

Thus, in conclusion, the proposed SE algorithm was predicted to be able to achieve the best overall subjective quality for most SNRs and all noise types considered when comparing with WT and KLT algorithms.

5.2 Performance of predicting speech intelligibility

The SNR_{LOSS} measure values obtained from each algorithm (include UP) were subjected to statistical analysis in order to assess their significant differences. A highly significant effect ($p < 0.005$) was found in all SNR levels and all types of noise by analysis of variance (ANOVA). Following the ANOVA, multiple comparison statistical tests according to Tukey's HSD test were done to assess the significance between algorithms. The difference was deemed significant if the p value was smaller than 0.05.

Table 3 gives the statistical comparisons of the SNR_{LOSS} measure between unprocessed noisy sentences (UP) and enhanced sentences by four SE algorithms (WT, KLT, Wiener_Clean, and proposed). At the same time, the comparisons between our proposed SE algorithm and the other three algorithms were also given. From Table 3, we know that when compared with the UP, our proposed algorithm was predicted by the SNR_{LOSS} measure to be able to improve the intelligibility in low SNRs for most noises tested (italicized). The R in the table gives the

Table 3 Statistical comparisons of the SNR_{LOSS} measure between unprocessed sentences and enhanced sentences by SE algorithms

		WT-UP		KLT-UP		Wiener_Clean-UP		Proposed-UP		Proposed-WT		Proposed-KLT		Proposed-Wiener_Clean	
Noise	**SNR (dB)**	***R*** **(%)**	***p*** **value**	***R*** **(%)**	***p*** **value**	***R*** **(%)**	***p*** **value**	***R*** **(%)**	***p*** **value**	***R*** **(%)**	***p*** **value**	***R*** **(%)**	***p*** **value**	***R*** **(%)**	***p*** **value**
Train	−8	1.30	0.000	0.98	0.011	−7.05	0.000	*−0.87*	*0.031*	*−2.15*	*0.000*	*-1.83*	*0.000*	6.64	0.000
	−5	1.61	0.000	1.26	0.004	−6.87	0.000	−0.91	0.076	*−2.47*	*0.000*	*−2.13*	*0.000*	6.41	0.000
	−2	1.93	0.000	1.43	0.005	−6.74	0.000	−0.96	0.127	*−2.83*	*0.000*	*−2.35*	*0.000*	6.20	0.000
	0	3.41	0.000	2.33	0.002	−9.83	0.000	−1.67	0.054	*−4.92*	*0.000*	*−3.92*	*0.000*	9.04	0.000
	5	3.44	0.001	2.12	0.084	−8.95	0.000	0.45	0.983	*−2.89*	*0.004*	−1.63	0.264	10.32	0.000
	10	2.96	0.024	2.47	0.091	−7.06	0.000	3.74	0.002	0.76	0.930	1.25	0.687	11.63	0.000
	15	3.54	0.034	4.96	0.001	−1.85	0.553	10.76	0.000	6.97	0.000	5.52	0.000	12.85	0.000
Babble	−8	1.61	0.000	0.73	0.128	−7.69	0.000	*−1.16*	*0.002*	*−2.73*	*0.000*	*−1.88*	*0.000*	7.07	0.000
	−5	2.25	0.000	1.21	0.010	−7.23	0.000	*−1.11*	*0.022*	*−3.28*	*0.000*	*−2.30*	*0.000*	6.60	0.000
	−2	2.42	0.000	1.22	0.047	−6.90	0.000	−1.06	0.118	*−3.39*	*0.000*	*−2.25*	*0.000*	6.28	0.000
	0	3.25	0.000	1.55	0.046	−9.57	0.000	*−1.85*	*0.009*	*−4.93*	*0.000*	*−3.34*	*0.000*	8.54	0.000
	5	4.88	0.000	2.76	0.002	−8.05	0.000	0.10	1.000	*−4.56*	*0.000*	*−2.59*	*0.004*	8.87	0.000
	10	6.11	0.000	4.74	0.000	−4.75	0.000	4.72	0.000	−1.31	0.684	−0.02	1.000	9.95	0.000
	15	6.84	0.000	7.72	0.000	1.91	0.687	12.76	0.000	5.54	0.001	4.67	0.006	10.65	0.000
Car	−8	1.18	0.000	0.12	0.984	−8.33	0.000	*−1.89*	*0.000*	*−3.04*	*0.000*	*−2.01*	*0.000*	7.09	0.000
	−5	1.90	0.000	0.46	0.424	−8.20	0.000	*−2.11*	*0.000*	*−3.94*	*0.000*	*−2.56*	*0.000*	6.64	0.000
	−2	2.25	0.000	0.56	0.321	−8.20	0.000	*−2.48*	*0.000*	*−4.62*	*0.000*	*−3.03*	*0.000*	6.23	0.000
	0	3.38	0.000	0.78	0.373	−11.23	0.000	*−3.45*	*0.000*	*−6.61*	*0.000*	*−4.19*	*0.000*	8.77	0.000
	5	3.46	0.000	−0.21	0.997	−10.70	0.000	*−2.33*	*0.001*	*−5.59*	*0.000*	*−2.12*	*0.005*	9.38	0.000
	10	5.32	0.000	1.90	0.345	−7.52	0.000	1.55	0.554	*−3.58*	*0.003*	−0.34	0.997	9.81	0.000
	15	5.10	0.000	2.78	0.075	−2.68	0.093	8.25	0.000	2.99	0.030	5.32	0.000	11.22	0.000
Exhibition Hall	−8	0.58	0.026	−0.42	0.201	−7.[illegible]9	0.000	*−1.44*	*0.000*	*−2.01*	*0.000*	*−1.03*	*0.000*	6.88	0.000
	−5	1.01	0.000	−0.17	0.947	−7.[illegible]0	0.000	*−1.61*	*0.000*	*−2.60*	*0.000*	*−1.45*	*0.000*	6.60	0.000
	−2	1.41	0.000	0.03	1.000	−7.48	0.000	*−1.83*	*0.000*	*−3.20*	*0.000*	*−1.86*	*0.000*	6.10	0.000
	0	4.00	0.000	0.77	0.738	−10.52	0.000	*−1.95*	*0.020*	*−5.72*	*0.000*	*−2.70*	*0.000*	9.58	0.000
	5	3.68	0.000	−0.62	0.889	−9.69	0.000	0.15	0.999	*−3.40*	*0.000*	0.77	0.779	10.90	0.000
	10	4.86	0.001	2.14	0.393	−6.57	0.000	4.34	0.004	−0.49	0.993	2.15	0.365	11.67	0.000
	15	6.17	0.000	4.88	0.003	−1.24	0.880	11.13	0.000	4.67	0.002	5.96	0.000	12.53	0.000

Table 3 Statistical comparisons of the SNR_{LOSS} measure between unprocessed sentences and enhanced sentences by SE algorithms *(Continued)*

Restaurant	−8	1.81	0.000	0.95	0.033	−7.42	0.000	*−1.14*	*0.005*	*−2.90*	*0.000*	*−2.07*	*0.000*	6.78	0.000
	−5	2.15	0.000	1.10	0.044	−7.09	0.000	−1.07	0.053	*−3.16*	*0.000*	*−2.15*	*0.000*	6.47	0.000
	−2	2.18	0.000	1.12	0.113	−6.67	0.000	−1.05	0.159	*−3.16*	*0.000*	*−2.14*	*0.000*	6.03	0.000
	0	4.76	0.000	3.21	0.000	−8.82	0.000	−0.88	0.746	*−5.38*	*0.000*	*−3.96*	*0.000*	8.70	0.000
	5	4.06	0.002	2.53	0.142	−7.90	0.000	0.83	0.939	*−3.10*	*0.028*	−1.65	0.525	9.49	0.000
	10	7.40	0.000	6.27	0.000	−3.21	0.053	6.53	0.000	−0.81	0.945	0.24	1.000	10.06	0.000
	15	8.27	0.000	9.01	0.000	−3.77	0.101	15.18	0.000	6.38	0.000	5.67	0.001	11.00	0.000
Street	−8	1.11	0.045	0.72	0.363	−7.26	0.000	*−1.27*	*0.015*	*−2.35*	*0.000*	*−1.98*	*0.000*	6.46	0.000
	−5	1.48	0.017	0.89	0.329	−7.24	0.000	*−1.32*	*0.044*	*−2.76*	*0.000*	*−2.19*	*0.000*	6.39	0.000
	−2	1.58	0.030	0.86	0.502	−7.12	0.000	−1.36	0.088	*−2.90*	*0.000*	*−2.20*	*0.001*	6.21	0.000
	0	4.76	0.000	2.90	0.004	−9.14	0.000	−1.13	0.622	*−5.62*	*0.000*	*−3.91*	*0.000*	8.82	0.000
	5	4.64	0.000	2.85	0.036	−8.59	0.000	0.74	0.944	*−3.73*	*0.001*	−2.05	0.210	10.20	0.000
	10	6.28	0.001	5.29	0.007	−5.61	0.003	5.36	0.006	−0.86	0.976	0.07	1.000	11.62	0.000
	15	7.55	0.000	7.49	0.000	0.47	0.998	12.22	0.000	4.34	0.014	4.41	0.012	11.70	0.000
Airport	−8	1.80	0.000	0.96	0.147	−7.89	0.000	*−1.51*	*0.004*	*−3.25*	*0.000*	*−2.44*	*0.000*	6.93	0.000
	−5	2.31	0.000	1.37	0.048	−7.42	0.000	*−1.37*	*0.048*	*−3.60*	*0.000*	*−2.70*	*0.000*	6.53	0.000
	−2	2.65	0.000	1.60	0.035	−6.78	0.000	−1.13	0.249	*−3.68*	*0.000*	*−2.68*	*0.000*	6.06	0.000
	0	4.53	0.000	1.72	0.225	−9.43	0.000	−1.45	0.392	*−5.72*	*0.000*	*−3.12*	*0.001*	8.81	0.000
	5	5.26	0.000	3.36	0.007	−7.69	0.000	0.60	0.972	*−4.42*	*0.000*	*−2.67*	*0.042*	8.99	0.000
	10	7.72	0.000	5.88	0.000	−2.90	0.113	6.58	0.000	−1.05	0.876	0.66	0.977	9.76	0.000
	15	7.76	0.000	7.43	0.000	2.61	0.270	13.81	0.000	5.62	0.000	5.95	0.000	10.92	0.000
Train station	−8	1.67	0.000	0.89	0.034	−8.37	0.000	*−1.91*	*0.000*	*−3.52*	*0.000*	*−2.78*	*0.000*	7.04	0.000
	−5	2.30	0.000	1.22	0.010	−7.89	0.000	*−1.90*	*0.000*	*−4.11*	*0.000*	*−3.09*	*0.000*	6.50	0.000
	−2	2.64	0.000	1.47	0.006	−7.53	0.000	*−1.88*	*0.000*	*−4.41*	*0.000*	*−3.30*	*0.000*	6.11	0.000
	0	3.19	0.000	0.58	0.857	−10.72	0.000	*−2.97*	*0.000*	*−5.98*	*0.000*	*−3.54*	*0.000*	8.67	0.000
	5	4.25	0.003	1.24	0.812	−9.47	0.000	−1.63	0.608	*−5.64*	*0.000*	−2.84	0.092	8.65	0.000
	10	5.43	0.000	2.38	0.192	−5.94	0.000	2.79	0.082	−2.50	0.115	0.41	0.995	9.29	0.082
	15	7.60	0.000	6.30	0.000	0.38	0.999	11.19	0.000	3.33	0.065	4.60	0.004	10.76	0.000

Four SE algorithms: WT, KLT, Wiener_Clean, and proposed. The comparisons between our proposed SE algorithm and the other three were also given. The results in italics show the SNR_{LOSS} measure by which our proposed SE algorithm is better than others, and the difference was significant ($p<0.05$).

percentage by which our algorithm is better than others; the value is negative because better performance gave smaller SNR_{LOSS} measure. Furthermore, our proposed SE algorithm was also compared with the WT and KLT algorithms and was proved to supply better performance for most conditions tested.

6 Conclusions

The main contribution of this paper was the introduction of a new SE algorithm based on imposing constraint on Wiener gain function. Experiments were done on NOIZEUS database for eight kinds of noise (AURORA database) across seven different SNRs ranging from −8 to 15 dB. The Wiener_Clean algorithm was taken as the ground truth. The performance of our proposed algorithm was compared with WT and KLT methods. The results were analyzed mainly by three composite measures and the SNR_{LOSS} measure to predict the performance on subjective quality and speech intelligibility, respectively. Through extensive experiments, we showed that when averaged over all eight kinds of noises, our proposed SE algorithm achieved the best results in terms of predicting signal distortion C_{sig} and overall quality C_{ovl} when SNR is no more than 10 dB. Furthermore, we investigated the individual performance on each noise type. Our proposed SE algorithm outperformed the KLT algorithm for all noise types tested in terms of both C_{sig} and C_{ovl}. On the other hand, the SNR_{LOSS} measure comparisons with both the UP and other SE algorithms predicted that our proposed algorithm was able to improve speech intelligibility for low SNR levels and outperform WT and KLT algorithms for most conditions examined.

It is important to point out that the three composite measures and the SNR_{LOSS} measure used in this paper are adopted for predicting the subjective quality and intelligibility of noisy speech enhanced by noise suppression algorithms because of their high correlation with real subjective tests [42,43]. Further subjective tests on both normal-hearing listeners and hearing-impaired listeners are needed to verify the effectiveness of the proposed algorithm on improving both subjective quality and speech intelligibility. It is also worth mentioning that depending on the nature of the application, some practical SE systems may require very high quality speech but can tolerate a certain amount of noise, while other systems may want speech as clean as possible even with some degree of speech distortion. Therefore, it should be noted that according to different applications, different SE algorithms should be chosen to meet the variant requirement.

Competing interests

The authors declare that they have no competing interests.

References

1. Y Hu, PC Loizou, A comparative intelligibility study of single-microphone noise reduction algorithms. J. Acous. Soc. Am. **122**(3), 1777–1786 (2007)
2. MR Schroeder, Apparatus for suppressing noise and distortion in communication signals. U.S Patent 3180936 (April27, 1965)
3. P Scalart, J Vieira-Filho, Speech enhancement based on a priori signal to noise estimation, in *Proceedings of the 21st IEEE International Conference on Acoustics, Speech, and Signal Processing (ICASSP-96)* (Atlanta Georgia, 1996), pp. 629–632
4. J Chen, J Benesty, Y Huang, New insights into the noise reduction wiener filter. IEEE Trans. Audio Speech Lang. Process. **14**(4), 1218–1234 (2006)
5. S Boll, Suppression of acoustic noise in speech using spectral subtraction. IEEE Trans. Acoust. Speech Signal Proces. **27**(2), 113–120 (1979)
6. M Berouti, R Schwartz, J Makhoul, Enhancement of speech corrupted by acoustic noise, in *IEEE International Conference on Acoustics, Speech, and Signal Processing (ICASSP)*, vol. 4, (1979), pp. 208–211
7. H Gustafsson, SE Nordholm, I Claesson, Spectral subtraction using reduced delay convolution and adaptive averaging. IEEE Trans. Speech Audio Proces. **9**(8), 799–807 (2001)
8. S Kamath, PC Loizou, A multi-band spectral subtraction method for enhancing speech corrupted by colored noise, in *Student Research Abstracts of Proc. IEEE International Conference On Acoustics, Speech, and Signal Processing (ICASSP)*, vol. 4 (Orlando, FL, USA, 2002), pp. IV-4164
9. Y Hu, PC Loizou, A generalized subspace approach for enhancing speech corrupted by colored noise. IEEE Trans. Speech Audio Proces. **11**(4), 334–341 (2003)
10. F Jabloun, B Champagne, Incorporating the human hearing properties in the signal subspace approach for speech enhancement. IEEE Trans. Speech Audio Proces. **11**(6), 700–708 (2003)
11. Y Ephraim, D Malah, Speech enhancement using a minimum-mean square error short-time spectral amplitude estimator. IEEE Trans. Acoust. Speech Signal Proces. **32**(6), 1109–1121 (1984)
12. Y Ephraim, D Malah, Speech enhancement using a minimum mean-square error log-spectral amplitude estimator. IEEE Trans. Acoust. Speech Signal Proces. **33**(2), 443–445 (1985)
13. PC Loizou, Speech enhancement based on perceptually motivated Bayesian estimators of the magnitude spectrum. IEEE Trans. Speech Audio Proces. **13**(5), 857–869 (2005)
14. JS Lim, AV Oppenheim, Enhancement and bandwidth compression of noisy speech. Proc. IEEE. **67**(12), 1586–1604 (1979)
15. JS Lim, AV Oppenheim, All pole modeling of degraded speech. IEEE Trans. Acoust. Speech Signal Proces. **26**(3), 197–210 (1978)
16. Y Hu, PC Loizou, Speech enhancement based on wavelet thresholding the multitaper spectrum. IEEE Trans. Speech Audio Proces. **12**(1), 59–67 (2004)
17. Y Hu, PC Loizou, Subjective comparison and evaluation of speech enhancement algorithms. Speech Commun. **49**(7), 588–601 (2007)
18. PC Loizou, *Speech Enhancement: Theory and Practice*. (CRC, Boca Raton, 2007)
19. PC Loizou, G Kim, Reasons why current speech-enhancement algorithms do not improve speech intelligibility and suggested solutions. IEEE Trans. Audio Speech Lang. Process. **19**(1), 47–56 (2011)
20. Y Hu, PC Loizou, Subjective comparison of speech enhancement algorithms, in *Proceedings of the IEEE International Conference on Acoustics, Speech and Signal Processing (ICASSP)*, vol. 1, (2006), pp. 153–156
21. JS Lim, Evaluation of a correlation subtraction method for enhancing speech degraded by additive white noise. IEEE Trans. Acoust. Speech Signal Proces. **26**(5), 471–472 (1978)
22. PC Loizou, A Lobo, Y Hu, Subspace algorithms for noise reduction in cochlear implants. J. Acoust. Soc. Am. **118**(5), 2791–2793 (2005)
23. G Kim, Y Lu, Y Hu, PC Loizou, An algorithm that improves speech intelligibility in noise for normal-hearing listeners. J. Acoust. Soc. Am. **126**(3), 1486–1494 (2009)
24. EW Healy, SE Yoho, Y Wang, DL Wang, An algorithm to improve speech recognition in noise for hearing-impaired listeners. J. Acoust. Soc. Am. **134**(4), 3029–3038 (2013)
25. K Hu, P Divenyi, D Ellis, Z Jin, BG Shinn-Cunningham, DL Wang, Preliminary intelligibility tests of a monaural speech segregation system, in *ISCA Tutorial and Research Workshop on Statistical and Perceptual Audition (SAPA)*, (2008), pp. 11–16

26. DJ Thomson, Spectrum estimation and harmonic analysis. Proc. IEEE. **70**(9), 1055–1096 (1982)
27. SM Kay, *Modern Spectral Estimation*. (Prentice-Hall, Englewood Cliffs, 1988)
28. KS Riedel, A Sidorenko, Minimum bias multiple taper spectral estimation. IEEE Trans. Signal Process. **43**(1), 188–195 (1995)
29. P Moulin, Wavelet thresholding techniques for power spectrum estimation. IEEE Trans. Signal Process. **42**(11), 3126–3136 (1994)
30. AT Walden, DB Percival, EJ McCoy, Spectrum estimation by wavelet thresholding of multitaper estimators. IEEE Trans. Signal Process. **46**(12), 3153–3165 (1998)
31. AC Cristan, AT Walden, Multitaper power spectrum estimation and thresholding: wavelet packets versus wavelets. IEEE Trans. Signal Process. **50**(12), 2976–2986 (2002)
32. G Wahba, Automatic smoothing of the log periodogram. J. Am. Stat. Assoc. **75**(369), 122–132 (1980)
33. DB Percival, *Spectral Analysis for Physical Applications: Multitaper and Conventional Univariate Techniques*. (Cambridge University Press, Cambridge, 1993)
34. MS Bartlett, DG Kendall, The statistical analysis of variance-heterogeneity and the logarithmic transformation. Suppl. J. R. Stat. Soc. **8**(1), 128–138 (1946)
35. SR Quackenbush, TP Barnwell, MA Clements, *Objective Measures of Speech Quality*. (Prentice-Hall, Englewood Cliffs, 1988)
36. DL Donoho, IM Johnstone, Ideal spatial adaptation by wavelet shrinkage. Biometrika. **81**(3), 425–455 (1994)
37. DL Donoho, De-noising by soft-thresholding. IEEE Trans. Inform. Theory. **41**(3), 613–627 (1995)
38. DL Donoho, IM Johnstone, Adapting to unknown smoothness via wavelet shrinkage. J. Am. Stat. Assoc. **90**(432), 1200–1224 (1995)
39. IM Johnstone, BW Silverman, Wavelet threshold estimators for data with correlated noise. J. Roy. Stat. Soc. B. **59**(2), 319–351 (1997)
40. SG Mallat, A theory for multiresolution signal decomposition: the wavelet representation. IEEE Trans. Pattern Anal. Mach. Intell. **11**(7), 674–693 (1989)
41. S Rangachari, PC Loizou, A noise-estimation algorithm for highly non-stationary environments. Speech Commun. **48**(2), 220–231 (2006)
42. Y Hu, PC Loizou, Evaluation of objective measures for speech enhancement, in *Proceedings of INTERSPEECH*, (2006), pp. 1447–1450
43. J Ma, PC Loizou, SNR loss: a new objective measure for predicting the intelligibility of noise-suppressed speech. Speech Commun. **53**(3), 340–354 (2011)
44. Y Hu, PC Loizou, Subjective comparison and evaluation of speech enhancement algorithms. Speech Commun. **49**(7), 588–601 (2007)
45. EH Rothauser, WD Chapman, N Guttman, MHL Hecker, KS Nordby, HR Silbiger, GE Urbanek, M Weinstock, IEEE recommended practice for speech quality measurements. IEEE Trans. Audio Electroacoust. **17**(3), 225–246 (1969)
46. International Telecommunication Union, P.862: Perceptual Evaluation of Speech Quality (PESQ): An Objective Method for End-to-End Speech Quality Assessment of Narrow-Band Telephone Networks and Speech Codecs. (ITU-T Recommendation P. 862, 2000)
47. H Hirsch, D Pearce, The AURORA experimental framework for the performance evaluation of speech recognition systems under noisy conditions, in *ASR-2000* (Paris, France, 2000), pp. 181–188
48. Y Hu, PC Loizou, Evaluation of objective quality measures for speech enhancement. IEEE Trans. Audio Speech Lang. Process. **16**(1), 229–238 (2008)
49. Y Hu, J Ma, PC Loizou, Objective measures for predicting speech intelligibility in noisy conditions based on new band-importance functions. J. Acoust. Soc. Am. **125**(5), 3387–3405 (2009)
50. D Klatt, Prediction of perceived phonetic distance from critical-band spectra: a first step, in *IEEE International Conference on Acoustics, Speech, and Signal Processing (ICASSP)*, vol. 7, (1982), pp. 1278–1281
51. International Telecommunication Union, ITU-T Recommendation P.835: Subjective Test Methodology for Evaluating Speech Communication Systems that Include Noise Suppression Algorithms. (ITU-T Recommendation P.835, 2003)
52. ANSI, Methods for Calculation of the Speech Intelligibility Index. Technical Report S3.5-1997, American National Standards Institute (1997)

14

The self-taught vocal interface

Bart Ons*, Jort F Gemmeke and Hugo Van hamme

Abstract

Speech technology is firmly rooted in daily life, most notably in command-and-control (C&C) applications. C&C usability downgrades quickly, however, when used by people with non-standard speech. We pursue a fully adaptive vocal user interface (VUI) which can learn both vocabulary and grammar directly from interaction examples, achieving robustness against non-standard speech by building up models from scratch. This approach raises feasibility concerns on the amount of training material required to yield an acceptable recognition accuracy. In a previous work, we proposed a VUI based on non-negative matrix factorisation (NMF) to find recurrent acoustic and semantic patterns comprising spoken commands and device-specific actions, and showed its effectiveness on unimpaired speech. In this work, we evaluate the feasibility of a self-taught VUI on a new database called DOMOTICA-3, which contains dysarthric speech with typical commands in a home automation setting. Additionally, we compare our NMF-based system with a system based on Gaussian mixtures. The evaluation favours our NMF-based approach, yielding feasible recognition accuracies for people with dysarthric speech after a few learning examples. Finally, we propose the use of a multi-layered semantic frame structure and demonstrate its effectiveness in boosting overall performance.

Keywords: Vocal user interface; Dysarthric speech; Non-negative matrix factorisation; Semantic frame description; Command and control

Introduction

Currently, modern voice control technology is available in many applications such as direct voice input (DVI) in aviation [1], information requests using Siri and speech driven home automation. Command-and-control (C&C) appliances afford hands-free control, thus enhancing the independence of the physically incapacitated. Unfortunately, speech commands are sometimes misinterpreted when words overstep lexical boundaries and word sequences do not fit the preset grammars. Moreover, C&C appliances frequently fail to interpret dialectic or impaired speech, often encountered with physically challenged people. Consequently, people with non-standard speech are increasingly excluded from the growing market of voice-driven applications. The goal of this work is to investigate a vocal user interface (VUI) model which is able to learn words and grammars from end users, improving accessibility of C&C applications.

Over the past decade, various approaches have been proposed to improve the usability of automatic speech recognition (ASR) for speech-impaired users. For example, in [2-4], speaker-independent acoustic models were adapted to speaker-dependent and speaker-adapted models, both providing better recognition of user-specific vocalizations. Besides adaptation, dysarthric speakers also improved the recognition likelihood of their words by training the consistency of their pronunciations [5-7]; thus, users can adapt their vocalizations in order to alleviate the ASR shortcoming to cope with severe vocal variability. In [8,9], the increased phonatory variability associated with dysarthric speech was addressed by a system enabling more suitable hidden Markov model (HMM) topologies for each phoneme in the speaker?s repertoire. Another example is [10], where user needs were surveyed and reflected in the design of a VUI for which an isolated word recognition system with a customizable command list and a built-in word prediction function was proposed to improve usability of typical services on mobiles and tablets. Although these approaches resulted in considerable improvements in usability, the accessibility of voice control technology still needs to widen to cater for users with non-standard or impaired speech (see [11,12]).

State-of-the-art ASR is typically based on HMM acoustic models developed with Gaussian mixture (GMM) continuous emission densities and context-dependent

*Correspondence: bart.ons@esat.kuleuven.be
Department ESAT-PSI, KU Leuven, Kasteelpark 10, 3001 Leuven, Belgium

bi- or triphone models with multiple states per model. These language-dependent models are trained on hundreds of thousands of recorded and annotated speech utterances. Some applications in voice-enabled automated home environments use ASR models together with a speaker adaptation procedure to improve ASR performance for specific users or user groups. For example, the DIRHA [13], SWEET-HOME [14] and HomeService [15] projects aim for voice-enabled assistive technology in home environments for people with a physical impairment. In the DIRHA and SWEET-HOME projects, maximum likelihood linear regression (MLLR) speaker adaptation is used starting from a speaker-independent ASR system. In the HomeService project, speaker-independent ASR models were obtained using normal or dysarthric speech followed by maximum a posteriori (MAP) speaker adaptation. These approaches require annotated language-dependent speech material in addition to annotated user-specific speech material. The advantage of the adaptation approach is that the amount of user-specific speech material composes only a fraction of the data required for building a speaker-dependent state-of-the-art speech recognizer. Speaker-dependent data usually requires an enrolment session and automated or non-automated transcriptive resources. Contrary to the adaptation approach, the basic approach here and in the (ALADIN) project (see [16] for an overview) is to build a VUI model that starts from scratch and learns from speech and demonstrations of the end user without transcription. Considering the VUI usability, the training procedure requires the ability to learn from a few examples and should be able to work with easily obtainable annotations such as content or context information. In our language-independent approach, the VUI learns to understand spoken commands by mining the speech input from the end user and the changes that are provoked on a device.

The first aim of the study is to test the feasibility of the learning procedure to construct speech patterns such as words from a few examples and content-related annotations. The speech of the user and the content information entered by the device are two sources of information that we combine by using non-negative matrix factorisation (NMF, see [17]). This procedure allows the VUI to learn co-occurring patterns from two information sources. In [18], we proposed a novel grammar induction technique based on HMM learning and semantic descriptions of commands guiding the learning process. Here, we propose multi-layered semantic structures and implement the semantic dependencies in a parse tree structure. The second aim of the study is to compare the new semantic structure with the ones employed in [18]. For this, we use two databases: one with recordings of normally speaking subjects playing a card game by voice, and another one with commands provoked in a virtual home automation setting for people diagnosed with dysarthria. The first database is referred to as PATCOR, whereas the second one is a new database called DOMOTICA-3. Besides the validation of new semantic structures, we will evaluate the NMF procedure as well by comparing our NMF-based framework with a Gaussian mixture model (GMM)-based baseline system.

The remainder of the paper is organised as follows. In the section ?Language learning in the vocal user interface?, we describe the learning framework, including the semantic and acoustic representations as well as the NMF learning procedure. In the section ?Reference model?, we describe a reference model employing GMMs instead of NMF. We proceed by describing the databases used for evaluation in the section ?Speech material?. Subsequently, we explain the semantic structure of spoken commands (cf. the section ?Hierarchical knowledge representation?) before conducting a series of experiments (cf. the section ?Experiments?) where we evaluate the feasibility of our approach and the effectiveness of more layered semantic structure. We present our conclusion and thoughts on a future work in the final section ?Conclusions?.

Language learning in the vocal user interface

A schematic overview of the learning framework is depicted in Figure 1. Here, two different types of data are processed: one processing stream is depicted in the upper part and builds up a *semantic representation*, while the other one, depicted in the lower part, builds up an *acoustic representation*. In the upper processing stream, device-specific functionality is parsed into a semantic *frame description*. The conversion is guided by a handcrafted *semantic frame structure* as indicated with the dotted arrow pointing towards the arrow leading to the block ?frame description? (cf. the section ?Semantic representation?). The frame description is turned into a *label vector* and passed on to the *NMF* module.

In the lower part of Figure 1 (cf. the section ?Acoustic representation?), spectro-temporal features are extracted and transformed into mel-frequency cepstral coefficients (MFCCs, cf. the section ?Feature extraction?). The MFCC features are converted into a *posteriorgram*, and the horizontal dotted arrow from the right leaving from the block *Codebook/Gaussians* indicates that, for this, intermediate procedures like *codebook* training and clustering are needed (cf. the section ?Codebook training?). The posteriorgram is then converted into an utterance-based representation by using histograms of acoustic co-occurrence (HAC, cf. the section ?Histogram of acoustic co-occurrence?), after which the NMF training takes place. The depicted matrices denoted by $\mathbf{H}$ contain column-wise entries for each learning example representing loads on recurrent patterns in the data matrices $\mathbf{V}_s$ and $\mathbf{V}_a$,

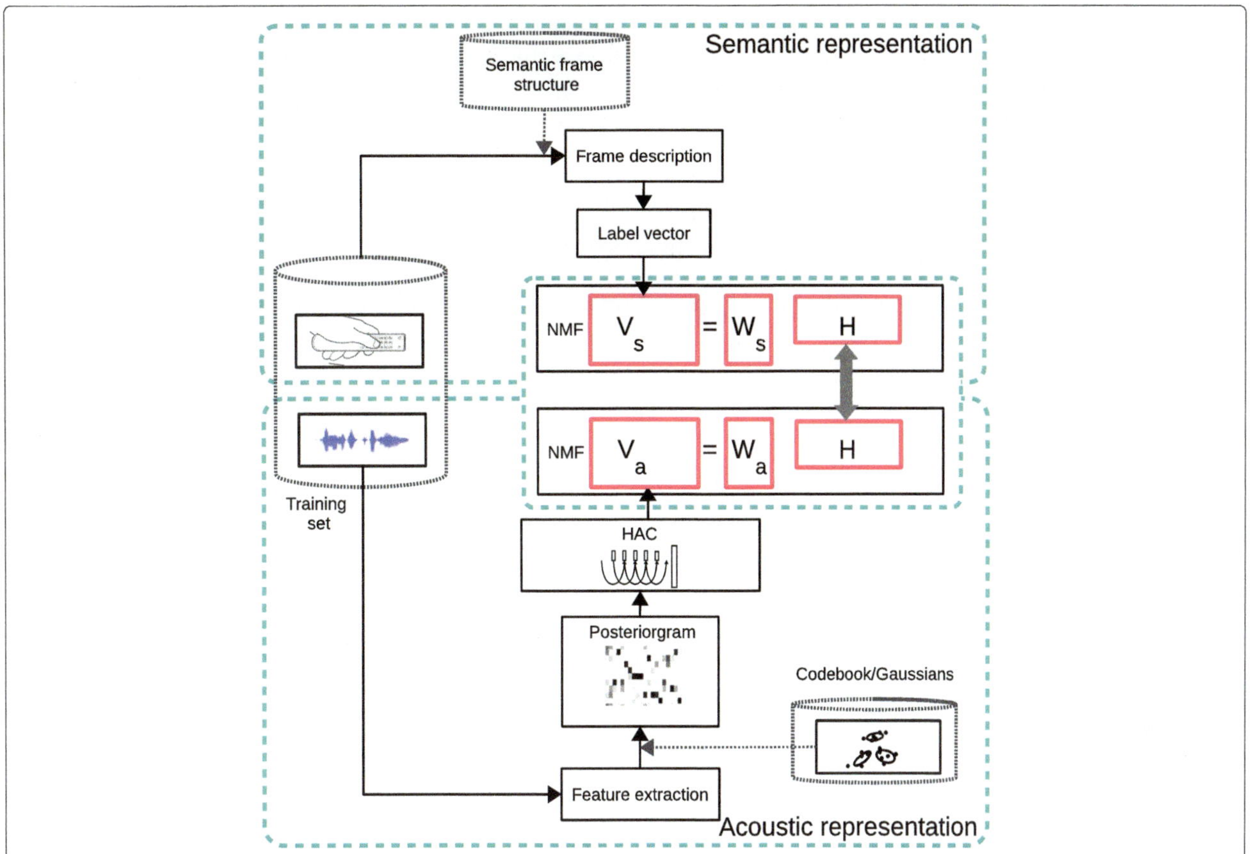

Figure 1 A schematic overview of the learning framework. Two data streams containing acoustic (lower part) and semantic (upper part) features are enhanced and processed in the direction of the arrows towards the centre where they are combined using NMF.

which are represented by the columns in the depicted matrices $\mathbf{W}_s$ and $\mathbf{W}_a$, with the subindex connoting the semantic or the acoustic stream, respectively. The large bi-directed arrow between the two matrices $\mathbf{H}$ indicates that a common matrix is sought for $\mathbf{H}$, and thus common loads on recurrent patterns which are co-occurring between the two data streams. The finding of recurrent patterns, co-occurring between the two data streams, lies at the heart of the learning procedure (cf. the section ?Non-negative matrix factorisation?), where idiosyncratic expressions are parsed and linked to operations on a device. The steps and algorithms are explained with more detail in the following sections.

Semantic representation

A *semantic frame* [19] is a data structure that represents the semantic concepts in a spoken utterance which users are likely to refer when they control a device by voice. Each semantic frame is composed of slots, which in turn contain slots or values. Different commands with a similar structure are represented by the same semantic frame structure but use different slot values. For example, the correspondence between commands like ?Switch off the kitchen light? and ?Switch on the bathroom light! could consist of a switching *action* on the *object*, here ?light?, at a particular*location*. A semantic frame with three slots, labelled by `< action >`, `< object >` and `< location >` is a possible generic structure for parsing such commands. The values in the slots relate to the concepts describing the intended setting. Each slot allows the selection of one value from a predefined list, such as `<on, off, ...>`, `<lights, ...>` and `<kitchen, bathroom, ...>` in this example. The values also relate to the functionality of the devices, and this can be understood as a place holder with the potential capacity to hold a spoken word or phrase referring to a relevant concept. For instance, the semantic frame structure in the example above also covers commands like ?Turn on the light in the kitchen?, where the spoken phrase ?Turn on? is related to the value `< on >`. The challenge is in learning to distinguish the semantic frame and filling in the correct values in the relevant slots, allowing the user to choose his own words and using his own pronunciations.

The semantic frame description of the nth utterance is converted into a binary *label vector*, denoted by the row vector $\mathbf{v}_{s,n}$, indicating the presence or absence of slot values collected in all frames and slots. It is a fixed-length vector with L entries equal to the total number of slot values. Note that multiple slot values are likely to be active in a single utterance and that their presence is highly correlated since the same slot values are usually active in different repetitions of the same command. Sorting all active label entries is a multi-label classification problem as multiple labels are decoded at the same time. For the collection of N utterances in the training set, the semantic representation is composed as $\mathbf{V}_s = \left[\mathbf{v}_{s,1}^T \mathbf{v}_{s,2}^T \ldots \mathbf{v}_{s,N}^T\right]$. A second utterance-based representation is built from the acoustic features as explained in the following section.

Acoustic representation

Feature extraction

The first steps in the feature extraction method are pre-emphasis and windowing followed by the fast Fourier transform. The obtained physical frequencies are rescaled to mel-frequencies which are believed to emulate the frequency scale of the human ear [20], which is approximately a linear frequency spacing below 1 kHz and a logarithmic spacing above 1 kHz. Mel-spectral magnitudes are logarithmically scaled as well and transformed into cepstral coefficients (MFCC features) by using the (inverse) discrete cosine transform. Other standard procedures in the preprocessing phase consists of voice activation detection to remove silence frames, and utterance-based mean and variance normalisation.

Codebook training

The acoustic frames are partitioned into clusters by using a codebook training procedure adopted from [17]. The procedure starts with one cluster and iteratively splits the cluster with the lowest frame sample density into subclusters. The clusters and the frames are repartitioned at each split iteration using k-means clustering. The Euclidean distance between frames was used as distance measure. The procedure continues until the requested number of K clusters is obtained. The codebook training procedure is followed by the estimation of a full-covariance Gaussian for each cluster. The set of clusters is denoted by Φ and $j = 1 \ldots K$ with K the cardinality of Φ. It is evidenced in [21] that speaker-dependent codebook training on smaller training sets is more effective than codebook training using larger training sets with speech pooled from different speakers. Therefore, we opted to use speaker-dependent codebooks for each speaker in this study.

Posteriorgram

A posteriorgram $\mathbf{P}_{t_i,\theta_j}$ is a two-dimensional data structure ($K \quad Q$) containing the posterior probabilities that the observation in the frame at time t_i, with $i = 1 \ldots Q$ and Q the number of frames, is drawn from the cluster θ_j under the assumption of a Gaussian distributed cluster membership. The posteriorgram provides a soft localization of the frame with respect to all the cluster locations in the feature space, and it contains positive values for which only a few are substantially different from zero.

Histogram of acoustic co-occurrence

The posteriorgram of an utterance has a variable length depending on the number of frames in the utterance while a fixed-length vector is required to compose a data matrix that is suitable for NMF. The aim of HAC [22] is twofold. HAC representations allow building fixed-length vectors for each utterance by accumulating the probability of observing the clusters (θ_a, θ_b) over two frames, shifted τ frames away from each other. The number of clusters is a constant; therefore, all possible $K \quad K$ co-occurring combinations for (θ_a, θ_b) are constant too. Secondly, the HAC of a posteriorgram is robust against small temporal variations because the HAC features consist of soft counts of co-occurring frames within small time delays (up to 20 frames in this study). Representations with an absolute time reference like posteriorgrams would be more prone to time-dependent variation, urging the use of time warping algorithms to compute the alignment between two time series. For the nth utterance spanning Q frames, the co-occurrence soft count over a time delay τ for the cluster pair (θ_a, θ_b) in $\Phi \quad \Phi$ is defined as follows (see [23]):

$$\left[\mathbf{v}_n^{\tau}\right]_{(\theta_a,\theta_b)} = \sum_{t_i=0}^{q \quad \tau} \mathbf{P}_{t_i,\theta_a} \mathbf{P}_{t_i+\tau,\theta_b} \tag{1}$$

and $\forall t_i, i = 1 \ldots Q, \sum_{\theta \in \Phi} \mathbf{P}_{t_i,\theta} = 1$.

The HAC is an accumulation of all the Gaussian co-occurrence probabilities denoted by the row vector $\mathbf{v}_n^{\tau}$. The information captured by the HAC depends largely on the chosen delay by which two frames are separated from each other in time. Therefore, multiple time aspects are incorporated by stacking HACs with shorter and longer delays to reach both within and across words and word boundaries. As a result, a large fixed-length column vector is built, denoted by the column vector $\mathbf{v}_{a,n} = \left[\mathbf{v}_n^{\tau_1} \mathbf{v}_n^{\tau_2} \ldots \mathbf{v}_n^{\tau_C}\right]^T$, where C represents the number of HACs. Analogous to the semantic denotation $\mathbf{V}_s$, the acoustic representation is composed as $\mathbf{V}_a = \left[\mathbf{v}_{a,1} \mathbf{v}_{a,2} \ldots \mathbf{v}_{a,N}\right]$ for the collection of N utterances in the training set.

Non-negative matrix factorisation

NMF decomposes a data matrix into the product of two low-rank matrices: one factor $\mathbf{W}$ represents latent structure, which are recurring patterns in the columns of $\mathbf{V}$, and the second factor $\mathbf{H}$ indicates which columns in $\mathbf{W}$ (patterns) are combined to approximate the columns in $\mathbf{V}$. In simultaneous NMF [24], data from different modalities are factorised simultaneously, leading to recurrent patterns in the columns of $\mathbf{W}$ consisting of pattern combinations over two or more sources that coincide with each other. Many names have been used for the same multimodal factorisation algorithm depending on the kind of source material, like for instance joint NMF in [25], or when one stream consists of supervision data, it has been referred to as semi-supervised NMF [26] or weakly supervised NMF [22].

Here, we jointly factorise the semantic and the acoustic representation in order to find the acoustic patterns that co-occur with the active label entries. The joint factorisation of $\mathbf{V}_s$ (cf. the section ?Semantic representation?) and $\mathbf{V}_a$ (cf. the section ?Acoustic representation?) is expressed as follows:

$$\begin{bmatrix} \mathbf{V}_s \\ \mathbf{V}_a \end{bmatrix} \approx \begin{bmatrix} \mathbf{W}_s \\ \mathbf{W}_a \end{bmatrix} \mathbf{H} \tag{2}$$

where $\mathbf{W} = [\mathbf{W}_s \mathbf{W}_a]^T$ and $\mathbf{H}$ are two matrices of lower rank. The co-occurring semantic and acoustic patterns are found in $\mathbf{W}_s$ and $\mathbf{W}_a$, respectively. The nth column in $\mathbf{H}$ describes which co-occurring patterns are active in the nth utterance. The inner dimension in the right half of Equation 2 determines the number of co-occurring patterns in which the dataset is decomposed. It is usually a low number in a small vocabulary task since it reflects the number of slot values L in the VUI. However, by increasing the inner dimension with a number D, columns are introduced in $\mathbf{W}$ to represent patterns for filler words, i.e. recurrent acoustic patterns such as ?please? or ?the? that are usually left out in the semantic representation.

The latent patterns are found by minimising the difference between both sides of Equation 2. Since $\mathbf{V}_s$ and $\mathbf{V}_a$ consist of (soft) count data, the Kullback-Leibler divergence [27] is preferred as loss function and is expressed as follows:

$$\left(\mathbf{H}^*, \mathbf{W}_a^*, \mathbf{W}_s^*\right) = \underset{(\mathbf{H},\mathbf{W}_a,\mathbf{W}_s)}{\arg\min} \, D_{KL}\left(\begin{bmatrix} \mathbf{V}_s \\ \mathbf{V}_a \end{bmatrix} || \begin{bmatrix} \mathbf{W}_s \\ \mathbf{W}_a \end{bmatrix} \mathbf{H}\right) \tag{3}$$

Iterative update rules for minimising a distance measure between the left- and the right-hand side can be found in [27]. It has been demonstrated that convergence is guaranteed towards a local optimum. Note that the loss function in Equation 3 can be seen as a regularisation in which acoustic patterns are preferred that correspond to the occurrences of slot values. Writing the loss function in Equation 3 as a regularised loss function results in

$$\begin{aligned} \left(\mathbf{H}^*, \mathbf{W}_a^*, \mathbf{W}_s^*\right) = \underset{(\mathbf{H},\mathbf{W}_a,\mathbf{W}_s)}{\arg\min} \, & [D_{KL}\left(\mathbf{V}_a || \mathbf{W}_a \mathbf{H}_a\right) \\ & + \lambda D_{KL}\left(\mathbf{V}_s || \mathbf{W}_s \mathbf{H}_s\right)] \end{aligned} \tag{4}$$

with $\lambda = 1$ and $\mathbf{H}_a = \mathbf{H}_s$ for equivalence with Equation 3. If $\mathbf{H}_a$ and $\mathbf{H}_s$ are allowed to be tied loosely, then an additional regularisation term should be added to minimise the difference between $\mathbf{H}_a$ and $\mathbf{H}_s$, which was pursued in [25].

Recognition

The aim of the VUI is to find the frame description for a spoken utterance. A schematic overview of the recognition phase is depicted in Figure 2. Speech processing of a command proceeds from the spectro-temporal representation in the lower part of Figure 2 to the HAC representation in the centre, after which NMF takes place in order to obtain the load matrix $\mathbf{H}_t$ using the learned patterns in $\mathbf{W}_a$ that were co-occurring with the semantic patterns in $\mathbf{W}_s$ in the training phase. $\mathbf{H}_t$ is then transferred to the upper part of Figure 2. The slot value activations $\mathbf{A}$ are found by using $\mathbf{H}_t$ and using $\mathbf{W}_s$ obtained in the learning phase. Finally, the arrow leaving from the box ?Semantic frame structure? indicates that the semantic structure is superimposed on slot value activations as a decision process where groups of slot values are compared and related to each other (cf. the section ?Decision process?). Knowing the correct frame description of the spoken command allows for the proper execution of the command.

Activation

We denote the data matrix and the load matrix in the test phase by $\mathbf{V}_t$ and $\mathbf{H}_t$. The data matrix $\mathbf{V}_t$ contains the processed speech signal, and $\mathbf{H}_t$ is found by minimising the Kullback-Leibler divergence between $\mathbf{V}_t$ and the matrix product of the acquired $\mathbf{W}_a^*$ and the unknown $\mathbf{H}_t$.

$$\mathbf{H}_t^* = \arg\min_{\mathbf{H}_t} D_{KL}\left(\mathbf{V}_t || \mathbf{W}_a^* \mathbf{H}_t\right) \tag{5}$$

Contrary to Equation 3, an optimal solution for $\mathbf{H}_t^*$, given $\mathbf{W}_a^*$, is guaranteed since the loss function in Equation 5 expresses a convex problem. The obtained matrix $\mathbf{H}_t^*$ and the acquired matrix $\mathbf{W}_s^*$ are used to provide the slot value activations in $\mathbf{A}$,

$$\mathbf{A} = \mathbf{W}_s^* \mathbf{H}_t^* \tag{6}$$

Note that the last step in Equation 6 allows the freedom to obtain slot value activation from different latent factors in $\mathbf{W}^* = \left[\mathbf{W}_s^* \mathbf{W}_a^*\right]^T$. A slot value activation can depend on one latent factor or a combination of latent factors in $\mathbf{W}^*$.

Decision process

The decision whether a particular slot value applies or not depends on the ensemble of activations signalling the presence of the related frame, slots and values. The interdependent relation between frames, slots and values is vital information that we use on top of the activations obtained from Equation 6. The whole semantic frame structure is taken into account by implementing a parse

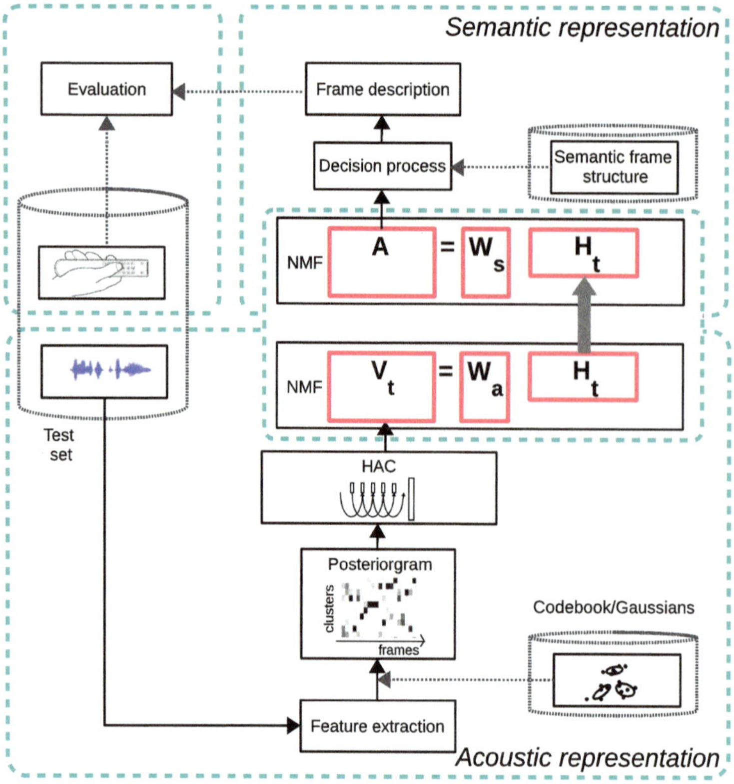

Figure 2 A schematic overview of decoding. Only acoustic data is available, and the processing proceeds from the bottom to the upper part where a decision process takes place to validate the interdependent activations of different words.

tree for each frame and a few activation spreading rules. In the parse tree (see Figure 3), the frames are considered root nodes, slots compose branch nodes, and slot values compose leaf nodes. Activations in leaf nodes correspond with the activations in the columns of **A**, denoted by the vector **a** and representing the NMF activations of one utterance. For each entry in **a**, there is one leaf node[a]. The activation spreading rules enable values in **a** to propagate to slot and frame levels. These activation spreading rules bear on child-parent relations to which we refer as ?exclusive? and ?selective? relations. In *exclusive* relation, only a predefined number of entries, denoted by the constant U, occur at the same time. For instance, the light can be switched on or off, but not both at the same time; therefore, a command can only be assigned one true value ($U = 1$) in the list <on, off>. Generally, if we denote the activation of a parent node by a_p and the activations of its child node by a_{ci}, with $i = 1, \ldots, z$ and $a_{c1} > a_{c2} > \ldots > a_{cz}$, then the following activation spreading rule applies for an exclusive relation: $a_p = \text{median}\left(\{a_{ci} | a_{ci} \geq a_{cU}\}\right)$. A selective relation differs from an exclusive one in the precognition of the number of valid child nodes. If the number of valid child nodes is unknown, then activations are compared against a threshold. The following general activation spreading rule applies for a selective relation: $a_p = \text{median}\left(\{a_{ci} | a_{ci} \geq a_0\}\right)$ and a_0 is a threshold determined by the pth percentile of all activations in **a**. Multiple frame descriptions are in competition, and the frame description with the highest activation in the root node is selected. The U highest activated slots and values for exclusive relations and the slots and values scoring higher than a_0 for selective relations are included in the predicted frame.

A toy example is depicted in Figure 3 where the first nine entries of **a** are set to $[0.4, 0, 0.1, 0, 0, 0, 0.5, 0, 0.1]^T$ corresponding to the frame description listed in the upper half of Table 1. Exclusive relations are depicted as solid arrows, and selective relations are depicted as dashed arrows. The first slot value has activation 0.4 and corresponds to the empty slot value of the frame ?increase heating?. The activation is propagated to the slot level and from there to the frame level. Exclusive relations are presumed for the slots <action> and <object> and their respective slot values. A preset value of $U = 1$ for both slots yields the propagation of the highest activation, $a_p = 0.1$ and $a_p = 0.5$, to the <action> and <object> slots, respectively. The relation between the ?open_close? frame and its slots is evaluated as selective, and by assuming a preset threshold beneath 0.1, their median, $a_p = 0.3$, is propagated to the frame level ?open_close?. Generally, the

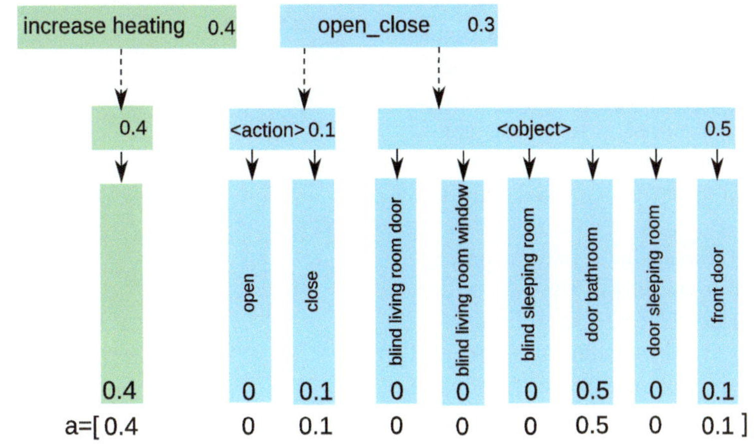

Figure 3 Parse tree and propagation of activation. A parse tree of the first two frame descriptions listed in Table 1 and the propagation of activation depending on exclusive and selective relations. See text for more explanations.

median is an unbiased measure for propagating activations when the number of slots and values differs between different frames. In this hypothetical example, the frame ?increase heating? and ?open_close? have activation 0.4 and 0.3, respectively; thus, the frame ?increase heating? and its selected slots and slot values are the predicted outcome of the spoken utterance yielding the activations **a** in this toy example.

Reference model

We compared NMF learning with GMMs. It is hard to preset the number of GMM components especially when datasets are small and have varying sizes among speakers (see the subsequent section ?Speech material?). Therefore, we investigated four GMMs having 10, 20, 40 or 80 components fixed for each speaker, respectively, with a diagonal covariance structure instead of a full one in order to limit the number of free parameters. The GMMs were embedded in the architecture of our framework in a similar way as the NMF learning module. At the front end, feature extraction was identical up to and including the posteriorgram step (see Figure 1), after which a scaling step was introduced using the logit function, $\log p/(1 \quad p)$, to map probabilities to the real line $\mathbb{R}$. Subsequently, utterances in the data with a common semantic entry, i.e. utterances with ?1? at a particular position in the label vector $\mathbf{v}_s$, were pooled together to compose the training set for GMM estimation of each respective slot value. Thus, for each label entry in $\mathbf{v}_s$, there is one GMM predicting the presence of the respective slot value in the decoding phase. Similar to the NMF activations (see the section ?Activation?), the posterior probabilities are committed to the same decision process (cf. the section ?Decision process?). It should be noted that GMMs do not capture temporal dependencies while HACs do. A GMM can be conceived as a HMM with one state per slot value. By using a HMM with multiple states and tuning transition probabilities, temporal relations among the acoustic features can be captured. However, it is evidenced in [28] that GMMs outperform HMMs in feasibility for small training sets.

Table 1 Semantic frame descriptions for DOMOTICA-3 - compositional

Frame (exclusive)	Slot (selective)	Value (exclusive)
Increase heating	-	-
Open_close	< action >	open, close
	< object >	1-6
Ranged	< *range* >	1,2,3
	< object >	1,2
on_off	< *action* >	on, off
	< object >	1-6

The numbers 1-6 refer to objects such as a kitchen lamp or a bathroom door.

Speech material

Similar to [18], two datasets are employed. The first dataset is PATCOR containing recordings of ten speakers playing a solitaire card game by voice. The second dataset is a recent recorded dataset called DOMOTICA-3 dubbing its precursor DOMOTICA-2 employed in [18]. The utterances consist of commands controlling a home automation system by voice.

PATCOR

The database PATCOR contains recordings of subjects playing the card game ?patience? on a computer, using only spoken commands. The database contains ten speakers with more than 2,000 commands. The data was collected from unimpaired subjects with non-pathological speech, speaking Belgian Dutch. As depicted in Table 2, six participants were females and the age ranged between 22 and 45 years old for almost all speakers except for speaker 9 who was 73 years old. All players played at least four games leading to 254 recorded utterances on average, except for the speaker in the second entry who played only two games.

In order to provoke commands in a natural human-machine-like interaction, a wizard-of-Oz setup was employed for five players as indicated in column 4 of Table 2. In a wizard-of-Oz setup, a subject is deceived to believe that the machine is able to commit responsive behaviour, while in reality, the administrator is taking care of the responsive actions of the machine. The five other players in PATCOR were committed to the same procedure; however, they were told that the administrator took care of all the actions.

The users were free to choose their own words and grammars allowing different expressions for the same card move. A typical utterance in PATCOR is ?Put the four of clubs on the five of hearts?. The standard frame structure of the utterances used in [18] is demonstrated in Table 3.

Table 3 Semantic frame descriptions for PATCOR-compositional

Frame (exclusive)	Slot (selective)	Value (exclusive)
Dealcard	-	-
Movecard	<from_suit>	c,d,h,s
	<from_value>	1-13
	<from_foundation>	1-4
	<from_column>	1-7
	<from_hand>	-
	<target_suit>	c,d,h,s
	<target_value>	1-13
	<target_foundation>	1-4
	<target_column>	1-7

Here, the letters c,d,h and s represent the suits clubs, diamonds, hearts and spades, respectively.

Domotica-3

The DOMOTICA-3 database contains Dutch dysarthric speech commands related to home automation. A typical DOMOTICA-3 utterance is ?turn on the kitchen light?. The dataset contains recordings of the speakers that participated in the collection of the DOMOTICA-2 dataset in [18].

In short, a two-phase data collection method was used in DOMOTICA-2. In the first phase, nine users were asked to command 27 distinct actions (see Table 4) in a 3D home environment on a computer, guided by a visualised and narrative scenario such as ?you enter the kitchen, but it is dark...?, in order to provoke an action, but to ensure an unbiased choice of words and grammar. Consequently, each user produced a list of natural induced commands; thus, nine different lists of commands controlling the same actions were created. Some participants missed out a few actions during the guidance of the narrative scenario, but never more than two. The lists were read repeatedly by multiple speakers in the DOMOTICA-2 data collection.

A selection of 27 actions from the DOMOTICA-2 collection were used in the new recordings of the DOMOTICA-3 database. A recording session lasted more or less a half hour in which the speaker read repeatedly the commands from one of the nine lists (see the fifth column in Table 5). To keep correspondence with the previous and future work, we refer to these speakers by unique IDs. For all adult speakers, speech intelligibility scores were obtained by analysing the recorded speech using the automated procedure in [29]. While a score above 85 is considered as normal speech intelligibility, a score equal to or below 70 is considered as severely impaired. Speaker characteristics are listed in Table 5. Speakers 31 and 37 were children and did not conduct an intelligibility test. Additionally, speakers 43, 44, 46, 47 and 48 were diagnosed as multiple sclerosis patients, and some of them demonstrated adequate speech intelligibility. They were recruited because the digressive nature in time of their speech ability would allow for speech-degenerating data collection in the future. Most speakers were able to generate six or more repetitions of the command lists, except for speakers 31 and 47 who were able to produce one and two repetitions, respectively. A few speakers received a reduced list with ten commands with at least ten repetitions. A larger number of repetitions allow us to investigate whether learning improvements proceed even further by adding more learning examples, or whether it levels off at a particulate stage. The number of utterances

Table 2 Participants in PATCOR

PID	Gender	Age (years)	Wizard-of-Oz	Number of games	Number of utterances
1	♀	33	Yes	6	274
-	♀	41	Yes	2	169
2	♂	45	Yes	4	260
3	♂	42	Yes	5	278
4	♀	23	No	4	222
5	♀	26	No	4	248
6	♂	24	No	4	223
7	♂	26	No	4	240
8	♀	73	No	5	235
9	♀	22	Yes	5	262

Table 4 Synoptic description of all actions in DOMOTICA-3, partitioned in columns according to frame type

Increase_heating	Open_close	Ranged	On_off
Increase_heating	close_blind_living_room_door	dimstate1_floor_lamp	off_light_living_room
	close_blind_living_room_window	dimstate2_floor_lamp	off_light_sleeping_room
	close_blind_sleeping_room	dimstate3_floor_lamp	off_lights
	close_door_bathroom	level1_head_bed	on_light_bathroom
	close_door_sleeping_room	level2_head_bed	on_light_kitchen
	close_front_door	level3_head_bed	on_light_living_room
	open_blind_living_room_door		on_light_sleeping_room
	open_blind_living_room_window		on_reading_light
	open_blind_sleeping_room		
	open_door_bathroom		
	open_door_sleeping_room		
	open_front_door		

is indicated in column 6 of Table 5. The frame description used in [18] is displayed in Table 1.

The database contains speech recorded in realistic environments with two microphones. One microphone was a head-worn set C520, and the other one was a R?DE M2 live condenser microphone, which was located in front of the speaker on top of a table at about 50 to 100 cm. The recordings were held in a room selected in the respective health care centre of the patient which ranged from quiet to some background speech. The recordings were carried out with a sampling rate of 48 kHz and a resolution of 24 bit for each channel, after which it was downgraded to a sampling rate of 16 kHz and stored as such in the corpus. The recordings of speakers 33 and 40 barely reached voice activation levels because the directed microphone of the headset was too far out of reach; however, the recordings on the second channel did succeed.

Hierarchical knowledge representation

An optimal structure depends on different factors like the number of decision steps in the recognition process (cf. the section ?Decision process?) and the complexity for each decision depending on the number of alternatives. These factors are not independent from each other. For instance,

Table 5 Participants in DOMOTICA-3

PID	Gender	Age (years)	Profile	Spoken list number	Number of utterances	Intelligibility score
17	♀	25	Spastic quadriparesis	6	347	88.6
28	♀	42	Severe nasal dysarthria	6	204	73.1
29	♂	44	Spastic quadriparesis	7	174	73.6
30	♂	33	Ataxic dysarthria	5	198	69
31	♂	11		8	225	
32	♀	43	Mild dysarthria and hyperkinetic speech	4	41	65.6
33	♂	33	Ataxic dysarthria, short phonation	3	113	66.2
34	♂	61	Multiple sclerosis	6	331	76.2
35	♀	25	Spastic quadriparesis	4	268	72.3
37	♂	10		8	156	
40	♂	55	Myotonic-flaccid dysarthria	1	184	85.5
41	♀	39	Dysarthria	2	144	64.2
43	♀		Multiple sclerosis	1	133	89.4
44	♂		Multiple sclerosis	9	164	89.2
46	♀	50	Multiple sclerosis	1	97	74.9
47	♂		Multiple sclerosis	7	64	73.4
48	♂		Multiple sclerosis	5	169	85.8

an ordered tree with more levels will induce more decisions, but with lower complexity. Semantic structures allowing for overall simple decisions improve slot value perplexity. In this study, we explore the influence of the semantic frame composition on the recognition performance by considering two different frame structures employing different hierarchical levels and decision rules.

We will investigate two approaches for the validation of frame structure on the database PATCOR; one is the compositional standard shown in Table 3 and employed in [18], and the second one is a hierarchical semantic frame structure with one additional level shown in Table 6. The decision rules for each layer are listed in the column headings. In the standard description, commands are decomposed into parts such as suits, values and columns. In the *compositional* structure, a selective rule is used to compare the activations of alternative slots against a threshold. Known information is left unexplored like for example the impossible co-occurrence of a card moved from the hand and from the foundation. A second structure is called *hierarchical* referring to more levels in which slots contain slots or values. Such a structure alleviates the decision step as the number of alternatives is limited in each layer with no more than two alternatives in selective decisions.

For the DOMOTICA-3 dataset, we employ even more distinctive structures on the semantic representation. The first structure entails the mapping of entire spoken commands to frames without slots or values, leading to a scenario where the machine learning problem reduces to a multi-class paradigm, that is one class for each possible command. Clearly, such a mapping is unattainable for sets with complex commands as in PATCOR, but for a small set of commands, modelling entire utterances is a viable option. Note that such a structure is less robust to

Table 6 Semantic frame descriptions for PATCOR - hierarchical

Frame (exclusive)	Slot (selective)	Slot (exclusive)	Slot (selective)	Value (exclusive)
Dealcard	-	-	-	-
Movecard	< from >	< card >	< suit >	c,d,h,s
			< value >	1-13
		< foundation >	-	1-4
		< column >	-	1-7
		< hand >	-	-
	< target >	< card >	< suit >	c,d,h,s
			< value >	1-13
		< foundation >	-	1-4
		< column >	-	1-7

Here, the letters c,d,h and s represent the suits clubs, diamonds, hearts and spades, respectively.

word order variation, and alternative expressions of the same command as utterances are learned in their entirety. We compare a semantic frame structure with commands modelled in their entirety with a compositional approach which parses commands into parts such as objects and actions [18]. This semantic frame structure is shown in Table 1. The values 1 to 6 refer to objects or devices such as kitchen lamp or bathroom door. Once again, we expect improved performance for the multi-layered frame structure since selective rules are used for layers holding only two slots, while multiple alternatives are gathered in levels with exclusive rules.

Experiments

The goal of the experiments is twofold: first, we test the feasibility of our VUI by evaluating the performance of the framework using the F-score on slot value recognition as defined in [18]; furthermore, we investigate the added value of using a more layered semantic frame structure on two datasets: PATCOR containing commands having a complex grammar and DOMOTICA-3 containing realistic recordings of commands from speech-impaired speakers in the setting of a virtual home automation system.

An important feasibility issue is the speed of learning, which we evaluate by tracking the gain in slot value recognition for incrementally increasing training sets. This procedure allows us to plot a learning curve, that is, the curve representing the average slot value recognition score in function of the average number of learning examples. The rate of learning is usually sharpest in the beginning and gradually evens out against an asymptotic level. We are especially interested in the initial and final phase of the learning curve; on the one hand, the speed of learning should be high so users gain interest in keeping on using the VUI, thus keeping on training the system; on the other hand, the learning curve should not level off to low, that is, the VUI should not get stuck in suboptimal functionality in the long run. Clearly, the speed of learning and the asymptotic performance are important attributes of a useful learning procedure.

Setup

Evaluation procedure

The data was partitioned in blocks containing approximately an equal number of slot values using an algorithm outlined in [18]. This algorithm minimises the Jensen-Shannon divergence between the slot value distributions over all blocks. Likewise [18], block creation was followed by the composition of a Latin square from which the first five rows were submitted to a fivefold cross-validation experiment. In each fold or row of the Latin square, the first x blocks were used as train set while the remaining Z x blocks were used as test set with Z as the total number of blocks. While the train sets increased incrementally with one block, $x = 1, \ldots, Z \quad 1$, the test sets decreased decrementally with one block. The incrementally increasing training sets allowed us to evaluate the learning performance at different time stamps in the learning process of the VUI. The slot values that appeared in each block at least once were used for scoring in the test sets. Note that the real performance of the vocal interface also depends on the interface?s ability to distinguish between commands and other utterances spoken in a domestic environment. Here, we focus on the rate of learning assuming a perfect classification of commands directed to the system against utterances that were not.

For the evaluation of the framework, we excluded the speaker without PID number in Table 2 in PATCOR and speakers 32 and 47 in DOMOTICA-3, due to data insufficiency for block creation. In addition, we created two groups for the DOMOTICA-3 corpus in order to evaluate the feasibility of the framework. Speakers $29, 30, 33, 41$ and 46 have an intelligibility score below 75 and uttered less than 200 commands. We refer to this group as *severe dysarthria*. Note that an intelligibility score higher than 85 is not considered pathologic. Speakers $17, 28, 31, 34$ and 35 were joined in another group because they uttered more than 200 commands allowing us to track the performance of the system in the long run.

Parameters

We used pre-emphasis ($\alpha = 0.97$, sampling rate at 16 kHz) and Hamming windowing with 30-ms frames in addition to a frame shift of 10 ms. Fourteen cepstral dimensions were retained, and the first and second order differences were appended, leading to 42 feature dimensions. Silence frames were removed before the codebook training started, aiming for $K = 100$ clusters from which posteriorgrams were obtained with 100 entries. The main portion of the probability mass in a frame seems to originate from only a few clusters; therefore, we retained only the three highest probabilities in each frame in order to gain computational efficiency by using sparse matrices HAC features. We stacked $C = 4$ HACs with delays $\tau = 2, 5, 9$ and 20 resulting in $4 \quad 100^2 = 40,000$ entries for each utterance-based acoustic representation $\mathbf{v}_a$.

$\mathbf{H}_{\text{init}}$ and $\mathbf{W}_{\text{init}}$ denote the initialisation of $\mathbf{H}$ and $\mathbf{W}$, respectively,

$$\mathbf{H}_{\text{init}} = \begin{bmatrix} \mathbf{V}_s + \lambda\mathbf{A}(R \quad N) \\ \mathbf{B}(D \quad N) + \gamma\mathbf{1}(D \quad N) \end{bmatrix} \tag{7}$$

$$\mathbf{W}_{\text{init}} = \begin{bmatrix} \mathbf{I}(R \quad R) + \lambda\mathbf{O}(R \quad R) & \mathbf{P}(R \quad D) + \theta\mathbf{1}(R \quad D) \\ \multicolumn{2}{c}{\mathbf{Q}(F \quad (D + R)))} \end{bmatrix} \tag{8}$$

with D being the largest integer smaller than $0.2 \quad R$; hence, by way of example, for $R = 40$ slot values, $D = 8$ extra columns were added to $\mathbf{W}$. This proportion was constant for all experiments. The parameters λ, γ and θ were set to $1e^{\ 4}, 0.1$ and 0.2, respectively. All entries in $\mathbf{A}, \mathbf{B}, \mathbf{O}, \mathbf{P}$ and $\mathbf{Q}$ are i.i.d samples from the uniform distribution $\mathcal{U}$ with boundaries $(0, 1)$. $\mathbf{I}$ is the identity matrix and $\mathbf{1}$ is a matrix with all ones. The columns of $\mathbf{W}$ were normalised to sum to one throughout the multiplicative updates to prevent drift towards large numbers reducing the cost function.

Results and discussion

Feasibility

Results on DOMOTICA-3 are shown in Figure 4 as a function of the number of learning examples in the training set. The depicted results concern recordings on the field microphone. The F-scores for the more severe dysarthric group are depicted in the upper panel. The plotted numbers are the PIDs of the speakers (see Table 5) with circle-shaped lighter and darker gray background colours indicating the NMF-based and the 80-component GMM approach, respectively. When we compare GMM-based learning with NMF-based learning in the upper panel, we observe steeper learning curves for the NMF-based learning for the group of severely dysarthric speakers, yielding an average improvement of 23% ($t_{(159)} = 30.2, p < 0.001$). Moreover, a similar trend can be observed in the group with more training material depicted in the lower panel of Figure 4, yielding an average improvement of 20.2% ($t_{(159)} = 38.5, p < 0.001$) by using the NMF-based approach.

We used the non-parametric method in [30,31] to estimate a smooth learning curve for each speaker using the locally weighted scatterplot smoothing (LOWESS) procedure. Optimal smoothness parameters were found by cross-validating different smoothness values between 0.4 and 0.8. We plotted the learning curve for the average speaker using a full blue-coloured and a dashed green-coloured line to indicate the NMF-based and the GMM-based scores, respectively. Furthermore, we constructed 95% confidence limits by bootstrapping the LOWESS procedure, and we indicated these bounds by dotted lines. The curves provide an indication of how the average speaker is expected to perform.

When comparing F-scores, the gain by using NMF learning is higher in the beginning of the learning curve, up to 40% absolutely on average, as can be seen from the difference between the full and the dashed-line averaging curve. For the group with severe dysarthria, we observe that the NMF-based approach yields a score close to 80% on average after only one repetition. For instance, speaker 33 has an intelligibility score of 66.2 and yields an F-score of 70% after one repetition and 96% after nine repetitions.

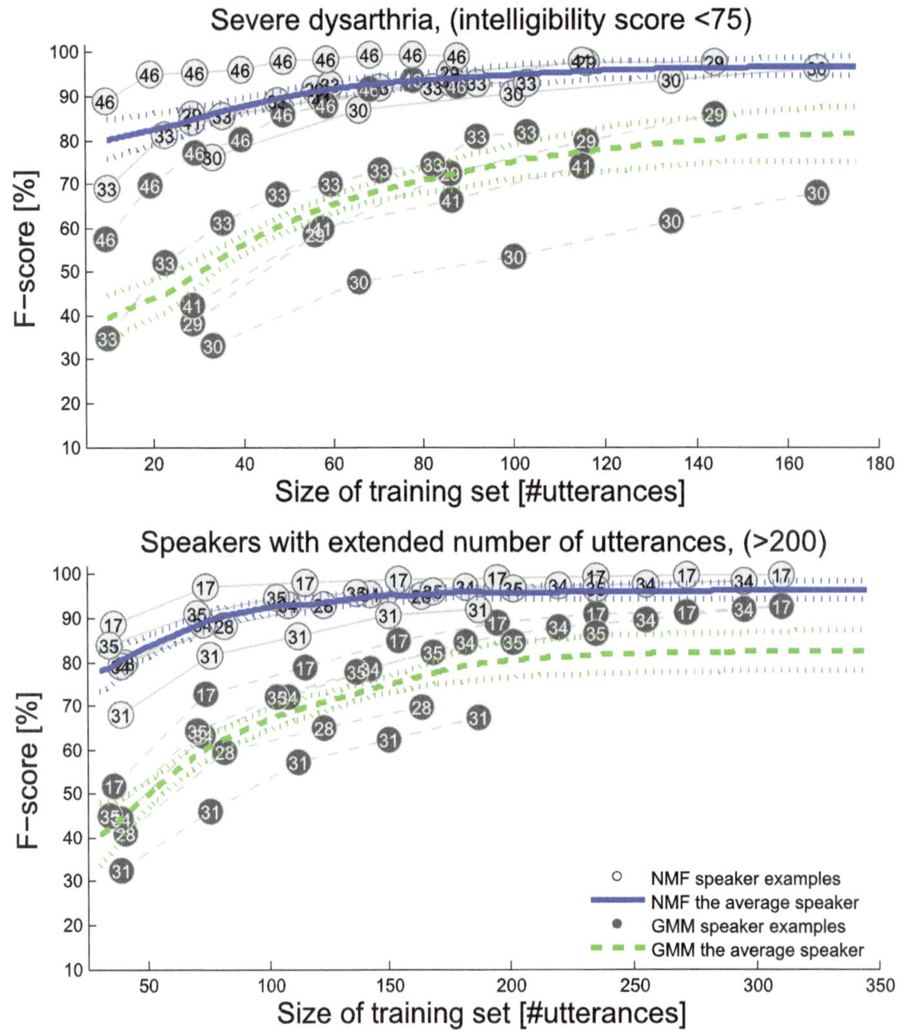

Figure 4 NMF-based learning against GMM-based learning. Upper part: for severe dysarthric speakers. Lower part: for speakers with extended training sets in the lower part. Numbered circles represent PID, and their locations indicate F-scores as a function of the number of utterances in the training sets. Furthermore, the smoothed curves are interpolations of the scattered F-scores using the LOWESS procedure, and they exemplify the performance of an average speaker.

Moreover, some speakers yield scores close to 100% after a few repetitions, like for instance speaker 17 depicted in the lower part of Figure 4 obtaining a score above 99% after four repetitions only. Note that the results using the headset recordings are similar as can be seen in Table 7. These results are very promising, especially for dysarthric speakers, as both the learning rate and the accuracy are already in a range that is usable for a vocal interface. Moreover, all learning curves that did not reach a ceiling performance at the end are still rising, indicating that with more learning examples, the accuracies will probably further improve.

Semantic structure

Here, we compare the results for NMF-based learning using two different semantic frame structures on the PATCOR database depicted in the upper panel and the DOMOTICA-3 database depicted in the lower panel of Figure 5. When comparing F-scores for the hierarchical and the compositional frame structure in Figure 5, we find a small but significant overall improvement for using a hierarchical frame structure instead of a compositional one, i.e. $t_{(179)} = 12.4, p < 0.001$, with an absolute average improvement in F-score of 3.3%. The improvements are fairly consistent among speakers despite the fact that the individual scores for the PATCOR database are wide ranged. The scores are wide ranged because speakers 3, 5, 7 and 8 frequently used the words ?red? and ?black? instead of ?hearts?, ?spades?, ?clubs? and ?diamonds?. While the use of colours such as ?red? and ?black? allows the VUI to distinguish ?clubs? and ?spades? from ?hearts? and ?diamonds?, it will not allow to learn the

Table 7 F-scores after 40 and 120 training utterances for DOMOTICA-3

				Speakers															Average
				17	**28**	**29**	**30**	**31**	**33**	**34**	**35**	**37**	**40**	**41**	**43**	**44**	**46**	**48**	
			GMM 10	60	51	53	41	43	58	52	53	62	43	53	88	45	83	77	57.5
			GMM 80	53	45	47	34	34	65	46	49	56	37	46	87	37	81	79	53.1
			NMF	90	69	83	82	71	87	76	83	84	73	77	99	75	96	98	82.9
	Compo-sitional		GMM 10	75	66	66	51	52	67	68	66	69	57	66	88	75	88	85	69.3
		$N = 120$	GMM 80	80	65	76	56	59	82	74	73	79	69	68	97	78	96	93	76.3
			NMF	99	88	90	93	86	93	91	94	94	92	96	100	99	97	99	94.1
DOMOTICA-3 R?DE M2																			
			GMM 10	43	37	43	23	25	41	38	43	48	29	35	78	62	73	66	45.6
		$N = 40$	GMM 80	24	19	27	18	12	45	23	31	42	16	28	84	57	75	74	38.3
			NMF	88	72	80	76	63	78	79	81	77	71	68	99	98	96	98	81.6
	Flat		GMM 10	65	55	65	39	39	48	61	57	51	51	48	91	88	82	81	61.4
		$N = 120$	GMM 80	65	50	74	49	35	67	59	61	65	53	60	97	97	85	87	66.9
			NMF	98	83	95	94	78	85	90	89	88	86	93	100	100	100	99	91.9
			GMM 10	58	45	54	42	42	40	45	53	60	35	56	87	59	86	82	56.3
		$N = 40$	GMM 80	54	42	49	37	33	51	46	49	58	31	50	87	54	82	82	53.7
	Compo-sitional		NMF	89	80	89	80	69	79	80	85	86	60	88	99	90	96	98	84.5
			GMM 10	74	56	69	48	52	43	63	69	65	46	69	87	74	90	88	66.2
		$N = 120$	GMM 80	79	65	79	57	58	70	74	75	79	55	75	97	89	96	92	76
			NMF	98	92	97	92	86	92	94	96	95	87	98	100	99	99	100	95
DOMOTICA-3 headset																			
			GMM 10	41	29	39	24	24	25	36	41	45	25	37	82	70	73	75	44.4
		$N = 40$	GMM 80	27	18	32	18	13	31	23	30	38	11	30	85	59	75	74	37.6
			NMF	88	79	90	78	63	66	83	87	79	52	84	98	98	98	99	82.8
	Flat		GMM 10	64	48	67	45	38	32	57	55	48	38	61	80	89	80	80	58.8
		$N = 120$	GMM 80	70	45	73	52	39	52	61	58	64	39	63	96	97	82	89	65.3
			NMF	98	88	97	94	82	92	94	92	92	74	98	100	100	100	100	93.4

The F-scores are interpolated using the LOWESS procedure.

difference between the two black or the two red card suits. Since 40% to 50% of the words in the move commands consisted of words referring to the card suits, a drop in overall F-score is observed because the incorrectly recognised card suits are counted as false positives despite the fact that the user did not provide this information in the VUI training. As can be seen in the upper part of Figure 5, there is a considerable gap between the learning curves of speakers 3, 5, 7 and 8 using the words ?red? and ?black? and speakers 2,4,6 and 9 who all preferred the consistent use of the words ?clubs?, ?spades?, ?hearts?, and ?diamonds?. Another reason for the wide-ranged performances is that some users tend to use a lot of synonyms, which we did not anticipate in the NMF-based approach here. More results on the PATCOR database, including results on GMMs, are listed in Table 8. Note that GMMs with more

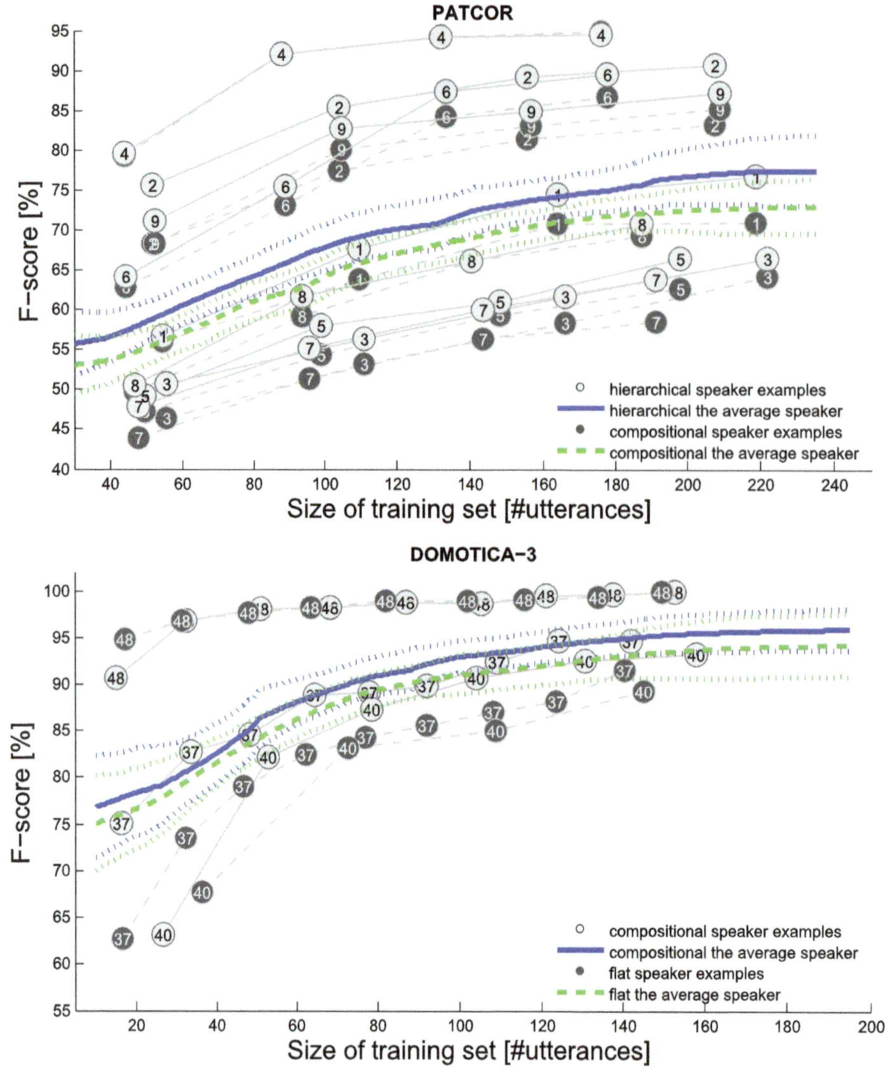

Figure 5 Hierarchical, compositional and flat structures. Hierarchical against compositional frame structure for PATCOR in the upper part, and the compositional against the flat structure for DOMOTICA-3 in the lower part. Numbered circles represent speaker ID, and their locations indicate F-scores as a function of the number of utterances in the training sets. Furthermore, the smoothed curves are interpolations of the scattered F-scores using the LOWESS procedure, and they exemplify the performance of an average speaker.

compounds perform better if there is enough data to adequately fit all free parameters as can be seen in Tables 7 and 8 when comparing GMM scores for small datasets ($N = 40$) against large datasets ($N = 120$ or $N = 175$). As can be seen in the tables, the GMM with 80 components demonstrates better performance than the GMM with 40 components for $N = 175$, but not for $N = 40$. These tables include GMM scores for 10 and 80 components. The GMM results for GMMs with 20 and 40 components are not reported here because these scores are similar to the 80-component GMM scores.

The corresponding results for the DOMOTICA-3 database are depicted in the lower panel of Figure 5, displaying a positive, though, non-significant statistical tendency in favour of the compositional frame structure, i.e. the more profound structure compared to the flat one. The average speaker plot represents scores of all 15 speakers in the database, though the varying range of results is exemplified by three speakers only for reasons of visibility. A considerable number of speakers yield high F-scores in the beginning while other speakers yield lower F-scores in the beginning, but a steeper rise towards the end, as demonstrated by speakers 48 and 37, respectively. The non-significant statistical tendency is probably caused by the ceiling effect, in which a considerable number of speakers have maximum scores for both conditions, making discrimination between conditions more difficult. We verified this explanation by running the same analyses

Table 8 F-scores after 40 and 175 training utterances for PATCOR

				Speakers								Average
				1	2	3	4	5	6	7	8	9	
PATCOR	Hierarchical	$N = 40$	GMM 10	38	46	35	49	36	45	34	36	51	41.1
			GMM 80	40	43	38	47	37	43	33	38	46	40.6
			NMF	55	73	50	79	47	64	47	49	69	59.2
		$N = 175$	GMM 10	38	46	35	49	36	45	34	36	51	41.1
			GMM 80	54	71	44	70	46	58	40	43	62	54.2
			NMF	78	91	63	95	63	90	62	68	87	77.4
	Compositional	$N = 40$	GMM 10	41	47	35	49	36	45	34	36	50	41.4
			GMM 80	39	43	38	48	37	42	32	38	46	40.3
			NMF	53	66	45	79	46	63	42	49	66	56.6
		$N = 175$	GMM 10	41	47	35	49	36	45	34	36	50	41.4
			GMM 80	54	66	44	70	45	62	40	47	61	54.3
			NMF	72	83	61	95	62	87	58	69	85	74.7

The F-scores are interpolated using the LOWESS procedure.

for the overall lower GMM scores, using the same blocks, speech material and semantic structures. When comparing the flat and compositional frame structures, we found a considerable average improvement of 19% after one training block, $t_{(74)} = 9.8, p < 0.001$, and 7% for the maximal number of training blocks, $t_{(74)} = 3.6, p < 0.001$.

We probably obtain a good performance using a flat semantic structure, because the NMF-based acoustic representation is sufficiently distinctive to set each command apart. As a consequence, the more elaborated semantic frame structure becomes redundant. However, when the GMM-based processing flow provides less distinctive representations, information contained in the semantic frame structure becomes vital to the decision process. Nevertheless, overall results are in favour of the hierarchical approach, confirming our hypothesis that using additional knowledge in the form of a hierarchical semantic frame structure is an effective method to boost performance.

Conclusions

This work presents results on the recently recorded dysarthric speech database DOMOTICA-3, with speech intelligibility ranging from normal to severe dysarthric levels. Our NMF-based framework yields 90% to 100% F-score for all speakers, with typically 70% F-score after a single example. These scores validate the use of NMF-based learning as the basis for a self-taught vocal interface for normal and dysarthric speech.

The results on PATCOR and DOMOTICA-3 demonstrate higher asymptotic F-scores by using a more advanced semantic frame structure. The lower scores for the patience card game players, using words like ?red? and ?black? instead of anticipated semantic suit concepts, further confirm the importance of using a semantic structure with more levels similar to categories used in humans. However, the mismatch in user concepts and the concepts that designers had in mind in their applications is considered a weak aspect in our framework in spite of its overall strength. Therefore, we will focus on generic procedures in future work to induce a proper semantic structure. Moreover, further improvements are expected from embedding an algorithm to detect synonyms as alternative referents to the device slot values.

The hierarchical semantic frame structure was superimposed by a decision tree dominating decoded NMF activations. This decision stage can be integrated into the NMF procedure by using group sparsity [32] which obviates the need for a back-end decision stage in future work. All these moderations will boost performance, bringing us one step further in the design process towards a self-taught non-standard speech interface.

Endnote

[a]Note that frames without slots should have a corresponding entry in **a**, which determines the activation scores of the empty slot value.

Competing interests
The authors declare that they have no competing interests.

Authors contributions
All authors made substantial contributions to conception and design, analysis and interpretation of the data, as well as drafting and revising the manuscript. All authors read and approved the final manuscript.

Acknowledgements
The research in this work is funded by IWT-SBO grant 100049.

References

1. G Zon, M Roerdink, *Using Voice to Control the Civil Flightdeck*. (Technical Report, NLR-TP-2006-720, National Aerospace Laboratory Amsterdam, Nederland, 200)
2. R Kuhn, J-C Junqua, P Nguyen, N Niedzielski, Rapid speaker adaptation in eigenvoice space. IEEE Trans. Speech Audio Process. **8**(6), 695?707 (2000)
3. G Potamianos, C Neti, in *AVSP 2001-International Conference on Auditory-Visual Speech Processing*. Automatic speechreading of impaired speech (Volterra, Italy, 2001), pp. 177?182
4. F Rudzicz, in *Proc SLPAT*. Acoustic transformations to improve the intelligibility of dysarthric speech (Association for Computational Linguistics Edinburgh, Scotland, 2011), pp. 11?21
5. P Green, J Carmichael, A Hatzis, P Enderby, MS Hawley, M Parker, in *Proc Interspeech*. Automatic speech recognition with sparse training data for dysarthric speakers (Geneva, Switzerland, 2003), pp. 1189?1192
6. M Parker, S Cunningham, P Enderby, M Hawley, P Green, Automatic speech recognition and training for severely dysarthric users of assistive technology: the stardust project. Clin. Linguist. Phon. **20**(2-3), 149?156 (2006)
7. AM Acrey, Speech recognition in individuals with dysarthria. PhD thesis, Texas Tech University (2012)
8. S-O Caballero-Morales, Estimation of phoneme-specific HMM topologies for the automatic recognition of dysarthric speech. Comput. Math. Methods Med. **2013** (2013). doi:10.1155/2013/297860
9. S-O Caballero-Morales, F Trujillo-Romero, Evolutionary approach for integration of multiple pronunciation patterns for enhancement of dysarthric speech recognition. Expert Syst. Appl. **41**(3), 841?852 (2014)
10. Y Hwang, D Shin, C-Y Yang, S-Y Lee, J Kim, B Kong, J Chung, S Kim, M Chung, in *Computers Helping People with Special Needs*. Lecture Notes in Computer Science, ed. by K Miesenberger, J Klaus, W Zagler, and A Karshmer. Developing a voice user interface with improved usability for people with dysarthria, vol. 7383 (Springer, Berlin Heidelberg, 2012), pp. 117?124
11. P Raghavendra, E Rosengren, S Hunnicutt, An investigation of different degrees of dysarthric speech as input to speaker-adaptive and speaker-dependent recognition systems. Augment. Altern. Commun. **17**(4), 265?275 (2001)
12. F Rudzicz, in *Proceedings of the 9th International ACM SIGACCESS Conference on Computers and Accessibility*. Assets ?07. Comparing speaker-dependent and speaker-adaptive acoustic models for recognizing dysarthric speech (ACM New York, NY, USA, 2007), pp. 255?256. doi:10.1145/1296843.1296899 http://doi.acm.org/10.1145/1296843.1296899
13. M Matassoni, R Astudillo, A Natsamanis, M Ravanelli, in *Proc. Interspeech*. The dirha-grid corpus: baseline and tools for multi-room distant speech recognition using distributed microphones (Singapore, 2014), pp. 1613?1617
14. B Lecouteux, M Cacher, F Portet, in *Proc Interspeech*. Distant speech recognition in a smart home: comparison of several multisource ASRs in realistic conditions (Florence, Italy, 2011), pp. 2273?2276
15. H Christensen, I Casanuevo, S Cunningham, P Green, T Hain, in *Proc SLPAT*. homeService: voice-enabled assistive technology in the home using cloud-based automatic speech recognition (Grenoble, France, 2013), pp. 29?34
16. J Gemmeke, B Ons, M Tessema, J van de Loo, G De Pauw, W Daelemans, J Huyghe, J Derboven, L Vuegen, B Van Den Broeck, H Van hamme, in *Proc Interspeech*. Self-taught assistive vocal interfaces: an overview of the ALADIN project (Lyon, France, 2013), pp. 2038?2043
17. J Driesen, Discovering words in speech using matrix factorization. PhD thesis, K.U.Leuven, ESAT, July 2012
18. B Ons, N Tessema, J van de Loo, JF Gemmeke, in *Proc SLPAT*. A self learning vocal interface for speech-impaired users (Grenoble, France, 2013), pp. 1?9
19. Y Wang, A Acero, Rapid development of spoken language understanding grammars. Speech Commun. **48**(3-4), 390?416 (2006)
20. SB Davis, P Mermelstein, Comparison of parametric representations for monosyllabic word recognition in continuously spoken sentences. IEEE Trans. Acoustics Speech Signal Process. **28**(4), 357?366 (1980)
21. B Ons, JF Gemmeke, H Van hamme, Fast vocabulary acquisition in an NMF-based self-learning vocal user interface. Comput. Speech Lang. **28**(4), 997?1017 (2014)
22. H Van hamme, in *Proc. Interspeech*. HAC-models: a novel approach to continuous speech recognition (Brisbane, Australia, 2008), pp. 255?258
23. M Van Segbroeck, H Van hamme, Unsupervised learning of time-frequency patches as a noise-robust representation of speech. Speech Commun. **51**, 1124?1138 (2009)
24. A Cichocki, R Zdunek, A-H Phan, S Amari, *Nonnegative Matrix and Tensor Factorizations: Applications to Exploratory Multi-way Data Analysis and Blind Source Separation*. (John Wiley & Sons, Ltd chichester, United Kingdom, 2009)
25. Z Akata, C Thurau, C Bauckhage, in *16th Computer Vision Winter Workshop*. Non-negative matrix factorization in multimodality data for segmentation and label prediction (Mitterberg, Austria, February 2?4 2011)
26. H Lee, J Yoo, S Choi, Semi-supervised nonnegative matrix factorization. Signal Process. Lett. IEEE. **17**(1), 4?7 (2010)
27. DD Lee, HS Seung, Learning the parts of objects by nonnegative matrix factorization. Nature. **401**, 788?791 (1999)
28. B Lize, D Katrien, FG Jort, H Van hamme, in *Proc SLPAT*. Comparing and combining classifiers for self-taught vocal interfaces (Grenoble, France, 2013), pp. 21?28
29. C Middag, Automatic analysis of pathological speech. PhD thesis. Ghent University, Belgium, 2012
30. WS Cleveland, Robust locally weighted regression and smoothing scatterplots. J. Am. Stat. Assoc. **74**(368), 829?836 (1979)
31. WS Cleveland, SJ Devlin, Locally weighted regression: an approach to regression analysis by local fitting. J. Am. Stat. Assoc. **83**(403), 596?610 (1988)
32. R Jaiswal, D Fitzgerald, E Coyle, S Rickard, in *Proc at 15th International Conference on Digital Audio Effects DAFx-12*. Shifted NMF with group sparsity for clustering NMF basis functions (York, UK, 2012), pp. 17?21

15

A cross-lingual adaptation approach for rapid development of speech recognizers for learning disabled users

Marek Bohac[1*†], Michaela Kucharova[1†], Zoraida Callejas[2†], Jan Nouza[1†] and Petr Červa[1†]

Abstract

Building a voice-operated system for learning disabled users is a difficult task that requires a considerable amount of time and effort. Due to the wide spectrum of disabilities and their different related phonopathies, most approaches available are targeted to a specific pathology. This may improve their accuracy for some users, but makes them unsuitable for others. In this paper, we present a cross-lingual approach to adapt a general-purpose modular speech recognizer for learning disabled people. The main advantage of this approach is that it allows rapid and cost-effective development by taking the already built speech recognition engine and its modules, and utilizing existing resources for standard speech in different languages for the recognition of the users' atypical voices. Although the recognizers built with the proposed technique obtain lower accuracy rates than those trained for specific pathologies, they can be used by a wide population and developed more rapidly, which makes it possible to design various types of speech-based applications accessible to learning disabled users.

Keywords: Automatic speech recognition; Cross-lingual adaptation; Assistive technology; Speech technology; Atypical voices; Learning disabled; Intellectual disability; Dysarthria

1 Introduction

Millions of individuals suffer from learning disabilities that also affect their speech production. These conditions result in atypical voices that are very difficult to understand even for human listeners, as they may affect one or more of the major language subsystems, including phonology, morphology, syntax and semantics. Focusing on phonology, impaired speech can affect voice timing, pitch, volume, fluency and articulation [1].

Different studies have focused on the nature of such mispronunciations and their impact in intelligibility. For example, [2] shows that impaired speakers have a good control of tone but a diminished discrimination between stressed and unstressed vowels, as well as abnormal production of extremely long or short vowels. In [3], the authors focus on how to measure the intelligibility of atypical voices objectively along different perceptual dimensions.

*Correspondence: marek.bohac@tul.cz
†Equal contributors
[1]Institute of Information Technology and Electronics, Technical University of Liberec, Studentska 2, 461 17 Liberec, Czech Republic
Full list of author information is available at the end of the article

Speech technology enhances the functional and affective experience of technology for many user groups, including people with reading difficulties, hearing and visually impaired, older adults, and people with learning disabilities [4]. One of the main applications of speech technology is voice therapy. For example, [5] presents the PreLingua tool, which aims to train skills such as intensity, tone, vocal onset, phonation time and vocalization.

However, when the phonological disorder is severe, it can be useful to complement the voice therapy with other applications for augmentative and alternative communication. For example, VIVOCA [6] is a voice-input voice-output augmentative communication aid for people with severe impairment. Joode et al. present a detailed study on assistive technologies for people with cognitive deficits including different uses of speech technology [7], and Lancioni et al. provide a review of speech generating devices for augmented communication [8]. Recently, there was a special issue on speech and language processing as assistive technologies [9], which shows the potential interest of this area.

Despite their high potential to help users, these technologies usually address specific disorders. For example, [10] studied the best configuration of parametrization, feature selection and classification techniques for the recognition of stuttered events, while [3] studied different measures of vocal quality, articulation, nasality and prosody of spastic dysarthria. This means that most of the systems have been tailored to specific population groups, which makes them more effective for those users but not so adequate for people suffering from other related disorders.

In this paper, we present an approach to develop speech recognizers for learning disabled users aimed at a general population. Also, we were particularly interested in defining a procedure which is rapid and cost-effective. The existence of affordable assistive technology is an effective mean to ensure full and equal enjoyment of all human rights and fundamental freedoms [11]. To make the development fast and efficient, and to speed up the transition from an experimental tool to a professional program, we decided to employ a modular automatic speech recognition (ASR) system designed at the Technical University of Liberec during the last decade. It can be easily adapted to various tasks, including on-line and off-line speech-to-text transcription [12] and robust voice-command control based on real-time keyword spotting [13]. The system has been originally developed for the Czech language, but later it was ported to other languages such as Slovak, Polish, Croatian or Russian, using a cross-lingual adaptation approach [14].

It is well known that ASR depends on large amounts of transcribed speech recordings in order to estimate the parameters of the acoustic model. Recording such large speech corpora is time-consuming and expensive; as a result, there do not exist sufficient quantities of data for disabled users.

Our proposal uses a cross-lingual adaptation approach in which we use most of the available resources in order to recognize atypical speech. This approach is based on an idea similar to the one used to recognize poorly resourced languages: to use data from a well-resourced source language to estimate the acoustic models for a recognizer in a poorly resourced target language [15,16]. In our case, to use data from typical voices in different languages in order to recognize impaired speech using a small amount of training data from disabled people.

Thus, the general idea is to achieve acceptable results reducing the cost of adapting a model to atypical voices. This can contribute to fostering the development of speech applications and help disabled users to be more actively involved in choosing their assistive technology from a wider range of options.

The rest of the paper is structured as follows. 'Related work' section presents related works, 'Proposed method' section presents our proposal for cross-lingual adaptation of speech technologies, the 'A case study for Spanish disabled users' section shows a case study developing our proposal in which we ported a speech recognizer from Czech to Spanish and then adapted it to Spanish disabled users. The experimental results with this example are discussed in 'Experimental evaluation' section. Finally, in 'Conclusions', we present the conclusions and propose future improvements which may increase the performance of the recognizers developed following our proposal.

2 Related work

As discussed in [6], traditional automatic speech recognition techniques are unsuitable for impaired speech for several reasons: the amount of training material is limited, the training samples are highly variable, and they are very different from voices corresponding to non-disabled users.

Due to their very reduced intelligibility, some authors have addressed the problem of automatic recognition of disabled users by carrying out in-depth studies of the most salient features of some types of atypical voices. For example, [17] studied the predictability of articulatory errors and trained a Bayesian network in order to build an augmented ASR system that considered the statistical relationships between vocal tract configurations and their acoustic consequences. Similarly, [18] focused on aspects of syllabic strength for moderate hypokinetic dysarthric speech.

Other work is not so much focused on the recognition itself, but on facilitating the correction of the errors that the recognizer will presumably make. This way, some authors focus on allowing the users to select between alternative word candidates using *n*-best lists [19], while others propose methods that use different approaches to compute the most probable mismatch taking into account the peculiarities of certain pathologies [20].

Despite their high performances, these approaches are focused on particular disabilities or demand a detailed study of the characteristics of the target users. Other authors have addressed more general-purpose approaches. For example, Hawley et al. proposed an incremental approach in which they collected an initial corpus from each user and employed it to train models of words in a reduced vocabulary of commands [6]. Then, they re-estimated the model using the initial training examples and subsequent examples collected from the users while they were employing an application. The advantage of this proposal is that it does not require expert knowledge about the pathologies or a high amount of atypical voice samples, for which the resources are very limited.

We believe that the challenge of limited resources for atypical voices is similar to the case of under-resourced languages: it is costly and time-consuming to gather and process speech material in both cases, which is one of the major limiting factors for speech-enabled application development [21]. Cross-language approaches allow the common exploitation of acoustical similarities between languages in order to be able to use resources available in different languages for the recognition of a less-resourced one. In fact, this approach has also been used to recognize other types of atypical voices such as non-native or accented speech [22]. There are different ways in which existing models can be used along with new data in a target language, mainly training on multilingual data [23], or cross-lingual adaptation of the acoustic [24] and language [25] models.

Our recent experiments show that it is possible to make a cost-efficient cross-lingual adaptation of speech recognition technologies [14]. The continuous speech recognizer may achieve an accuracy between 65% and 75% when using an acoustic model of related languages (e.g. recognition of Croatian using Slovak, Polish, Russian or Czech acoustic models) and between 80% and 85% when the acoustic model is enriched by (semi-automatically obtained) training data of the target language (e.g. recognition of Croatian using Czech + Croatian mixed acoustic model). In this paper, we propose a method to exploit the benefits of cross-lingual adaptation of speech technologies for the recognition of the atypical speech of learning disabled users.

Our previous work in the development of assistive speech technologies for motor handicapped users [13] showed that assistive speech technologies can improve the living conditions of disabled people (and even help them to find job opportunities). Thus, the availability of methods for rapid and cost-efficient prototyping of speech applications, such as the one proposed in this paper, can be of great help for people who suffer from communication disorders.

3 Proposed method

As the development of speech recognition technologies starting from scratch is a very time- and resource-consuming process, we propose to avoid these costs by means of cross-lingual adaptation. The general idea is to use the resources already created for a source language to ease and accelerate the production of the target language resources. The same idea can be used to adapt existing models for common speakers to the needs of handicapped speakers.

3.1 Cross-lingual adaptation

The proposed method for cross-lingual adaptation consists of three partially-independent steps: grapheme-to-phoneme conversion (G2P), building the acoustic model, and building a vocabulary and the corresponding language model. A scheme of the procedure is shown in Figure 1, where the source language (SL) is the one for which the

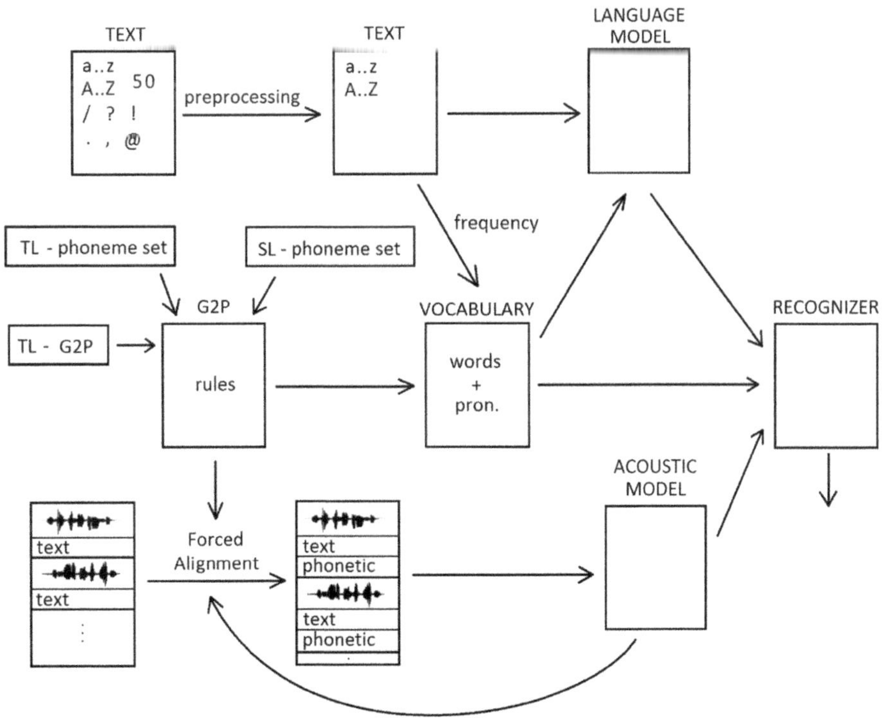

Figure 1 Proposed steps for cross-lingual adaptation from a target language (TL) to a source language (SL).

main resources are available, and the target language (TL) is to be ported to the source language in order to use such resources.

As can be observed, the inputs to the ASR are as follows: the language model in the target language, the (adapted) acoustic model, and a vocabulary in the target language.

G2P conversion determines how the target language words sound in terms of the source language phonetic inventory (*SL - phoneme set* in the figure). It is carried out in two main stages. Firstly, we convert between the target language orthographic form and target language phonetic form (this process is denoted as *TL - G2P* in the figure). Secondly, we decide how to map between the target language phoneme inventory (*TL - phoneme set*) and that of the source language. This involves determining which phoneme pairs (or maybe phoneme groups) are the most similar and possibly which phonemes remain unused.

When the G2P conversion is defined, a fitting acoustic model can be prepared. There exist three alternatives: i) to use the data already available in the SL and mix the model from one or more source languages, ii) mix the SL recordings with some amount of the TL data, and iii) use the TL data only. In the latter case, we can exploit the SL for the development of support technologies in order to lower the demands of expert work (e.g. forced alignment of the TL data using SL models).

Finally, a language model can be built. To do so, it is necessary to gather and pre-process a sufficient amount of textual data in the target language and analyse it to choose a suitable vocabulary. Part of this vocabulary is the phonetic form of the items obtained by the G2P conversion, which can be manually corrected.

3.2 Acoustic and language model

The preparation of training data for the acoustic model is the most time-consuming phase of the adaptation. To reduce the time requirements, we propose to use forced alignment, which consists in assigning time stamps to every word in the input orthographic transcription according to the corresponding audio recording. Forced alignment has many other applications, such as indexing a spoken document for searching, timing subtitles automatically or testing data preparation [26-28]. Processing a document with a forced alignment algorithm requires an audio recording, its corresponding text transcription, an acoustic model (fitting the phonetic inventory used in the phonetic transcription) and a vocabulary containing all words in the document or a G2P conversion module. Eventually, there can be some more input resources to process more complicated tasks (e.g. processing of numerals, physical units, degrees and titles, and special symbols like @, &, or %).

The forced alignment tool that we use is described in [26] and is based on the continuous speech recognizer in the source language. The language model is very constrained as all words must appear strictly in the correct order. They can either fit one of its acoustic forms or be skipped. After the recording is 'recognized' a special post-processing takes place which corrects the different errors that may arise (e.g. when the textual transcript somehow differs from the audio content). As this is a complex process, we encourage the readers to see the details in [26].

Forced alignment is particularly advantageous for the adaptation to atypical voices, as the quality of the acoustic model required can be much lower than the model demanded by the continuous speech recognizer and still it is possible to obtain very accurate results. Moreover, our approach was developed to process inaccurate transcriptions as it can handle 'low quality' acoustic models, i.e. non-robust models trained with a small amount of data [26].

The preparation of the language model does not demand much manual work. The most sensible step is to find a suitable source for the text in the language model. Also, usually there are many characters that are not appropriate for training a language model, there are several ways in which they can be handled, for example, some of them may be erased (e.g. '.', '?'...), others can be rewritten (e.g. numerals) and some of them can be unified and then rewritten (e.g. brackets). Once the text has been processed, a bigram statistical language model is automatically created.

3.3 Adaptation to atypical voices

Every application for disabled users demands a high level of adaptation and customization. Some of these enhancements can be done by the system developers, while others demand active cooperation between the final user and the developers (or a trained assistant). The main enhancements we can enlist include the following: i) training the acoustic model using the data recorded with a subset of final users, ii) general changes in the G2P so it covers some typical speech disorders, iii) adaptation of the acoustic model for the concrete speaker or environment, and iv) reasonable vocabulary limitation (usually context-dependent).

Some users are able to cooperate further (e.g. motor handicapped) but some are not (e.g. severe intellectual disability). If the users are able to cooperate, it is possible to enhance the described adaptation scheme by recording 10 to 60 min of additional recordings in the environment in which they will more frequently use the application and redefine the keywords so that they are easier to pronounce [13]. Finally, users can be offered some training to help them master the assisting technology and improve the performance [29].

Once the cross-lingual adaptation is carried out, a second step is performed to adapt the recognizer trained with common voices to detect impaired speech. Manually sorting and transcribing the recordings is very time-consuming. There is no other choice in the case of longer utterances (sentences), but we propose an automated approach to choose and prepare the isolated word recordings (even if it may imply losing some data).

Our solution requires the audio recording, expected transcription of the recording, and a G2P tool. The solution must be robust to face different phenomena. For example, as some of the users may have reading difficulties, they can be prompted by an assistant to repeat a word that he/she has previously read aloud. This may lead to recording the annotated word more than once. Also, the observed pronunciation may strongly differ from the G2P one. All these obstacles are solved by setting the forced alignment tool properly. We suggest changing the forced aligner language model so the words may be repeated. The phonetic inventory of the forced aligner can also be enhanced by a set of rules modelling the most common pronunciation distortions observed in the data set, as well as the totally mispronounced words (e.g. when omitting groups of phonemes).

Once the recordings are aligned, we select the data suitable for training. If the detected pronunciation differs slightly from the presumed one, we use it or we do not use it otherwise. The similarity is checked using the minimum edit distance (MED) algorithm [30]. The MED algorithm aligns the reference and result sequences in the terms of hits, substitutions, deletions or insertions needed to transform one sequence to the other. We set the rule where a word containing N phonemes in the reference transcription ($N > 3$) has to reach at least $N - 2$ hits to be accepted for training, a heuristic that we found to be appropriate from our previous work [31]. Once the data has been selected, they are added to the training set and the whole procedure is repeated in a second iteration.

4 A case study for Spanish disabled users

In previous work [14], we have successfully ported different ASR models from different Slavic languages, such as Polish, Croatian, Slovak or Russian, to Czech. In this paper, we propose to use a similar approach to build a Spanish model with our ASR system using the resources available for Czech (source language) and Spanish (target language) and adapt it for the recognition of impaired speech (adaptation of the target language).

Although Czech and Spanish are not as similar as the Slavic languages considered in our previous works, we can exploit the Czech resources to speed-up the preparation of the Spanish training data (which is a very time-consuming task). Once we have sufficient amount of the Spanish data, we can leave out the original Czech data and use the Spanish resources only.

The ASR system we have used was originally proposed for processing Czech, a highly inflective Slavic language [12,32], so it supports large vocabularies (it can operate with 500,000 vocabulary items in the on-line mode). The inputs are converted via the FFmpeg codec to the standard 16 kHz pulse-code modulation (PCM) wave format (16 bits per sample), this way the system supports most audio (and video) input formats. The parametrization uses standard 39-MFCC vectors computed on 20 ms frames with 10 ms overlap, and the feature vector is processed with the cepstral mean subtraction (CMS) normalization.

We cover 42 phonemes and 8 non-speech events (e.g. click, breathe, silence, hesitation). The output of the recognizer comprises the written form of the detected word, the detected phonetic form (words usually have several phonetic alternatives), and the time stamps (beginning and end of each word and noise). Additionally, it can be used on-line, when it can also run a post-processing that formats the output as it was pronounced.

Continuous speech recognition is very powerful, but it can also be very demanding for disabled users who might not be able to pronounce a whole sentence correctly but still be able to say it word by word. That is why we employ a keyword spotter (KWS), which is a speech recognition technology used for the detection of isolated words of interest (keywords) from an audio stream. Typical applications include smart homes, industrial enhancements, making audio-archives accessible, or security purposes. The specific implementations and their accuracies may differ between on-line and off-line applications as shown in [31].

As the algorithms mostly rely on the acoustic similarity between keywords and features of the audio stream, we must pay attention when choosing the keywords. If we were interested in detecting acoustically similar words or words that are substrings (one word is included in the other) the system will raise many false alarms or confuse the output. This problem is particularly significant in Slavic languages where the words often differ in the ending only [33].

As we demand the ability of on-line response, we use our KWS system derived from the continuous Czech speech recognition system described previously. It can be used in both on-line and off-line modes. Another advantage is that this KWS uses the same parametrization and acoustic model as the continuous speech recognizer. The difference is in the language model and the vocabulary. It employs phoneme-based n-grams to model speech and build the filler model. Such defined fillers compete with keywords from the vocabulary and with the non-speech events and noises. The performance is controlled by language model parameters and penalties so we can tune the ratio between false alarms and missing detections.

As indicated in 'Adaptation to atypical voices' section, we propose to constrain the minimal length of a keyword to three phonemes and the minimum difference between keywords in at least two phonemes. This ensures there will be no substrings in the vocabulary but still there can be false alarms caused by a combination of two (or more) words in the audio stream that form together one of the keywords. This problem does not appear in the case of isolated word utterances.

4.1 Cross-lingual adaptation from Czech to Spanish

To build the Spanish acoustic model we used the Albayzin [34] corpus. The corpus comprises two sub-corpora with 6,800 utterances each: one based on texts extracted from novels and the other based on queries to a geography database. The utterances were recorded under good acoustic conditions (quiet offices, with the same set of professional microphones) and were pronounced by 304 speakers (152 female, 152 male), whose age varied from 18 to 55 years. The Albayzin corpus represents 12 h 52 min of annotated speech in 13,600 gender and phonetically balanced sentences.

To obtain the training data, we carried out the following preparations. The first step was the conversion of the original audio data from original .ses format to the .wav format - we used (16 kHz, mono 16 bit per sample PCM). From the transcriptions, we removed the punctuation marks, replaced the numbers with their word forms (e.g. 512 = 'quinientos doce'), and processed some special symbols (mainly units of areas or distances). To annotate noises, we employed our forced aligner module (see 'Acoustic and language model' section), which was able to detect and annotate the noises and select the best alternative phonetic representation for each word.

For the cross-lingual adaptation, we employed the proposed G2P conversion in two stages. The first stage was the conversion from the Spanish orthographic form to the Spanish phonetic form. In order to carry out the conversion, we used several rules that varied from unigrams to trigrams. The converter sequentially parsed the orthographic form finding the longest fitting rule. Then, the output was slightly modified to reflect the voicing assimilation and some more phenomena by a set of regular expressions.

The second stage was a substitution of the phonetic inventories: we substituted the phonemes of the target language (Spanish) with the phonemes contained in the phonetic set of the source language (Czech). As discussed before, we had already done this between Slavic languages [14] where the phonetic inventories were quite similar. As shown in Table 1, for the adaptation between Spanish and Czech, we substituted 30 Spanish phonemes with 28 Czech phonemes. So, not all the Czech phonemes are used, and a few Spanish phonemes were substituted by the same Czech phoneme. The phonetic inventory can be optimized after the prototyping phase is finished.

Table 1 The proposed mapping between Spanish and Czech phonemes

Spanish phoneme [adapted X-SAMPA] → Czech phoneme [PAC]				
z→s	d→d	ñ→ň	J→j	á→á
k→k	f→f	p→p	é→é	u→u
C→č	j→X	r→r	í→í	o→o
g→g	l→l	s→s	ó→ó	i→i
y→Č	m→m	t→t	ú→ú	e→e
b→b	n→n	x→ks	0→0	a→a

For example, the conversion of the Spanish sentence 'Guillermo y Yolanda practicaban ciclismo con Jaime' to Czech phonetic form is as follows:

- ESP_{text}: guillermo y yolanda practicaban ciclismo con Jaime
- ESP_{phon}: giyermo i yolanda praktikaban ziklismo kon jaJme
- ESP_{PAC}: giČermo i Čolanda praktikaban siklismo kon Xajme

Where the rules used for the first three items in the sentence are as follows: 'gui' → 'gi' ; 'll' → 'y' ; 'e' → 'e' ; 'r' → 'r' ; 'm' → 'm' ; 'o' → 'o' ; ' y' → 'i' ; 'y' → 'y' ; 'o' → 'o' ; 'l' → 'l' ; 'a' → 'a' ; 'n' → 'n' ; 'd' → 'd' ; 'a' → 'a' ; ' ' → ' '. As can be observed, rules may differ if the beginning or end of a word is encountered.

To build a language model and vocabulary for continuous Spanish recognition, it was necessary to retrieve and process a large amount of Spanish texts. As we were interested in rapid development, we used daily Spanish and international news from different web pages. We downloaded 11.7 GB of texts from http://elpais.com/, http://www.20minutos.es/, and http://spanish.news.cn/.

As the text corpus was gathered from downloaded articles, we carried out a careful post-processing to prepare the corpus for training the target language model (statistical bigram model of Spanish). This way, we used different scripts to remove all HTML tags, foreign (non-Spanish) characters, English words, and other parts of the text that were not suitable for our purpose (e.g. currency rates, information from stock exchange and sports results). We also replaced all numbers with their orthographic transcription. For example, instead of 'in 1926' we had 'in nineteen hundred and twenty six' (in Spanish: 'en 1926' - 'en mil novecientos veintiséis').

Once the data was processed, we computed the bigram language model. For preparing the vocabulary, we employed all words that occurred more than 10 times in the corpus. We decided to use collocations (several words that usually go together - for example 'Los Ángeles') for definite and indefinite articles (e.g. 'el pan', 'un profesor') as short items are disadvantageous for speech recognition. Using this approach, we generated a vocabulary with 54,217 words and word collocations.

4.2 Acoustic models generated

As we trained and compared several acoustic models, we decided to list them here together with their features. We have quantified the amount and sources of training data, so they can be easily compared. In all the models, the Spanish data are used twice - first with floating CMS, then with CMS computed over all the recordings.

AM_CZ is the model made from 200 hours of Czech recordings already available from our previous work [14] (as previously discussed, we consider Czech our source language).

AM_cz&ES_cross denotes five different models. All consist of 1.45 h of Czech recordings (chosen to cover the Czech phonemes not used in Spanish phonetic inventory) and approximately 10 h of Spanish data from the Albayzin corpus (for details, see 'Cross-lingual adaptation from Czech to Spanish' section).

AM_CZ&es denotes the acoustic model consisting of 133 h 24 min of Czech recordings, 12 h 52 min of continuous Spanish speech (whole Albayzin corpus), and 69 min of isolated words uttered by disabled people.

AM_cz&ES consists of 1 h 27 min Czech data and all the Spanish data mentioned in *AM_CZ&es*. The Czech training data mostly covers phonemes unused in the Spanish vocabulary (as we experimented with the best pairing of Czech and Spanish phonemes). This can be considered equivalent to a 'Spanish only' model.

AM_cz&ES_2 is similar to *AM_cz&ES* but including more isolated words uttered by the disabled people (140 min).

AM_cz&ES_sent consists of the *AM_cz&ES_2* and 28 min of continuous speech uttered by disabled people.

4.3 Adaptation to Spanish atypical voices

To gather a corpus of impaired speech, we have worked together with two associations of people with learning disabilities in Southeastern Spain: JABALCÓN[a] and APAFA[b]. Both associations are based in mainly rural areas and have around 100 affiliated persons. They work for the social integration of the learning disabled through different programs of professional development such as wood workshops. People in these associations are mainly adults from the towns and villages nearby who visit the centres during the day.

Due to the characteristics of the users, the recordings had to be carried out in special conditions. Firstly, we had several meetings with the professionals who work with the users on a daily basis, in order to select the individuals that would participate in the recordings. The selection was carried out according to the following criteria:

- To select subjects with a wide range of phoniatric problems.
- To select only subjects for which the participation in the recordings would not impact their wellness, as certain disabilities imply that a change in the person's agenda can be very disruptive.
- To select only subjects who were willing to participate voluntarily with the consent of their families and/or caregivers.

Following this approach, 42 subjects were selected. As the subjects could not participate in long recording sessions, we had to make several visits to the associations to make the recordings. During these visits, we carried out three types of recordings: single words, sentences, and conversations. The first group of recordings were frequent words from their daily activities from a vocabulary that was agreed with their caregivers and categorized in to the following six scenarios: 'street' (street, coin, house…), home (bed, sofa, table…), food (apple, meat, fork…), 'family' (father, mother, sister…), 'dressing' (trousers, jersey, coat…), and 'me' (cold, happy, hungry…). The second group were basic sentences containing words in this vocabulary (e.g. 'The fork is on the table', 'I am cold', 'Open the door'…). Finally, the third group was comprised of open conversations about their daily activities: activities inside the association, visits around the village, sports (especially football), summer activities and family.

During the recording sessions, there was an assistant from our team, a caregiver from the association, and the subject being recorded. Although the sessions were planned to be adapted to the subjects, they could be interrupted if the subject was tired or was not willing to continue with it, or if the caregiver decided for some reason to stop the session. In the first two groups of recordings, the assistant or the caregiver (depending on the subject being recorded) would read the word or sentence aloud and the subject would repeat it. Our aim was to record only the subject, but in some cases they would start speaking while the other person was reading aloud, which forced us to process the corpus. In the last group, the assistant asked them questions that they could respond to without restrictions. In this case, the questions were not always the same, as the idea was to facilitate conversation-asking questions about the topics for which each subject seemed to be more talkative. The recordings were not cut and the whole conversation was recorded, from which we discarded the bits not corresponding to the disabled speakers. Table 2 shows a summary with the number of recordings of each type.

The recordings took place in the associations, so that the participants did not have to travel and the recordings contained the environmental noise that would also surround the users when employing the generated recognizers. Due to the high number of activities carried out by the associations, the same rooms were not always available during the recording sessions, also the recordings were done at different times of the day, so the levels of acoustic noise vary and in some cases, there appear some events that produced louder noises such as opening/closing doors. We believe that these situations are desirable to build acoustic models that consider the noise that will be present during the usage of the speech recognizer. State-of-the-art databases of impaired speech (some of them are described in [35]), are usually recorded in laboratory conditions, which makes it more difficult to employ the recognizers trained with them in real settings.

With respect to the single words, the accepted data (chosen by the automatic classification) were used for training a new acoustic model, and we repeated this procedure with the adapted model in several iterations. In the first round, we used 6,156 recordings to obtain 2,268 words suitable for training. The second round (with the improved acoustic model) chose 921 more words for training. Then, from the other 6,900 recordings, we chose other additional 3,300 samples.

With respect to the sentences, they consist of 45 min of annotated speech in 940 utterances. We automatically prepared the phonetic annotations using the speech, its orthographic transcription and the G2P technique described in 'Cross-lingual adaptation' section, and corrected them as well as the orthographic transcriptions in the cases in which they were inaccurate (i.e. they did not correspond to what was pronounced). For adapting the acoustic model, we employed 618 utterances with 28 min of speech.

Finally, the conversations were split by speakers. We used only the parts with disabled speakers. From 47 min of conversation between disabled speakers and moderators, we separated 16 annotated min for test purposes.

Table 2 Number of elements and impaired subjects recorded

Group of recordings	Number of recordings	Number of users
Single words	13,056 words	42 users (30 male, 12 female)
Sentences	940 utterances (45 min)	18 users (13 male, 5 female)
Conversations	47 min	18 users (13 male, 5 female)

5 Experimental evaluation

To evaluate our proposal, we have carried out different experiments corresponding to the recognition of Spanish disabled users with Czech as the source language (the case study described in the 'A case study for Spanish disabled users' section). Concretely, we have studied two speech recognition technologies (keyword spotter and continuous speech recognizer) using the different acoustic models described in the 'Acoustic models generated' section and also varying the vocabulary and language models.

5.1 Evaluation metrics

We have used four evaluation metrics: ACC, DR, FA, and CORR. ACC stands for accuracy (Eq. 1), DR stands for detection rate (Eq. 2), FA stands for false alarm rate as the number of false positives per keyword per hour (Eq. 3), and CORR stands for correctness (Eq. 4). In the following equations, TP means true positives (correctly detected items), FP means false positives (false alarms), FN means false negatives (items that were not detected), N_{kw} denotes the number of keywords in the vocabulary, Dur stands for the total duration of recordings, and N_{rec} is the number of words in the reference transcription that really appear in each audio recording.

$$\mathrm{ACC} = \frac{\mathrm{TP}}{\mathrm{TP} + \mathrm{FP} + \mathrm{FN}} \times 100\ [\%] \tag{1}$$

$$\mathrm{DR} = \frac{\mathrm{TP}}{\mathrm{TP} + \mathrm{FN}} \times 100\ [\%] \tag{2}$$

$$\mathrm{FA} = \frac{\mathrm{FP}}{N_{\mathrm{kw}} \times \mathrm{Dur}}\ [1/\mathrm{kw/h}] \tag{3}$$

$$\mathrm{CORR} = \frac{\mathrm{TP}}{N_{\mathrm{rec}}} \times 100\ [\%] \tag{4}$$

5.2 Keyword spotter for common speakers

This experiment was the first to be done. As we needed to decide which ratio between Czech and Spanish training data was the best for training the acoustic models, we made a comparative experiment between the *AM_CZ*, *AM_CZ&es* and *AM_cz&ES_cross* acoustic models (Table 3). The latter represents a group of five models that were cross-verified: we split the Albayzin corpus into five equivalent subparts and used one for testing and the remaining four for training. From the text transcription, we chose all the words without substring occurrence (every vocabulary item had to have at least three phonemes and differ from each other in two phonemes at least). The main idea behind this rule is that we need to distinguish completely mispronounced words and somehow distorted words. Given a word that should

Table 3 Comparison of acoustic models with different balance of Czech and Spanish training data evaluated over typical Spanish voices

Acoustic model	ACC (%)	DR (%)	FA [1/kw/h]
AM_cz&ES_cross	54.10	85.88	1.94
AM_CZ&es	52.30	81.42	2.03
AM_CZ	45.75	57.00	0.46

be pronounced as x and was pronounced as x', the N-2 threshold means that there are only N-2 out of N correctly pronounced phonemes in x'. This usually implies that x' (e.g. 'boleto'/ticket) is a completely different word from x (e.g. 'coleta'/ponytail) and thus, it is not as restrictive as it may appear.

In average, there were 350 items chosen for the spotting. In this case, we spot the words from a continuous speech so we also checked if the combination of words could substitute a vocabulary item. The phonetic forms were generated automatically by the G2P module (only one alternative for each item). As our work is focused on disabled speakers who have troubles with spoken communication, we set the system for high DR (even if it implies possibly higher FA). Although false alarms may raise errors, it is possible to recover from them in the application that employs the ASR by establishing interaction contexts or scenarios, and also providing *N*-best lists from which the users may select the most appropriate response. This may be more suited for disabled users than having a recognizer with a lower DR that gives the impression to be not responding to the user's inputs.

From all points of view, the *AM_cz&ES_cross* achieved the best results. This was a surprise for us because when we were porting Slavic languages, it was advantageous to use more data (even from relative languages and not the source or target one) as it guaranteed the robustness of the model. In this case, when porting a Roman language, the mapped phonemes were so different that it was better to use the Czech data only for uncovered phonemes (phonemes not used in Spanish). The *AM_CZ* model had low FA, but the DR was insufficient. The results of this experiment (with almost no Czech data) are very promising in comparison with our former work with Czech KWS [31] so we decided to minimize the usage of Czech acoustic data in the acoustic model training.

5.3 Keyword spotter for disabled speakers

We made two groups for these experiments: pack1, with 131 items in the lexicon, and pack2 with 177 items. We launched the KWS using the *AM_cz&ES* acoustic model and created two lexicons for each group: the initial lexicon (*pack1_base* and *pack2_base*) and a lexicon with alternative phonetics automatically obtained with a set of rules based on expert observation of the disabled users (*pack1_alter* and *pack2_alter*). Phonetic conflicts and similarities were revealed by the MED algorithm [30] and corrected. As there was a large portion of data that was not marked as suitable for training (approximately one half), we decided to test if there is a difference between the recognition of suitable (*suit*) and unsuitable (*unsuit*) data. The results are shown in Table 4.

As can be observed, the results are not very encouraging but we must realize there are only 69 min (approximately 35 min of pure speech) to adapt the acoustic model to disabled speakers. However, we can state that the algorithm that chooses the training data was correct, as it was able to correctly discriminate the unsuitable data from the suitable samples. We can also clearly see the impact of the alternative phonetics in the vocabulary (labelled *alter*). The only drawback is the increase of the false alarm rate. But as stated before, in this domain, it is preferable to have a higher detection rate than to lower the false alarm rate.

Even though the ratio between suitable and unsuitable data is practically the same in *pack1* and *pack2*, the results of *pack2* are better. The reason lies in the vocabulary (although there are 35% more items in *pack2*, there are less phonetic conflicts). That is why we decided to split the whole vocabulary into the six scenarios described in the 'Adaptation to Spanish atypical voices' section so only a subset of the vocabulary is used each time (street, home, food, dressing, me, or family). The results are shown in Table 5. As can be observed, narrowing the vocabulary helped to obtain much better recognition results.

We also wanted to verify that increasing the amount of the training data pronounced by the disabled speakers improves the KWS performance. We ran the KWS with *AM_cz&ES_2* over the *pack2* recordings and the accuracy was 25.30% when using one phonetic only and 29.60% when using the alternative phonetics. This shows the negative impact that the reduced amount of utterances by disabled speakers had in the previously discussed experimental results.

Table 4 Impact of alternative phonetics and recordings' pronunciation quality of the atypical voices

Test data set	ACC (%)	DR (%)	FA [1/kw/h]
$pack1\_base_{all}$	07.60	08.10	2.75
$pack1\_base_{suit}$	13.60	14.90	3.80
$pack1\_base_{unsuit}$	00.80	00.80	1.70
$pack1\_alter_{all}$	08.80	10.10	5.98
$pack1\_alter_{suit}$	15.60	18.30	7.25
$pack1\_alter_{unsuit}$	01.20	01.40	4.71
$pack2\_base_{all}$	11.80	12.40	1.56
$pack2\_alter_{all}$	14.10	15.80	4.15

Table 5 Impact of dividing vocabulary into scenarios - atypical voices

Scenario	ACC (%)	DR (%)	N_{kw}
Street	23.20	25.27	47
Home	25.00	28.54	78
Food	24.40	27.54	77
Family	30.70	30.77	6
Dressing	23.10	26.07	65
Me	21.70	22.96	22

Our last experiment with the KWS was the evaluation of the possible improvements gained via speaker (and simultaneously channel) adaptation, as the parameters of the recording differed between the Albayzin corpus and the devices used to record the disabled speakers. We used the *pack2* data which were found suitable for training to make the adaptation and compared the results when using *AM_cz&ES* with and without the adaptation.

Maximum likelihood linear regression (MLLR) [36] was employed for adaptation. We adapted the constrained version of this method, known as constrained MLLR (CMLLR) [37], where the transformation matrix applied for adaptation of means has to be the same as the one used for adaptation of variances. Therefore, the adaptation can also be performed in the feature space and the adapted feature vector $\hat{\mathbf{o}}$ can be expressed as

$$\hat{\mathbf{o}} = \mathbf{W}\xi \tag{5}$$

where $\mathbf{W}$ is the extended transformation matrix, $\xi = \left[\, \omega \; o_1 \; o_2 \; \ldots \; o_n \,\right]^\top$ is the extended vector of features, n is the dimension of data and ω represents a bias offset.

The matrix $\mathbf{W}$ has to be calculated within an iterative process, where the likelihood of adaptation data with known transcription is maximized [37]. Note that in our case, only one global transform was estimated for all Gaussian components of the system. Hence, it was not necessary to include the Jacobian of the transformation in the likelihood calculation.

There exist two basic approaches on how to perform adaptation for a target speaker, supervised and unsupervised [38]. The former utilizes adaptation data that is annotated manually by a human expert. The latter employs a speech recognizer, which creates these transcripts automatically.

We have employed a supervised adaptation approach. At first, the manual orthographic transcripts of adaptation data were available. Then, we performed forced alignment to adapt the utterances using a speech recognizer operating with the baseline speaker independent model and the lexicon containing all pronunciation variants of all words occurring in the orthographic transcripts. As a result of this process, we obtained accurate phonetic transcripts with labelled noises produced by speakers (breathing, various hesitation sounds, cough, lip-smack, etc.). Finally, general speaker-specific transformations were estimated using this annotated adaptation data.

As the impact of adaptation differs between speakers, it is not reasonable to measure an average improvement. For speakers who were badly processed by the baseline, the gain varied. For some of them ACC decreased from 11.37% to 8.75%; for others, increased from 2.92% to 6.07%. For better recognized speakers, the behaviour was much more predictable: ACC increased by approximately 20% (e.g. 40.19% increased to 58.52%, and 59.85% increased to 73.14%). Thus, the speaker-channel adaptation can be very useful but, on the other hand, it is necessary to identify the user.

5.4 Continuous speech recognition

For testing the language model and vocabulary, we randomly chose 1,500 testing utterances from the Albayzin corpus. These utterances contained 1 h and 24 min of continuous speech. We did not use texts from these utterances for preparation of the language model or for the vocabulary. Thus, there could appear out of vocabulary (OOV) words. Accounting for the impact of the number OOV words is very important, as if it is high, it indicates that the vocabulary should be bigger and there is space for improving the recognition results. Table 6 shows the results from experiments in which we used the language model and vocabulary with several acoustic models. The worst results were obtained with the *AM_CZ* model that consisted only on Czech training data. The inclusion of Spanish data increased the results to a great extent, obtaining a maximum accuracy of 63.29% with *AM_cz&ES_2*. The number of OOV words was very high, so there is possibly a good opportunity to improve our vocabulary and reach even better accuracy. In the case of the non-disabled speakers, a part of Albayzin was obtained using a geography database, the OOV were mainly substantives describing geographical accidents and proper names of Spanish areas, rivers and mountains. In the case of the learning disabled, the OOV were mainly related to the way they talk, which is very informal and also affected by the way of speaking in the

Table 6 Results for continuous speech recognition with different acoustic models - typical voices

Acoustic model	CORR (%)	ACC (%)	OOV (%)
AM_CZ	31.59	29.39	23.18
AM_cz&ES_sent	65.08	63.03	23.18
AM_CZ&es	64.55	62.45	23.18
AM_cz&ES	65.44	63.24	23.18
AM_cz&ES_2	65.44	63.29	23.18

location of the associations, where many diminutives are employed.

The results of continuous speech recognition with the acoustic model for disabled speakers is shown in Table 7. We used 16 min of impaired speech recorded as a conversation (see more in the 'Adaptation to Spanish atypical voices' section). We performed several experiments: normal language model and vocabulary (*LM_ES*), vocabulary with multiple pronunciations (*LM_ES_multiVOC*), and adding test sentences to the language model once (*LM_ES_DS_multiVOC*) or several times (*LM_ES_nDS_multiVOC*). As can be observed, the results are much worse for disabled speakers than for average speakers. This is not only because of their pronunciation but also because the recordings did not take place under ideal conditions as in the Albayzin corpus, which was gathered with noise isolation and professional equipment.

However, as the continuous speech recognizer accuracy is higher than 60%, it is possible to use this recognizer to semi-automatically obtain new data for training. In order to do this, it is possible to use the approach that we described in [14], specially if we have at least partially transcribed speech. For example, it is possible to gather this type of data from radio broadcasts, as on their web pages there are usually some abstracts about the recordings and in a lot of cases there are some rewritten parts also. With our recognizer, we can automatically transcript the recordings and compare them with the text from the web page. If we find matchings that are long enough (usually a few words), we can extract these parts and use them to build more accurate models in a short time.

6 Conclusions

Speech technology can be a very valuable help for learning disabled users, whose varied conditions result in different dysfunctions of their speech system. Different efforts have been made by the scientific community in order to propose different approaches to the implementation of automatic speech recognizers for these users. However, many of the state-of-the-art approaches focus on particular pathologies or imply an in-depth study of specific dimensions such as certain articulatory features. This makes the development of automatic speech recognizers costly, as there is a need for a large amount of specific data samples to train the recognizer.

In this paper, we have presented a cross-lingual approach for the development of speech recognizers for disabled people. Our main aim was to study mechanisms to take the most of the resources already available for average users in different languages for the recognition of atypical voices. Although the generality of the method reduces its performance compared to recognizers trained on large databases of impaired speech, it allows rapid and cost-efficient development, which makes it suitable for fast development of assistive speech applications.

We have evaluated the proposed model preparing a KWS and CSR for Spanish using Czech resources, and adapting the Spanish resources obtained with the cross-lingual approach to fit learning disabled speakers.

The experimental results show that the KWS for common Spanish speakers achieves results comparable to those reached by the source KWS when used for Czech. The CSR obtains accuracies over 60%. Although it does not seem very satisfactory, we would like to emphasize that there were more than 20% OOV words. The high OOV rate demonstrates we have to increase the vocabulary for CSR (and concurrently retrain the language model). Taking all these aspects into account, the results for the recognition of common Spanish speakers are very promising, and when the new vocabulary is ready, we should be able to launch the almost automatic improvement procedure proposed in [14]. This procedure will give us new data to improve the robustness of the acoustic model for Spanish.

Both technologies (KWS and CSR) obtained lower accuracy rates with the disabled speakers. In the case of CSR, we believe that limiting the vocabulary and changing the system behaviour from continuous speech recognition to recognition of isolated words would provide several benefits. On the one hand, the speakers (who are not used to speak for long periods of time) will have the opportunity to relax their vocal tract and pronounce the words better and on the other hand isolated word dictation is more reliable than CSR. In the case of KWS, although the results are not sufficient for real operation, we have shown the positive impacts of the tested enhancement techniques. We show the importance of a careful choice of keywords as well as context-dependent limitation of the vocabulary, together with the use of proper alternative phonetics. We have also demonstrated the improvement achieved through the supervised speaker (and channel) adaptation.

We have proved that for common speakers a cost-efficient cross-lingual adaptation can be done even with a training dataset smaller than the usual databases for training speech recognizers. Especially, the forced alignment tool has proven to be very useful. In the case of disabled speakers, the task itself is challenging.

Table 7 Results for continuous speech recognition with different language models - atypical voices

Language model	CORR (%)	ACC (%)	OOV (%)
LM_ES	2.75	2.30	11.25
LM_ES_multiVOC	3.32	2.70	11.25
LM_ES_DS_multiVOC	3.72	2.97	11.25
LM_ES_nDS_multiVOC	5.32	4.52	11.25

However, we succeeded in the automatic elimination of the data which was not suitable for training, and generally, we can say that the proposed method leads to significant reduction of time-consuming expert work.

For future work, we plan to improve the quality of phonetic alternatives using a data-driven weighted finite state transducer (WFST)-based approach. The WFST can be trained directly from a vocabulary where the input data consists of (*Spanish word - Czech phonetic*) pairs [39]. Along with the general-purpose approach, it is also possible to adapt the vocabularies to fit the speakers individually, this WFST-based system should be able to 'learn' the rules created by an expert and to propose alternatives replacing the actual G2P module.

Another promising guideline comes from our ongoing work. We are currently making experiments replacing the physical state decoder of the speech recognizer by a neural network. This network has the advantage that it uses seven subsequent parametrized frames to classify the middle one, so it uses more information. The experiments show promising improvement especially for the data with low initial recognition score. Another advantage of this approach is that it uses the same training data as the current HMM-based decoder. We want to apply this change for the existing speech recognizer and also to prepare another KWS based on these neural networks which output would be processed directly by the weighted finite state transducers.

Although the number of recordings in our database of atypical voices is in the same order as other state-of-the-art corpora (see a review of existing corpora in [35] and [40]), it would be desirable to record new data and/or manually annotate the data marked as 'not suitable for training'. The latter solution has the drawback of requiring a large amount of expert work, but we can presume it would help to recognize the more severely impaired speakers.

As the emphasis in the paper has been to take advantage of the existing resources to help to develop new assistive technologies for disabled people, we offer our collaboration and resources to interested researchers. In the near future, we plan to test the adapted recognizer under real conditions integrating it in an application for tablets, the results of this new stage of our research will also be at the disposal of the scientific community.

Endnotes

[a] Asociación pro discapacitados psíquicos Jabalcón, www.jabalcon.org.

[b] Asociación de padres, familiares y amigos de personas con discapacidad intelectual del norte de Almería, www.apafa.es.

Competing interests
The authors declare that they have no competing interests.

Authors' contributions
All authors have read and approved the manuscript.

Acknowledgements
This research was supported by the project 'Favoreciendo la vida autónoma de discapacitados intelectuales con problemas de comunicación oral mediante interfaces personalizados de reconocimiento automático del habla', financed by the Centre of Initiatives for Development Cooperation (Centro de Iniciativas de Cooperación al Desarrollo, CICODE), University of Granada, Spain. This research was supported by the Student Grant Scheme 2014 (SGS) at the Technical University of Liberec. The authors want to thank the staff from the associations JABALCON and APAFA and the participants in the recordings for their contribution to make this work possible.

Author details
[1] Institute of Information Technology and Electronics, Technical University of Liberec, Studentska 2, 461 17 Liberec, Czech Republic. [2] Department of Languages and Computer Systems, University of Granada, CITIC-UGR, Periodista Daniel Saucedo Aranda s/n, 18071 Granada, Spain.

References

1. J Sigafoos, RW Schlosser, GE Lancioni, MF O'Reilly, VA Green, NN Singh, GE Lancioni, NN Singh, in *Assistive Technology for People with Communication Disorders*. Autism and Child Psychopathology Series (Springer, New York, 2014), pp. 77–112
2. O Saz, J Simón, W-R Rodríguez, E Lleida, C Vaquero, Analysis of acoustic features in speakers with cognitive disorders and speech impairments. EURASIP J. Adv. Signal Process. **2009**, 159–234 (2009)
3. TH Falk, W-Y Chan, F Shein, Characterization of atypical vocal source excitation, temporal dynamics and prosody for objective measurement of dysarthric word intelligibility. Speech Commun. **54**, 622–631 (2012)
4. MA Neerincx, AHM Cremers, JM Kessens, DA van Leeuwen, KP Truong, Attuning speech-enabled interfaces to user and context for inclusive design: technology, methodology and practice. Univers. Access. Inform. Soc. **8**, 109–122 (2009)
5. WR Rodríguez, O Saz, E Lleida, A prelingual tool for the education of altered voices. Speech Commun. **54**, 583–600 (2012)
6. MS Hawley, SP Cunningham, PD Green, P Enderby, R Palmer, S Sehgal, P O'Neill, A voice-input voice-output communication aid for people with severe speech impairment. IEEE Trans. Neural Syst. Rehabil. Eng. **21**, 23–31 (2013)
7. Ed Joode, Cv Heugten, F Verhey, Mv Boxtel, Efficacy and usability of assistive technology for patients with cognitive deficits: a systematic review. Clin. Rehabil. **24**, 701–714 (2010)
8. GE Lancioni, NN Singh, MF O'Reilly, J Sigafoos, D Oliva, in *Assistive Technology for People with Severe/Profound Intellectual and Multiple Disabilities*. Autism and Child Psychopathology Series (Springer, New York, 2014), pp. 277–313
9. KF McCoy, JL Arnott, L Ferres, M Fried-Oken, B Roark, Speech and language processing as assistive technologies. Comput. Speech Lang. **27**, 1143–1146. (2013-09)
10. O Chia Ai, M Hariharan, S Yaacob, L Sin Chee, Classification of speech dysfluencies with MFCC and LPCC features. Expert Syst. Appl. **39**, 2157–2165 (2012)
11. J Borg, S Larsson, P Östergren, The right to assistive technology: for whom, for what, and by whom? Disabil. Soc. **26**, 151–167 (2011)
12. J Nouza, K Blavka, P Červa, J Zdansky, J Silovsky, M Bohac, J Prazak, Making czech historical radio archive accessible and searchable for wide public. J. Multimed. **7**, 159–169 (2012)
13. P Červa, J Nouza, in *Proceedings of the Conference of the International Speech Communication Association Interspeech: 27-31 August 2007; Antwerp, Belgium, (ISCA, France)*. Design and development of voice controlled aids for motor-handicapped persons, (2007), pp. 2521–2524

14. J Nouza, P Červa, M Kucharová, Cost-efficient development of acoustic models for speech recognition of related languages. Radioengineering. **22**, 866–873 (2013)
15. P Lal, S King, Cross-lingual automatic speech recognition using tandem features. **21**, 2506–2515 (2013)
16. L Besacier, E Barnard, A Karpov, T Schultz, Automatic speech recognition for under-resourced languages: a survey. Speech Commun. **56**, 85–100 (2014)
17. F Rudzicz, Production knowledge in the recognition of dysarthric speech. PhD thesis, University of Toronto (2011)
18. SA Borrie, MJ McAuliffe, JM Liss, GA O'Beirne, TJ Anderson, A follow-up investigation into the mechanisms that underlie improved recognition of dysarthric speech. J. Acoust. Soc. Am. **132**, 102–108 (2012)
19. J-P Hosom, T Jakobs, A Baker, S Fager, in *Proceedings of the 11th Conference of the International Speech Communication Association (Interspeech): 26-30 September 2010; Makuhari, Japan (International, Speech Communication Association, France)*, ed. by T Kobayashi, K Hirose, and S Nakamura. in *Automatic speech recognition for assistive writing in speech supplemented word prediction*, (2010), pp. 2674–2677
20. WK Seong, JH Park, HK Kim, in *Dysarthric Speech Recognition Error Correction Using Weighted Finite State Transducers Based on Context-Dependent Pronunciation Variation*. LNCS, ed. by Miesenberger K, A Karshmer, P Penaz, and W Zagler (Springer, Heidelberg, 2012), pp. 475–482
21. I Kraljevski, G Strecha, M Wolff, O Jokisch, S Chungurski, R Hoffmann, in *Cross-Language Acoustic Modeling for Macedonian Speech Technology Applications*, ed. by S Markovski, M Gusev. Advances in Intelligent Systems and Computing (Springer, Berlin, 2013), pp. 35–45
22. D Imseng, H Bourlard, J Dines, PN Garner, M Magimai-Doss, Applying multi- and cross-lingual stochastic phone space transformations to non-native speech recognition. IEEE Trans. Audio Speech Lang. Process. **21**, 1713–1726 (2013)
23. T Schultz, K Kirchhoff, *Multilingual Speech Processing*. (Academic Press, USA, 2006)
24. D Imseng, P Motlicek, PN Garner, H Bourlard, in *Proceedings of the IEEE Workshop on Automatic Speech Recognition and Understanding (ASRU): 8-12 December 2013; Olomouc, Czech Republic (IEEE, USA)*. Impact of deep MLP architecture on different acoustic modeling techniques for under-resourced speech recognition, (2013), pp. 332–337
25. P Xu, P Fung, Cross-lingual language modelling for low-resource speech recognition. IEEE Trans. Audio Speech Lang. Process. **21**, 1134–1144 (2013)
26. M Bohac, K Blavka, Text-to-speech alignment for imperfect transcriptions. LNCS: Text, Speech Dialogue. **8082**, 536–543 (2013)
27. J Zhang, F Pan, Y Yan, in *Proceedings of the 4th International Conference on Intelligent Human-Machine Systems and Cybernetics (IHMSC): 26-27 August 2012; Nanchang, China (IEEE, U.S.A.)* An LVCSR based automatic scoring method in English reading tests, (2012), pp. 34–37
28. DP Córdova Lucero, DT Toledano, in *Proceedings of the Joint 7th Spanish Speech Technology Workshop and the Iberian SLTech Workshop: 21-23 November 2012; Madrid, Spain (Springer, Germany)*. Preliminary results of alignment of text and audio in news and songs, (2012), pp. 59–68
29. J Nouza, P Červa, J Chaloupka, in *Proceedings of the International Conference on Health Informatics (HEALTHINF - BIODEVICES): 26-29 January 2011; Rome, Italy (SciTePress, U.K.)* Rainbow bridge: Training center based on voice technology for people with physical disabilities, (2011), pp. 529–533
30. RA Wagner, MJ Fischer, String-to-string correction problem. J. ACM. **21**, 168–173 (1974)
31. M Bohac, in *Proceedings of the 54th International Symposium ELMAR: 12-14 September 2012; Zadar, Croatia (IEEE, USA)*. Performance comparison of several techniques to detect keywords in audio streams and audio scene, (2012), pp. 215–218
32. J Nouza, J Zdansky, P Červa, J Silovsky, Challenges in speech processing of Slavic languages (case studies in speech recognition of Czech and Slovak). LNCS. **5967**, 225–241 (2010)
33. M Bohac, J Nouza, K Blavka, Investigation on most frequent errors in large-scale speech recognition applications. LNCS: Text, Speech Dialogue. **7499**, 520–527 (2012)
34. Albayzin corpus in the European Language Resources Association. http://catalog.elra.info/product_info.php?products_id=746. Accessed 10 October 2014
35. D-L Choi, B-W Kim, Y-W Kim, Y-J Lee, Y Um, M Chung, in *Proceedings of the 8th International Conference on Language Resources and Evaluation (LREC): 23-25 May 2012; Istanbul, Turkey (IEEE, USA)*, ed. by N Calzolari (Conference Chair), K Choukri, T Declerck, MU Dogan, B Maegaard, J Mariani, Odijk J, and S Piperidis. Dysarthric speech database for development of QoLT software technology, (2012), pp. 47-50
36. MJF Gales, PC Woodland, Mean and variance adaptation within the MLLR framework. Comput. Speech Lang. **10**, 249–264 (1996)
37. MJF Gales, Maximum likelihood linear transformations for HMM-based speech recognition. Comput. Speech Lang. **12**, 75–98 (1998)
38. P Červa, J Nouza, Supervised and unsupervised speaker adaptation in large vocabulary continuous speech recognition of Czech. LNCS (including subseries Lecture Notes in Artificial Intelligence and Lecture Notes in Bioinformatics). **3658**, 203–210 (2005)
39. M Bohac, J Malek, K Blavka, in *Proceedings of the 36th International Conference on Telecommunications and Signal Processing (TSP): 2–4 July 2013; Brno, Czech Republic (IEEE, U.S.A.)* Iterative grapheme-to-phoneme alignment for the training of WFST-based phonetic conversion, (2013), pp. 474–478
40. F Rudzicz, Using articulatory likelihoods in the recognition of dysarthric speech. Speech Commun. **54**, 430–444 (2012)

New aliasing cancelation algorithm for the transition between non-aliased and TDAC-based coding modes

Jeongook Song and Hong-Goo Kang*

Abstract

This paper proposes a new aliasing cancelation algorithm for the transition between non-aliased coding and transform coding with time domain aliasing cancelation (TDAC). It is effectively utilized for unified speech and audio coding (USAC) that was recently standardized by the Moving Picture Experts Group (MPEG). Since the USAC combines two coding methods with totally different structures, a special processing called forward aliasing cancelation (FAC) is needed at the transition region. Unlike the FAC algorithm embedded in the current standard, the proposed algorithm does not require additional bits to encode aliasing cancelation terms because it appropriately utilizes adjacent decoded samples. Consequently, around 5% of total bits are saved at 16- and 24-kbps operating modes in speech-like signals. The proposed algorithm is sophisticatedly integrated on the decoding module of the USAC common encoder (JAME) for performance verification, which follows the standard process exactly. Both objective and subjective experimental results confirm the feasibility of the proposed algorithm, especially for contents that require a high percentage of mode switching.

1 Introduction

Unified speech and audio coding (USAC; ISO/IEC 23003-3) standardized in early 2012 shows the best performance for speech, music, and mixed type of input signals [1]. Verification tests confirmed the superiority of quality, especially at low bit rates [2]. In an initial stage of designing the coding structure, it was not possible to acquire high-quality output to all input contents because only a single type of traditional audio or speech coding structure was adopted. The best result could be obtained by simultaneously running two types of codecs: Adaptive Multi-rate Wideband plus (AMR-WB+ [3]) for speech signals and high-efficiency advanced audio coding (HE-AAC [4]) for audio signals. In case of encoding signals with mixed characteristics, one of two coding modes is chosen depending on the characteristic of input contents. Although this approach improves the quality of all types of contents, many problems occur at transition frames where mode switching is needed between entirely different types of codecs. For example, the segment of perceptually weighted signal encoded by speech codec needs to be smoothly combined with that of the signal encoded by audio codec. Since the characteristic of speech and audio codec is different, however, the overlapped segment between two codecs must not be similar to the input signal. How to determine the encoding mode for the various types of input signal is also important. The problems are mostly solved by adopting novel technologies such as a signal classifier, frequency domain noise shaping (FDNS), and forward aliasing cancelation (FAC) technique [5].

The FAC algorithm is one of the key technologies in USAC, which enables the successful combination of two different types of codecs, especially at transition frames. To remove the aliasing terms caused by cascading different types of codecs, FAC additionally generates the aliasing cancellation signals, and then they are quantized and transmitted to the decoder. In the earlier version of USAC that had not introduced the FAC technique, the frame boundary of transition frame was variable; thus, a special windowing operation was needed for compensating the aliased signal in the overlap region. Although FAC

*Correspondence: hgkang@yonsei.ac.kr
School of Electrical and Electronic Engineering, Yonsei University, Seoul 120-749, South Korea

somewhat solves the problem, it still requires additional bits.

This paper proposes a new aliasing cancelation algorithm that does not need any additional bits, which uses the decoded signal of the adjacent frames. At first, the algorithm generates the relevant aliasing cancelation part by considering the error that occurred by the encoding mode switching. Then, the output signals are reconstructed by adding the generated aliasing cancelation part to the decoded signal and by normalizing the weight caused by the encoding window. In the overall process, the most important thing is how to obtain the aliasing cancelation part by properly utilizing the adjacent signal.

The aliasing cancelation process of the proposed algorithm is conceptually similar to that of the block switching compensation scheme proposed for low delay advanced audio coding (AAC-LD [6,7]). In the literature, the scheme introduced time domain weightings applicable as a post processing in the decoder in order to remove a look-ahead delay inevitable for a window transition from the long window to the short window. This is similarly considered as an aliasing cancellation signal described in this paper. However, its application and the resulting aliasing form are different.

A new aliasing cancelation algorithm is sophisticatedly integrated in the decoding module of the USAC common encoder (JAME) [8], which has been designed by our team as an open source paradigm. Objective and subjective test results show that the proposed method has comparable quality to the FAC algorithm while saving the bits for encoding the aliasing signal component in the FAC algorithm.

Section 2 describes the overview of USAC techniques and FAC algorithm. In Section 3, the proposed algorithm is explained in detail. In Section 4, experiments and evaluation results are also described.

2 USAC overview and FAC algorithm

2.1 Overview

USAC, recently standardized codec by MPEG, provides high quality for speech, audio and mixed signals even in very low bit rates [2]. Figure 1 shows a block diagram of the encoding process that consists of frequency domain (FD) and time domain (TD) coding modules. At first, the encoding mode is determined by analyzing the spectral information of input signal in the signal classifier block [9]. The FD coder transforms the time domain input signal into frequency spectrum by taking the modified discrete cosine transform (MDCT) [10], then calculates the perceptual entropy of each frequency band using a psychoacoustic model [11,12]. The number of allocated bits to each band is determined by considering the distribution of perceptual entropy. In the TD coding module, an input signal is encoded by either algebraic code-excited linear

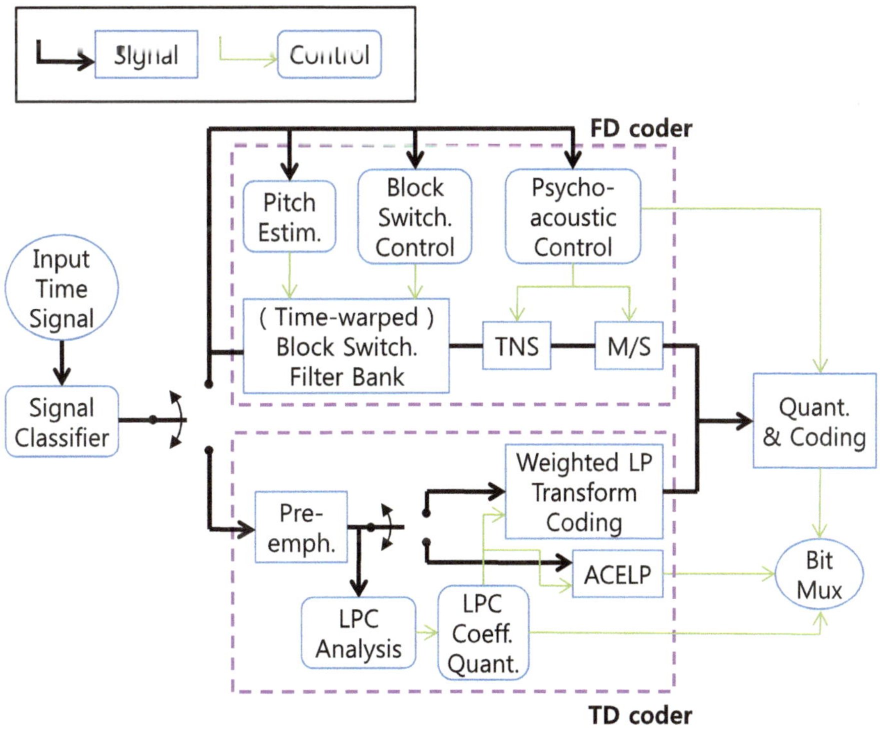

Figure 1 Block diagram of the encoding process of USAC.

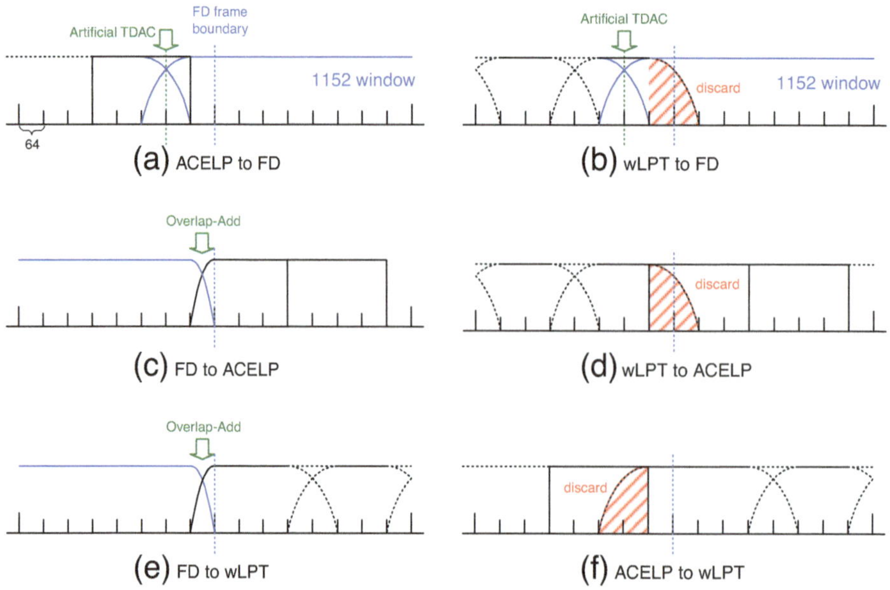

Figure 2 Synthesis process in the transition frame of an initial version of USAC. **(a)** ACELP to FD, **(b)** wLPT to FD, **(c)** FD to ACELP, **(d)** wLPT to ACELP, **(e)** FD to wLPT, and **(f)** ACELP to wLPT.

prediction (ACELP) or weighted linear prediction transform coding (wLPT) similar to the AMR-WB+ codec. The wLPT is a modified version of transform coded excitation (TCX) mode that the residuals of LPC filter are encoded in the frequency band using the MDCT method [13]. Note that its quantizer is the same as the one used for the FD coder to keep compatibility and efficiency. Finally, the quantized spectrum is encoded by context adaptive arithmetic coding (CAAC), which has a higher coding efficiency than the Huffman coding [14].

2.2 Forward aliasing cancelation algorithm

Since the USAC consists of two different types of coding methods, it is very important to handle the transition frame where the encoding mode is switched from FD codec to TD codec or vice versa. Note that the MDCT

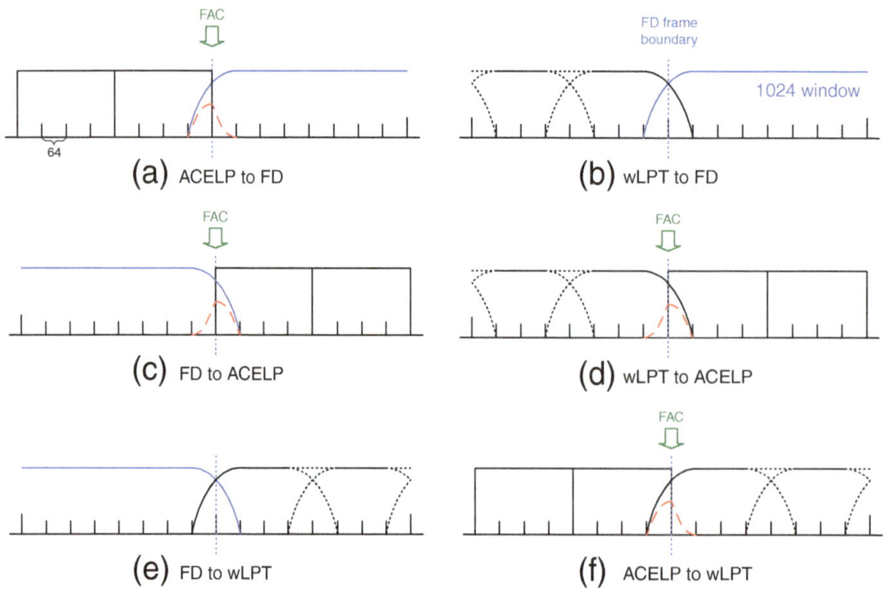

Figure 3 Synthesis process in the transition frame using FAC algorithm. **(a)** ACELP to FD, **(b)** wLPT to FD, **(c)** FD to ACELP, **(d)** wLPT to ACELP, **(e)** FD to wLPT, and **(f)** ACELP to wLPT.

removes the aliasing part of the current frame by combining the signal decoded at the following frame. However, if the encoding mode of the next frame is TD codec, the aliasing term must not be generally canceled. In an initial version of USAC, this problem was solved by discarding the aliased signal and using inconsistent frame length. When the frame length of TD codec is decreased due to aliased signal, the following frame length is increased for synchronizing the starting position of FD codec [15].

Figure 2 describes the synthesis process in the transition frame of an initial version of USAC. The synthesized signals in the overlapped region between wLPT and other coding methods are discarded as given in Figure 2b,d,f. In case the encoding mode is changed from FD codec to TD codec, the signals decoded by ACELP are windowed to perform an overlap-add processing with the FD output. Since the frame encoded by TD codec starts at the front position of the frame boundary, the starting point of the long frame of FD mode needs to be compensated by decreasing the length, which allows the early start of the TD codec mode. Since the frame size is inconsistent, therefore, a new type of window should be designed [15].

The forward aliasing cancelation algorithm is proposed to solve the awkward frame structure mentioned above. Figure 3 shows the FAC algorithm [5]. All transitions are made at the same position in each frame boundary. Note that the FAC is needed for the ACELP transition frames given in Figure 3a,c,d,f. Since the decoded output of ACELP mode cannot cancel out the aliased outputs

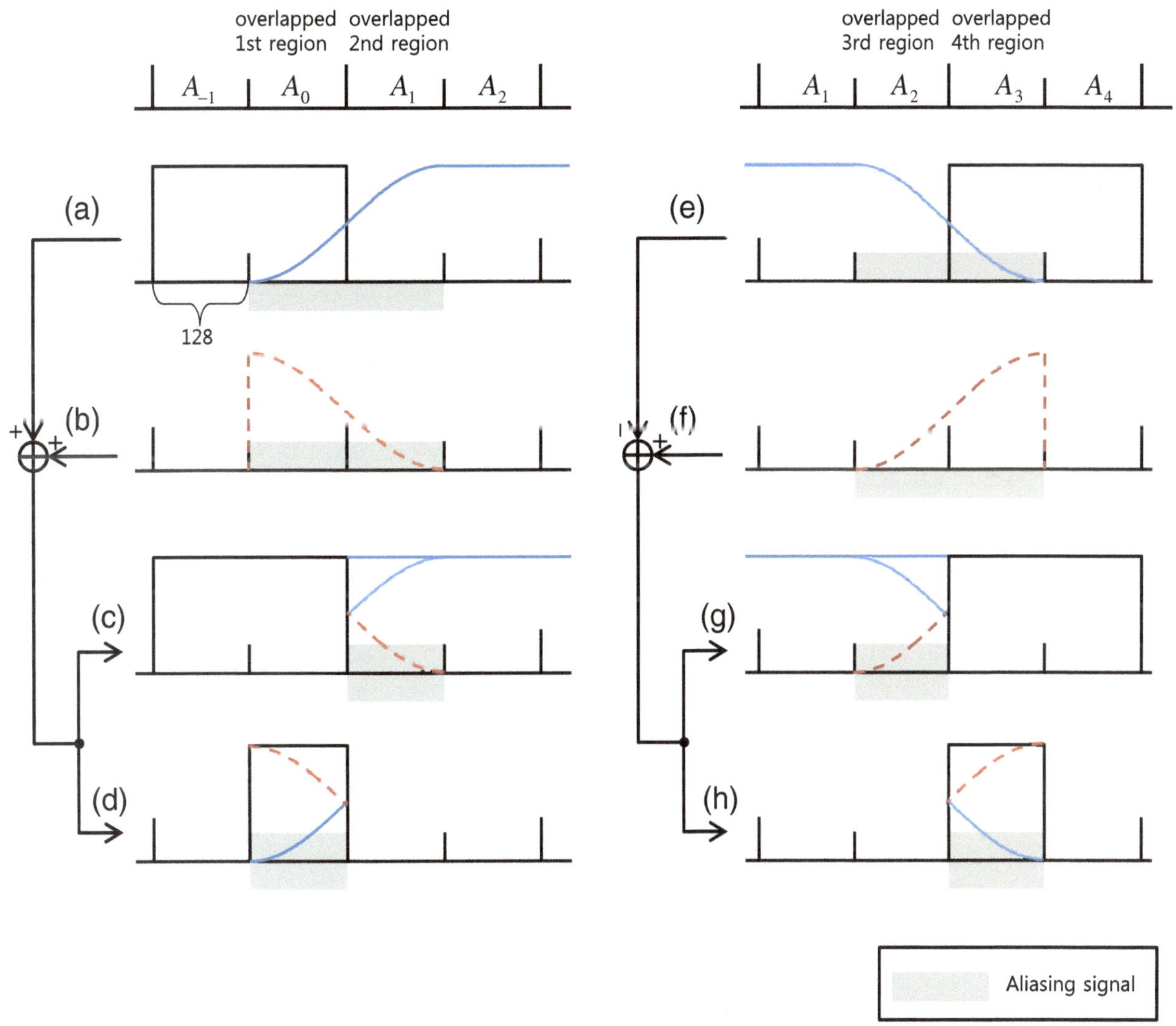

Figure 4 Aliasing cancelation processes using two different types of FAC signals. (a) Aliased signal in the first case, **(b)** FAC signal in the first case, **(c)** total signal in the first case, **(d)** dummy signal in the first case, **(e)** aliased signal in the second case, **(f)** FAC signal in the second case, **(g)** total signal in the second case, and **(h)** dummy signal in the second case.

decoded by FD or wLPT codec modes, the FAC algorithm artificially generates the additional signals for canceling the aliasing component. The generated signals are mixed with the quantization error portion of wLPT or FD coder, and then they are quantized by the adaptive vector quantization (AVQ) tool [9]. The AVQ tool consists of three parts: FAC gain, two codebook indices, and 16 Voronoi extension indices for AVQ refinement. Seven bits are allocated to the FAC gain, and the bits for other indices are variable due to adopting a unary coding. For example at 24 kpbs, around 130 bits per one frame is used to encode the FAC parameters. It corresponds to the 11% of the average frame bits.

3 Proposed aliasing cancelation algorithm

As is shown in Figure 4, the FAC algorithm happens to be applied in two different cases depending on the order of coding modules, i.e., whether the transition is made from ACELP to other coding modes (wLPT or FD) or vice versa. The first case given in Figure 4a,b,c,d describes the way of removing aliasing signals from ACELP to other coding modes. The second case given in Figure 4e,f,g,h does the reverse direction. The aliased signals given in Figure 4a,e are compensated by adding the FAC signal. The FAC signal given in Figure 4b,f consists of an aliasing cancelation component and a symmetric windowed signal. Note that the aliasing cancelation term in the FAC signal plays a key role in designing the proposed algorithm later. The dummy signal is simply generated by adding the FAC signals and aliased signals in the decoding stage. Since the aliasing signal depicted in Figure 4d is canceled out, the sum of the remained signals becomes the output signal marked with the black rectangular shape. From now on, it is called 'dummy signal'. Assuming that there is no quantization error, dummy signals are equivalent to ACELP signals in the same position in the time domain. Similarly, dummy signals in Figure 4h are also equivalent to the first 128 samples of ACELP signals. Since those ACELP signals are available in the decoder, the dummy signals do not need to be sent as it does in the FAC algorithm. In other words, the region located on dummy signals can be directly decoded by the synthesized signal obtained from the ACELP scheme, i.e., it is regarded as a non-aliased part. Please also note that the method requires additional bits to quantize FAC signals. This paper proposes a new aliasing cancelation algorithm that does not need any additional bits while successfully removing the aliasing parts.

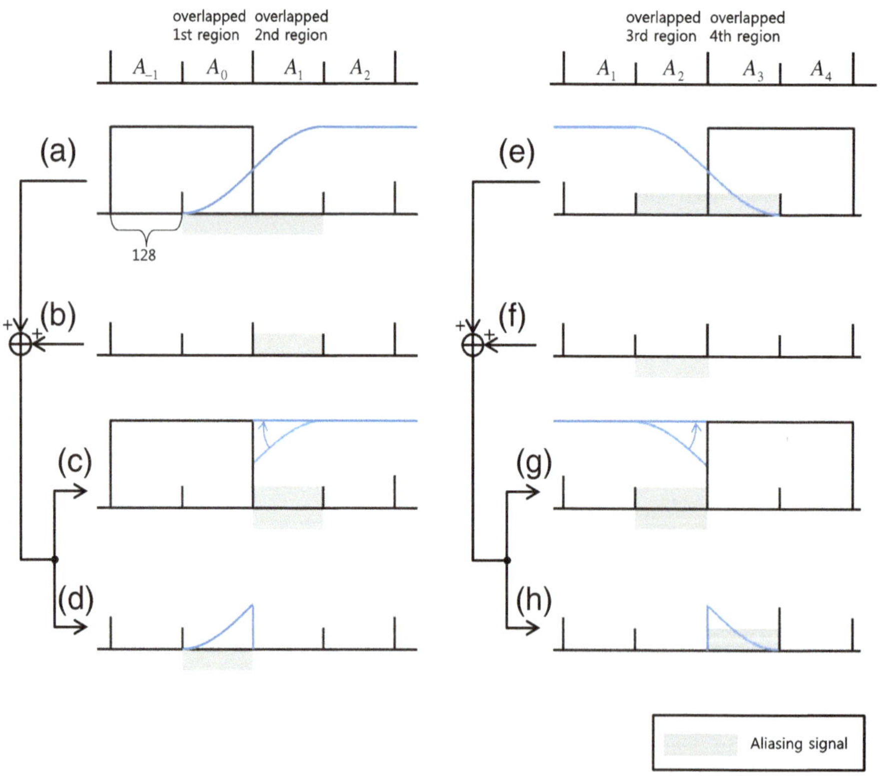

Figure 5 Aliasing cancelation processes using two different types of the proposed AC signals. (a) Aliased signal in the first case, **(b)** proposed AC signal in the first case, **(c)** total signal in the first case, **(d)** unused signal in the first case, **(e)** aliased signal in the second case, **(f)** proposed AC signal in the second case, **(g)** total signal in the second case, and **(h)** unused signal in the second case.

Table 1 Test items for the evaluation of the proposed algorithm

Item number	Class	Label	Item
1	Music	m1	salvation
2	Music	m2	te15
3	Music	m3	Music_1
4	Music	m4	Music_3
5	Music	m5	Phi7
6	Speech	s6	Es01
7	Speech	s7	louis_raquin_15
8	Speech	s8	Wedding_speech
9	Speech	s9	te1_mg54_speech
10	Speech	s10	Arirang_speech
11	Mixed	×11	twinkle_ff51
12	Mixed	×12	SpeechOverMusic_1
13	Mixed	×13	SpeechOverMusic_4
14	Mixed	×14	HarryPotter
15	Mixed	×15	Lion

Figure 5 shows the schematic diagram of the proposed algorithm. As is shown in Figure 5b,f, the proposed algorithm generates signals for canceling the aliasing components. After aliasing cancelation (AC) signal is added to the aliased output of the decoder, the combined signal becomes unaliased as given in Figure 5c,g. The signals given in Figure 5d,h are simply disregarded because the region can be reconstructed by the ACELP output only as already described in Figure 4.

Hereinafter, we further derive the relationship of the FAC signals (Figure 4b,f) and the aliasing cancelation signals (Figure 5b,f) by utilizing the specific relation between the formulae of MDCT and DCT-IV. Note that the MDCT is a modified form of DCT-IV that is suitable for saving the bits. MDCT spectral coefficient, $X_M(k)$, and DCT-IV spectral coefficient, $X_D(k)$, are respectively defined as follows [10]:

$$X_M(k) = \sum_{n=0}^{2N-1} x(n)\Phi_k\left(n+\frac{N}{2}\right), \tag{1}$$

$$X_D(k) = \sum_{n=0}^{N-1} x(n)\Phi_k(n),$$
$$\Phi_k(n) = \cos\left(\frac{\pi}{N}\left(n+\frac{1}{2}\right)\left(k+\frac{1}{2}\right)\right), \tag{2}$$

where N is the frame length.

$$\begin{aligned}\Phi_k(2N+n) &= \cos((2k+1)\pi+\alpha) = -\Phi_k(n),\\ \Phi_k(2N-1-n) &= \cos((2k+1)\pi-\alpha) = -\Phi_k(n),\\ \alpha &= \frac{\pi}{2N}(n+\frac{1}{2})(2k+1).\end{aligned} \tag{3}$$

Table 2 Actual achieved bit rates by each item in the operating mode

Item	12-kbps mode		16-kbps mode		24-kbps mode	
	Conv. (kbps)	Prop.-B (kbps)	Conv. (kbps)	Prop.-B (kbps)	Conv. (kbps)	Prop.-B (kbps)
m1	12.09	12.09	16.18	16.18	24.71	24.71
m2	12.15	12.13	16.22	16.19	24.75	24.73
m3	12.20	11.94	16.19	15.96	24.69	24.69
m4	12.14	12.11	16.24	16.20	24.76	24.58
m5	11.85	11.83	15.94	15.91	24.52	24.50
s6	11.83	11.64	15.74	14.58	24.29	22.96
s7	12.20	11.99	16.19	15.24	24.70	23.34
s8	11.95	11.90	15.50	14.76	24.08	22.53
s9	11.60	11.45	15.37	14.65	23.93	22.52
s10	11.76	11.61	15.41	14.76	23.94	22.71
×11	12.12	12.11	16.20	16.18	24.73	24.73
×12	11.81	11.76	15.94	15.91	24.49	24.41
×13	12.20	11.58	16.27	15.30	24.77	23.67
×14	12.21	12.08	16.19	15.65	24.71	23.47
×15	11.89	11.68	15.76	14.85	24.30	23.06
Music	12.08	12.02	16.15	16.09	24.69	24.64
Speech	11.87	11.72	15.64	14.80	24.19	22.81
Mixed	12.04	11.84	16.07	15.58	24.60	23.87
Total	12.00	11.86	15.96	15.49	24.49	23.77

By utilizing the cosine property given in Equation 3, the MDCT spectral coefficient can be represented by the form of DCT-IV:

$$X_M(k) = \sum_{n=0}^{N/2-1} (-x(\frac{3}{2}N - 1 - n) - x(\frac{3}{2}N + n))\Phi_k(n) + \sum_{n=N/2}^{N-1} (x(n - \frac{N}{2}) - x(\frac{3}{2}N - 1 - n))\Phi_k(n) = \sum_{n=0}^{N-1} c(n)\Phi_k(n). \quad (4)$$

Equation 4 informs us that the MDCT spectral values transformed by $2N$ consecutive inputs are exactly equivalent to DCT-IV spectral values transformed by N inputs, which are folded at the $\frac{1}{2}N$ position and $\frac{3}{2}N$ position. Since the DCT-IV should be invertible, we know that the folded signals are the aliased parts generated by taking an inverse MDCT [16]. Two parts of the folded signals are

$$\begin{cases} -x\left(\frac{3}{2}N - 1 - n\right) - x\left(\frac{3}{2}N + n\right) & , 0 \le n < \frac{N}{2} \\ x\left(n - \frac{N}{2}\right) - x\left(\frac{3}{2}N - 1 - n\right) & , \frac{N}{2} \le n < N. \end{cases} \quad (5)$$

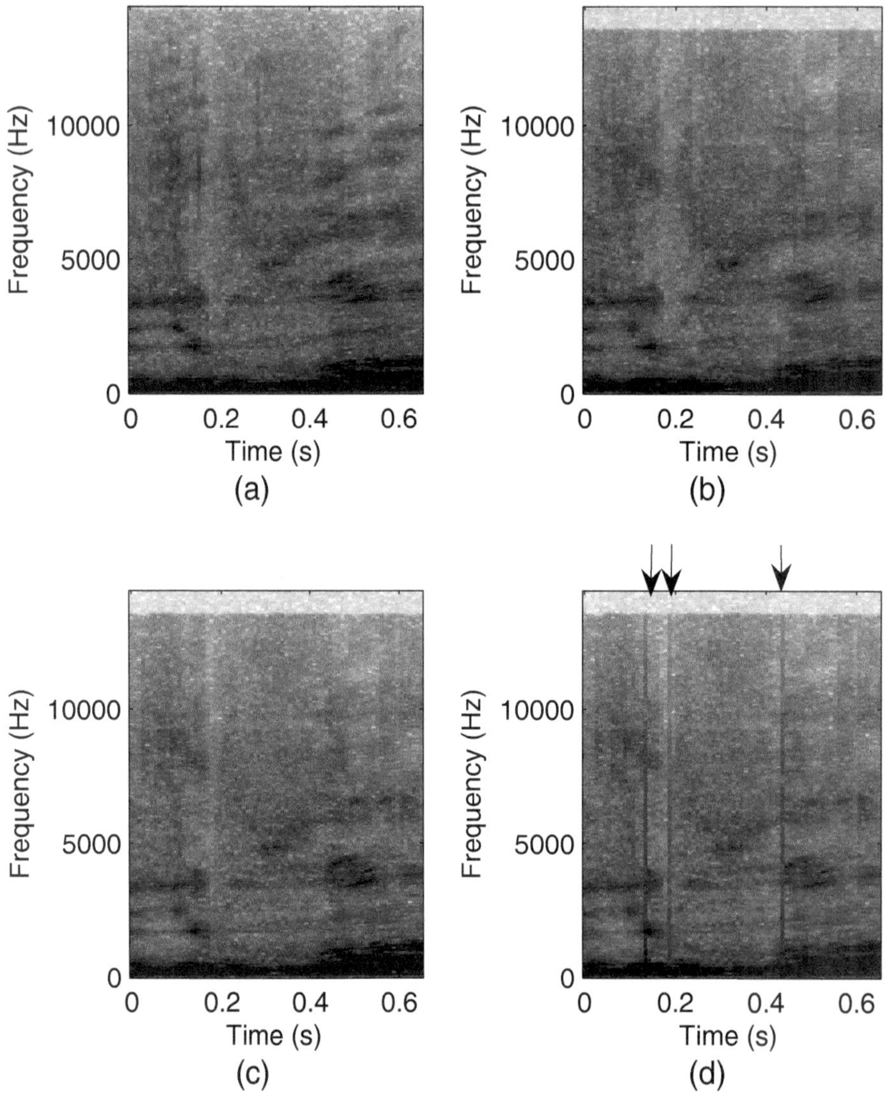

Figure 6 An example of speech spectrogram. **(a)** Original spectrogram, **(b)** synthesized spectrogram using proposed algorithm, **(c)** synthesized spectrogram using FAC algorithm, and **(d)** aliased spectrogram without FAC signal.

Let $x(n)$ be input samples and A_m be the $\left(1 \times \frac{N}{2}\right)$ vector of input samples:

$$A_m = \left[\, x\left(m\tfrac{N}{2}\right) \; x\left(m\tfrac{N}{2}+1\right) \; \cdots \; x\left((m+1)\tfrac{N}{2}-1\right) \right]. \quad (6)$$

Equation 5 is reformulated as

$$\begin{cases} S(A_0, A_1) = A_0 - A_1 R, \\ U(A_2, A_3) = -A_2 R - A_3, \end{cases} \quad (7)$$

where R depicts an $\left(\frac{N}{2} \times \frac{N}{2}\right)$ reverse identity matrix:

$$R = \begin{bmatrix} 0 & \cdots & 0 & 1 \\ & & 1 & 0 \\ & \iddots & & \vdots \\ 1 & & & 0 \end{bmatrix}. \quad (8)$$

Practically, windowing is introduced to remove the side-lobe artifacts. By introducing the windowing to Equation 7, the first and second folded signals can be expressed as

$$\begin{cases} S(A_0 \circ W_0, A_1 \circ W_1) = A_0 \circ W_0 - A_1 \circ W_1 R, \\ U(A_2 \circ W_2, A_3 \circ W_3) = -A_2 \circ W_2 R - A_3 \circ W_3, \end{cases} \quad (9)$$

where the operator '∘' denotes the Hadamard product [17] and W_k depicts a $(1 \times \frac{N}{2})$ window matrix:

$$W_k = \left[w\left(\frac{\pi}{4}k\right) \quad w\left(\frac{\pi}{4}k + \frac{\pi}{2N}\right) \quad \cdots \quad w\left(\frac{\pi}{4}k + \frac{\pi}{2N}\left(\frac{N}{2}-1\right)\right) \right]. \quad (10)$$

The window matrix, W_k, must be symmetric and satisfy the Princen-Bradley condition for perfect reconstruction [10]:

$$W_0 R = W_3, \qquad W_1 R = W_2, \quad (11)$$
$$W_k \circ W_k + W_{k+2} \circ W_{k+2} = [\, 1 \; 1 \; \cdots \; 1 \,].$$

Note that the aliased signals in the overlap region of Figures 4 and 5 are equivalent to the folded signals through the analysis of MDCT transform as is given in Equation 9. Therefore, the time signals in the overlapped regions can be synthesized perfectly using the aliasing cancelation terms and windowing property. FAC signals in Figure 4b,f are respectively defined as

$$\begin{cases} U(A_0 \circ W_2, A_1 \circ W_3) = -A_0 \circ W_2 R - A_1 \circ W_3, \\ S(A_2 \circ W_0, A_3 \circ W_1) = A_2 \circ W_0 - A_3 \circ W_1 R. \end{cases} \quad (12)$$

We can obtain the dummy signals in Figure 4d,h by

$$\begin{cases} S(A_0 \circ W_0, A_1 \circ W_1) \circ W_0 - U(A_0 \circ W_2, A_1 \circ W_3) R \circ W_2 = A_0, \\ -U(A_2 \circ W_2, A_3 \circ W_3) \circ W_3 - S(A_2 \circ W_0, A_3 \circ W_1) R \circ W_1 = A_3. \end{cases} \quad (13)$$

Note that there is no difference between dummy signals and adjacent ACELP signals if they have the same quantization error or do not have any quantization error. The synthesized signals in Figure 4c,g are calculated as follows:

$$\begin{cases} -S(A_0 \circ W_0, A_1 \circ W_1) R \circ W_1 - U(A_0 \circ W_2, A_1 \circ W_3) \circ W_3 = A_1, \\ -U(A_2 \circ W_2, A_3 \circ W_3) R \circ W_2 + S(A_2 \circ W_0, A_3 \circ W_1) \circ W_0 = A_2. \end{cases} \quad (14)$$

Actually, the aliasing parts in Equation 9 are $-A_0 \circ W_0$ and $-A_3 \circ W_3$. As previously mentioned in Figure 5, it is clear that outputs are perfectly synthesized if these terms are removed. The new algorithm generates the aliasing cancelation terms from the adjacent ACELP signals such as

$$\begin{cases} (-S(A_0 \circ W_0, A_1 \circ W_1) R + A_0 \circ W_0 R) \circ W_1^{-1} = A_1, \\ (-U(A_2 \circ W_2, A_3 \circ W_3) R - A_3 \circ W_3 R) \circ W_2^{-1} = A_2. \end{cases} \quad (15)$$

Theoretically, if there is no quantization error, the FAC algorithm and new aliasing cancelation algorithm are able to perfectly reconstruct the original signal in the transition frame. Practically, since the quantization error is generated by several passes of non-linear filters in the time and frequency domain, it is very difficult to mathematically model the impact of the error. However, it is clear that the FAC method has a quantization error in the frequency domain, while the proposed algorithm includes

Table 3 SNR at 12-, 16-, and 24-kbps operating modes

Category	Mode	Prop.-B (dB)	Conv. (dB)
Music	12 kbps	10.662	10.682
	16 kbps	12.452	12.544
	24 kbps	14.530	14.531
Speech	12 kbps	10.794	10.840
	16 kbps	12.033	12.026
	24 kbps	12.934	12.842
Mixed	12 kbps	9.831	9.846
	16 kbps	11.678	11.703
	24 kbps	13.456	13.479
Total	12 kbps	10.429	10.456
	16 kbps	12.054	12.091
	24 kbps	13.640	13.617

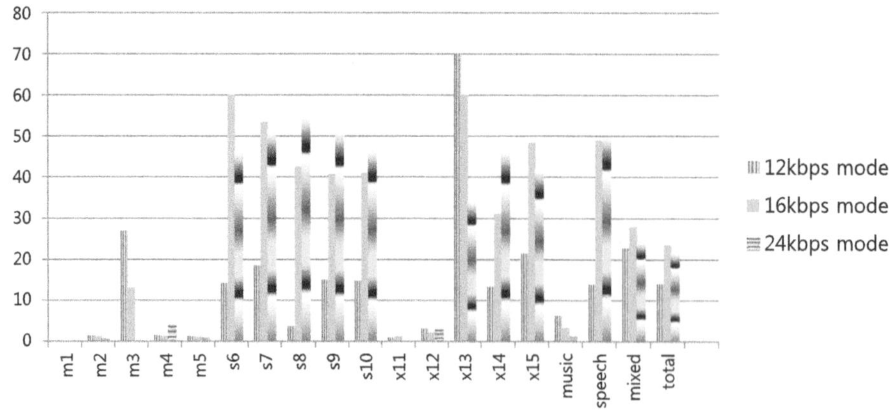

Figure 7 FAC frame rate of each test item and each category.

the error caused by ACELP encoding and inverse windowing. Accordingly, the amount of quantization error can be evaluated and compared by measuring signal-to-noise ratio (SNR) values. As will be shown from the experimental results given in the next section, there is no difference between the proposed and the conventional FAC algorithm. Subjective listening test also confirms the result.

4 Performance evaluation

4.1 Simulation setup and implementation

To verify the performance of the proposed algorithm, the USAC common encoder (JAME) is used as a baseline. The JAME developed by ourselves is officially released by MPEG as an open source [8], and its decoder module generates the bit-exact output set by the standardization process. In the recent verification test [18], the JAME encoder shows significantly better quality than the reference model encoder (RME) and comparable quality to the state-of-the-art reference quality encoder (RQE). Since the RQE is not publicly available, the JAME is a good baseline system for implementing the proposed algorithm. Table 1 summarizes the 15 test items used for the USAC standardization process, which are selected for performance evaluation in this paper. Both objective and subjective tests are performed to evaluate the performance of the proposed algorithm.

Note that USAC is designed to have a capability of dynamic bit allocation in each frame. Therefore, the achieved average bit rate of each test item in each operating mode needs to be measured. Two methods are implemented for evaluation. First is the conventional method using FAC algorithm (Conv.), and second is the proposed method using new aliasing cancelation algorithm (Prop. -B). Table 2 shows the actual achieved bit rates of two methods in operating modes of 12, 16, and 24 kbps. The bit rates of the proposed algorithm (Prop.-B) are much less than those of the conventional algorithm (Conv.) because it does not need bits for encoding FAC signal. As shown in Table 2, we attached the symbol (-B) into the label of the proposed output (Prop.-B) for emphasizing not to use additional bits.

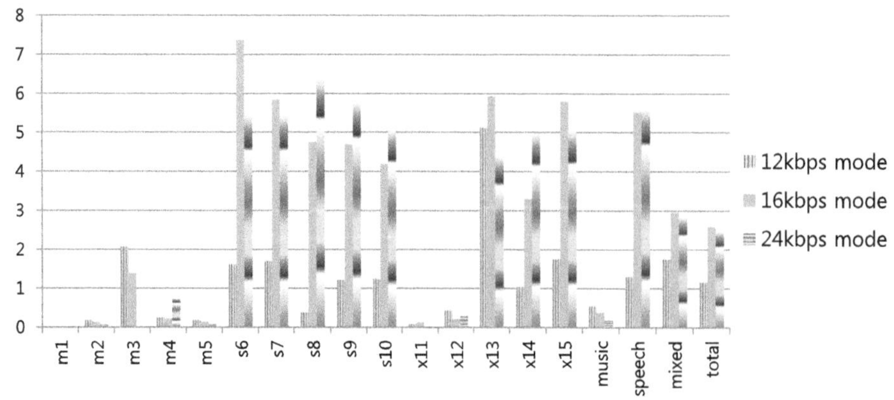

Figure 8 FAC bit ratio of each test item and each category.

Table 4 Subjective test environment

Feature	Description
Methodology	MUSHRA
Number of subjects	8
Headphones	Sennheiser HD600
Systems under the test	ref : Hidden reference
	lp35 : 3.5 kHz Low-pass anchor
	Conv. : JAME with FAC
	Prop.-B : JAME with New AC
Modes	12, 16, and 24 kbps mono

4.2 Objective test

Figure 6 shows an example of speech spectrogram that includes mode transition frames. If there is no aliasing cancelation algorithm, the output has severe distortion as shown in Figure 6d. Since the distortion is spread out to all frequency bands, it is heard as strong click noise. These perceptually annoying noises exist more frequently in speech and mixed signals because more transition frames occur in the contents. To clarify the effectiveness of the proposed algorithm, the signal-to-noise ratio is measured at 12-, 16-, and 24-kbps operating modes.

Table 3 summarizes the results. The SNR of the proposed algorithm (Prop.-B) is similar to that of the FAC algorithm (Conv.). Note that the proposed method does not need any additional bits compared to FAC algorithm; thus, the transmitted bits for encoding FAC frames can be saved. To measure the number of bits to be saved, the FAC frame rate is computed in each test item and each category. The FAC frame rate, α, is calculated as

$$\alpha\ (\%) = \frac{N_{\text{fac}}}{N} \times 100, \qquad (16)$$

where N_{fac} is the number of FAC frames, and N is the number of total frames.

The FAC bit ratio, β, is obtained as

$$\beta\ (\%) = \frac{1}{\bar{B}} \sum_{i=1}^{N} B_{i,\text{fac}} \times 100, \qquad (17)$$

where $B_{i,\text{fac}}$ is the FAC bits for ith frame and $\bar{B}$ is the number of total bits.

Figures 7 and 8 depict the results. The FAC frame rate at the 12-kbps operating mode is lower than those at the 16- and 24-kbps operating modes because the allocated bits for the FAC frame are insufficient. Since music contents generally do not use the ACELP coding mode, it hardly has any FAC frame. On the contrary, FAC rates of speech at 16- and 24-kbps operating modes are around 50%. In case of mixed signal, the speech-dominant content has many FAC frames. The FAC bit ratio of the speech-like signals at 16- and 24-kbps operating modes are over 5%. The rate at the 12-kbps operating mode is lower than others due to the insufficient amount of available bits.

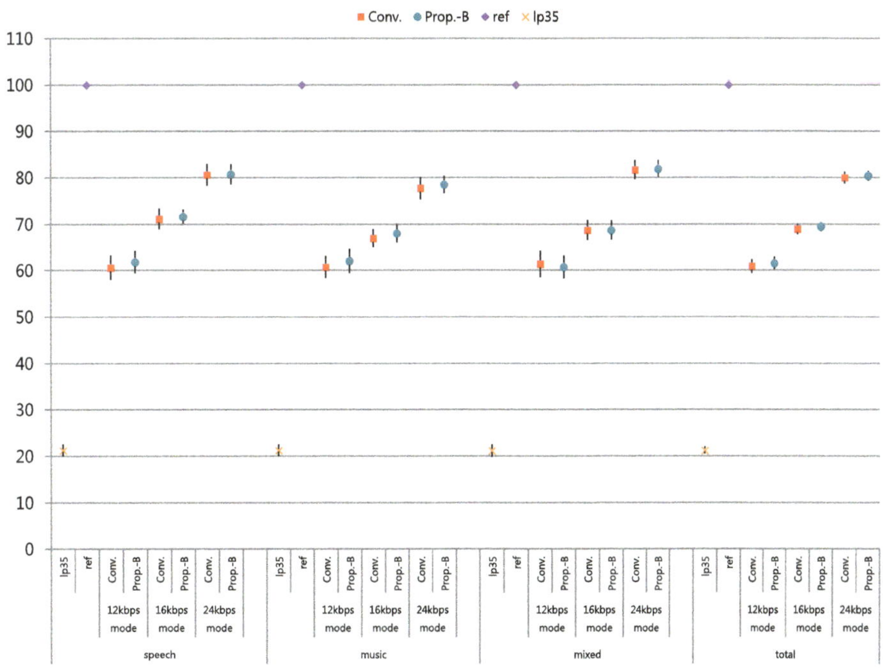

Figure 9 MUSHRA results at 12, 16, and 24 kbps operating modes.

4.3 Subjective test

Through the measurement of SNR and FAC bit ratio, it shows that the proposed algorithm has comparable performance to the USAC standard while it does not need any additional bits for FAC frames as given in Table 2. To verify the performance in terms of perceptual quality, listening tests are performed. Table 4 summarizes the test environment. Eight trained listeners participated in the Multiple Stimuli with Hidden Reference and Anchor (MUSHRA) [19] test for the contents encoded and decoded at 12-, 16-, and 24-kbps operating modes. Results given in Figure 9 denote mean values and 95% confidence levels of test scores, and they used the same achieved bit rates given by Table 2.

The synthesized signal using the proposed algorithm (Prop.-B) has comparable performance to the FAC algorithm (Conv.). Note again that the proposed method does not need additional bits to remove the aliasing term as we have explained before.

5 Conclusions

Although the FAC algorithm solves the switching problem caused by combining two heterogeneous types of coders, i.e., time domain coder and frequency domain coder, it needs additional bits to cancel out the aliasing components at every transition frame. The proposed new aliasing cancelation algorithm does not need additional bits because it efficiently utilizes decoded signals in the adjacent frames. The proposed algorithm is sophisticatedly integrated into the recently released open-source platform. In case of speech-like signals, it saves over 5% of the total bits compared with the conventional FAC algorithm. Both subjective listening tests and objective tests confirmed that the proposed algorithm showed comparable quality to the conventional FAC algorithm, but it does not require any additional bits for FAC encoding.

Competing interests

The authors declare that they have no competing interests.

Authors' information

JS received his B.S. and M.S. degrees in electrical and electronic engineering from Yonsei University, Seoul, South Korea, in 2004 and 2008, respectively. He is currently pursuing his Ph.D. degree at Yonsei University. His research interests include speech coding, unified speech and audio coding, spatial audio coding, and 3D audio. HGK (M94) received his B.S., M.S., and Ph.D. degrees in electronic engineering from Yonsei University, Seoul, South Korea, in 1989, 1991, and 1995, respectively. He was a Senior Member of the Technical Staff at AT&T, Labs-Research, from 1996 to 2002. In 2002, he joined the Department of Electrical and Electronic Engineering, Yonsei University, where he is currently a professor. His research interests include speech signal processing, array signal processing, and pattern recognition.

Acknowledgements

The authors would like to thank the reviewers for their suggestions which have contributed a lot to the great improvement of the manuscript.

References

1. M Neuendorf, M Multrus, N Rettelbach, G Fuchs, J Robilliard, J Lecomte, S Wilde, S Bayer, S Disch, C Helmrich, R Lefebvre, P Gournay, B Bessette, J Lapierre, K Kjörling, H Purnhagen, L Villemoes, W Oomen, E Schuijers, K Kikuiri, T Chinen, T Norimatsu, CK Seng, E Oh, M Kim, S Quackenbush, B Grill, MPEG unified speech and audio coding - the ISO/MPEG standard for high-efficiency audio coding of all content types, in *130th AES Convention* (Budapest, 26–29 April 2012)
2. ISO/IEC JTC1/SC29/WG11, *Unified Speech and Audio Coding Verification Test Report N12232*. (ISO/IEC JTC 1, New York, 2011)
3. J Makinen, B Bessette, S Bruhn, P Ojala, R Salami, A Taleb, AMR-WB+: a new audio coding standard for 3RD generation mobile audio services. IEEE Int. Conf. Acoustics Speech Signal Process. (ICASSP '05). **2**, 1109–1112 (2005)
4. M Wolters, K Kjorling, D Homm, H Purnhagen, Closer look into MPEG-4 high efficiency AAC, in *115th AES Convention* (Jacob K Javits Convention Center New York, 10–13 October 2003)
5. ISO/IEC JTC1/SC29/WG11, *Proposal for Unification of USAC Windowing and Frame Transitions M17020*. (ISO/IEC JTC 1, New York, 2009)
6. D Virette, B Kövesi, P Philippe, Adaptive time-frequency resolution in modulated transform at reduced delay. IEEE Int. Conf. Acoustics Speech Signal Process. (ICASSP '08). **2**, 3781–3784 (2008)
7. ISO/IEC JTC1/SC29/WG11, *Proposed Core Experiment for Enhanced Low Delay AAC M14237*. (ISO/IEC JTC 1, New York, 2007)
8. ISO/IEC JTC1/SC29/WG11, *Unified Speech and Audio Coder Common Encoder Reference Software N12022*. (ISO/IEC JTC 1, New York, 2011)
9. ISO/IEC JTC1/SC29/WG11, *ISO/IEC 23003-3/FDIS, Unified Speech and Audio Coding N12231*. (ISO/IEC JTC 1, New York, 2011)
10. JP Princen, AB Bradley, Analysis/synthesis filter bank design based on time domain aliasing cancellation. IEEE Trans. Acoustics Speech Signal Process. **34**(5), 1153–1161 (1986)
11. K Brandenburg, M Bosi, Overview of MPEG audio: current and future standards for low-bit-rate audio coding. J. Audio Eng. Soc. **45**(1–2), 4–21 (1997)
12. JD Johnston, Estimation of perceptual entropy using noise masking criteria. IEEE Int. Conf. Acoustics Speech Signal Process. (ICASSP '98). **5**, 2524–2527 (1998)
13. G Fuchs, M Multrus, M Neuendorf, R Geiger, MDCT-based coder for highly adaptive speech and audio coding, in *European Signal Processing Conference (EUSIPCO 2009)* (Glasgow, August 2009), pp. 24–28
14. G Fuchs, V Subbaraman, M Multrus, Efficient context adaptive entropy coding for real-time applications, in *IEEE International Conference on Acoustics Speech Signal Process. (ICASSP '11)* (IEEE Piscataway, 2011), pp. 493–496
15. J Lecomte, P Gournay, R Geiger, B Bessette, M Neuendorf, Efficient cross-fade windows for transitions between LPC-based and non-LPC based audio coding, in *126th AES Convention* (Munich, 7–10 May 2009)
16. C-M Liu, W-C Lee, Unified fast algorithm for cosine modulated filter banks in current audio coding standards. J. Audio Eng. Soc. **47**(12), 1061–1075 (1999)
17. RA Horn, The Hadamard product. Symp. Appl. Math. **40**, 87–169 (1990)
18. ISO/IEC JTC1/SC29/WG11, *Verification Test Report on USAC Common Encoder, JAME N13215*. (ISO/IEC JTC 1, New York, 2012)
19. ITU, *Recommendation ITU-R BS.1534-1. Method for the Subjective Assessment of Intermediate Quality Level of Coding Systems 2001–2003*. (International Telecommunication Union, Geneva, 2003)

17

The Ethnic Lyrics Fetcher tool

Rafael P Ribeiro, Murilo AP Almeida and Carlos N Silla Jr*

Abstract

The task of automatic retrieval and extraction of lyrics from the web is of great importance to different Music Information Retrieval applications. However, despite its importance, very little research has been carried out for this task. For this reason, in this paper, we present the Ethnic Lyrics Fetcher (ELF), a tool for Music Information Researchers, which has a novel lyrics detection and extraction mechanism. We performed two experiments to evaluate ELF's lyrics extraction mechanism and its performance as a lyrics fetcher tool. Our experimental results show that ELF performs better than the current state of the art, using website-specific lyrics fetchers or manually retrieving the lyrics from the web.

Keywords: Lyrics retrieval; Ethnic lyrics; Music information retrieval

1 Introduction

Song lyrics are used in many different types of music information retrieval (MIR) applications. Examples of MIR applications with song lyrics are Music Mood Identification [1,2], Music Genre [3] and Mood [4] Classification, Artist Similarity [5,6], Artist Style Identification [7], Enhancing Query By Humming Systems [8,9], Automatic Recognition of Lyrics in Singing [10], among others [11-13]. Furthermore, the authors of [14] have shown that lyrics (at least in part) are the second most used in a MIR system and are also the second most desired information on such system.

The need for automated lyrics retrieval and extraction systems has been reported in several research papers. For example, in [15], the authors point out that due to some limitations on how they received their dataset (a list of song artist and song titles without blank spaces, e.g. ironmainden - fearofthedark), they were limited to using 5,631 lyrics out of 10,000 songs they had available.

In [16], the authors used a specialized online lyrics library (Leo's Lyrics); however, they mentioned that in their experiments, direct searching did not always work for cover songs. For example, the system would only find the lyrics for the song 'Smells Like Teen Spirit' if the song artist provided was 'Nirvana'. It would not find the lyrics if they provided 'Tori Amos' (who has recorded a cover of the song) as the song artist.

In the work of [17], the authors noted that their simple lyrics mining procedure failed to filter out all the non-text lyrics. In their approach, they examined the line of each html page to determine whether it contained lyrics-like text or not and then selected the longest sequence of lyrics-like text lines in the page. Any lyrics that were less than three lines or over 200 lines long were discarded.

In the work of [18], the authors created a lyrics dataset manually for the task of music genre classification. However, despite having available a private collection of about 12,000 songs, they only selected (randomly) 30 to 45 songs from each of the ten music genres available because of the inexistence of reliable automated tools to retrieve the lyrics for the whole collection.

Despite the importance of song lyrics to the different types of MIR applications and systems and the problems reported by several authors [15-18], to the best of the authors' knowledge (as we shall see in Section 2), very little research has been carried out in the area of the automated lyrics retrieval from the web [19,20].

The main contributions of this paper are the Ethnic Lyrics Fetcher (ELF) System[a] (presented in Section 3) and its novel lyrics detection and extraction procedure (presented in Section 4). It should be noted that our system is named as Ethnic Lyrics Fetcher in the same sense the word 'ethnic' is used in Computational Ethnomusicology [21], i.e., our system is potentially able to retrieve lyrics for all of the world's music. In order to evaluate ELF,

*Correspondence: carlosjunior@utfpr.edu.br
Computer Music Technology Laboratory at the Federal University of Technology of Paraná, Av. Alberto Carazzai, 1.640, Cornélio Procópio 86300-000, Brazil

we have performed two experiments. In the first experiment (Section 5), we evaluate the effectiveness of ELF's novel lyrics extraction algorithm. In the second experiment (Section 6), we evaluate the effectiveness of ELF as an automated tool to retrieve lyrics from the web. Finally, in Section 7, we present the conclusions of this work and future research directions.

2 Related work

The task of automatically retrieving lyrics from the web presents two problems: The first problem is how to detect whether or not a given webpage contains the lyrics of a given song. The second problem is how to extract only the lyrics content from a webpage that contains the lyrics of a given song.

It is possible to avoid these two problems by using an online lyrics website that has an internal lyrics database such as LeosLyrics [22]. However, this approach has one main limitation, i.e., no matter how big the lyrics collection of a given website is, it will still have a limited number of song lyrics which may not contain the song lyrics the user is interested in. This is particularly true for new and/or non-mainstream songs.

Another approach that has been used by some researchers, in order to avoid these two problems, is to develop a website-specific automated lyrics fetcher, such as jLyrics [23]. This approach avoids the issue of manually retrieving and storing the desired lyrics but it is still limited to retrieving only the lyrics available in the website it was designed for. Also, in order to develop such an automated lyrics fetcher, it is necessary to know the website structure in order to correctly locate and extract the lyrics content from the website.

To the best of the authors' knowledge, the only reported studies that are related to dealing with the two problems related to the task of automatic lyrics retrieval from the web are [19,20]. The work of Knees et al. [20] is possibly the first reported work that deals with the problem of retrieving lyrics from the web. However, the main objective of their work was to use sequence alignment algorithms to produce one robust (i.e., typo-free) lyric for a given song. In [20], the authors used the Google search engine with queries containing the following information: song artist, song title, and the keyword 'lyric'. In their pre-processing phase, they removed all HTML tags and links. Since, the authors were dealing with a different task, they could retrieve a collection of webpages and use co-relation methods in order to determine which webpages might have contained the lyrics of a given song.

In the work of [19], their main goal is to retrieve the lyrics from a given song. In order to search for lyrics, the authors also use the Google search engine and store the first 40 retrieved results of the query. Then, for each webpage, they use a regular expression-based approach to extract the lyrics content from the webpages.

It should be noted that in both works [19,20], the authors used the same database of 258 songs, and although both papers mention the Evil Lyrics software [24], they did not make any experimental comparisons with it. The Evil Lyrics software is a plug-in used by several digital music players such as Winamp and iTunes. Although all the implementation details of the software are not available, the existing information about the software allows us to classify it as a hybrid approach to retrieve lyrics from the web. In the version of the software used in this work, the Evil Lyrics plug-in had a list of 437 websites which we assume that it knows their structure. If the lyrics are not found in one of these websites, it then searchers the web for the lyrics; however, no information is available about how this web search is performed by the software.

3 ELF system overview

Figure 1 presents an overview of the ELF system. The ELF system works as follows:

1. In the first step, the ELF system needs information about the song lyrics the user wants to retrieve. The ELF system requires that the user provides the song title information (mandatory) and song artist information (optional).
2. In the second step, the ELF system pre-processes the song title and song artist information by applying case folding (e.g.,'É' becomes 'é') and accent removal (e.g., 'é' becomes 'e').
3. In the third step, the ELF system creates a query for the search engine. In this paper, we used the Google search engine, and the search queries have been constructed by concatenating the song title and artist information and by replacing all the white spaces in the resulting string with the plus (+) symbol. We also add to the search query a special keyword which is the the word 'lyric' in the language of the lyric the user is searching for.
4. In the fourth step, the ELF system process the JSON string returned by the Google application programming interface (API) to create a list of websites which may contain the lyrics the user is interested in.
5. In the fifth step, the ELF system uses its novel lyrics detection and extraction mechanism in order to identify if a given website contains a song lyrics or not.
   If the website contains the song lyrics, the ELF system returns it to the user. Otherwise, it analyzes the next website on the list created in the fourth step.

Figure 1 Overview of the Ethnic Lyrics Fetcher (ELF) system.

Note that the ELF system can work with lyrics in any language that can be encoded using UTF-8.

4 ELF's novel lyrics detection and extraction procedure

The ELF system employs a novel lyric detection and extraction procedure. The usual approach behind current lyrics extraction is to assume that the lyrics of a song are somewhere within the webpage where there are ends of line tags (i.e., <BR> tags) [19]. One of the differences between ELF novel lyrics extraction procedure and the existing approaches is the way ELF uses all the html tags (including the <BR> tags) to locate the lyrics within any webpage.

Before attempting to extract the lyrics from the webpage, the system verifies if the webpage does not come from the list of websites that are known to not have lyrics.

If this is not the case, ELF's lyric extraction procedure works as follows:

- The first step is to store the retrieved webpage into a string.
- The second step is to analyze the string looking for opening html tags with the exception of '<br' (line break), '<p' (paragraph), '</' (closing tag), and '<!-' (html comment). For each opening html tag found, do the following:

  1. Create a new position in the lyric extraction vector;
  2. Add all the content of the webpage until a new opening html tag is found.

- After the webpage has been converted into the lyric extraction vector, ELF analyzes the lyric extraction vector counting the number of '<BR' tags in each position of the vector.
- In order to decide what to output, the ELF lyric extraction procedure has threshold (θ), and it considers that any position in the lyric extraction vector that has a number of '<BR' tags higher than θ is the song lyrics and output it to the user.
- If the song lyric to be shown to the user is empty, ELF considers that the lyric was not detected on the webpage given as input.

Figure 2 presents an example of the lyrics extraction procedure used by ELF with different θ values for the song 'To rindo a toa' by the Brazilian band 'Falamansa'.

5 Evaluating ELF's lyrics extraction procedure

In this section, we are interested in answering the following questions by using controlled experiments: (a) How well does ELF's lyric extraction procedure (with different θ values) performs when compared against other approaches? (b) What value should be used as the θ value of ELF's lyric extraction procedure?

5.1 Experimental settings

In order to evaluate ELF's lyric extraction procedure, we manually verified 12 different websites which have lyrics for songs of the Latin Music Database (LMD) [25]. Each of the websites has a different way of marking the song lyrics content. Table 1 presents the list of websites we used in this experiment along with specific tag delimiters each website uses to identify the song lyrics content within their webpages.

In order to evaluate ELF's lyric extraction procedure, we provide ELF with the full html page and compare its output (using different θ values) with the gold standard of each website. The gold standard of each website is obtained by creating a website-specific lyrics extractor which knows beforehand the special delimiter tags used in each website.

In order to measure how similar the output of ELF's lyric extraction procedure (using different θ values) is with that of the gold standard in each website, we use the document similarity measure from the information retrieval field, known as cosine similarity [26]. The cosine similarity measure computes how similar two documents (lyrics in our case) are. The measure provides a value between 0 and 1. The higher the value, the more similar the two documents are. We also use the method presented in [19], which is, to the best of the authors' knowledge, the reported state of the art and the lyrics obtained by using the Evil Lyrics software [24].

5.2 Experimental results

Table 2 presents the experimental results of the lyrics extraction mechanism used by ELF with different θ values, [19] and [24] for the 12 different websites presented in Table 1.

The analysis of Table 2 shows some interesting results. First is that for higher θ values, a higher value of overall similarity is obtained. This can be explained by the fact that a low value will simply return all the webpage content that has a <BR> tag in the same position of the lyric extraction vector.

Second, we manually inspected the lyrics with higher θ values in order to understand why the similarity was not 1.00. Interestingly, we observed that some of the websites have among their structured lyrics some kind of promotional message such as 'Ringtone - Send this ringtone to your cell phone!' which was removed by the lyric extraction procedure used by ELF. In our experiments, this happened with the lyricsmania.com, 1songlyrics.com, lyricshall.com, and lyricsreg.com websites. Another observation that we made was that some websites such as lyricsoncall.com and 6lyrics.com have in the middle of the song lyrics the name of their website. This information was also removed by the lyric extraction procedure used by ELF. In the case of the vounessa.com.br website, it included the fields composer and music genre at the end of the lyrics which were also filtered out by ELF's lyric extraction procedure.

Third, based on our experimental results, it seems that there is a trade-off between using a low value or a high value for θ. A low value might return everything, while a high value may filter out more than it should. Therefore, for the experiments in the next section, we employed the value of $\theta > 3$. Fourth, despite the value of θ used, the novel lyrics detection and extraction procedure used by ELF is better than the method

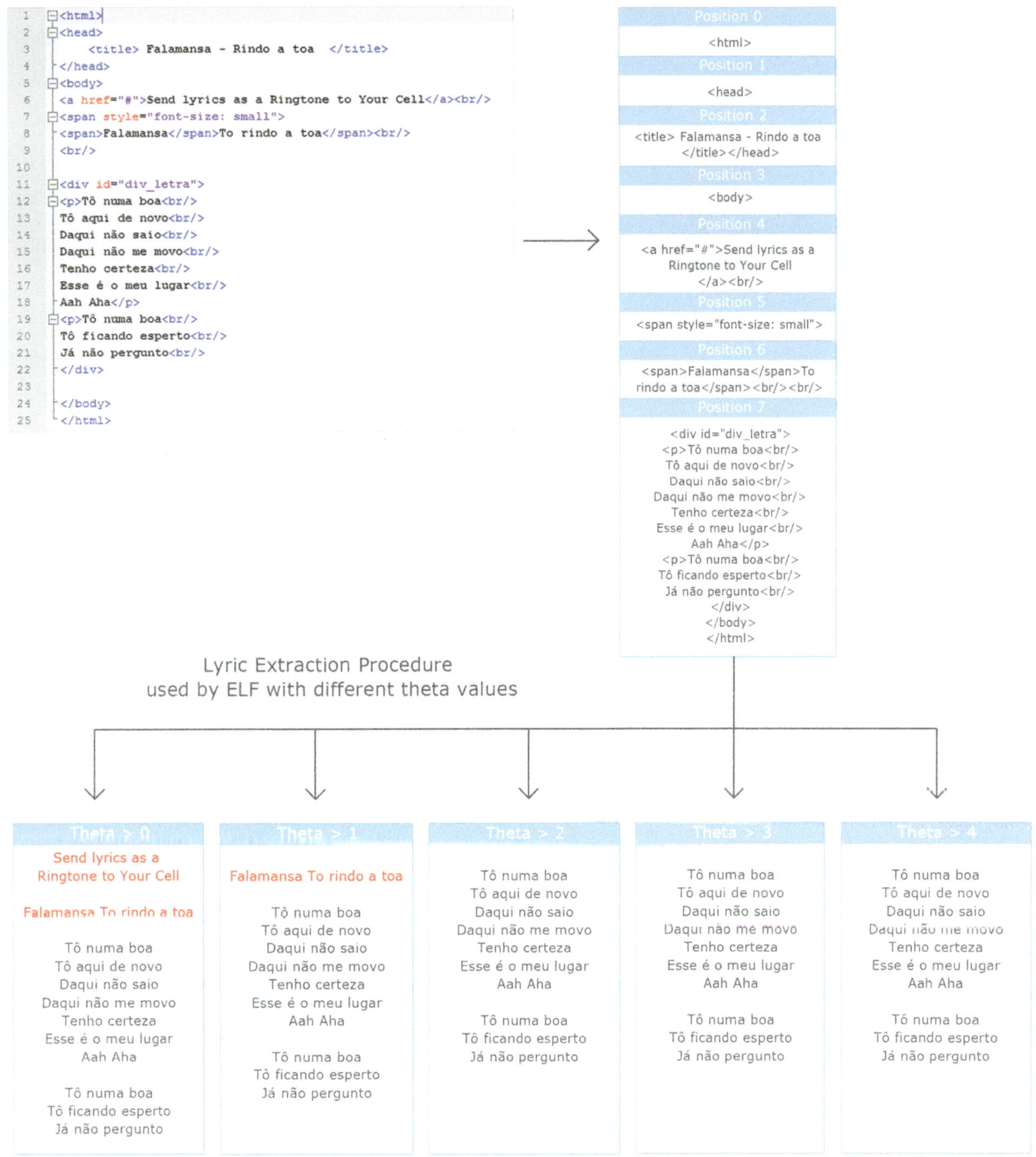

Figure 2 **An example of the lyric extraction procedure used by ELF with different theta values.**

proposed in [19] or the method used by the Evil Lyrics software.

6 Evaluating ELF as a lyrics retrieval system

In this section, we are interested in evaluating how well ELF performs as an online lyrics retrieval system. That is, we want to evaluate how well it performs by using a search engine API and its lyrics detection mechanism.

6.1 Experimental settings

In order to evaluate ELF as a online lyrics retrieval system, we have provided ELF with the song artist and song title information for all the songs available in the LMD [25]. The motivation behind using the LMD is that it contains songs with lyrics in English, Spanish, Brazilian Portuguese, and Spanglish (i.e., lyrics with words in English and Spanish).

Table 1 Lyrics websites and their lyrics identifier tags

Website	Lyrics-content starting tag	Lyrics-content enclosing tag
1songlyrics.com	`<div id="wrapper">`	`<p><b>`
6lyrics.com	`<span id="ctl00_ContentPlaceHolder1_lbl Content "style="display: block">`	`</span></p>`
lyricmania.com	`<div id="wrapper">`	`<div id="wrapper2">`
www.lyricsoncall.com	`<p><strong>`	`<div id="info">`
www.moron.nl	`<td colspan=2 style="line -height:15px; size:14px;"></td>`	
vounessa.com.br	`<div class="bloco-c">`	`<!- Fim div bloco c ->`
letras.mus.br	`<div id="main_cnt">`	`</div>`
www.vagalume.com.br	`<div id=lyr_original class="left originalOnly">`	`</div>`
lyricshall.com	`<div class="correct-button">`	`<h3 class="smalltitle">`
lyricspedia.com	`<!- AddThis Button END ->`	`</a></b></div>`
lyricsreg.com	`</div><div style="text -align:center;"><g:plusone>`	
themusic-world.com	`<BLOCKQUOTE>`	`</BLOCKQUOTE>`

In this experiment, we have used the Google API as ELF's search engine for querying possible websites that may contain the lyrics for the song the user is looking for. As a baseline for ELF, we have implemented 12 website-specific lyrics fetchers (those listed in Table 1) that have a website-specific URL encoding and the lyrics-specific information. Table 3 presents the website structure for each of the 12 lyrics websites.

We also evaluate two versions of the ELF system. One version uses the song title and song artist information, referred to as ELF_{ST+A}, and the other version uses only the song title information, referred to as ELF_{ST}.

It should be noted that in the experiments reported in this section, the following websites have been listed in ELF as non-lyrics websites: 'blogspot', 'wikipedia', and 'youtube'.

Additionally, in order to compare ELF with the current state-of-the-art technology in online lyrics fetching, we also present the performance of the Evil Lyrics software [24]. To make a fair comparison with ELF, we have also run the experiments with Evil Lyrics using song title and song artist information (referred to as $EvilLyrics_{ST+A}$) or using only song title information (referred to as $EvilLyrics_{ST}$).

6.2 Experimental results

Let us now evaluate ELF as a lyrics retrieval system by comparing the total number of lyrics retrieved by it with the other approaches. Table 4 presents the experimental results (broken down by music genre) when using ELF_{ST+A}, ELF_{ST}, the other 12 lyrics websites, and the two configurations of the Evil Lyrics software to retrieve lyrics for the Latin Music Database.

Table 2 Evaluation of ELF's and the lyrics extraction approaches used in [19] and [24]

Website	ELF with $\theta > 0$	ELF with $\theta > 1$	ELF with $\theta > 2$	ELF with $\theta > 3$	ELF with $\theta > 4$	[19]	[24]
letras.mus.br	0.9226	1.0	1.0	0.9998	0.9998	0.7199	0.3556
vagalume.com.br	0.9151	0.9191	1.0	1.0	1.0	0.7202	0.3434
songlyrics.com	0.9753	0.9761	0.9767	0.9780	0.9780	0.6389	0.3753
6lyrics.com	0.9141	0.9800	0.9972	0.9972	0.9966	0.5808	0.3721
vounessa.com.br	0.9996	0.9996	0.9995	0.9996	0.8960	0.6379	0.3409
lyricmania.com	0.9552	0.9617	0.9618	0.9617	0.9617	0.6295	0.3703
lyricshall.com	0.9702	0.9706	0.9706	0.9706	0.9162	0.8781	0.1893
lyricsoncall.com	0.9668	0.9666	0.9666	0.9787	0.9787	0.5932	0.3232
lyricspedia.com	0.9974	0.9967	0.9920	0.9920	0.9905	0.6725	0.3756
lyricsreg.com	0.9533	0.9686	0.9689	0.9730	0.9730	0.7305	0.3474
moron.nl	0.8847	0.9890	1.0	1.0	1.0	0.8088	0.3907
themusic-world.com	0.9905	0.9890	0.9922	0.9923	0.9922	0.7096	0.3646
Average	0.9537	0.9764	0.9855	0.9869	0.9736	0.6933	0.3457

Table 3 Fixed structure of the 12 lyrics websites used as baselines

Website	Structure
letras.mus.br	`http://letras.mus.br/$song artist$/$song title$`
vagalume.com.br	`http://www.vagalume.com.br/$song artist$/$song title$.html`
moron.nl	`http://www.moron.nl/lyrics/$song artist$/$song title$-lyrics.html`
lyricsoncall.com	`http://www.lyricsoncall.com/lyrics/$song artist$/$song title$-lyrics.html`
lyricsreg.com	`http://www.lyricsreg.com/lyrics/$song artist$/$song title$`
vounessa.com.br	`http://www.vounessa.com.br/musicas/$song title$`
6lyrics.com	`http://www.6lyrics.com/$song artist$-lyrics-artista.aspx`
lyricshall.com	`http://www.lyricshall.com/lyrics/$song artist$/$song title$`
lyricmania.com	`http://www.lyricmania.com/$song artist$-$song title$-lyric.html`
1songlyrics.com	`http://www.1songlyrics.com/$first letter of song artist$/$song artist$/$song title$.html`
lyricspedia.com	`http://www.lyricspedia.com/$song artist$/$song title$-lyrics/`
themusic-world.com	`http://www.themusic-world.com/$song artist$/lyrics/$song title$`

In Table 4, from row 1 to row 12, we present the results of using the individual lyrics websites and their lyrics page structure to retrieve the song lyrics from the LMD. The analysis of these results shows that the Brazilian websites vagalume.com.br (1,569 lyrics retrieved) and vounessa.com.br (1,944 lyrics retrieved) contain the highest number of lyrics for songs from the LMD among the individual websites. To some extent, this is an expected result since six out of the 10 music genres in the LMD are Brazilian music genres.

However, we were also interested in knowing if these websites could be combined into one lyrics retrieval system in order to obtain a higher number of lyrics. The result of this combination is shown in Table 4 row 13. The combined approached of the individual websites retrieved 2,480 lyrics for the songs in the LMD. It should be noted

Table 4 Number of lyrics retrieved by the different approaches for each music genre in the LMD

Row number	Approach	A	Ba	Bo	F	G	M	P	Sa	Se	T	Total
1	1songlyrics.com	28	94	71	7	0	53	2	102	0	24	381
2	6lyrics.com	17	107	112	2	0	70	9	96	2	101	516
3	letras.mus.br	132	98	127	229	126	39	168	140	14	168	1,241
4	lyricmania.com	21	98	72	8	0	52	3	109	0	23	386
5	lyricshall.com	0	0	2	0	0	0	0	0	0	0	2
6	lyricsoncall.com	46	80	67	10	6	42	36	85	27	23	422
7	lyricspedia.com	13	43	61	11	0	35	1	77	0	22	263
8	lyricsreg.com	11	23	18	1	0	6	0	32	0	4	95
9	moron.nl	0	12	6	1	0	3	0	0	7	0	29
10	themusic-world.com	0	23	66	1	0	18	0	0	35	0	143
11	Vagalume.com.br	204	125	136	208	105	68	224	129	257	113	1,569
12	Vounessa.com.br	247	210	239	259	179	171	0	230	143	266	1,944
13	All 12 websites	268	238	251	277	208	185	237	253	275	288	2,480
14	Manually retrieved	292	251	268	296	178	101	299	172	310	336	2,503
15	EvilLyrics_{ST+A}	289	283	275	311	272	292	288	257	298	298	2,863
16	ELF_{ST+A}	298	247	233	307	272	227	301	267	309	386	2,847
17	EvilLyrics_{ST}	294	272	292	289	298	289	278	255	289	379	2,935
18	ELF_{ST}	303	303	301	313	300	291	301	299	308	397	3,116
19	LMD total	304	308	302	315	306	307	301	303	310	404	3,160
20	LMD instrumental songs	0	0	0	6	24	0	2	0	0	54	86

A, Axé; Ba, Bachata; Bo, Bolero; F, Forró; G, Gaúcha; M, Merengue; P, Pagode; Sa, Salsa; Se, Sertaneja; T, Tango.

that among these results (individual or the combined approach), the webpages would be extracted using the known webpage structure and, as discussed in section 5.2, may contain some additional content besides the song lyrics (such as some type of advertisement).

In Table 4 row 14, we present the results of the lyrics being manually searched (using the Google search engine) and retrieved by the authors of this paper. It should be noted that the authors are familiar with all the languages present in the songs of the LMD and that the method used for manually gathering the lyrics consisted on, first, using well-known websites for the particular music genres presented in the LMD such as letras.mus.br and then, on a second stage, the authors searched for lyrics using song title and artist information using Google search engine. This process has led to the retrieval of 2,503 lyrics of songs from the LMD.

In Table 4 row 15, we present the results for the Evil Lyrics software using song artist and song title information (EvilLyrics$_{ST+A}$). This version of Evil Lyrics has retrieved 2,863 lyrics for songs from the LMD. In order to evaluate if all the lyrics retrieved were the actual lyrics of the songs, we have manually inspected each one of them. After this analysis, we verified that 422 out of the 2,847 retrieved lyrics were not actual lyrics but some other type of information such as artist's biographies and discographies. Therefore, in total, the EvilLyrics$_{ST+A}$ correctly retrieved 2,425 lyrics.

In Table 4 row 16, we present the results for ELF$_{ST+A}$ (with $\theta > 3$). This version of ELF has managed to retrieve 2,847 lyrics from songs in the LMD. This is a very interesting result, as these lyrics were retrieved automatically by the system. In order to evaluate if all the lyrics retrieved were the actual lyrics of the songs, we have manually inspected each one of them. After this analysis, we verified that 204 out of the 2,847 retrieved lyrics were not actual lyrics but some other type of information. Therefore, in total, the ELF$_{ST+A}$ correctly retrieved 2,643 lyrics.

Another aspect that we analyzed was which lyrics websites were being used by ELF$_{ST+A}$ to retrieve lyrics. In total, the ELF$_{ST+A}$ has gathered lyrics from 157 different websites. Due to space limitations, it is not viable to list all the websites here, but it is interesting to note that the top five websites used by ELF$_{ST+A}$ were letras.mus.br (1,490 lyrics), www.todotango.com (279 lyrics), www.musica.com (224 lyrics), www.vagalume.com.br (91 lyrics), and www.mp3lyrics.org (76 lyrics). Out of these websites, only two of them were known to the authors of this paper before these experiments.

In Table 4 row 17, we present the results for the Evil Lyrics software using only song title information (EvilLyrics$_{ST}$). This version of Evil Lyrics has retrieved 2,935 lyrics for songs in the LMD. In order to evaluate if all the lyrics retrieved were the actual lyrics of the songs, we have manually inspected each one of them. After this analysis, we verified that 215 out of the 2,935 retrieved lyrics were not actual lyrics but some other type of information. Therefore, in total, the EvilLyrics$_{ST}$ correctly retrieved 2,720 lyrics.

In Table 4 row 18, we present the results for using ELF$_{ST}$ (with $\theta > 3$), i.e., using only song title information. This version of ELF has managed to retrieve 3,116 lyrics for songs in the LMD. As with the ELF$_{ST+A}$, we manually verified if all the lyrics were the actual lyrics of the songs. Our analysis has shown that for ELF$_{ST}$, only 60 of the retrieved lyrics were not the actual lyrics of the songs. Therefore, the ELF$_{ST}$ has successfully retrieved 3,056 lyrics of the LMD using 110 different websites. The top five websites used were letras.mus.br (1,454 lyrics), www.musica.com (760 lyrics), www.todotango.com (274 lyrics), www.vagalume.com.br (114 lyrics), and www.lyricsmode.com (38 lyrics).

In Table 4 row 19, we present the total number of songs available within each music genre of the LMD, and in row 20, we present the number of instrumental songs in the LMD. After the analysis of this information, it becomes clear that the maximum number of lyrics that could be retrieved correctly for the songs in the LMD is 3,074.

Let us now analyze the results under the information retrieval metrics of precision, recall, and F-measure presented in Table 5. The analysis of Table 5 shows some interesting results. First, it should be noted that for the

Table 5 Evaluation of the different approaches for retrieving lyrics from the web

Approach	Precision	Recall	*F*-measure
lyricshall.com	1	0.000632911	0.001265022
moron.nl	1	0.009177215	0.018187520
lyricsreg.com	1	0.030063291	0.058371736
themusic-world.com	1	0.045253165	0.086587950
lyricspedia.com	1	0.083227848	0.153666375
1songlyrics.com	1	0.120569620	0.215193448
lyricmania.com	1	0.122151899	0.217710096
lyricsoncall.com	1	0.133544304	0.235622557
6lyrics.com	1	0.163291139	0.280739935
letras.mus.br	1	0.392721519	0.563962736
vagalume.com.br	1	0.496518987	0.663565236
vounessa.com.br	1	0.615189873	0.761755486
EvilLyrics$_{ST+A}$	0.847013622	0.906012658	0.875520318
All 12 websites	1	0.784810127	0.879432624
Manually retrieved	1	0.792405063	0.884180791
ELF$_{ST+A}$	0.928345627	0.900949367	0.914442348
EvilLyrics$_{ST}$	0.926746167	0.928797468	0.927770684
ELF$_{ST}$	0.980744544	0.986075949	0.983403021

specific websites, although their precision is always 1.0, there is a need to know their website structure and their webpage structure in order to correctly fetch lyrics which is unfeasible in practice.

Second, with the exception of the vagalume.com.br and vounessa.com.br websites, all the specific websites provided low recall values as they only retrieved a small portion of the number of lyrics requested. Third, the Evil Lyrics software using both song artist and song title information achieved an F-measure similar to the baseline 12-website combined approach and the manually retrieval of the songs. This might have happened because the Evil Lyrics software is a hybrid approach to web-based lyrics retrieval, since it has a list of websites which we assume it knows their structure. In the version of the software used in the experiments, the websites letras.mus.br and www.vagalume.com.br were present in the list of known websites. For this reason, we believe that it might have used this websites first when searching for the lyrics, since it already knew their structure, hence its similar performance with the 12-website combined approach.

Fourth, both versions of the Ethnic Lyrics Fetcher system have achieved high F-measure rates, being 0.9144 for the ELF_{ST+A} and 0.9834 for the ELF_{ST}. The difference between the performance of ELF_{ST+A} and the ELF_{ST} lies in the fact that in Latin music, many authors perform covers of songs performed by well-known artists. Therefore, in many cases when the song in the LMD is sung by a less famous band, the ELF_{ST+A} may not find the lyrics, even though the lyrics is widely available under the famous artist name. This might also explain why the $EvilLyrics_{ST}$ was the second best performing approach in our experiments.

Fifth, the F-measure achieved by the ELF_{ST} is higher than those of all the other approaches: manually searching for the lyrics (0.8841), using all the 13 lyrics websites together (0.8794), or using the $EvilLyrics_{ST}$ (0.9277). This demonstrates the feasibility of using the ELF system to automatically retrieve lyrics for different music information retrieval tasks and research experiments.

7 Conclusions

The task of automatic lyrics retrieval and extraction from the web is a non-trivial task. Despite its importance to several music information retrieval applications, very little research has been carried out so far. For this reason, in this paper, we have presented the ELF tool and its novel lyrics detection and extraction mechanism.

In order to evaluate ELF's novel lyric extraction method from any webpage, we have evaluated its performance against 12 lyrics websites with known structure for delimiting the lyrics content and also against the method proposed by [19]. Our experimental results show that ELF's novel lyrics extraction procedure is better than the method prosed by [19] and that it should be used with a θ value higher than 3 ($\theta > 3$).

We have also evaluated ELF as a lyrics retrieval system. In our experiments, we compared its performance with 12 site-specific lyrics fetchers, the manual retrieval of the lyrics, and also with two versions of the Evil Lyrics software [24]. In these experiments, two versions of ELF were used: the first version (ELF_{ST+A}) uses both the song artist and song title information, while the second version (ELF_{ST}) uses only the song title information. Our experiments have shown that both versions of ELF can be used as lyrics retrieval tools, but that the ELF_{ST} has a better performance (measured by the F-measure) because it is capable of correctly fetching lyrics for cover songs. Furthermore, ELF_{ST} has outperformed all the other approaches, even the Evil Lyrics software which is currently one of the most well-known and used plug-ins in several digital music player softwares. As future research, we intend to perform experiments with the Ethnic Lyrics Fetcher system on other music databases with songs in other languages.

Endnote

[a]Available online at http://music.cp.utfpr.edu.br/elf/.

Abbreviations

API: Application Programming Interface; ELF: Ethnic Lyrics Fetcher; ELF_{ST}: Ethnic Lyrics Fetcher using only the song title information; ELF_{ST+A}: Ethnic Lyrics Fetcher using the song title and artist information; MIR: music information retrieval.

Competing interests

The authors declare that they have no competing interests.

Acknowledgements

We thank the anonymous reviewers for their insightful and valuable feedback on the earlier versions of this manuscript.

References

1. X Hu, JS Downie, AF Ehmann, Lyric text mining in music mood classification, in *Proceedings of the 10th International Society for Music Information Retrieval Conference (ISMIR)* (Kobe, Japan, 26–30 Oct 2009), pp. 411–416
2. X Hu, JS Downie, When lyrics outperform audio for music mood classification: a feature analysis, in *Proceedings of the 11th International Society for Music Information Retrieval Conference (ISMIR)* (Utrecht, Netherlands, 9–13 Aug 2010), pp. 619–624
3. R Mayer, A Rauber, Musical genre classification by ensembles of audio and lyrics features, in *Proceedings of the 12th International Society for Music Information Retrieval Conference (ISMIR)* (Miami, USA, 24–28 Oct 2011), pp. 675–680
4. B den Brinker, R van Dinther, J Skowronek, Expressed music mood classification compared with valence and arousal ratings. EURASIP J. Audio Speech Music Process. **2012**, Article ID 24 (2012)
5. S Baumann, O Hummel, Using cultural metadata for artist recommendations, in *Proceedings of the 3rd International Conference on Web Delivering of Music* (Leeds, UK, 15–17 Sept 2003), pp. 138–141
6. B Logan, A Kositsky, P Moreno, Semantic analysis of song lyrics, in *Proceedings of the IEEE International Conference on Multimedia and Expo* (Taipei, Taiwan, 27–30 June 2004), pp. 827–830

7. T Li, M Ogihara, Music artist style identification by semi-supervised learning from both lyrics and content, in *Proceedings of the 12th Annual ACM International Conference on Multimedia* (New York, USA, 10–16 Oct 2004), pp. 364–367
8. Z Guo, Q Wang, G Liu, J Guo, Y Lu, A music retrieval system using melody and lyric, in *Proceedings of the IEEE International Conference on Multimedia and Expo Workshops* (Melbourne, Australia, 9–13 July 2012), pp. 343–348
9. C-C Wang, J-SR Jang, W Wang, An improved query by singing/humming system using melody and lyrics information, in *Proceedings of the 11th International Society for Music Information Retrieval Conference (ISMIR)* (Utrecht, Netherlands, 9–13 Aug 2010), pp. 45–50
10. A Mesaros, T Virtanen, Automatic recognition of lyrics in singing. EURASIP J. Audio Speech Music Process. **2010**, Article ID 546047 (2010)
11. S Funasawa, H Ishizaki, K Hoashi, Y Takishima, J Katto, Automated music slideshow generation using web images based on lyrics, in *Proceedings of the 11th International Society for Music Information Retrieval Conference (ISMIR)* (Utrecht, Netherlands, 9–13 Aug 2010), pp. 63–68
12. JPG Mahedero, A Martinez, P Cano, M Koppenberger, F Gouyon, Natural language processing of lyrics, in *Proceedings of the 13th Annual ACM International Conference on Multimedia* (Singapore, 6–11 Nov 2005), pp. 475–478
13. T O'Hara, N Schuler, Y Lu, DE Tamir, Inferring chord sequence meanings via lyrics: process and evaluation, in *Proceedings of the 13th International Society for Music Information Retrieval Conference (ISMIR)* (Porto, Portugal, 8–12 Oct 2012), pp. 463–468
14. JS Downie, SJ Cunningham, Toward a theory of music information retrieval queries: system design implications, in *Proceedings of the 3rd International Conference on Music Information Retrieval* (Paris, France, 13–17 Oct 2002), pp. 299–300
15. MV Zaanen, P Kanters, Automatic mood classification using tf*idf based on lyrics, in *Proceedings of the 11th International Society for Music Information Retrieval Conference (ISMIR)* (Utrecht, Netherlands, 9–13 Aug 2010), pp. 75–80
16. DA Shamma, B Pardo, KJ Hammond, Musicstory: a personalized music video creator, in *Proceedings of the 13th Annual ACM International Conference on Multimedia* (Singapore, 6–11 Nov 2005), pp. 563–566
17. R Macrae, S Dixon, Ranking lyrics for online search, in *Proceedings of the 13th International Society for Music Information Retrieval Conference (ISMIR)* (Porto, Portugal, 8–12 Oct 2012), pp. 362–366
18. R Mayer, R Neumayer, A Rauber, Rhyme and style features for musical genre classification by song lyrics, in *Proceedings of the 9th International Society for Music Information Retrieval Conferece (ISMIR)* (Philadelphia, USA, 14–18 Sept 2008), pp. 337–342
19. G Geleijnse, J Korst, Efficient lyrics extraction from the web, in *Proceedings of the 7th International Society for Music Information Retrieval Conference (ISMIR)* (Victoria, Canada, 8–12 Oct 2006), pp. 371–372
20. P Knees, M Schedl, G Widmer, Multiple lyrics alignment: automatic retrieval of song lyrics, in *Proceedings of the 6th International Society for Music Information Retrieval Conference (ISMIR)* (London, UK, 11–15 Sept 2005), pp. 564–569
21. G Tzanetakis, A Kapur, WA Schloss, M Wright, Computational ethnomusicology. J. Interdiscip. Music Stud. **1**(2), 1–24 (2007)
22. Leo's Lyrics, http://www.leoslyrics.com/. Accessed 14 Oct 2013
23. jLyrics, http://jmir.sourceforge.net/jLyrics.html. Accessed 14 Oct 2013
24. Evil Lyrics, http://www.evillabs.sk/evillyrics/. Accessed 10 Dec 2013
25. CN Silla Jr, AL Koerich, CAA Kaestner, The Latin Music Database, in *Proceedings of the 9th International Conference on Music Information Retrieval* (Philadelphia, USA, 14–18 Sept 2008), pp. 451–456
26. G Salton, C Buckley, Term weighting approaches in automatic text retrieval. Inf. Process. Manag. **24**(5), 513–523 (1998)

18

Evaluation of linguistic and prosodic features for detection of Alzheimer's disease in Turkish conversational speech

Ali Khodabakhsh*, Fatih Yesil, Ekrem Guner and Cenk Demiroglu

Abstract

Automatic diagnosis and monitoring of Alzheimer's disease can have a significant impact on society as well as the well-being of patients. The part of the brain cortex that processes language abilities is one of the earliest parts to be affected by the disease. Therefore, detection of Alzheimer's disease using speech-based features is gaining increasing attention. Here, we investigated an extensive set of features based on speech prosody as well as linguistic features derived from transcriptions of Turkish conversations with subjects with and without Alzheimer's disease. Unlike most standardized tests that focus on memory recall or structured conversations, spontaneous unstructured conversations are conducted with the subjects in informal settings. Age-, education-, and gender-controlled experiments are performed to eliminate the effects of those three variables. Experimental results show that the proposed features extracted from the speech signal can be used to discriminate between the control group and the patients with Alzheimer's disease. Prosodic features performed significantly better than the linguistic features. Classification accuracy over 80% was obtained with three of the prosodic features, but experiments with feature fusion did not further improve the classification performance.

Keywords: Alzheimer's disease; Speech processing; Linguistic features; Prosodic features; Machine learning

1 Introduction

As the worldwide elderly population increases, the incidence of Alzheimer's disease is becoming more widespread. It is estimated that 7% of the world's population over 65 years old has Alzheimer's or a related dementia [1]. Moreover, only one in four patients has been diagnosed [1]. Because there is no treatment to cure the disease, years of healthcare costs are becoming a significant economic burden on governments as well as patients and their families. The global cost of Alzheimer's and dementia is estimated to be $605 billion, which is equivalent to 1% of the entire world's gross domestic product [2]. The problem intensifies each year with the aging world population. Thus, simplifying healthcare processes and reducing costs through the use of automated systems can make a significant socio-economic impact.

Diagnosis of the disease is costly and difficult. Moreover, even if the disease is diagnosed correctly, monitoring the progression of the disease by a clinician over time further increases the cost. Thus, patients cannot visit clinicians frequently and what happens between the visits is largely unknown to clinicians.

Typically, clinicians use tests such as Mini-Mental State Examination (MMSE) and linguistic memory tests [3]. Linguistic memory tests are based on the recall rates of word lists and narratives, and they are typically more effective than the MMSE. Moreover, individual's medical and family histories are used, along with MRI scans to test for other brain-related conditions, such as stroke. Biomarkers showing the level of beta-amyloid accumulation in the brain or the neurons that are injured or actually degenerating can also be used in combination with the other tests [4].

None of those typical practices consider the speech signal in diagnosing the disease even though the part of the brain cortex that processes language abilities is one of the earliest parts to be affected by the disease [5]. For example,

*Correspondence: ali.khodabakhsh@ozu.edu.tr
Electrical and Computer Engineering Department, Ozyegin University, Orman Street, 34794, Istanbul, Turkey

narrative retelling ability is found to be strongly correlated with the disease [6]. Similarly, linguistic features derived from the transcription of a narrative retelling task were found to be significantly correlated with primary progressive aphasia, which is a type of dementia [7]. Similarly, analysis of the speech signal has been shown to be useful for Alzheimer's detection in [8,9]. However, in those works, the speech signal is recorded during the administration of standard clinical tests. Moreover, most of the focus is on the high-level structural processing of spoken language for a specific language. For example, in [10], features such as moderate word finding difficulty, reduced phrase length, and reduced comprehension are manually tagged by humans and shown to contain information complementary to standardized tests. Correlation of linguistic capability with Alzheimer's disease was also shown in [11]. Speech-based features are investigated in [5] to detect fronto-temporal lobar degeneration with promising results. Speech is recorded in a semi-structured interview setting in [5]. The frequency and ratio of syntactic categories such as pronouns and adverbs are found to be markers of the disease.

In addition to natural language processing (NLP) features, speech acoustics have also been studied and reported in the literature. A limited study with one patient with a focus on the prosodic features of speech such as stress, intonation, and emotion is reported in [12]. Problems of speech production that are related to central nervous system problems are also noted in [13]. In [14], dysfluency cycles in speech are measured using the length and frequency of hesitations in speech. Subjects with dementia were found to have patterns different from those of control subjects.

Here, we focus on extracting an extensive set of acoustic and linguistic features from spoken language to detect Alzheimer's disease. Because the patients with Alzheimer's are usually not able to take automated tests or to carry on a structured conversation, data collection is done during unstructured conversational speech. In this way, a subject's speech can be recorded in the most natural and effortless way by a person with minimal technical or clinical skills. Semi-structured conversational data has been investigated in [15,16], but only linguistic features are analyzed, and speech features are not considered. Similarly, conversational data has been investigated for limited sets of linguistic and speech dysfluency features by [5,17], who measured the correlation of those features with the disease and attempt to use the features for diagnosis.

In our work, we focused on evaluating the effectiveness of a large set of features for detecting Alzheimer's disease in unstructured conversations. The data was collected in Turkish, which has not been studied to the extent of languages such as English. We propose 20 prosodic features extracted automatically from the recordings and 18 linguistic features derived from the transcriptions of patients and control subjects. We have investigated the predictive power of each feature as well as combination of features using support vector machines (SVM), nearest neighbor (NN) classifiers, naive Bayesian classifiers, and classification trees (CTree). Our results indicate that some of the investigated features are strong predictors of the disease with high statistical significance independent of the age, education, and gender of the subjects. Prosodic features were more successful than the linguistic features. In fact, only two of the linguistic features were found to be significant. Accuracy of greater than 80% was obtained with three of the prosodic features. Silence ratio, which is defined as the rate of silences in speech regardless of their durations, was found to be the most useful feature. Feature fusion did not improve the performance, which indicates that the features are not complementary to one another.

2 Linguistic features

A list of the linguistic features that are extracted from the transcriptions of the recorded conversations with the test subjects is shown in Table 1. The features are geared towards detecting problems with the flow of the conversation and measuring how well the subject can understand the question or carry on the conversation without getting confused. Recordings are manually transcribed. Each recording is first split into conversation turns. Then, the turns where the subjects speak are further split into utterances that are segments where the subjects talk without interruption by the interviewer and without long silences. The splitting mechanism is shown in Figure 1 where a voice activity detector is used for detecting silences. Only the turns of subjects are used in feature extraction.

2.1 Hesitation and confusion features

During recording, we found that patients tend to hesitate more, forget what they were talking about, and have a harder time finding the right words or remembering details about their pasts. They also sometimes get confused about why they cannot remember the details or forget the context of the conversation. Those observations led us to propose features that will be able to capture those patterns in transcriptions.

2.1.1 Question ratio

Patients are more likely to forget details in the middle of conversation, to not understand the questions, or to forget the context of the question. In those cases, they tend to ask the interviewer to repeat the question or they get confused, talk to themselves, and ask further questions about the details. The question words such as 'which,' 'what,' etc. are tagged automatically in each conversation. The full list of question tags that were used here is shown in Table 2. The question ratio of a subject is computed by dividing

Table 1 Lists of linguistic and prosodic features and their IDs

	Category	ID	Features
Linguistic features	Hesitation and puzzlement features	1.1	Question ratio
		1.2	Filler ratio
		1.3	Incomplete sentence ratio
	POS-based features	2.1	Verb freq.
		2.2	Noun freq.
		2.3	Pronoun freq.
		2.4	Adverb freq.
		2.5	Adjective freq.
		2.6	Particle freq.
		2.7	Conjunction freq.
		2.8	Pronoun-to-noun ratio
	Intelligibility	3	Unintelligible word ratio
	Complexity features	4.1	Standardized word entropy
		4.2	Suffix ratio
		4.3	Number ratio
		4.4	Brunet's index
		4.5	Honore's statistic
		4.6	Type-token ratio
Prosodic features	Voice activity-related features	5.1	Response time
		5.2	Response length
		5.3	Silence ratio
		5.4	Silence to utt. ratio
		5.5	Long silence ratio
		5.6	Avg. silence count
		5.7	Silence rate
		5.8	Cont. speech rate
		5.9	Avg. cont. word count
	Articulation-related features	6.1	Avg. abs. delta energy
		6.2	Dev. of abs. delta energy
		6.3	Avg. abs. delta pitch
		6.4	Dev. of abs. delta pitch
		6.5.1	Avg. abs. delta formant 1
		6.5.2	Avg. abs. delta formant 2
		6.5.3	Avg. abs. delta formant 3
		6.5.4	Avg. abs. delta formant 4
		6.6	Voicing ratio
	Rate of speech-related features	7.1	Phoneme rate
		7.2	Word rate

the total number of question words by the number of utterances spoken by the subject.

2.1.2 Filler ratio

Filler sounds such as 'ahm' and 'ehm' are used by people in spoken language when they think about what to say next. We hypothesize that they may be used more frequently by the patients because of slow thinking and memory recall processes. Patients tend to forget what they are talking about and to use fillers more often than the control subjects. The filler ratio is computed by dividing the total number of filler words by the total number of utterances spoken by the subject.

2.1.3 Incomplete sentence ratio

One of our observations of the patients is their inability to complete sentences. They seem to either forget what they were going to say or to completely change the context and start talking about a different topic. Incomplete sentences are manually labeled for each conversation. To compute this feature, the ratio of incomplete sentences to the total number of the sentences is calculated.

2.2 POS-based features

Part of speech (POS) tags can be used to extract markers for detecting the disease. For example, frequent adjectives can indicate more colorful and descriptive use of language, while frequent adverbs can indicate the ability to relate different utterances to each other. The frequency of each POS tag can also be a useful identifier of patients with Alzheimer's disease.

POS tags are added automatically to each word using a Turkish stemmer [18]. In cases where a word can have multiple alternative POS tags, equal weights are given to all possibilities. For instance, if a word can be either a noun or an adverb, depending on the sentence, that word is counted as half adverb and half noun in computation. The following POS tag frequencies are used as features:

- Verb frequency
- Noun frequency
- Pronoun frequency
- Adverb frequency
- Adjective frequency
- Particle frequency
- Conjunction frequency
- Pronoun-to-noun ratio

Frequency of a POS tag is computed by dividing the total number of words with that tag by the total number of words spoken by the subject in the recording. Pronoun-to-noun ratio is the ratio of the total number of pronouns to the total number of nouns.

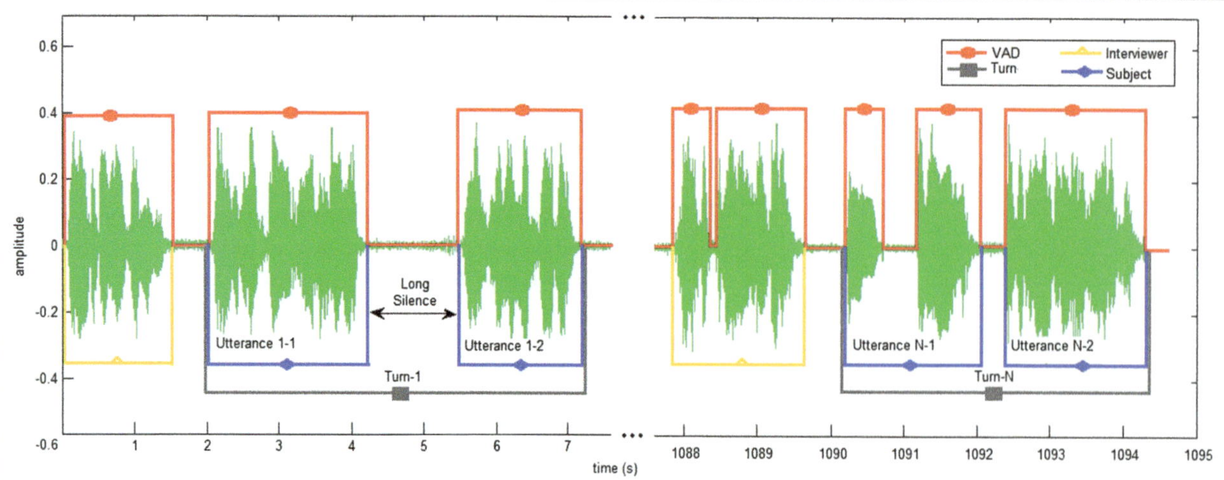

Figure 1 Conversation turns between a patient and the interviewer are shown. Each conversation turn contains one or more utterances. Voice activity detector (VAD) is used to detect the long and short silences.

2.3 Unintelligible word ratio

During the conversations, some of the words spoken by the patients were unintelligible. These are mostly because patients could not produce the words correctly, they mumbled, or they were thinking while talking, which reduced intelligibility. Unintelligible word ratio is the ratio of unintelligible words to all words spoken by the subject.

Annotation of unintelligible words was done manually by three listeners for each conversation. A word was marked as unintelligible only when at least two of the three listeners could not understand it.

2.4 Complexity features

2.4.1 *Standardized word entropy*

One of the earliest parts of the brain to be damaged by Alzheimer's disease is the part of the brain that deals with language ability [5]. We hypothesize that this may cause a degradation in the variety of words and word combinations that a patient uses. Standardized word entropy, i.e., word entropy divided by the log of the total word count, is used to model this phenomenon. Because the aim is to compute the variety of word choice, stemming is done, and only the stems of the words are considered.

Table 2 Question tags that were used in computing the question ratio

Word	Translation
Efendim	Excuse me
Hangi[si]	Which [one]
Hani	So where's/Why ... not .../You remember
Kaç[ı]	How many/much [of]
Kim[in]	Who[se]
M[i|ı|u|ü]	Question suffix
Nasıl	How
Ne[suffix+]	What
Nere[suffix+]	Where
Niye	Why

Some of the question tags occur with one or more additional suffixes. Those are indicated with [.] symbol.

2.4.2 *Suffix ratio*

The standardized word entropy feature focuses on the variety of the stem words while ignoring the suffixes. However, suffixes can also be strong indicators of the complexity of a sentence. Turkish, in particular, has a rich and complex morphological structure [19]. Hundreds of different words can be generated from the same stem word by appending suffixes to it. Thus, we investigated whether the patients tend to construct simpler words than the control subjects by analyzing the suffixes they used. The suffix ratio of a subject is calculated by dividing the total number of suffixes by the total number of words spoken by the subject.

2.4.3 *Number ratio*

During conversations, subjects give details about their birth dates, how many kids they have, and other numerical information. Such use of numbers in a sentence can be a measure of recall ability. The number ratio feature is calculated by dividing the total count of numbers by the total count of words the subject used in the conversation.

2.4.4 *Brunet's index*

Brunet's index (W) quantifies lexical richness [20]. It is calculated as $W = N^{V^{-0.165}}$, where N is the total text length and V is the total vocabulary. Lower values of W correspond to richer texts. As with standardized word entropy, stemming is done on words and only the stems are considered.

2.4.5 Honore's statistic

Honore's statistic [21] is based on the notion that the larger the number of words used by a speaker that occur only once, the richer his overall lexicon is. Words spoken only once (V_1) and the total vocabulary used (V) have been shown to be linearly associated. Honore's statistic generates a lexical richness measure according to $R = 100 \times \log(N/(1 - V_1/V))$, where N is the total text length. Higher values correspond to a richer vocabulary. As with standardized word entropy, stemming is done on words and only the stems are considered.

2.4.6 Type-token ratio

A pattern that we noticed in the recordings of the Alzheimer's patients is the frequency of repetitions in conversation. Patients tend to forget what they have said and to repeat it elsewhere in the conversation. The metric that we used to measure this phenomenon is type-token ratio [22]. Type-token ratio is defined as the ratio of the number of unique words to the total number of words. In order to better assess the repetitions, only the stems of the words are considered in calculations.

3 Prosodic features

A total of 20 prosodic features were extracted and evaluated for detecting Alzheimer's disease. A list of all prosodic features used here is shown in Table 1. Descriptions of the prosodic features are given below. All prosodic feature computations are performed over the locution of the subject. Locution is the total response period of the subject which is the sum of all of the subject's speech turns. Each speech turn includes utterances, long silences, and short silences, as shown in Figure 1.

3.1 Voice activity-related features

Silence and speech segments are automatically labeled in each conversation with a voice activity detector (VAD). The VAD used here is based on the distribution of the short-time frame energy of the speech signal. Because there is both silence and speech in the recordings, the energy distribution has two modes, both of which can be modeled with a Gaussian distribution. The bimodal distribution of silence and speech is trained using the expectation-maximization (EM) algorithm. The mode that has a lower mean is used to represent silence, and the mode that has a higher mean is used to represent speech.

Energy of each short-time speech frame in the recording is classified as either speech or silence using the likelihood ratio test (LRT). Because the test treats each frame independently, a second processing step is used where silence and speech segments that were shorter than four frames are removed.

The transcriptions of recordings were available and could be used for VAD through forced alignment using an automatic speech recognition system. However, the VAD described above worked well and more sophisticated VAD techniques were not required.

3.1.1 Response time

When the interviewer asks a question, it often takes some time before the subject gives a response. It is hypothesized that this time can be an indicator of the disease since it is expected to be related to cognitive processes such as attention and memory. The time it takes the subject to answer a question is calculated in each segment as the response time measure.

3.1.2 Response length

Response length is the average length of a subject's response in seconds to the interviewer's question. Beginning and trailing silences are removed.

3.1.3 Silence ratio

The plan-and-execute cycle in speech production was found to be distinctly different in patients compared to control subjects as noted in [14]. In our data, we also observed that patients tend to stop more in the middle of sentences to think about what to say next. The silence ratio is computed by dividing the total number of silences in the whole locution by the total number of words in the locution. Dividing by the number of words, we reduce the variability that arises from different speaking rates.

3.1.4 Silence-to-utterance ratio

Silence-to-utterance ratio is the ratio of the total number of silences to the total number of utterances. Similar to silence ratio, it is a measure of the hesitation rate of the subject.

3.1.5 Long silence ratio

Patients sometimes pause for a long time while answering a question. They do not use fillers during these long periods, and the interviewers did not interrupt these periods of silence. We hypothesized that these pauses may correspond to moments when the subject is retrieving information which is expected to be longer for the Alzheimer's patients. Similarly, confusion may also lead to long silences. The rate of such long hesitation events, defined as silences longer than approximately one second, is used to detect the disease. This feature is computed as the ratio of the total count of long silences to the total number of words.

3.1.6 Average silence count

This feature specifies the average number of silences produced by a speaker in one second of speech. It is calculated by dividing the total number of silences by the duration of the locution.

3.1.7 *Silence rate*
The silence rate measures the silence as a proportion of the whole locution. It is computed by dividing the total duration of all silence segments by the duration of the locution.

3.1.8 *Continuous speech rate*
This feature measures how long the subject speaks until the next long silence, which is considered to be a thinking or recalling state. It is defined as the average duration of continuous speech segments over the whole locution.

3.1.9 *Average continuous word count*
As mentioned above, the thinking process longer for patients than for the control subjects. The silence rate features discussed above try to exploit this long thinking process. Another way to measure it is to compute the average number of consecutive words that are spoken without intervening silences. First, the number of words for each continuous segment is computed. Then, the mean of these counts is used as the feature.

3.2 Articulation-related features
The voice activity-related features discussed above are related to cognitive thought processes. However, it is also important to measure how the subject uses his or her voice articulations during speech. For example, if the subject becomes emotional, significant changes in the fundamental frequency (pitch) can be expected. Similarly, changes in the resonant frequencies (formants) of speech can be a strong indicator of the subject's health. If the formants do not change fast enough or are not distinct enough, sounds may become harder for listeners to identify, leading to the perception of mumbling. In order to see the impact of these effects on classification of the disease, pitch and formant trajectories are extracted, and the following features are derived over the whole locution.

3.2.1 *Average absolute delta energy*
Energy variations can convey information about the mood of the subject. Changing energy significantly during speech may indicate a conscious effort to stress words that are semantically important or a change in mood related to the content of the speech. The absolute value of each frame-to-frame energy change is measured, and the average of these changes over the whole locution is computed.

3.2.2 *Deviation of absolute delta energy*
In addition to changes in energy, changes in the delta energy, which is the acceleration of energy, can be used. The standard deviation of the average absolute delta energy is used to further investigate the possible impacts of the disease on the energy change rate.

3.2.3 *Average absolute delta pitch*
The average of the absolute delta pitch shows the rate of variations in pitch. This feature is highly correlated with the emotions carried through the speech signal.

3.2.4 *Deviation of absolute delta pitch*
The standard deviation of the absolute delta pitch is also used as a feature to further analyze the possible impacts of the disease on the pitch change rate. A monotonic increase or decrease in the pitch may simply be related to routine changes in sentence structure. However, acceleration of pitch, measured with the standard deviation of absolute delta pitch, can capture unusual pitch events in speech.

3.2.5 *Average absolute delta formants*
The average of the absolute delta formant frequencies indicates the rate of change in the formant features. Formants are related to the positions of the vocal organs such as the tongue and lips. Reduction of control over these organs related to damage in the brain caused by Alzheimer's disease can create speech impairments such as mumbling. In this case, formants do not change quickly and speech becomes less intelligible [23]. Changes in the first four formants are used as features in this research.

3.2.6 *Voicing ratio*
Another speech impairment is the loss of voicing in speech. In this case, the subject loses the ability to control the vibrations of the vocal cords, which results in breathy and noisy speech. The ratio of the total duration of voiced speech to the total duration of speech in the locution is used to detect any potential impairment in the vocal cords.

3.3 Rate of speech-related features
3.3.1 *Phoneme rate*
A basic identifier of rate of speech is the average number of phonemes spoken per second. The phoneme rate of a subject is computed by dividing the number of phonemes by the duration of the locution.

3.3.2 *Word rate*
Similar to phoneme rate, word rate is used to measure the rate of speech at the word level. Word rate is computed by dividing the number of words by the duration of the locution.

4 Experiments
Conversational speech recordings of 32 patients and 51 age and education-matched control subjects were collected and manually transcribed. Recordings from four

patients were neglected because they were either unintelligible for the most part or they did not talk much. Thus, recordings from a total of 28 patients were used in experiments. The Alzheimer's patients and the control subjects were recruited from the same healthcare facility, but the control subjects were receiving treatment for injuries or illnesses other than Alzheimer's disease. Gender, age, and education details of the subjects are shown in Table 3. The age range is between 60 and 90 in both control subjects and patients.

Unstructured conversations were carried with the subjects where questions were asked depending on the flow of the conversation. Thus, different topics and different questions were used to make the subjects feel comfortable. The transcriptions were produced by one person and then reviewed by another person. The transcribers and the subjects were native Turkish speakers. In order to annotate unintelligible words, a third person also listened to the recordings.

The data was collected at elderly healthcare facilities in Istanbul. For each subject, approximately 10 min of conversation was recorded using a high-quality microphone. The recording was then manually segmented into speech turns between the interviewer and subject. In each speech turn, only the subject or the interviewer speaks. Segments of speech where both the subject and the interviewer talk were not used in the analysis.

After linguistic and prosodic features are extracted, SVM, NN classifiers, naive Bayesian classifiers, and CTree are used for classification. A linear Kernel is used for the SVM. For the NN classifier, Euclidean distance is used. For the CTree, nodes are split to minimize within-node impurity. Impure nodes that contain samples both from patients and control subjects are split only if they have more than nine samples.

There is more data available for the control subjects than for the Alzheimer's patients because the number of subjects that were in the healthcare facilities and willing to provide data was larger. Even though equal amounts of data from both groups could be used in the experiments to have a balance, all of the available data was used with special care while training the classifiers as discussed below.

Table 3 Gender, age, and years of education for the patient and control subjects

	AD (n = 28)	Control (n = 51)
Male/Female	18/10	31/20
Age	75 (6)	75.9 (6.4)
Education	11.6 (4.9)	11.4 (6)

Age and education data are presented in mean (standard deviation) format.

Data imbalance can become a problem for the SVM, NN, and decision tree algorithms, where the data points are used directly, as opposed to the naive Bayes approach, where the distribution of the data is used. For the NN, SVM, and decision tree classifiers, a random subsampling approach is used, in which a subset of the control subjects is randomly selected such that there is an equal number of data points for the control subjects and the patients. For each test case, the subsampling procedure is repeated ten times, and the average performance is reported.

In the first phase of testing, each feature is tested separately to assess the classification power of individual features. Then, in the second phase, combinations of features are used to increase the classification power of the algorithms. Features are normalized to have zero-mean and unit variance.

Because there is a limited number of subjects in the data set, a leave-one-out evaluation strategy is used, in which one of the subjects is left out and the classifier is trained with the rest of the subjects and tested on the left-out subject.

5 Results and discussion

The age, education level, and gender of the subjects can significantly affect performance in classification tests. Therefore, initial testing is done to control for the effect of age, education, and gender on the performance. Only features that have significant performance in all three control tests are reported as significant markers. Significance is measured using the paired t-test. A given feature may not have significant performance with all classifiers. In this case, the feature is reported as significant if it can pass the significance test with at least one of the classifiers.

Age-, education-, and gender-controlled experimental results are discussed below. Analysis and discussion of the features and combination of features with statistically significant discriminative power are reported in Section 5.4.

5.1 Age-controlled experiments

The age-controlled linguistic features are shown in Table 4. All features related to POS tags other than nouns and pronouns were found to be insignificant with this control variable. Incomplete sentences and unintelligible word ratios were found to be age related and not disease related. Similarly, all features that are related to the complexity of the language also became insignificant when age was used as a control variable.

The age-controlled prosodic features are shown in Table 5. The significance of these features was found to be less related to age compared to the linguistic features. In particular, formant and voicing features that are related to articulation were found to be age related and not disease related. Patients in the older age group had a harder time controlling their vocal cords and other articulatory

Table 4 Accuracy (%) of each classifier using the linguistic features in age-controlled experiments

		Age between 60 and 75 ($p = 15, c = 25$)					Age between 76 and 90 ($p = 13, c = 26$)				
ID	Features	SVM	Bayes	CTree	NN	*p* value	SVM	Bayes	CTree	NN	*p* value
1.1	Question ratio	***70.0***	***67.5***	***70.0***	60.0	**0.011**	***66.7***	35.0	38.5	43.6	**0.037**
1.2	Filler ratio	52.5	65.0	57.5	47.5	0.058	48.7	38.5	46.2	53.9	0.631
1.3	Incomplete sentence ratio	60.0	57.5	52.5	60.0	0.206	61.6	***66.7***	59.0	64.1	**0.037**
2.1	Verb freq.	37.5	62.5	47.5	50.0	0.114	***69.2***	61.5	53.9	59.0	**0.016**
2.2	Noun freq.	***67.5***	65.0	60.0	60.0	**0.027**	59.0	***71.8***	53.9	59.0	**0.006**
2.3	Pronoun freq.	***72.5***	***72.5***	65.0	47.5	**0.004**	***69.2***	***66.7***	56.4	59.0	**0.016**
2.4	Adverb freq.	52.5	62.5	52.5	52.5	0.114	59.0	41.0	38.5	43.6	0.262
2.5	Adjective freq.	57.5	47.5	45.0	42.5	0.343	51.3	43.6	51.3	51.3	0.873
2.6	Particle freq.	52.5	60.0	52.5	55.0	0.206	48.7	61.5	***77.0***	***66.7***	**0.001**
2.7	Conjunction freq.	62.5	62.5	65.0	52.5	0.058	53.9	56.4	56.4	46.2	0.423
2.8	Pronoun-to-noun ratio	***70.0***	***70.0***	***67.5***	62.5	**0.011**	***71.8***	***66.7***	61.5	46.2	**0.006**
3	Unintelligible word ratio	60.0	62.5	55.0	55.0	0.114	***82.1***	***84.6***	***84.6***	***82.1***	**<0.001**
4.1	Standardized word entropy	55.0	65.0	45.0	57.5	0.058	***74.4***	64.1	59.0	56.4	**0.002**
4.2	Suffix ratio	45.0	52.5	55.0	60.0	0.206	***69.2***	***74.4***	64.1	53.9	**0.002**
4.3	Number ratio	***70.0***	***67.5***	***70.0***	65.0	**0.011**	48.7	61.5	51.3	48.7	0.150
4.4	Brunet's index	62.5	57.5	***67.5***	65.0	**0.027**	48.7	59.0	61.5	56.4	0.150
4.5	Honore's statistic	40.0	45.0	65.0	47.5	0.058	43.6	61.5	61.5	***71.8***	**0.006**
4.6	Type-token ratio	47.5	45.0	42.5	45.0	1.000	***79.5***	***79.5***	64.1	***76.9***	**<0.001**

Number of patients (*p*) and control subjects (*c*) in each group is given in parenthesis. Classifiers with significant results are denoted in italics. Features with *p* values less than 0.05 are considered significant and are shown in bold. *p* value of the best classifier is reported for each feature.

Table 5 Accuracy (%) of each classifier using the prosodic features in age-controlled experiments

		Age between 60 and 75 ($p = 15, c = 25$)					Age between 76 and 90 ($p = 13, c = 26$)				
ID	Features	SVM	Bayes	CTree	N	p-value	SVM	Bayes	CTree	NN	p-value
5.1	Response time	60.0	60.0	52.5	45.0	0.206	61.5	59.0	48.7	48.7	0.150
5.2	Response length	62.5	57.5	52.5	65.0	0.058	***66.7***	41.0	51.3	61.5	**0.037**
5.3	Silence tatio	***70.0***	***70.0***	62.5	***72.5***	**0.004**	***94.9***	***94.9***	***94.9***	***84.6***	**<0.001**
5.4	Silence to Utt. ratio	***67.5***	65.0	60.0	62.5	**0.027**	***76.9***	***71.8***	***69.2***	59.0	**0.001**
5.5	Long silence ratio	57.5	***70.0***	57.5	55.0	**0.011**	***74.4***	***74.4***	***69.2***	***69.2***	**0.002**
5.6	Avg. silence count	***82.5***	***80.0***	***82.5***	***80.0***	**<0.001**	***74.4***	***74.4***	64.1	59.0	**0.002**
5.7	Silence rate	***67.5***	***75.0***	62.5	***72.5***	**0.002**	***74.4***	***79.5***	***74.4***	59.0	**<0.001**
5.8	Cont. speech rate	***80.0***	***80.0***	***75.0***	***75.0***	**<0.001**	***74.4***	***71.8***	***66.7***	64.1	**0.002**
5.9	Avg. cont. word count	***67.5***	62.5	60.0	55.0	**0.027**	***87.2***	***92.3***	***94.9***	***84.6***	**<0.001**
6.1	Avg. abs. delta energy	***67.5***	***75.0***	***67.5***	57.5	**0.002**	***71.8***	***71.8***	***79.5***	64.1	**<0.001**
6.2	Dev. of abs. delta energy	***75.0***	***72.5***	62.5	55.0	**0.002**	***69.2***	***71.8***	41.0	51.3	**0.006**
6.3	Avg. abs. delta pitch	***67.5***	62.5	55.0	47.5	**0.027**	61.5	***82.1***	59.0	64.1	**<0.001**
6.4	Dev. of abs. delta pitch	45.0	55.0	55.0	***72.5***	**0.004**	59.0	***69.2***	56.4	56.4	**0.016**
6.5.1	Avg. abs. delta formant 1	62.5	57.5	55.0	62.5	0.114	***74.4***	***71.8***	***69.2***	61.5	**0.002**
6.5.2	Avg. abs. delta formant 2	55.0	62.5	52.5	52.5	0.114	***69.2***	***66.7***	***66.7***	***71.8***	**0.006**
6.5.3	Avg. abs. delta formant 3	52.5	47.5	62.5	55.0	0.114	***66.7***	***66.7***	***76.9***	59.0	**0.001**
6.5.4	Avg. abs. delta formant 4	45.0	65.0	55.0	57.5	**0.002**	***74.4***	***76.9***	***66.7***	***69.2***	**0.001**
6.6	Voicing ratio	***75.0***	***75.0***	62.5	***75.0***	**0.002**	59.0	59.0	53.9	43.6	0.262
7.1	Phoneme rate	***67.5***	***75.0***	***77.5***	60.0	**0.001**	***82.1***	***82.1***	***76.9***	58.0	**<0.001**
7.2	Word rate	***72.5***	***72.5***	***67.5***	60.0	**0.004**	56.4	61.6	***66.7***	48.7	**0.037**

The number of patients (*p*) and control subjects (*c*) in each group is given in parenthesis. Classifiers with significant results are denoted in italics. Features with *p* values less than 0.05 are considered significant and are shown in bold. *p* value of the best classifier is reported for each feature.

organs, but this was not a significant discriminative factor for the younger group of subjects.

5.2 Education-controlled experiments

As in the age-controlled experiments, linguistic features performed poorly in the education-controlled experiments, shown in Table 6. Most notably, pronoun frequency and pronoun-to-noun ratio were significant indicators of the disease independent of the education level. Patients use pronouns more often than nouns. This is surprising since we hypothesized that patients would use pronouns less often since they are used to refer to nouns mentioned earlier in the conversation which we assumed would require more cognitive effort. Analyzing the transcripts in more detail, we have found that patients use pronouns without necessarily referring to a specific noun. Sometimes it is hard, even impossible, for the interviewer to understand what a pronoun is referring to. Patients seem to prefer using pronouns instead of actual nouns, which are not always specified in the conversation.

Prosodic features were less dependent on the education level compared to age, as shown in Table 7. Response length was found to be insignificant in education-controlled experiments. Some of the patients with higher education either do not talk much or talk significantly more than the control group. However, such speakers do not exist in the lower education group. Hence, on average, response length was not found to be significant.

The average absolute delta pitch and average absolute delta formant-2 features were also found to be dependent on the education level. These two features are particularly interesting since they also have high correlation with the display of mood and depression [23]. Patients in the high education group sometimes displayed exaggerated emotions which increased the pitch variability. Interestingly, some of the subjects in the same group tend to have lower second formant deviations, which can be a sign of depression. Those two patterns, however, were not observed in the younger patients. Thus, they were not found to be significant markers of the disease.

5.3 Gender-controlled Experiments

Gender-controlled experiments were performed to measure the performance of features for each gender separately. Results are shown in Tables 8 and 9. There are three features that performed well in the age-controlled and

Table 6 Accuracy (%) of each classifier using the linguistic features in education-controlled experiments

		High school and below ($p = 18, c = 24$)					College and above ($p = 10, c = 27$)				
ID	Features	SVM	Bayes	CTree	NN	*p* value	SVM	Bayes	CTree	NN	*p* value
1.1	Question ratio	54.8	38.1	59.5	45.2	0.217	***73.0***	***75.7***	***70.3***	62.2	**0.002**
1.2	Filler ratio	40.5	35.7	35.7	38.1	1.000	62.2	59.5	***70.3***	62.2	**0.014**
1.3	Incomplete sentence ratio	52.4	54.8	57.1	59.5	0.217	56.8	59.5	64.9	***67.6***	**0.033**
2.1	Verb freq.	61.9	61.9	61.9	47.6	0.123	64.9	54.1	64.9	***73.0***	**0.005**
2.2	Noun freq.	57.1	40.5	59.5	50.0	0.217	***78.4***	***78.4***	***86.5***	***73.0***	**<0.001**
2.3	Pronoun freq.	***76.2***	***71.4***	61.9	54.8	**0.001**	***75.7***	***73.0***	64.9	51.4	**0.002**
2.4	Adverb freq.	38.1	35.7	50.0	40.5	1.000	***70.3***	***81.1***	59.5	62.2	**<0.001**
2.5	Adjective freq.	52.4	45.2	54.8	57.1	0.355	54.1	51.4	56.8	56.8	0.411
2.6	Particle freq.	54.8	47.6	61.9	50.0	0.123	43.3	62.2	54.1	62.2	0.139
2.7	Conjunction freq.	42.9	35.7	35.7	42.9	1.000	***73.0***	***73.0***	***75.7***	64.9	**0.002**
2.8	Pronoun-to-noun ratio	***73.8***	***71.4***	57.1	59.5	**0.002**	***78.4***	***73.0***	51.4	62.2	**0.001**
3	Unintelligible word ratio	64.3	57.1	57.1	57.1	0.064	***78.4***	***78.4***	62.2	***67.6***	**0.001**
4.1	Standardized word entropy	64.3	***69.1***	57.1	57.1	**0.014**	***70.3***	56.8	59.5	56.8	**0.014**
4.2	Suffix ratio	57.1	57.1	59.5	57.1	0.217	62.2	37.8	54.1	62.2	0.139
4.3	Number ratio	57.1	50.0	50.0	35.7	0.355	59.5	56.8	56.8	43.3	0.250
4.4	Brunet's index	40.5	61.9	54.8	52.4	0.123	***67.6***	54.1	48.7	62.2	**0.033**
4.5	Honore's statistic	40.5	52.4	45.2	54.8	0.537	56.7	56.8	54.1	48.7	0.411
4.6	Type-token ratio	59.5	64.3	45.2	50.0	0.064	***67.6***	37.8	59.5	56.8	**0.033**

The number of patients (*p*) and control subjects (*c*) in each group is given in parenthesis. Classifiers with significant results are denoted in italics. Features with *p* values less than 0.05 are considered significant and are shown in bold. *p* value of the best classifier is reported for each feature.

Table 7 Accuracy (%) of each classifier using the prosodic features in education-controlled experiments

		High school and below ($p = 18, c = 24$)					College and above ($p = 10, c = 27$)				
ID	Features	SVM	Bayes	CTree	NN	*p* value	SVM	Bayes	CTree	NN	*p* value
5.1	Response time	52.4	61.9	54.8	***69.1***	**0.014**	***70.3***	59.5	62.2	54.1	**0.014**
5.2	Response length	47.6	42.9	38.1	42.9	1.000	***73.0***	51.4	46.0	***67.6***	**0.005**
5.3	Silence ratio	***73.8***	***71.4***	64.3	***73.8***	**0.002**	***91.9***	***97.3***	***94.6***	***91.9***	**<0.001**
5.4	Silence to utt. ratio	***69.1***	***69.0***	57.1	59.5	**0.014**	***83.8***	***81.1***	***73.0***	***75.7***	**<0.001**
5.5	Long silence ratio	64.3	***66.7***	59.5	***66.7***	**0.031**	***70.3***	***78.4***	54.1	51.4	**0.001**
5.6	Avg. silence count	***83.3***	***85.7***	***83.3***	***71.4***	**<0.001**	***75.7***	***70.3***	***73.0***	***67.6***	**0.002**
5.7	Silence rate	***66.7***	***73.8***	***83.3***	***76.2***	**<0.001**	***75.7***	62.2	***78.4***	59.5	**0.001**
5.8	Cont. speech rate	***83.3***	***83.3***	***78.6***	***73.8***	**<0.001**	***75.7***	***70.3***	59.5	***73.0***	**0.002**
5.9	Avg. cont. word count	59.5	***71.4***	57.1	59.5	**0.005**	***89.2***	***91.9***	***91.9***	***83.8***	**<0.001**
6.1	Avg. abs. delta energy	***69.1***	57.1	64.3	61.9	**0.014**	***78.4***	***78.4***	***75.7***	48.7	**0.001**
6.2	Dev. of abs. delta energy	***69.1***	***66.7***	50.0	42.9	**0.014**	***73.0***	***75.7***	59.5	56.8	**0.002**
6.3	Avg. abs. delta pitch	45.2	45.2	38.1	40.5	1.000	64.9	***83.8***	***73.0***	56.8	**<0.001**
6.4	Dev. of abs. delta pitch	38.1	50.0	47.6	42.9	1.000	56.8	62.2	54.1	***67.6***	**0.033**
6.5.1	Avg. abs. delta formant 1	61.9	64.3	61.9	61.9	0.064	***67.6***	64.9	48.7	46.0	**0.033**
6.5.2	Avg. abs. delta formant 2	61.9	64.3	54.8	47.6	0.064	***67.6***	***70.3***	***73.0***	***78.4***	**0.001**
6.5.3	Avg. abs. delta formant 3	57.1	47.6	35.7	38.1	0.355	***75.7***	62.2	64.9	62.3	**0.002**
6.5.4	Avg. abs. delta formant 4	***69.1***	***66.7***	57.1	***66.7***	**0.014**	54.1	62.2	***70.3***	62.3	**0.014**
6.6	Voicing ratio	64.3	64.3	61.9	***69.0***	**0.014**	***67.6***	64.9	64.9	56.8	**0.033**
7.1	Phoneme rate	***66.7***	***66.7***	47.6	42.9	**0.031**	***83.8***	***86.5***	***83.8***	***70.3***	**<0.001**
7.2	Word rate	***66.7***	64.3	61.9	***66.7***	**0.031**	64.9	***73.0***	***81.1***	64.9	**<0.001**

The number of patients (*p*) and control subjects (*c*) in each group is given in parenthesis. Classifiers with significant results are denoted in italics. Features with *p* values less than 0.05 are considered significant and are shown in bold. *p* value of the best classifier is reported for each feature.

Table 8 Accuracy (%) of each classifier using the linguistic features in gender-controlled experiments

		Male ($p = 18, c = 31$)					Female ($p = 10, c = 20$)				
ID	Features	SVM	Bayes	CTree	NN	p-value	SVM	Bayes	CTree	NN	p-value
1.1	Question ratio	47.1	50.0	***69.0***	60.8	**0.007**	***70.7***	***68.0***	58.7	61.0	**0.022**
1.2	Filler ratio	43.7	40.2	37.1	39.4	1.000	***66.7***	51.0	63.7	***69.0***	**0.031**
1.3	Incomplete sentence ratio	46.1	43.9	54.5	52.9	0.295	53.7	47.7	47.7	52.0	0.379
2.1	Verb freq.	51.4	55.5	45.5	51.0	0.242	46.7	54.0	52.0	48.0	0.335
2.2	Noun freq.	48.6	46.7	55.3	***64.7***	**0.031**	***80.7***	***70.7***	***76.0***	***69.7***	**0.001**
2.3	Pronoun freq.	59.2	59.7	***63.6***	52.8	**0.016**	59.0	***71.5***	65.5	50.0	**0.012**
2.4	Adverb freq.	47.1	46.7	50.8	48.0	0.458	57.0	45.0	44.7	41.7	0.183
2.5	Adjective freq.	53.7	52.2	54.7	***65.3***	**0.030**	36.0	42.0	44.0	50.7	0.473
2.6	Particle freq.	***65.3***	***66.9***	***66.7***	59.2	**0.010**	39.0	38.0	41.7	41.0	1.000
2.7	Conjunction freq.	45.3	***65.3***	50.8	49.2	**0.021**	53.0	***71.0***	64.7	***69.7***	**0.010**
2.8	Pronoun-to-noun ratio	***65.0***	61.1	39.4	39.4	**0.032**	***67.0***	***72.0***	65.0	***69.5***	**0.014**
3	Unintelligible word ratio	***66.7***	***66.1***	58.0	62.2	**0.022**	59.7	58.0	57.0	***72.0***	**0.009**
4.1	Standardized word entropy	57.1	60.8	42.2	46.7	0.092	56.0	53.7	52.7	58.7	0.148
4.2	Suffix ratio	56.9	48.6	46.1	42.4	0.196	***69.0***	***73.0***	***66.7***	58.7	**0.008**
4.3	Number ratio	***65.1***	***65.3***	59.4	55.3	**0.031**	51.7	58.7	60.0	49.0	0.120
4.4	Brunet's index	41.7	52.9	51.6	62.4	0.061	48.0	51.0	42.0	41.7	0.459
4.5	Honore's statistic	47.1	53.3	55.1	52.9	0.249	39.7	40.7	39.0	41.0	1.000
4.6	Type-token ratio	***64.1***	62.0	59.0	51.6	**0.041**	51.7	46.7	54.0	51.7	0.308

The number of patients (*p*) and control subjects (*c*) in each group is given in parenthesis. Classifiers with significant results are denoted in italics. Features with *p* values less than 0.05 are considered significant and are shown in bold. *p* value of the best classifier is reported for each feature.

Table 9 Accuracy (%) of each classifier using the prosodic features in gender-controlled experiments

		Male ($p = 18, c = 31$)					Female ($p = 10, c = 20$)				
ID	Features	SVM	Bayes	CTree	NN	*p* value	SVM	Bayes	CTree	NN	*p* value
5.1	Response time	***68.6***	58.4	***69.0***	***64.5***	**0.005**	50.7	50.0	44.7	42.7	0.468
5.2	Response length	50.0	46.1	***63.9***	***64.1***	**0.033**	51.0	50.7	59.0	50.0	0.114
5.3	Silence ratio	***86.1***	***83.3***	***83.9***	***80.8***	**<0.001**	66.0	***68.0***	***73.7***	***72.7***	**0.001**
5.4	Silence to utt. ratio	60.2	62.9	54.1	***66.9***	**0.017**	***83.0***	***82.0***	***71.7***	53.7	**<0.001**
5.5	Long silence ratio	***63.9***	***63.3***	55.9	48.0	**0.048**	51.7	63.0	47.0	53.7	0.077
5.6	Avg. silence count	***79.0***	***79.4***	***73.9***	***72.9***	**<0.001**	***76.0***	***72.0***	***80.0***	62.7	**<0.001**
5.7	Silence rate	***63.7***	***66.3***	***70.0***	***63.1***	**0.003**	59.0	60.7	58.0	***78.7***	**<0.001**
5.8	Cont. speech rate	***82.9***	***79.0***	***83.9***	***69.2***	**<0.001**	***73.0***	***72.0***	***71.7***	47.0	**0.007**
5.9	Avg. cont. word count	***86.1***	59.0	***84.1***	***77.1***	**<0.001**	***68.7***	62.7	***68.0***	50.7	**0.014**
6.1	Avg. abs. delta energy	***69.4***	***70.0***	***81.0***	***84.1***	**<0.001**	53.0	46.7	36.0	41.0	0.380
6.2	Dev. of abs. delta energy	***69.4***	***69.8***	49.2	51.0	**0.008**	55.7	57.0	42.7	38.7	0.176
6.3	Avg. abs. delta pitch	***67.8***	***74.7***	***63.1***	58.4	**0.001**	60.7	45.0	63.7	45.0	0.060
6.4	Dev. of abs. delta pitch	***76.1***	***72.2***	***63.9***	55.5	**<0.001**	48.0	38.7	***70.7***	***67.7***	**0.007**
6.5.1	Avg. abs. delta formant 1	59.4	***66.3***	***64.5***	53.9	**0.016**	53.0	52.0	46.7	48.0	0.374
6.5.2	Avg. abs. delta formant 2	55.9	60.2	57.8	52.0	0.108	39.0	38.7	42.7	55.7	0.259
6.5.3	Avg. abs. delta formant 3	***66.1***	***65.9***	60.2	50.8	**0.022**	52.7	42.7	***69.7***	48.0	**0.007**
6.5.4	Avg. abs. delta formant 4	***69.2***	***67.1***	***64.1***	58.6	**0.009**	42.7	53.7	56.7	52.7	0.209
6.6	Voicing ratio	***75.3***	***75.1***	***70.2***	***69.8***	**0.001**	57.7	41.0	58.7	50.7	0.111
7.1	Phoneme rate	***63.9***	***71.0***	***65.1***	***67.8***	**0.004**	***71.7***	***68.0***	48.0	46.0	**0.017**
7.2	Word rate	***70.0***	***70.2***	53.3	45.5	**0.005**	35.0	***66.7***	48.0	46.7	**0.014**

The number of patients (*p*) and control subjects (*c*) in each group is given in parenthesis. Classifiers with significant results are denoted in italics. Features with *p* values less than 0.05 are considered significant and are shown in bold. *p* value of the best classifier is reported for each feature.

Table 10 Overall accuracy (%), missed detection (%), and false alarm (%) rates of features with statistically significant performance

		Accuracy				Missed detection				False alarm			
ID	Features	SVM	Bayes	CTree	NN	SVM	Bayes	CTree	NN	SVM	Bayes	CTree	NN
2.3	Pronoun freq.	***73.4***	72.2	59.5	54.4	***39.3***	39.3	46.4	57.1	***19.6***	21.6	37.3	39.2
2.8	Pronoun-to-noun ratio	***72.2***	70.9	65.8	58.2	***39.3***	39.3	50.0	60.7	***21.6***	23.5	25.5	31.4
5.3	Silence ratio	***83.5***	81.0	79.8	78.5	***35.7***	35.7	28.6	25.0	***5.9***	9.8	15.7	19.6
5.4	Silence to utt. ratio	***73.4***	69.6	62.0	69.6	***35.7***	42.9	42.9	39.3	***21.6***	23.5	35.3	25.5
5.6	Avg. silence count	***81.0***	78.5	69.6	68.4	***21.4***	17.9	32.1	39.3	***17.7***	23.5	29.4	27.5
5.7	Silence rate	***73.4***	72.2	72.1	68.4	***64.3***	67.9	35.7	39.3	***5.9***	5.9	23.5	27.5
5.8	Cont. speech rate	***78.5***	76.0	68.4	62.0	***28.6***	25.0	25.0	39.3	***17.7***	23.5	35.3	37.4
5.9	Avg. cont. word count	76.0	***82.3***	76.0	70.9	28.6	***32.1***	28.6	35.7	21.6	***9.8***	21.6	25.5
7.1	Phoneme rate	69.6	***79.8***	60.8	57.0	32.1	***46.4***	46.4	50.0	29.4	***5.9***	35.3	39.2
7.2	Word rate	62.0	***65.8***	55.7	50.6	25.0	***32.1***	50.0	53.6	45.1	***35.3***	41.2	47.1

Features with statistically significant performance in age-, education-, and gender-controlled experiments. Performance of the classifier with highest accuracy is shown in italics for each feature. Note that because the number of patients and control subjects are not equal, sum of average error, which is (missed detection+false alarm)/2, and accuracy is not 100%.

Table 11 Correlation of features with statistically significant accuracies in age-, education-, and gender-controlled experiments

ID	Features	2.3	2.8	5.3	5.4	5.6	5.7	5.8	5.9	7.1	7.2
2.3	Pronoun freq.	—									
2.8	Pronoun-to-noun ratio	0.94	—								
5.3	Silence ratio	0.12	0.21	—							
5.4	Silence to utt. ratio	0.24	0.29	0.60	—						
5.6	Avg. silence count	−0.30	−0.36	−0.55	−0.38	—					
5.7	Silence rate	0.17	0.22	0.54	0.39	−0.68	—				
5.8	Cont. speech rate	0.33	0.34	0.38	0.31	−0.87	0.45	—			
5.9	Avg. cont. word count	0.04	−0.01	−0.67	−0.45	0.41	−0.40	−0.27	—		
7.1	Phoneme rate	−0.01	−0.14	−0.70	−0.44	0.52	−0.61	−0.23	0.49	—	
7.2	Word rate	−0.00	−0.06	−0.59	−0.38	0.47	−0.56	−0.19	0.42	0.90	—

education-controlled experiments but not in the gender-controlled experiments. Those features are: deviation of absolute delta energy, average absolute delta energy, and long silence ratio.

All three features performed well for males but not for females. In the recordings, we have found that males displayed less emotion which resulted in less expressive speech compared to male subjects in the control group. However, that pattern was not as strong with the female speakers. Moreover, the number of males is significantly larger than the number of females which makes it easier to get statistically significant results in classification experiments for the male subjects.

5.4 Analysis of significant features

Features that have significant performance in education-, age-, and gender-controlled tests are shown in Table 10, along with missed detection and false alarm rates. SVM and naive Bayes classifiers always outperform the CTree and NN classifiers. SVM has the best accuracy among all classifiers. In particular, SVM classifier with the silence ratio feature has the highest accuracy among all features and classifiers.

Missed detection rates are significantly higher than the false alarm rates in the best performing classifiers, as shown in Table 10. Even though more data from the control group is available, the subsampling method is used to

Table 12 Accuracy, missed detection, and false alarm rates of the best performing features

ID		Accuracy	Missed detection	False alarm
1 feature				
5.3	SVM	83.5%	35.7%	5.9%
		(73.5 to 90.9)	(18.6 to 55.9)	(1.2 to 16.2)
7.1	Bayes	79.8%	46.4%	5.9%
		(69.2 to 88.0)	(27.5 to 66.1)	(1.2 to 16.2)
2.3	SVM	73.4%	39.3%	19.6%
		(62.3 to 82.7)	(21.5 to 59.4)	(9.8 to 33.1)
2 features				
5.3 and 7.1	SVM	83.5%	35.7%	5.9%
		(73.5 to 91.0)	(18.6 to 55.9)	(1.2 to 16.2)
2.3 and 5.3	Bayes	82.3%	32.1%	9.8 %
		(72.1 to 90.0)	(15.9 to 52.4)	(3.3 to 21.4)
2.3 and 7.1	Bayes	78.5%	39.3%	11.8%
		(67.8 to 87.0)	(21.5 to 59.4)	(4.4 to 23.9)

Performance of best performing single feature in each feature category, as well as performance of combinations of these features. Results are reported only for the best classifier. Note that because the number of patients and control subjects are not equal, sum of average error, which is (missed detection + false alarm)/2, and accuracy is not 100%. Confidence intervals are shown in parenthesis.

ensure an equal number of patient and control subjects in the training datasets, as discussed in Section 4. Thus, the results show that significantly more patients were classified as healthy compared to control subjects classified as patients. It also indicates that features extracted from some patients are significantly different from some of the other patients and most of the control subjects, which helps in classification.

Voice activity-related features are particularly good at identifying the disease as shown in Table 10. Interestingly, features 5.3 and 5.6, which are related to the rate of silences, independent of the silence duration, were found to be more powerful discriminators than 5.4 and 5.7, which are related to long silences and duration of silences. Thus, frequency of silences during speech was found to be more important than the duration of silences. Features 5.8 and 5.9 indicate how long the subject can talk without a long silence. These two features are also highly correlated with the silence rate features, as shown in Table 11, and they had good performance in the classification experiments. Similarly, phoneme rate was strongly correlated with the rate of silences, and it performed well in experiments. Even though word rate is strongly correlated with the phoneme rate, it is not as strongly correlated with the silence rate as the phoneme rate, and it was not as successful in prediction of the disease.

Linguistic features did not perform as well as the prosodic features, as discussed in the previous section. Only pronoun rate and pronoun-to-noun ratio were strong indicators of the disease, but their prediction powers are not as strong as the prosodic features. However, their missed detection and false alarm rates are more balanced compared to prosodic features.

Features within each feature category are strongly correlated with each other, as shown in Table 11. Performances of the features with highest accuracy from each category are compared with confidence intervals in Table 12.

Feature fusion is used in an attempt to further boost the performance. In that approach, classifiers were trained with two features instead of a single feature. Because of high within-category correlations, feature fusion experiments were done by using the best performing feature in each category. Results are shown in Table 12. Not only statistically significant improvement over the best single feature could not be achieved but also performance slightly degraded with feature fusion.

Scatter diagrams of the features and decision boundaries for classification are shown in Figure 2. Silence ratio had better discrimination power than the other features. Unfortunately, the information in the other features failed to correct the errors made with silence ratio. Similarly, phoneme rate was found to be a powerful feature, but the pronoun frequency could not correct the errors it generates, as shown in the Figure 2C.

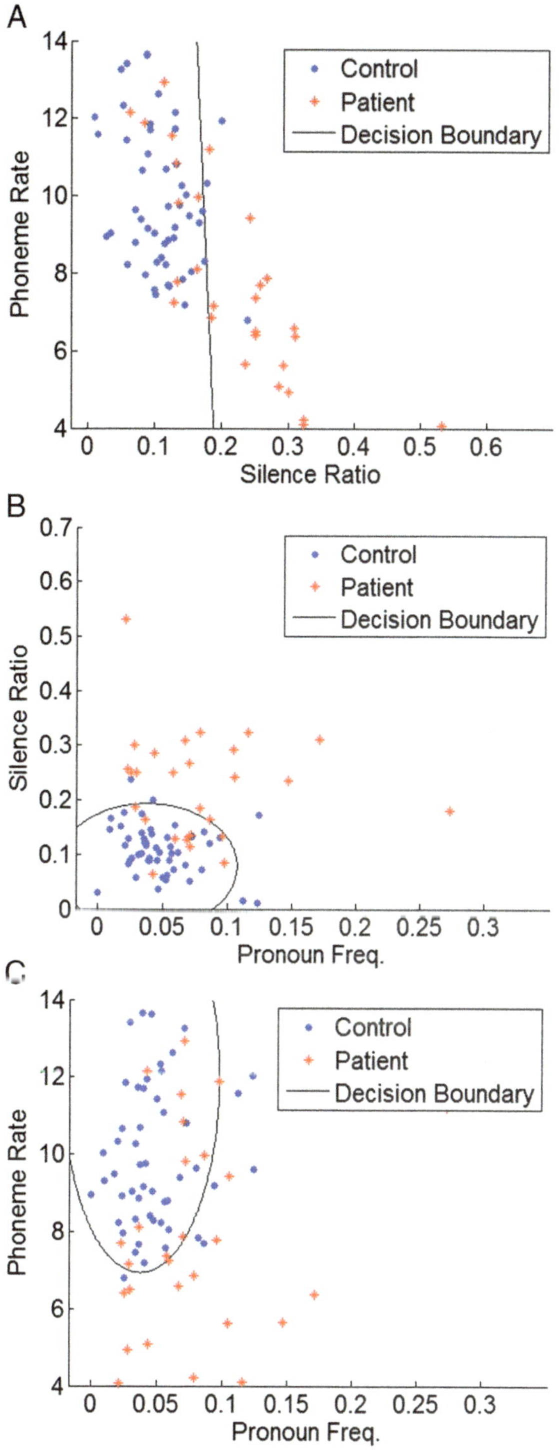

Figure 2 Scatter diagram of features in Table 12. Decision boundary of the best performing classifier is shown in each figure. Note that even though the features were normalized before classification, they are not normalized in these figures to make interpretation easier. **(A)** Scatter diagram for silence ratio vs. phoneme rate features. **(B)** Scatter diagram for pronoun frequency vs. silence ratio features. **(C)** Scatter diagram for pronoun frequency vs. phoneme rate features.

Note that increasing the size of feature vectors can in fact degrade the performance of classifiers due to the curse of dimensionality that occurs when there is not enough training data and the classifier cannot generalize and perform well on test data. That effect may be partly responsible for not observing an improvement with the feature fusion approach. For the same reason, feature fusion with larger number of features was not investigated.

6 Conclusions

We have investigated an extensive set of features derived from the speech signal and transcriptions of Alzheimer's patients and control subjects. It is already known that the part of the brain cortex that deals with linguistic abilities is one of the first to deteriorate with the onset of the disease. Our work explored how that deterioration is reflected in the patient's speech prosody and spoken language, and whether there are markers that can be effectively detected using machine learning techniques. Our results indicate that a prediction accuracy higher than 80% can be obtained with high confidence using the proposed features, independent of the age, education level, and gender of the subjects. Prosodic features were substantially better than the linguistic features. In fact, only two of the linguistic features were found to be strong markers of the disease.

Classification experiments were also done with combinations of features. However, using more than one feature did not outperform the best single feature. This may be a result of limited amounts of data used in training the classifiers which causes generalization problems when the number of features increases.

Our experiments are with late-stage patients, and the effectiveness of the markers that we have found should be measured with early-stage patients, where the signals are more subtle and more subjects may be needed to reach statistically significant results. However, our experimental results and manual observations from the data are encouraging, and we will start collecting data from early-stage patients in the near future.

Another topic that we will investigate in the future work is a cross-lingual study of the proposed features. Features that are independent of language can provide important clues about the neural degeneration process during the disease or perhaps can enable deeper understanding of neural networks in the brain that are responsible from cognition of language and speech production.

Competing interests

The authors declare that they have no competing interests.

Acknowledgments

The authors would like to thank the anonymous reviewers for their invaluable comments and suggestions to improve the quality of the paper. They are also grateful to Prof. Emily T. Prud'hommeaux of Rochester Institute of Technology for her invaluable comments during the revision of this paper.

References

1. M Prince, M Guerchet, M Prina, World Alzheimer report 2013: Journey of caring: an analysis of long-term care for dementia. http://www.alz.co.uk/research/world-report-2013 Accessed 2015-03-13
2. R Schmelzer, Roche Joins The Global CEO Initiative on Alzheimer's Disease. http://www.ceoalzheimersinitiative.org/node/71 Accessed 2014-08-20
3. MF Folstein, SE Folstein, PR McHugh, "Mini-mental state": a practical method for grading the cognitive state of patients for the clinician. J. Psychiat. Res. **12**(3), 189–198 (1975)
4. LM Bloudek, DE Spackman, M Blankenburg, SD Sullivan, Review and meta-analysis of biomarkers and diagnostic imaging in Alzheimer's disease. J. Alzheimer's Dis. **26**(4), 627–645 (2011)
5. RS Bucks, S Singh, JM Cuerden, GK Wilcock, Analysis of spontaneous, conversational speech in dementia of Alzheimer type evaluation of an objective technique for analysing lexical performance. Aphasiology. **14**(1), 71–91 (2000)
6. ET Prud'hommeaux, B Roark. Extraction of narrative recall patterns for neuropsychological assessment. Proceedings of the 12th Annual Conference of the International Speech Communication Association (Interspeech) (Florence, Italy, 2011), pp. 3021–3024
7. KC Fraser, JA Meltzer, NL Graham, C Leonard, G Hirst, SE Black, E Rochon, Automated classification of primary progressive aphasia subtypes from narrative speech transcripts. Cortex. **55**, 43–60 (2014)
8. B Roark, M Mitchell, JP Hosom, K Hollingshead, J Kaye, Spoken language derived measures for detecting mild cognitive impairment. IEEE Trans. Audio Speech Lang. Process. **19**(7), 2081–2090 (2011)
9. G Tosto, M Gasparini, GL Lenzi, G Bruno, Prosodic impairment in Alzheimer's disease: assessment and clinical relevance. J. Neuropsychiat. Clin. Neurosci. **23**(2), 21–23 (2011)
10. DS Knopman, S Weintraub, VS Pankratz, Language and behavior domains enhance the value of the clinical dementia rating scale. Alzheimers Dement. **7**(3), 293–299 (2011)
11. SV Pakhomov, GE Smith, S Marino, A Birnbaum, N Graff-Radford, R Caselli, B Boeve, DS Knopman, A computerized technique to assess language use patterns in patients with frontotemporal dementia. J. Neurolinguistics. **23**(2), 127–144 (2010)
12. V Iliadou, S Kaprinis, Clinical psychoacoustics in Alzheimer's disease central auditory processing disorders and speech deterioration. Ann. Gen. Hosp. Psychiat. **2**(1), 12 (2003)
13. I Hoffmann, D Nemeth, CD Dye, M Pakaski, T Irinyi, J Kalman, Temporal parameters of spontaneous speech in Alzheimer's disease. Int. J. Speech Lang. Pathol. **12**(1), 29–34 (2010)
14. SV Pakhomov, EA Kaiser, DL Boley, SE Marino, DS Knopman, AK Birnbaum, Effects of age and dementia on temporal cycles in spontaneous speech fluency. J. Neurolinguistics. **24**(6), 619–635 (2011)
15. C Thomas, V Keselj, N Cercone, K Rockwood, E Asp, in *Mechatronics and Automation 2005 IEEE International Conference*. Automatic detection and rating of dementia of Alzheimer type through lexical analysis of spontaneous speech, vol. 3 (Niagara Falls, Canada, 2005), pp. 1569–15743
16. LEE H, F Gayraud, F Hirsch, M Barkat-Defradas. Speech dysfluencies in normal and pathological aging: a comparison between Alzheimer patients and healthy elderly subjects. the 17th International Congress of Phonetic Sciences (ICPhS) (Hong Kong, China, 2011), pp. 1174–1177
17. DA Snowdon, SJ Kemper, JA Mortimer, LH Greiner, DR Wekstein, WR Markesbery, Linguistic ability in early life and cognitive function and Alzheimer's disease in late life. Findings from the Nun Study. JAMA. **275**(7), 528–532 (1996)
18. K Oflazer, S Inkelas, in *Proceedings of the EACL Workshop on Finite State Methods in NLP*. A finite state pronunciation lexicon for Turkish, vol. 82 (Budapest, Hungary, 2003), pp. 900–918
19. K Oflazer, Two-level description of Turkish morphology. Literary Linguist. Comput. **9**(2), 137–148 (1994)
20. v Brunet. Le Vocabulaire De Jean Giraudoux : Structure Et évolution : Statistique Et Informatique Appliquées à L'étude Des Textes à Partir Des

Données Du Trésor De La Langue Française. Le Vocabulaire des grands écrivains français (Genève, Slatkine, 1978). ASIN: B0000E99PZ
21. A Honore, Some simple measures of richness of vocabulary. Assoc. Literary Linguistic Comput. Bull. **7**, 1979
22. D Biber, S Conrad, G Leech, *The Longman student grammar of spoken and written English*, (Harlow: Longman, 2002). ISBN: 0 582 237262
23. E Moore, MA Clements, JW Peifer, L Weisser, Critical analysis of the impact of glottal features in the classification of clinical depression in speech. Biomed. Eng. IEEE Trans. **55**(1), 96–107 (2008)

Percussive/harmonic sound separation by non-negative matrix factorization with smoothness/sparseness constraints

Francisco Jesus Canadas-Quesada[1*], Pedro Vera-Candeas[1], Nicolas Ruiz-Reyes[1], Julio Carabias-Orti[2] and Pablo Cabanas-Molero[1]

Abstract

In this paper, unsupervised learning is used to separate percussive and harmonic sounds from monaural non-vocal polyphonic signals. Our algorithm is based on a modified non-negative matrix factorization (NMF) procedure that no labeled data is required to distinguish between percussive and harmonic bases because information from percussive and harmonic sounds is integrated into the decomposition process. NMF is performed in this process by assuming that harmonic sounds exhibit spectral sparseness (narrowband sounds) and temporal smoothness (steady sounds), whereas percussive sounds exhibit spectral smoothness (broadband sounds) and temporal sparseness (transient sounds). The evaluation is performed using several real-world excerpts from different musical genres. Comparing the developed approach to three current state-of-the art separation systems produces promising results.

1 Introduction

The separation of percussive and harmonic sounds remains a challenging problem in music research. Percussive sound pertains to drum instruments, whereas the harmonic sound pertains to pitched instruments. We develop a method to separate monaural music signals (for which spatial information is unavailable), motivated by the large number of one-channel music recordings such as live performances or old recordings (from before the 1960s). In a musical context, a listener can effortlessly distinguish between percussive and harmonic sounds; therefore, these two types of sounds must have significantly different characteristics. Ono et al. [1,2] used Harmonic/Percussive Sound Separation (HPSS) to separate harmonic and percussive sounds by exploiting the anisotropy of harmonic and percussive sounds in a maximum a posteriori (MAP) framework. The authors considered a spectrogram, assuming anisotropic smoothness [1], i.e., percussive sounds have a structure that is vertically smooth in frequency, whereas harmonic sounds are temporally stable and have a structure that is horizontally smooth in time.

Over the last decade, several approaches have been developed to separate percussive/harmonic sounds from monaural polyphonic music [3-6]. The method developed here offers the following advantages over the method outlined in [3]: (i) The method is robust for various sources and not only for flat spectrum sources, (ii) threshold choices are not required, (iii) hand-tuning is not necessary, (iv) the method is quite fast (e.g., the developed method can factorize an input signal lasting 30 s in approximately 18 s). Unlike the methods given in [4,5], no labelled data is required by the proposed method because the percussive/harmonic information is obtained from the spectro-temporal features used in the factorization stage. Other recently published state-of-the-art techniques are presented in [7,8]. In [7], anisotropy [1] is used in Median Filtering-based Separation (MFS) under two assumptions: the harmonics are considered to be outliers in a temporal slice that contains a mixture of percussive and pitched instruments, and the percussive onsets are considered to be outliers in a frequency slice. A median operator is used to remove these outliers because median filtering has been used extensively in image processing for removing salt and pepper noise from images [9]. In this manner, the

*Correspondence: fcanadas@ujaen.es
[1] Telecommunication Engineering Department, University of Jaen, Linares, Jaen, Spain
Full list of author information is available at the end of the article

extraction of percussive sounds can be seen as the removal of outliers (overtones from harmonic sounds) in a time frame of a spectrogram while the extraction of harmonic sounds can be seen as the removal of outliers (onsets from percussive sounds) in a frequency bin of a spectrogram. In [8], drum source separation is performed using non-negative matrix partial co-factorization (NMPCF). In NMPCF, the input spectrogram and a drum-only matrix (which is picked up from *a priori* drum recordings) are simultaneously decomposed. The shared basis vectors in this co-factorization are associated with the drum characteristics, which are used to extract drum-related components from the musical signals.

In this paper, we develop a percussive/harmonic sound separation approach, which we apply to monaural instrumental music (the singing voice is not considered). We do not consider vocals for the following reasons: (i) Voiced vocals can show harmonic features modelled by the developed method (e.g., smoothness in time for sustained sounds and sparseness in frequency for harmonic sounds) but can also exhibit harmonic features that are not modelled by the developed method (e.g., non-smoothness in time, as in the vibrato effect); (ii) vocals may be voiced (harmonic sounds) and unvoiced (percussive sounds), and the developed method has not been designed to distinguish between percussive music instruments and unvoiced vocals. The novelty of this work lies in modeling a percussive/harmonic mixture signal using a modified non-negative matrix factorization (NMF) approximation that can automatically distinguish between percussive and harmonic bases. That is, we decompose the mixture signal using an objective function to integrate spectro-temporal features, anisotropic smoothness and sparseness into the decomposition process. Anisotropic smoothness is related to the difference in the directions of continuity between the spectrograms of harmonic and percussive sounds. The spectrograms of harmonic sounds are quasi-stationary and are therefore typically smooth in time, whereas the spectrograms of percussive sounds are impulsive and are typically smooth in frequency [1,10]. However, anisotropic sparseness is also related to the difference in the directions of sparseness between the spectrograms of harmonic and percussive sounds. The spectrograms of harmonic sounds are typically sparse in frequency, as in narrowband sounds, whereas the spectrograms of percussive sounds are impulsive and are therefore typically sparse in time. These features enable us to model harmonic sounds using the sparseness in frequency (for spectral peaks) and smoothness in time (for amplitudes that vary slowly in time), whereas percussive sounds can be modelled using the smoothness in frequency (i.e., the energy slowly decreases in frequency) and the sparseness in time (i.e., most of the signal energy is concentrated over short time intervals), as seen in Figure 1. Therefore, the signal spectrogram can be reconstructed as the sum of two different spectrograms that are characterized by specific percussive/harmonic bases and gains. Our formulation does not require information about the number of active sound sources neither prior knowledge about the instruments nor supervised training to classify the bases.

The developed approach is practically useful in the field of audio engineering applications for music information retrieval, where the percussive/harmonic separation task can be used as a preprocessing tool. The extraction of a harmonic sound source can be used to enhance music transcription [11] and chord detection [12]. Extracting a percussive sound source can also enhance onset detection [2]. Extracting both harmonic and percussive sound sources is useful for remixing and for audio to score alignment [13].

We implement an unsupervised approach in which imposed smoothness and sparseness constraints are used to automatically discriminate between percussive and harmonic signals in a NMF framework. Our specific contribution is the inclusion of sparseness criteria in a NMF framework for percussive/harmonic separation. Compared to methods that require some training (semi-supervised or supervised), our approach provides a more robust source-to-distortion ratio (as we will show in Section 3) because the separation process does not depend on a supervised training. In Section 3, we show that the developed method produces promising results in comparison with two unsupervised (i.e., untrained) approaches (HPSS and MFS) and a supervised (i.e., trained) approach (NMPCF) dedicated to percussive/harmonic separation.

The remainder of this paper is organized as follows: In Section 2, we describe our novel method. In Section 3, the results are evaluated and compared. Finally, conclusions and future work are presented in Section 4.

2 Developed method

Non-negative matrix factorization (NMF) has been widely used in the field of digital image processing [14-17] in recent years and has also been successfully applied to music analysis [18,19]. Following [5,6,8], we apply NMF to percussive/harmonic separation, motivated in part by the aforementioned promising results.

Lee and Seung [20] developed standard NMF, a technique for multivariate data analysis in which an input magnitude spectrogram, represented by a matrix X, is decomposed into the product of two non-negative matrices W and H:

$$X \approx WH \tag{1}$$

where the i-th column of matrix W is a frequency basis that represents the spectral pattern of a component that is active in the spectrogram. Additionally, the i-th row

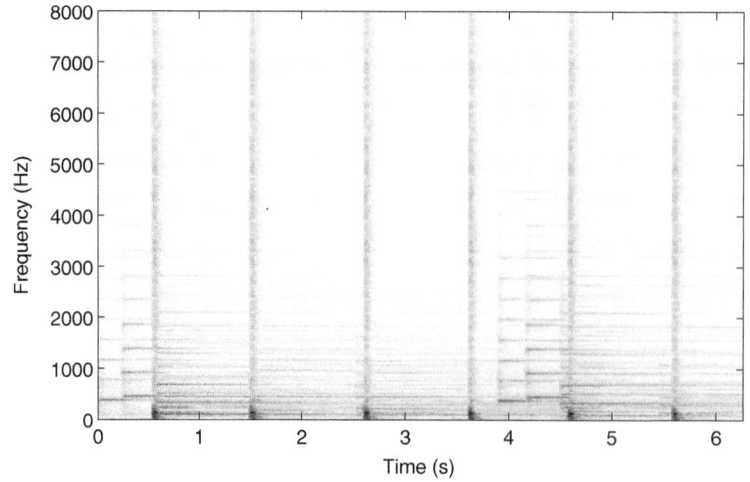

Figure 1 Magnitude spectrogram of an approximately 6-s-long excerpt of monaural mixture signal composed of percussive and harmonic sounds. The percussive sounds form vertical lines because of their smoothness in frequency and sparseness in time; the harmonic sounds form horizontal lines because of their sparseness in frequency and smoothness in time; the grey level represents the energy of each frequency.

of matrix H is a temporal gain or activation and represents the time interval over which the i-th frequency basis is active. Standard NMF cannot be used to distinguish between a percussive or harmonic frequency basis. Standard NMF can only ensure convergence to local minima, which enables the reconstruction of the signal spectrogram but cannot discriminate between percussive and harmonic frequency bases.

For clarity, the term *source* refers to a musical instrument, and the term *component* is used to select a specific frequency basis.

2.1 Signal representation

The magnitude spectrogram X of a music signal $x(t)$ is composed of T frames, F frequency bins and a set of time-frequency units $X_{f,t}$. Each $X_{f,t}$ is defined by the f-th frequency bin at the t-th frame and is calculated from the magnitude of the short-time Fourier transform (STFT) using a N-sample Hamming window $w(n)$ and a time shift J. A normalization process is necessary to adequately perform percussive/harmonic separation, for which the algorithm is independent of the norm of the input signal. Thus, the normalized magnitude spectrogram X_{n_β} is computed taking into account its dependence on the number of frames, the number of frequency bins and the constant β used in the β-divergence cost (see subsection 2.2.1).

$$X_{n_\beta} = \frac{X}{\left(\frac{\sum_{f=1}^{F}\sum_{t=1}^{T} X_{f,t}^{\beta}}{FT}\right)^{\frac{1}{\beta}}} \tag{2}$$

2.2 A modified non-negative matrix factorization for percussive/harmonic separation

Our formulation attempts to overcome the primary problem of the standard NMF approach by distinguishing between percussive and harmonic bases in the factorization process. For this purpose, an objective function is defined to decompose a mixture spectrogram X_{n_β} into two separate spectrograms, X_P (a percussive spectrogram) and X_H (a harmonic spectrogram) [5,6,8]. Each separated spectrogram exhibits specific spectro-temporal features for percussive or harmonic sounds. The factorization model is given in Equation 3 below:

$$X_{n_\beta} \approx X_P + X_H = W_{P_{F,R_p}} H_{P_{R_p,T}} + W_{H_{F,R_h}} H_{H_{R_h,T}} \tag{3}$$

where X_P, X_H, W_P, H_P, W_H and H_H are non-negative matrices. The parameter R_p denotes the number of percussive components, and the parameter R_h denotes the number of harmonic components used in the factorization process.

Next, we detail the decomposition process. This process adapts the concept of anisotropy which was initially appropriated by [1,2] to a NMF framework. Anisotropy is used to estimate W_P, H_P, W_H and H_H by minimizing a global objective function that depends on a β-divergence cost, a percussive cost and a harmonic cost.

2.2.1 *β-divergence cost*

Two different spectrograms X_P and X_H are constructed as to minimize the β-divergence cost $d_\beta(x|y)$ [21] compared to the input (known and fixed) normalized spectrogram X_{n_β}. The Euclidean distance ($\beta = 2$), the Kullback-Leibler (KL) divergence ($\beta = 1$) and the Itakura-Saito (IS) divergence ($\beta = 0$) are defined as functions of the parameter β:

$$d_\beta(x|y) = \begin{cases} \frac{1}{\beta(\beta-1)} \sum_{f=1}^{F} \sum_{t=1}^{T} \left(x_{f,t}^{\beta} + (\beta-1) y_{f,t}^{\beta} - \beta x_{f,t} y_{f,t}^{\beta-1} \right) & \beta \epsilon (0,1) \cup (1,2] \\ \sum_{f=1}^{F} \sum_{t=1}^{T} x_{f,t} log \frac{x_{f,t}}{y_{f,t}} - x_{f,t} + y_{f,t} & \beta = 1 \\ \sum_{f=1}^{F} \sum_{t=1}^{T} \frac{x_{f,t}}{y_{f,t}} - log \frac{x_{f,t}}{y_{f,t}} - 1 & \beta = 0 \end{cases} \quad (4)$$

where $x = X_{n_\beta}$ and $y = X_P + X_H$.

This factorization only considers the β-divergence cost and cannot automatically determine whether a basis belongs to a percussive signal or a harmonic signal. To overcome this drawback, the factorization process is modified to include specific spectro-temporal features related to percussive and harmonic sounds. Therefore, our contribution to the analysis of monaural signals is the development of a percussive/harmonic separation using an unsupervised NMF approach. This NMF approach models the mixture signal using an objective function that considers the β-divergence cost and common spectro-temporal features from the percussive/harmonic signals. Unlike other methods [4-6,8] that have been developed for percussive/harmonic separation, our approach does not use any labelled data to train the NMF bases maintaining competitive SDR results.

2.2.2 Percussive cost

The percussive cost is used to model percussive sounds by assuming smoothness in frequency and sparseness in time. Percussive sounds are typically represented as broadband signals, which have an energy that is confined to short time intervals. We define spectral smoothness (*SSM*), which is associated with the matrix W_P here, in the same way as continuity is defined in [22]:

$$\text{SSM} = \frac{T}{R_p} \sum_{r_p=1}^{R_p} \frac{1}{\sigma_{W_{P_{r_p}}}^{2}} \sum_{f=2}^{F} \left(W_{P_{f-1,r_p}} - W_{P_{f,r_p}} \right)^2 \quad (5)$$

where a high cost is assigned to large changes in the frequency between the bases $W_{P_{f,r_p}}$ and $W_{P_{f-1,r_p}}$ in adjacent frequency bins. Normalization is used to make the global objective function independent of the signal norm, i.e., the bases W_P are normalized by σ_{W_P}. The value σ_{W_P} of each percussive component r_p is calculated as $\sigma_{W_{P_{r_p}}} = \sqrt{\frac{1}{F} \sum_{f=1}^{F} W_{P_{f,r_p}}^2}$. To ensure that each percussive or harmonic cost has the same weight in the global objective function, each cost is normalized. Taking *SSM* into account, it is normalized by a factor equal to $\frac{T}{R_p}$. Thus, we avoid scaling problems that can arise from the number of frames, the number of frequency bins or the number of components considered.

Another percussive restriction can be applied to the temporal distribution of activations. Temporal sparseness has been previously used in [22,23] as the L1 norm of the activation gains to penalize solutions with nonzero gains. The concept of temporal sparseness (*TSP*) is defined following [22] but is applied to the matrix H_P. That is, a high cost is assigned to nonzero gains assuming that percussive sounds can be represented as transients with energies that are concentrated over short time intervals.

$$\text{TSP} = \frac{F}{R_p} \sum_{r_p=1}^{R_p} \sum_{t=1}^{T} \left| \frac{H_{P_{r_p,t}}}{\sigma_{H_{P_{r_p}}}} \right| \quad (6)$$

Similarly to the W_P treatment, the activations H_P are normalized by σ_{H_P}. The value σ_{H_P} of each percussive component r_p is calculated as $\sigma_{H_{P_{r_p}}} = \sqrt{\frac{1}{T} \sum_{t=1}^{T} H_{P_{r_p,t}}^2}$. As mentioned above, the TSP is normalized by a factor $\frac{F}{R_p}$ such that the percussive costs are equally weighted.

2.2.3 Harmonic cost

The harmonic cost is used to model harmonic sounds by assuming smoothness in time and sparseness in frequency. These sounds can be considered to be stable sounds that exhibit a slow variation in amplitude over time with most of their energy being concentrated at the spectral peaks. We define temporal smoothness (TSM), which is associated with the H_H matrix here, in the same way as temporal continuity is defined in [22]. This smoothness is used to assign a high cost to large changes in time between the gains $H_{H_{r_h,t-1}}$ and $H_{H_{r_h,t}}$ in adjacent frames.

$$\text{TSM} = \frac{F}{R_h} \sum_{r_h=1}^{R_h} \frac{1}{\sigma_{H_{H_{r_h}}}^{2}} \sum_{t=2}^{T} \left(H_{H_{r_h,t-1}} - H_{H_{r_h,t}} \right)^2 \quad (7)$$

Normalization is used to make the global objective function independent to the signal norm; Thus, the H_H gains are normalized by σ_{H_H}. The value σ_{H_H} of each harmonic component r_h is calculated as $\sigma_{H_{H_{r_h}}} = \sqrt{\frac{1}{T} \sum_{t=1}^{T} H_{H_{r_h,t}}^2}$. The harmonic cost TSM is normalized by a factor $\frac{F}{R_h}$.

Spectral sparseness (SSP) is defined following [22] but is applied to the matrix W_H. SSP assigns a high cost to nonzero bases assuming that the harmonic sounds can be represented as a set of overtones in the frequency.

$$\mathrm{SSP} = \frac{T}{R_h} \sum_{r_h=1}^{R_h} \sum_{f=1}^{F} \left| \frac{W_{H_{f,r_h}}}{\sigma_{W_{H_{r_h}}}} \right| \tag{8}$$

Similarly to the treatment for H_H, the bases W_H are normalized by σ_{W_H}. The value σ_{W_H} of each harmonic component r_h is calculated as $\sigma_{W_{H_{r_h}}} = \sqrt{\frac{1}{F}\sum_{f=1}^{F} W_{H_{f,r_h}}^2}$. The harmonic cost SSP is normalized by a factor $\frac{T}{R_h}$.

2.2.4 Global percussive/harmonic NMF algorithm

The global objective function D, including the β-divergence, percussive and harmonic costs, is formulated as follows:

$$\begin{aligned} D = d_\beta\left(X_{n_\beta} \mid (X_P + X_H)\right) + K_{\mathrm{SSM}}\mathrm{SSM} + K_{\mathrm{TSP}}\mathrm{TSP} \\ + K_{\mathrm{TSM}}\mathrm{TSM} + K_{\mathrm{SSP}}\mathrm{SSP} \end{aligned} \tag{9}$$

Preliminary results do not show significant differences from initializing K_{SSM} and K_{TSM} with different values; thus, these parameters are set equal to each other (i.e., $K_{\mathrm{SSM}} = K_{\mathrm{TSM}}$). The parameters K_{TSP} and K_{SSP} are treated in the same way (i.e., $K_{\mathrm{TSP}} = K_{\mathrm{SSP}}$) because a similar behaviour is observed for these parameters in the preliminary studies. To determine if both sparseness and smoothness affect the performance of the separation process, we define the parameter $K_{\mathrm{SP}} = K_{\mathrm{TSP}} = K_{\mathrm{SSP}}$ to represent the sparseness terms and the parameter $K_{\mathrm{SM}} = K_{\mathrm{TSM}} = K_{\mathrm{SSM}}$ to represent the smoothness terms.

Following [20], we use the so-called multiplicative update rules (see Equation 10), such that D is non-increasing while ensuring non-negativity of the bases and the gains. These rules can be implemented using a gradient descent algorithm with an appropriate choice of the step size and are estimated for each scalar parameter Z by expressing the partial derivatives of the objective function $\frac{\partial D}{\partial Z}$ as the division between two positive terms $\left[\frac{\partial D}{\partial Z}\right]^-$ and $\left[\frac{\partial D}{\partial Z}\right]^+$:

$$Z = Z \odot \frac{\left[\frac{\partial D}{\partial Z}\right]^-}{\left[\frac{\partial D}{\partial Z}\right]^+} \tag{10}$$

where $\odot$ is the element-wise product operator and the fraction is the element-wise division operator.

Substituting the percussive basis W_P and the percussive gain H_P into Equation 10, the multiplicative update rules of the percussive sounds are formulated in Equations 11 and 12. The equations of each term $\left[\frac{\partial d_\beta}{\partial W_P}\right]^\pm$, $\left[\frac{\partial \mathrm{SSM}}{\partial W_P}\right]^\pm$, $\left[\frac{\partial d_\beta}{\partial H_P}\right]^\pm$ and $\left[\frac{\partial \mathrm{TSP}}{\partial H_P}\right]^\pm$ can be found in the Appendix.

$$W_P = W_P \odot \frac{\left[\frac{\partial d_\beta}{\partial W_P}\right]^- + K_{\mathrm{SSM}}\left[\frac{\partial \mathrm{SSM}}{\partial W_P}\right]^-}{\left[\frac{\partial d_\beta}{\partial W_P}\right]^+ + K_{\mathrm{SSM}}\left[\frac{\partial \mathrm{SSM}}{\partial W_P}\right]^+} \tag{11}$$

$$H_P = H_P \odot \frac{\left[\frac{\partial d_\beta}{\partial H_P}\right]^- + K_{\mathrm{TSP}}\left[\frac{\partial \mathrm{TSP}}{\partial H_P}\right]^-}{\left[\frac{\partial d_\beta}{\partial H_P}\right]^+ + K_{\mathrm{TSP}}\left[\frac{\partial \mathrm{TSP}}{\partial H_P}\right]^+} \tag{12}$$

Substituting the harmonic basis W_H and the harmonic gain H_H into Equation 10, the multiplicative update rules of the harmonic sounds are formulated in Equations 13 and 14. The equations of each term $\left[\frac{\partial d_\beta}{\partial W_H}\right]^\pm$, $\left[\frac{\partial \mathrm{SSP}}{\partial W_H}\right]^\pm$, $\left[\frac{\partial d_\beta}{\partial H_H}\right]^\pm$ and $\left[\frac{\partial \mathrm{TSM}}{\partial H_H}\right]^\pm$ can be found in the Appendix.

$$W_H = W_H \odot \frac{\left[\frac{\partial d_\beta}{\partial W_H}\right]^- + K_{\mathrm{SSP}}\left[\frac{\partial \mathrm{SSP}}{\partial W_H}\right]^-}{\left[\frac{\partial d_\beta}{\partial W_H}\right]^+ + K_{\mathrm{SSP}}\left[\frac{\partial \mathrm{SSP}}{\partial W_H}\right]^+} \tag{13}$$

$$H_H = H_H \odot \frac{\left[\frac{\partial d_\beta}{\partial H_H}\right]^- + K_{\mathrm{TSM}}\left[\frac{\partial \mathrm{TSM}}{\partial H_H}\right]^-}{\left[\frac{\partial d_\beta}{\partial H_H}\right]^+ + K_{\mathrm{TSM}}\left[\frac{\partial \mathrm{TSM}}{\partial H_H}\right]^+} \tag{14}$$

By iteratively updating the matrices W_P, H_P, W_H, H_H using *maxIter* iterations, our scheme can automatically distinguish between the bases belonging to the percussive or harmonic sounds.

2.2.5 Signal reconstruction

The percussive signal $x_p(t)$ is synthesized by using the magnitude percussive spectrogram X_P (see Equation 3) computed as the product of the factorized bases W_P and the activations H_P. To ensure that the reconstruction process is conservative, a percussive mask M_P is generated using Wiener filtering [24]. The effect of a percussive mask is to scale every frequency bin with a ratio that explains how much the percussive source contributes in the mixed spectrogram. The phase information related to the percussive signal is computed by multiplying the percussive mask M_P by the complex spectrogram X_c related to the mixed signal $x(t)$. The harmonic mask M_H is taken into account to similarly compute the harmonic signal $x_h(t)$.

$$M_P = \frac{X_P^2}{X_P^2 + X_H^2} \tag{15}$$

$$M_H = \frac{X_H^2}{X_P^2 + X_H^2} \tag{16}$$

$$x_p(t) = \mathrm{IDFT}\left(M_P \cdot X_c\right) \tag{17}$$

$$x_h(t) = \mathrm{IDFT}\left(M_H \cdot X_c\right) \tag{18}$$

The developed percussive/harmonic sound separation is detailed in Algorithm 1.

Algorithm 1 The developed percussive/harmonic sound separation

1. Compute the normalized magnitude spectrogram X_{n_β} of the input signal $x(t)$
2. Initialize W_P, H_P, W_H, H_H with random nonnegative values
3. **for** iteration=1:*maxIter* **do**
4. Update W_P using the multiplicative update rule (see Equation 11)
5. Update W_H using the multiplicative update rule (see Equation 13)
6. Update H_P using the multiplicative update rule (see Equation 12)
7. Update H_H using the multiplicative update rule (see Equation 14)
8. **end for**
9. Reconstruction of the percussive signal $x_p(t)$ (see Equation 17)
10. Reconstruction of the harmonic signal $x_h(t)$ (see Equation 18)

3 Evaluation and comparison

3.1 Test data

The databases are composed of monaural real-world music excerpts. Each music excerpt contains percussive and pitched instruments but does not contain a vocal track. The development database E (Table 1) is taken from the Guitar Hero game [25,26] and is composed of five commercial excerpts, each lasting 30 s. The first test database T1 (Table 2), taken from the Guitar Hero game [25,26], is composed of 20 commercial excerpts, each lasting 30 s. A pseudo-random process, using the standard uniform distribution, is used to select a starting time followed by a 30-s excerpt. The second test database T2 (Table 3) is a public database taken from SiSEC 2010 [27] and consists of four professionally produced music recordings lasting between 14 and 24 s. All of the signals were converted from stereo to mono and sampled at 16 kHz with a 16-bit resolution.

In summary, the dataset is composed of three databases. The development database E (five excerpts) is used to optimize the parameters (β, K_{SM}, K_{SP}, R_p and R_h) of the developed method. Two test databases T1 (20 excerpts) and T2 (4 excerpts) are then used to evaluate the performance of the separation process. Note that the development database E is not a part of the test databases T1 and T2.

Table 1 Identifier, title and artist of the files of the development database E

Identifier	Title	Artist
*E*_01	Two minutes to midnight	Iron Maiden
*E*_02	Bullet with butterfly wings	Smashing Pumpkins
*E*_03	Gamma ray	Beck
*E*_04	Go your own way	Fleetwood Mac
*E*_05	Hotel California	Eagles

3.2 Experimental setup

The quality of the audio separation using different frame sizes N, time shifts J and *maxIter* iterations is evaluated in preliminary studies. The experimental results show approximately the same performance using $N > 1024$ samples and $J < 512$ samples; therefore, we use the values $N = 1024$ and $J = 512$ because these values provide the best trade-off between the performance and the computational cost. The convergence of the algorithm is empirically observed. In fact, in all the performed simulations the convergence is achieved after 100 iterations. For this reason, we choose $maxIter = 100$. The values R_p and R_h are initialized to 150 because we initially supposed that this number of components would be a representative number of percussive and harmonic spectral

Table 2 Identifier, title and artist of the files of the first test database T1

Identifier	Title	Artist
*T*1_01	In my place	Coldplay
*T*1_02	La bamba	Los Lobos
*T*1_03	Livin' on a prayer	Bon Jovi
*T*1_04	No one to depend on	Santana
*T*1_05	Ring of fire	Johnny Cash
*T*1_06	Rooftops	Lost prophets
*T*1_07	So lonely	The Police
*T*1_08	Song 2	Blur
*T*1_09	Sultans of swing	Dire Straits
*T*1_10	Under pressure	Queen
*T*1_11	Are you gonna go my way	Lenny Kravitz
*T*1_12	Feel the pain	Dinosaur Jr
*T*1_13	Hollywood nights	Bob Seger & The Silver Bullet Band
*T*1_14	Hurts so good	John Mellencamp
*T*1_15	Kick out the jams	MC5's Wayne Kramer
*T*1_16	Make it wit chu	Queens Of The Stone Age
*T*1_17	One way or another	Blondie
*T*1_18	Shiver	Coldplay
*T*1_19	Shout it out loud	Kiss
*T*1_20	Sympathy for the devil	The Rolling Stones

Table 3 Identifier, title and artist of the files of the second test database T2

Identifier	Title	Artist
T2_01	Roads	Bearlin
T2_02	The_ones_we_love	Another Dreamer
T2_03	Remember_the_name	Fort Minor
T2_04	Tour	Ultimate NZ Tour

patterns. However, these parameters will be later analyzed (subsection 3.4.1).

Sound source separation applications, based on NMF approaches, usually adopt logarithmic frequency discretization (e.g., uniformly spaced sub-bands on the equivalent rectangular bandwidth (ERB) scale [28]). As harmonic signals are organized in a chromatic scale and using this scale, the musical notes are defined with semitone resolution, the developed method uses a 1/4 semitone resolution. Therefore, the time-frequency representation is obtained by integrating the STFT bins corresponding to each 1/4 semitone interval. To obtain the separated signals, the frequency resolution of the percussive and harmonic masks defined in Equations 15 and 16 must be extended to the resolution of the STFT. Taking into account that each bin of the STFT belongs to a value in the 1/4 semitone resolution, each bin of the masks with the STFT resolution takes the value that belongs to in the 1/4 semitone resolution. Consequently, all bins belonging to the same 1/4 semitone have the same mask value. Percussive and harmonic masks with the resolution of the STFT are then obtained and the inverse transform can be computed following Equations 17 and 18.

3.3 Algorithms for comparison

We use three recent state-of-the-art percussive/harmonic sound separation methods to evaluate the developed method: HPSS [1], MFS [7] and NMPCF [8]. HPSS and MFS are implemented in this study, whereas the separation results from NMPCF have been provided directly by the authors.

3.4 Results

Three metrics are used to assess the performance of the developed method [29,30]: (1) the source-to-distortion ratio (SDR), which provides information on the overall quality of the separation process; (2) the source-to-interferences ratio (SIR), which is a measure of the presence of harmonic sounds in the percussive signal and vice versa; and (3) the source-to-artifacts ratio (SAR), which provides information on the artifacts in the separated signal from separation and/or resynthesis.

As previously mentioned, the output of each method is composed of two signals, a percussive signal $x_p(t)$ (i.e., harmonic sounds have been attenuated or removed) and a harmonic signal $x_h(t)$ (i.e., percussive sounds have been attenuated or removed). In the database T1, the percussive average (Perc) is computed using the mean of all of the separated percussive signals. The harmonic average (Harm) is computed using the mean of all of the separated harmonic signals. In the same way, the overall average (Overall) is computed using the mean of all of the separated percussive and harmonic signals.

3.4.1 Parameters optimization

Figure 2 shows the optimization of the parameters β, K_{SP} and K_{SM} that are used to analyze the effect of the β-divergence cost and the weight of the smoothness and sparseness constraints in the development database E. Standard NMF (smoothness and sparseness constraints are disabled, i.e., $K_{SM} = K_{SP} = 0$) achieves a SDR value approximately equal to 3 dB. This fact suggests that standard NMF does not properly separate percussive and harmonic sounds because each separated signal, using standard NMF, is composed of percussive and harmonic sounds but its energy is approximately half of the energy of the mixed input signal. The results show that the maximum SDR is obtained using $\beta = 1.5$ because the SDR is maximized at the best trade-off between high and low energy changes in the frequency, unlike the Euclidean distance (which corresponds to $\beta = 2.0$ and is more sensitive to high energy changes) or the Itakura-Saito divergence (which corresponds to $\beta = 0$ and is more sensitive to low energy changes). This optimization produces a significant improvement of approximately 2.7 dB over the standard NMF showing that a percussive separated signal is composed of sounds that exhibit percussive features in which harmonic sounds have been attenuated, and the harmonic separated signal is composed of sounds that exhibit harmonic features where percussive sounds have been attenuated. As a consequence, the factorized spectrograms exhibit time-frequency energy distributions such as can be found in real-world percussive or harmonic sounds. In order to obtain the maximum SDR, the percussive and harmonic costs are found to be equally significant ($K_{SM} = K_{SP} = 0.1$) but must be sufficiently small compared to the β-divergence cost to reconstruct the signal correctly. As a result of this fact, the developed method fails for K_{SM} >>1 or K_{SP} >>1 because NMF does not prioritize signal reconstruction under these conditions.

Figure 2 has shown the optimization of β, K_{SP} and K_{SM} using a fixed dictionary size ($R_p = R_h = 150$). In order to analyze the dependence of the parameters of the developed method, Figures 3, 4 and 5 are shown.

Figure 3 shows the optimization of the parameters K_{SP} and K_{SM} using $\beta = 1.5$ and different number of percussive R_p and harmonic R_h components in order to analyze the effect of different dictionary sizes. Results

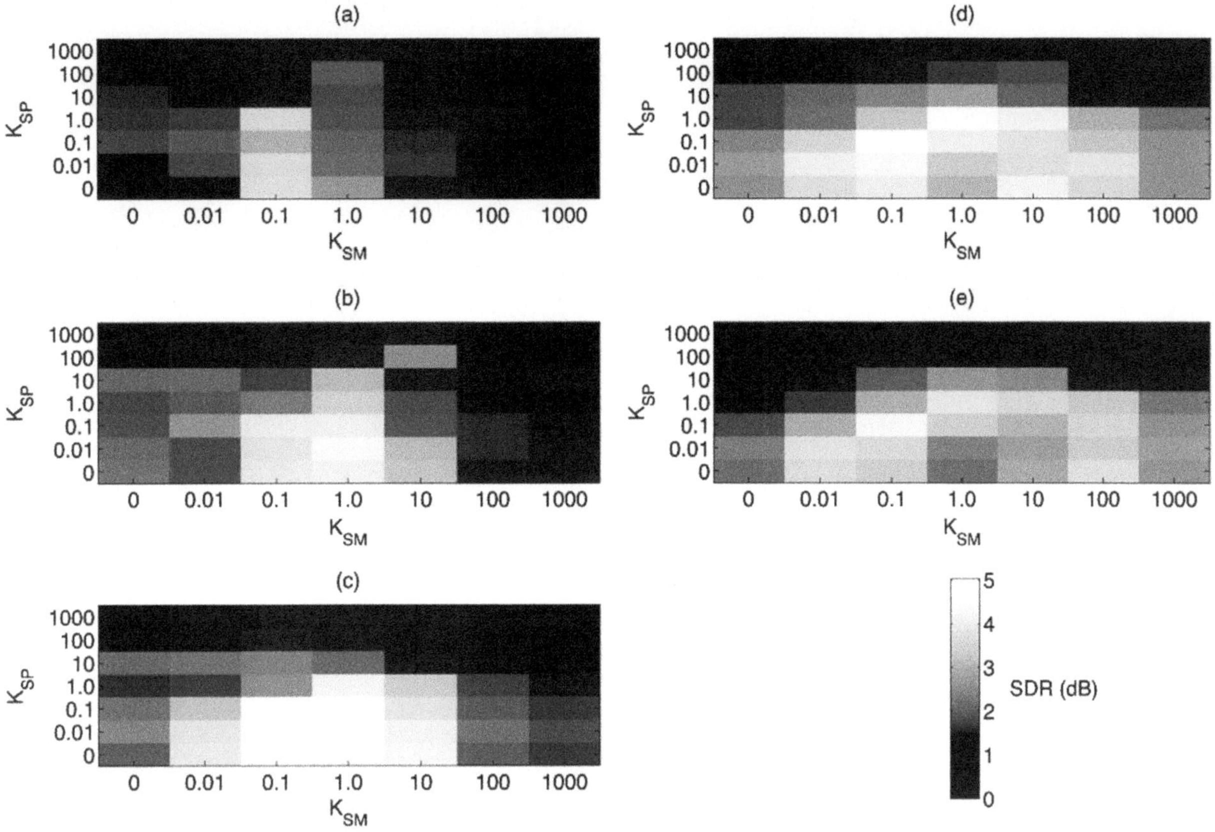

Figure 2 Optimization of β, K_{SP} and K_{SM} parameters using the development database E. (a) $\beta = 0$, **(b)** $\beta = 0.5$, **(c)** $\beta = 1.0$, **(d)** $\beta = 1.5$ and **(e)** $\beta = 2.0$. The number of percussive R_p and harmonic R_h spectral patterns is equal to 150. It can be observed the effect of the β-divergence cost in the separation process when varying the contribution of the smoothness and sparseness constraints. The goal of this analysis was to find the optimum value of parameters β, K_{SP} and K_{SM} looking for the value of the combination of them that maximizes the SDR measure for the development database E. The maximum SDR value is obtained using $\beta = 1.5$. This fact may be due to the best trade-off between high and low energy changes in frequency.

show the optimal values of the parameters K_{SP} and K_{SM} are obtained using $K_{SP} = K_{SM} = 0.1$ for all the dictionary sizes evaluated. This fact suggests that the dictionary size could not affect the optimal values of K_{SP} and K_{SM} because these values provide the maximum SDR for each dictionary size evaluated.

Figure 4 shows the optimization of the parameter β using the optimal $K_{SP} = K_{SM} = 0.1$ and different number of percussive and harmonic components in order to analyze the effect of a different dictionary sizes. For each value β, results show that SDR performance exhibits the similar behaviour independently of the dictionary size, increasing from $\beta = 0$ to $\beta = 1.5$ and decreasing from $\beta = 1.5$ to $\beta = 2$. As occurred in Figure 2, the value β that maximizes SDR is achieved using $\beta = 1.5$, but the differences, compared to the other results using different dictionary sizes, are not significant.

Figure 5 shows the optimization of the smoothness K_{TSM}-K_{SSM} and sparseness K_{TSP}-K_{SSP} parameters. Figure 5a shows that although different values of K_{TSM} and K_{SSM} have been used, the maximum SDR performance is obtained using $K_{TSM} = K_{SSM} = 0.1$ as occurred when these parameters were initialized with the same values (see Figure 2d). This fact suggests that the smoothness parameters K_{TSM} and K_{SSM} could be initialized using equal values $K_{SM} = K_{TSM} = K_{SSM}$ since they do not show significant differences from initializing them with different values in order to obtain the best SDR performance. A similar behaviour can be observed from the sparseness K_{TSP}-K_{SSP} parameters (see Figure 5b) which obtain the maximum SDR performance using a sparseness parameter equal to each other $K_{SP} = K_{TSP} = K_{SSP} = 0.1$.

Figure 6 shows how SDR results can be improved by analyzing a more accurate range of the parameters K_{SM} and K_{SP} around of the previous optimum value 0.1. The maximum SDR is obtained using a higher value of smoothness constraints than sparseness constraints, specifically $K_{SM} = 0.2$ and $K_{SP} = 0.1$, evaluating the development database E. This fact can be observed by

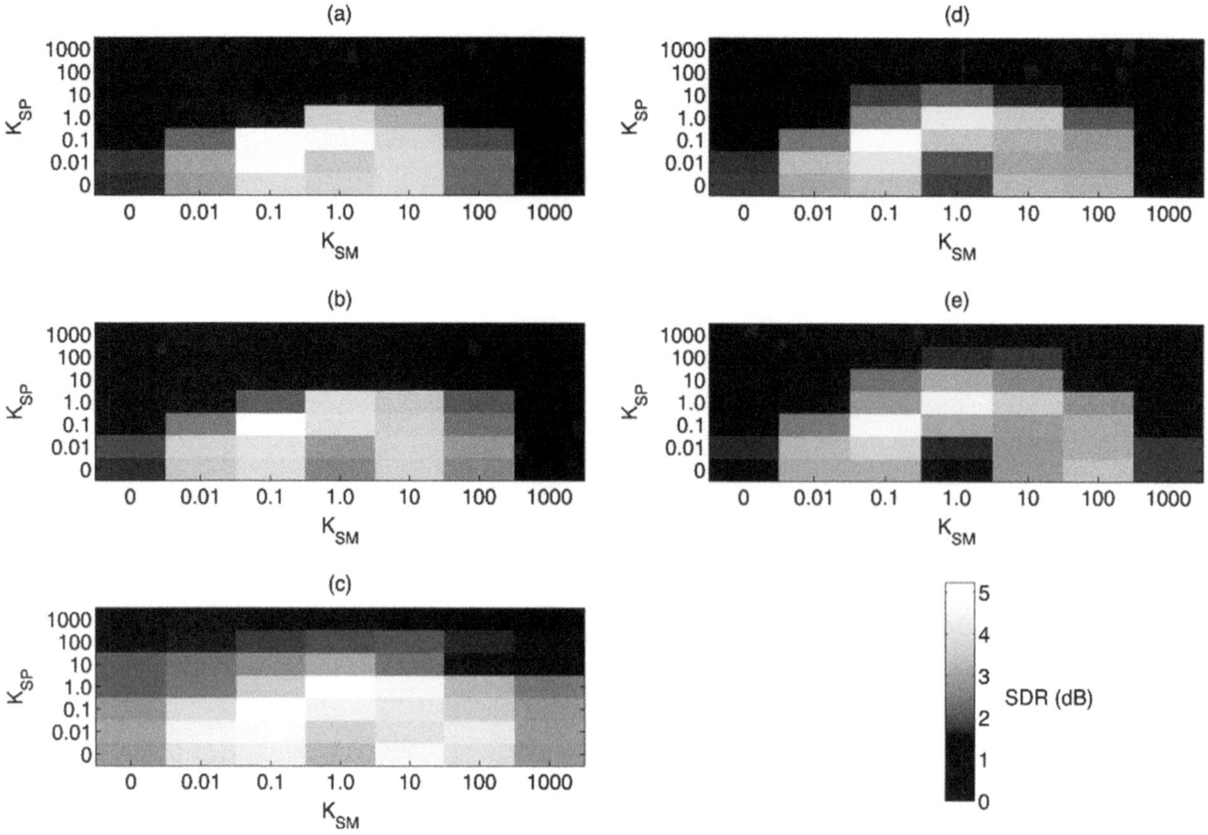

Figure 3 Optimization of parameters K_{SP} and K_{SM}. Using $\beta = 1.5$ and different dictionary sizes composed of the same number of percussive and harmonic components (**(a)** 50, **(b)** 100, **(c)** 150, **(d)** 250, and **(e)** 500 components) using the development database E.

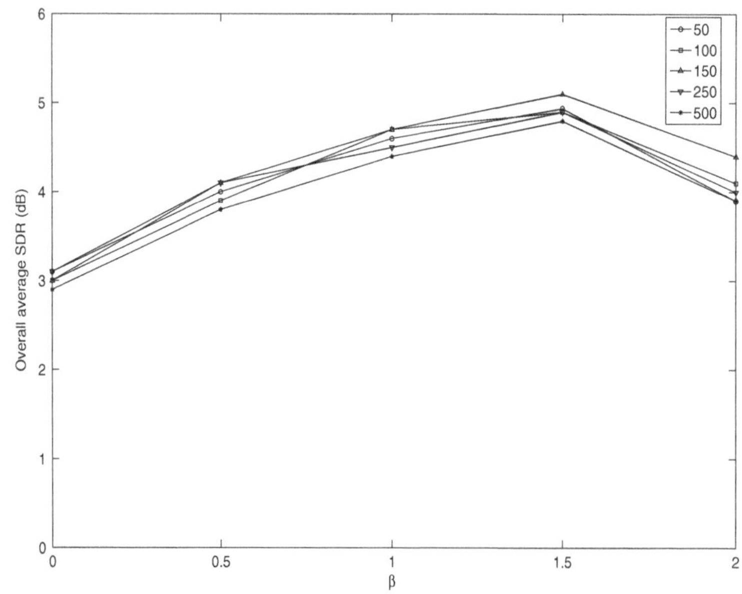

Figure 4 Optimization of the parameter β. Using the optimal $K_{SP} = K_{SM} = 0.1$ and different dictionary sizes composed of the same number of percussive and harmonic components (50, 100, 150, 250 and 500 components) in the development database E.

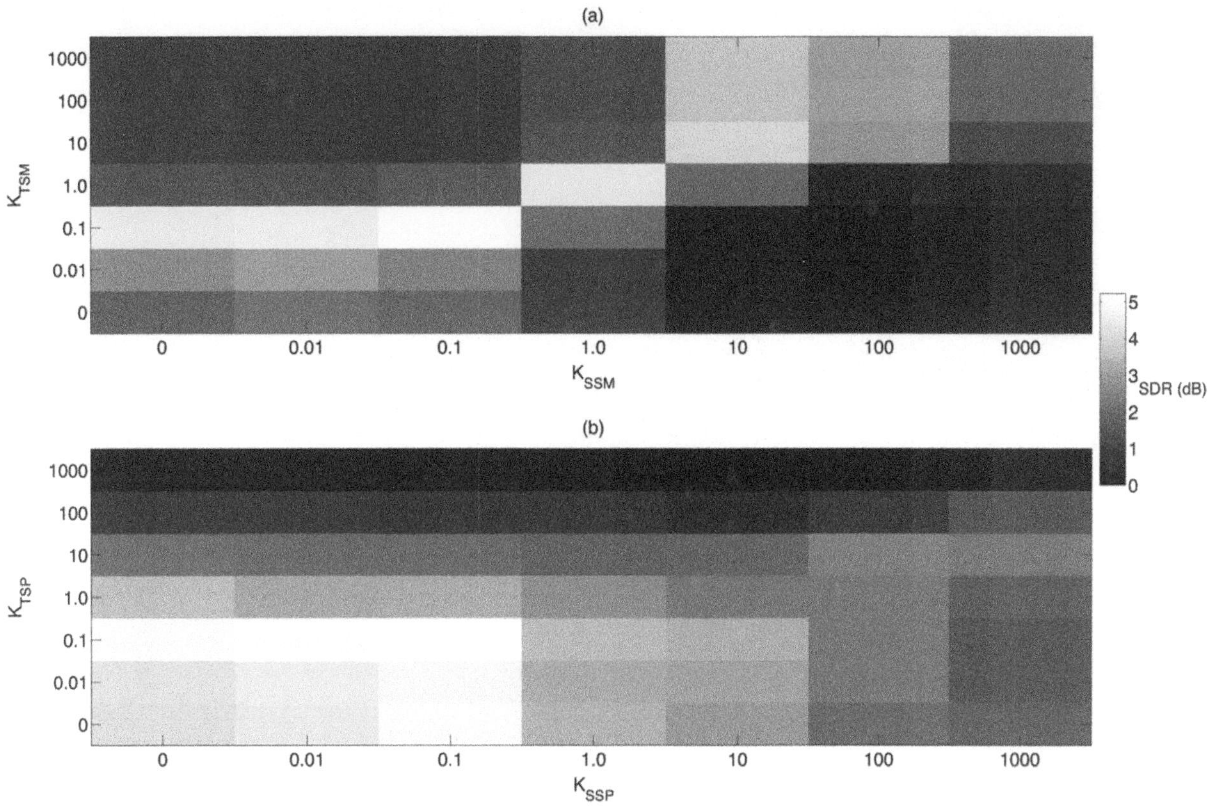

Figure 5 Optimization of (a) the smoothness and (b) sparseness parameters. (a) Smoothness parameters K_{TSM} and K_{SSM} using the values $\beta = 1.5, K_{TSP} = K_{SSP} = 0.1$ and $R_p = R_h = 150$ components in the development database E; **(b)** sparseness parameters K_{TSP} and K_{SSP} using the values $\beta = 1.5, K_{TSM} = K_{SSM} = 0.1$ and $R_p = R_h = 150$ components in the development database E.

analyzing the diagonals in Figure 6. Thus, using a K_{SM} value that is twice the K_{SP} value (i.e., see the lower diagonals from the main diagonal from left to right) improves the SDR results and slows the decrease in the SDR. However, when K_{SP} is twice K_{SM} (i.e., see the upper diagonals from the main diagonal from left to right), the SDR results are generally worse, and the SDR decreases more rapidly.

Once β, K_{SM} and K_{SP} are optimized ($\beta = 1.5, K_{SM} = 0.2$ and $K_{SP} = 0.1$), the number of the percussive R_p and harmonic R_h components is optimized (see Figure 7). A slight better separation is obtained when it is used $R_p = 250$ and $R_h = 500$. It seems that a wider variety of harmonic spectral patterns could improve SDR results; however, the improvement, about $0.3dB$, compared to $R_p = R_h = 150$ is not significant. In fact, a higher number of components, i.e., $R_p > 250$ and $R_h > 500$, reduces the SDR performance (in a similar way as occurred in Figure 3). This reduction of SDR may be explained by using a large dictionary size (usually $R_p + R_h$ is chosen to be smaller than F or T, so that $F(R_p + R_h) + (R_p + R_h)T << FT$ in a NMF framework [14,21]).

3.4.2 *Performance evaluation*

Figures 8, 9 and 10 show SDR, SIR and SAR results evaluating the database T1 for the proposed method and the three aforementioned state-of-the-art percussive/harmonic sound separation methods. Each box represents 20 data points, one for each excerpt of the test database: each data point in the blue box represents the average value of the percussive separation results; each data point in the red box represents the average value of the harmonic separation results; and each data point in the black box represents the average value of the overall separation results (the average value considering the percussive and harmonic separation results). The lower and upper lines of each box show the 25th and 75th percentiles for the database. The line in the middle of each box represents the mean value of the dataset. The lines extending above and below each box show the extent of the rest of the samples, excluding outliers. Outliers are defined as points that are over 1.5 times the interquartile range from the sample median, which are shown as crosses.

Figure 8 shows that the developed method obtains, on average, the best quality performance in terms of the percussive, harmonic and overall SDR for the separation

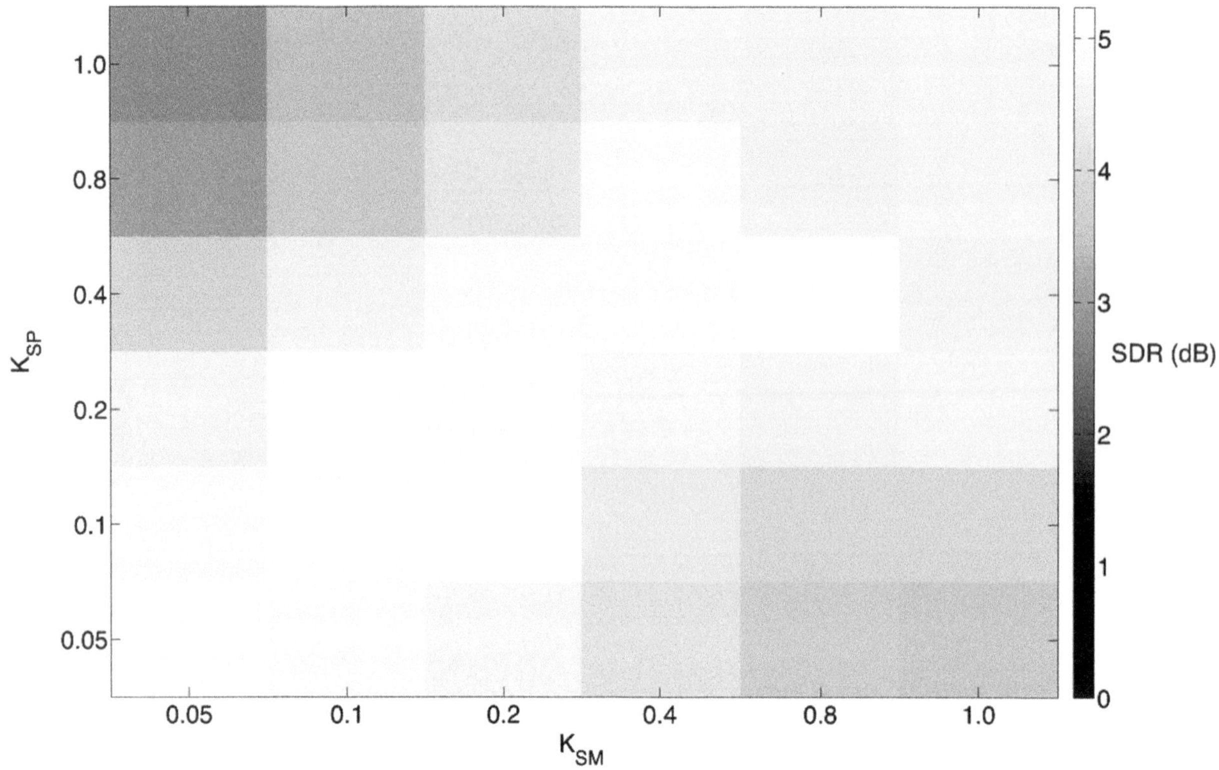

Figure 6 SDR average results. Obtained by using $\beta = 1.5$ and varying the smoothness term K_{SM} and the sparseness term K_{SP} for the development database E. The number of percussive R_p and harmonic R_h spectral patterns is equal to 150.

process relative to the three state-of-the-art separation methods. MFS and HPSS can still be considered to be competitive methods, unlike NMPCF (this method will be analyzed later). The proposed method significantly outperforms HPSS and NMPCF in percussive/harmonic SDR results but not compared to MFS which is confirmed by a one-sided paired t-test (see Table 4). A possible strength of the developed method, which is not exhibited by the other methods, seems to be its robustness in evaluating different databases (databases T1 and T2). This robustness is in the fact of including smoothness and sparseness constraints into the factorization process because the main difference between the develop method and HPSS and MFS is the use of sparseness constraints to achieve time-frequency energy distributions as can be found in real-world percussive or harmonic sounds. In both databases, the developed method produces nearly identical overall SDR results, 6.3 dB, independent of the database evaluated. Moreover, the lower line of each box of the developed method is above the mean obtained from using HPSS and NMPCF. Thus, more reliable SDR results are obtained in comparison with those methods when evaluating different types of sounds used in different music genres.

Figure 9 shows that HPSS produces the best overall SIR results but the SIR performance of the developed method is nearly identical to that obtained using HPSS. It can be confirmed using a one-sided paired t test that indicates HPSS does not significantly outperform SIR results compared to the developed method neither in percussive nor harmonic sounds (see Table 4). However, Table 4 shows that the developed method improves significantly SIR results compared to MFS and NMPCF for both percussive and harmonic sounds. Both HPSS and the proposed one: (i) enable most of the harmonic content to be removed while maintaining the quality of the percussive signal and vice versa and (ii) capture polyphonic richness. The developed method produces better percussive quality for the SIR than MFS because it uses information to model percussive sounds that is not used by MFS. That is, the developed method models percussive sounds using smoothness in the frequency and sparseness in the time, whereas MFS models percussive sounds using only smoothness in the frequency (by removing outliers in a temporal slice).

As previously mentioned, HPSS produces the best overall SIR on average. However, these results are obtained at the expense of introducing more artifacts and lead to greater overall distortion. This fact can be observed in Figure 10 in which the worst percussive and harmonic SAR results are obtained by HPSS. Considering SAR

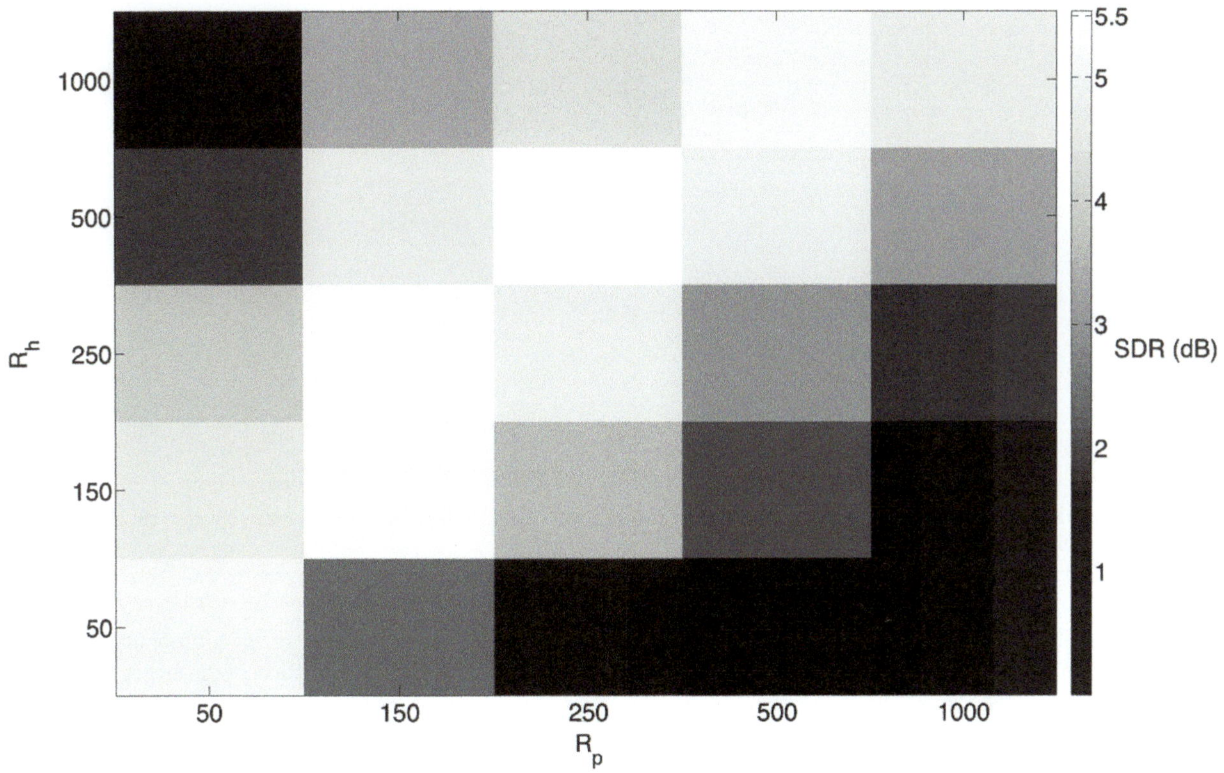

Figure 7 SDR average results (β = 1.5, K_{SM} = 0.2 and K_{SP} = 0.1). As a function of the number of percussive/harmonic components applied to the development database E.

results and using a one-sided paired *t* test (see Table 4), the developed method significantly outperforms HPSS and NMPCF in percussive sounds and the three state-of-the-art methods in harmonic sounds. The developed method also offers the advantage of producing the best SAR results (excluding NMPCF, which will be discussed in the next paragraph) because the artifacts in the reconstruction signal are minimized.

For the case of NMPCF, SDR and SIR results exhibit the worst separation performance, and therefore, this method

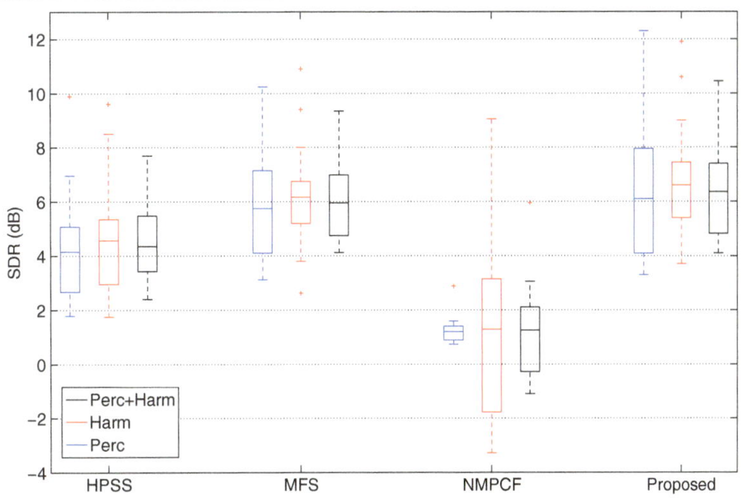

Figure 8 Comparative SDR results. Evaluating database T1 for the proposed method and the three aforementioned state-of-the-art percussive/harmonic sound separation methods.

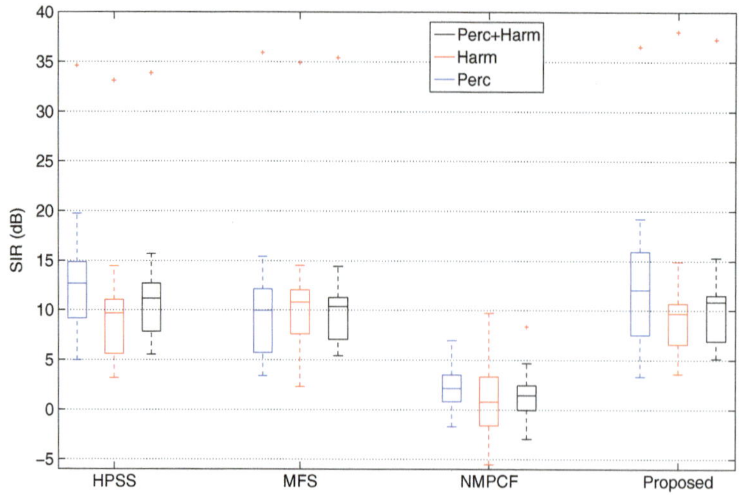

Figure 9 Comparative SIR results. Evaluating the database T1 for the proposed method and the three aforementioned state-of-the-art percussive/harmonic sound separation methods.

always ranks last. The poor performance of NMPCF may be attributed to its high dependence on the drum-only matrix used in the decomposition process. This drum-only matrix is obtained training with drum sounds, but these trained drum features may be sufficiently different from the percussive features evaluated in the test database T1. The harmonic signal, provided by NMPCF, is composed of most of the original harmonic and percussive sounds. It causes a high harmonic SAR because the proportion of artifacts is too small compared to the proportion of the target (harmonic) and the interference (percussive) sounds. However, the percussive signal, provided by NMPCF, is composed of a residual part of the original percussive and harmonic sounds. It causes a low percussive SAR because the proportion of artifacts is too high compared to the proportion of the target (percussive) and the interference (harmonic) sounds.

To illustrate some of the strengths and weaknesses of the developed method, Table 5 shows the SDR, SIR and SAR results for each individual music excerpt from database T2. Results show that the developed method produces the best average percussive, harmonic and overall SDR and SAR results. The main strength of the developed method is its high separation performance in evaluating purely harmonic or percussive sounds (e.g., SAR (dB) bass or drums). However, HPSS obtains the better SIR results on average with more than 1 dB above the proposed method and MFS. For all the methods evaluated, the

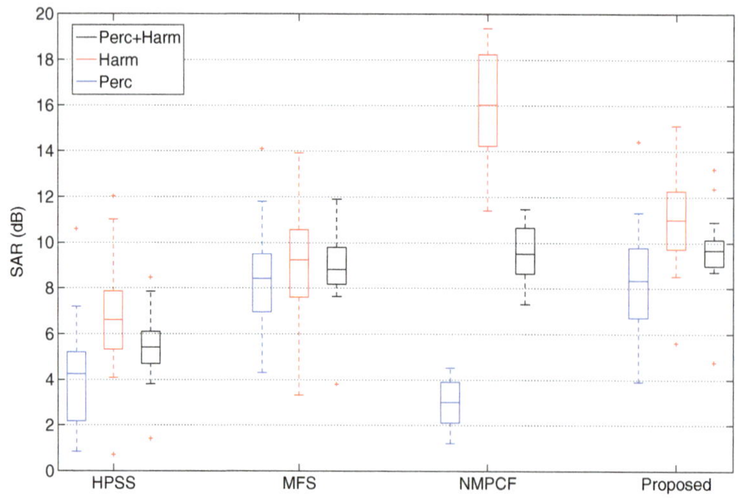

Figure 10 Comparative SAR results. Evaluating the database T1 for the proposed method and the three aforementioned state-of-the-art percussive/harmonic sound separation methods.

Table 4 Analysis of the statistical significance of the percussive/harmonic SDR-SIR-SAR results

	Percussive			Harmonic		
	SDR	SIR	SAR	SDR	SIR	SAR
HPSS	$< 10^{-8}$	0.2	$< 10^{-11}$	$< 10^{-9}$	0.9	$< 10^{-14}$
MFS	0.2	$< 10^{-3}$	0.8	0.1	$< 4 \cdot 10^{-2}$	$< 10^{-6}$
NMPCF	$< 10^{-7}$	$< 10^{-4}$	$< 10^{-7}$	$< 10^{-6}$	$< 10^{-4}$	$< 10^{-5}$

Comparing the developed method with the three state-of-the-art separation methods using a one-sided paired *t* test in the database T1. Each cell show the parameter *p* that represents the probability of setting a statistically significant result. Considering a confidence interval of 95%, small values of $p<0.05$ indicate that there exists statistical significance of the results evaluated.

main weaknesses of harmonic/percussive separation are the following: (i) It is not effective in separating harmonic onsets, that is, the transients of harmonic sounds played by harmonic instruments (e.g., the initial milliseconds of a note played by guitar) because they exhibit a percussive behaviour. In these cases, harmonic onsets are not separated in the harmonic signal as can be seen in Figure 11. The reason is because a harmonic onset exhibits spectro-temporal features that have been modelled as percussive sounds in the factorization process; (ii) it cannot effectively separate audio effects, e.g., a synthesizer can generate a harmonic sound that exhibits spectro-temporal features (e.g., the vibrato effect shows non-smoothness in time) which are not modelled in the factorization process.

To illustrate the separation performance of the developed method, audio examples (from the T1 and T2 databases) have been uploaded to a web page. Each audio example (mixed track, separated-percussive track and separated-harmonic track) has been evaluated using HPSS, MFS and the developed method. The web page can be found at https://dl.dropboxusercontent.com/u/22448214/PercHarmFeb2014/index.html.

4 Conclusions

This paper presents an unsupervised learning process for separating percussive and harmonic sounds from monaural instrumental music. Our formulation is based on a modified NMF approach that automatically distinguishes between percussive and harmonic bases by integrating spectro-temporal features, such as anisotropic smoothness or time-frequency sparseness, into the factorization

Table 5 Percussive, harmonic and overall SDR, SIR and SAR results for each excerpt of the database T2

	HPSS			MFS			Proposed		
	SDR	SIR	SAR	SDR	SIR	SAR	SDR	SIR	SAR
Percussive separation									
Identifier									
T2_01	2.6	13.2	1.1	−0.2	−1.5	8.4	4.0	6.5	5.7
T2_02	2.4	10.2	3.4	3.1	8.0	4.9	5.2	8.5	7.3
T2_03	2.6	6.9	4.0	2.5	2.1	12.3	2.8	2.6	11.1
T2_04	5.5	11.5	6.5	6.2	9.6	8.0	7.5	10.3	10.3
Average	3.2	*10.5*	3.8	2.9	4.6	8.4	*4.9*	7.0	*8.7*
Harmonic separation									
Identifier									
T2_01	9.8	13.8	11.9	7.1	13.8	11.5	11.0	14.8	13.9
T2_02	4.8	6.3	9.8	5.5	16.2	11.6	7.5	9.3	12.1
T2_03	4.8	8.7	6.3	4.6	11.0	8.0	5.0	9.1	8.6
T2_04	5.6	11.5	6.7	6.2	9.3	8.7	7.5	10.6	10.5
Average	6.3	10.1	8.7	5.9	*12.6*	10.0	*7.8*	11.0	*11.3*
Overall separation									
Identifier									
T2_01	6.2	13.5	6.5	3.5	6.2	10.0	7.5	10.7	9.8
T2_02	3.6	8.3	6.6	4.3	12.1	8.3	6.4	8.8	9.8
T2_03	3.7	7.8	5.2	3.6	6.6	10.2	3.9	5.9	9.9
T2_04	5.6	11.5	6.6	6.2	9.5	8.4	7.5	10.5	10.4
Average	4.8	*10.3*	6.2	4.4	8.6	9.2	*6.3*	9.0	*10.0*

Each row titled Average shows three italic values. Each italic value represents the best percussive, harmonic or overall SDR, SIR or SAR result comparing all methods evaluated in this table.

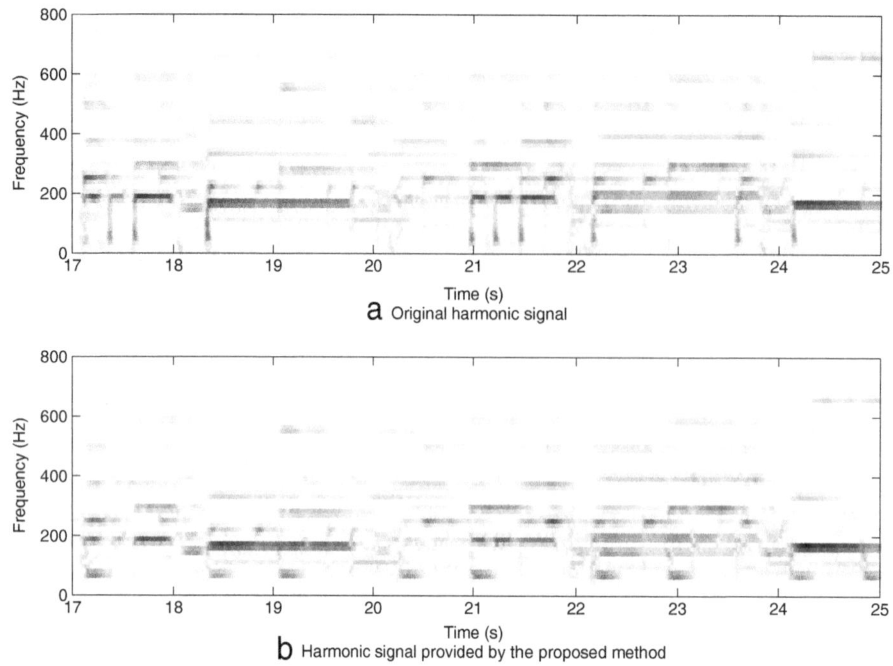

Figure 11 A harmonic excerpt, played by guitar instrument, extracted from the file *T2_02* in Table 3. It can be observed that the harmonic onsets (vertical lines at the beginning of some notes) in the original harmonic signal **(a)** do not appear in the harmonic signal **(b)** output of the proposed method.

process. The developed method exhibits the following advantages: (i) prior knowledge of the number of instruments playing the music excerpt is not required, and (ii) neither prior information about the musical instruments nor a supervised training are required to classify the bases.

Different experiments are performed to optimize the parameters of the developed method. The results show that (i) the value $\beta = 1.5$ provides the maximum SDR since this measure has been computed at the best trade-off between high and low energy changes in the frequency; (ii) The maximum SDR is achieved using a higher value of smoothness constraints compared to sparseness constraints evaluating the databases T1 and T2 and (iii) a higher number of components, i.e., $R_p > 250$ and $R_h > 500$, reduces the SDR performance.

The analysis of the dependence of the parameters of the developed method shows that (i) the dictionary size could not affect the optimal values of K_{SP} and K_{SM} because they provide the maximum SDR for each dictionary size evaluated; (ii) SDR performance obtains the maximum SDR using $\beta = 1.5$ independently of the dictionary size and (iii) the smoothness K_{TSM}-K_{SSM} and sparseness K_{TSP}-K_{SSP} parameters could be initialized using equal values $K_{SM} = K_{TSM} = K_{SSM}$ and $K_{SP} = K_{TSP} = K_{SSP}$ since they do not show significant differences from initializing them with different values in order to obtain the maximum SDR performance.

Evaluating the database T1 shows that the developed method obtains the best quality performance in terms of the percussive, harmonic and overall SDR for the separation process in relation to the three state-of-the-art separation methods. The proposed method significantly outperforms the other methods taking into account most of the percussive/harmonic SDR, SIR or SAR results which is confirmed by a one-sided paired t test. A significant strength of the developed method is its robustness in evaluating different databases.

Evaluating the database T2 illustrates some of the strengths and weaknesses of the developed method. An interesting strength shown by the developed method is its successful separation performance in evaluating purely harmonic or percussive sounds. However, harmonic onsets and audio effects are not successfully separated because their spectro-temporal features have not been modelled in the factorization process.

Future work will focus on three topics. First, we will try to improve the quality of the separated signals by defining a new spectral distance that integrates novel spectro-temporal features of the percussive and harmonic sounds. Second, a novel constraint based on the vibrato effect will be investigated to extend this method to singing-voice signals. Finally, in order to improve the singing-voice signals, a set of novel percussive constraints will be analyzed to distinguish between percussive music instruments and unvoiced vocal sounds.

Appendix

Detailed terms of the multiplicative update rules of percussive sounds

Here, each of the terms $\left[\frac{\partial d_\beta}{\partial W_P}\right]^{\pm}$, $\left[\frac{\partial \mathrm{SSM}}{\partial W_P}\right]^{\pm}$, $\left[\frac{\partial d_\beta}{\partial H_P}\right]^{\pm}$ and $\left[\frac{\partial \mathrm{TSP}}{\partial H_P}\right]^{\pm}$ belonging to the percussive multiplicative update rules are detailed:

$$\left[\frac{\partial d_\beta}{\partial W_P}\right]^{-} = \left[(W_P H_P + W_H H_H)^{\beta-2} \odot X_{n_\beta}\right] H_P^T \quad (19)$$

$$\left[\frac{\partial d_\beta}{\partial W_P}\right]^{+} = \left[(W_P H_P + W_H H_H)^{\beta-1}\right] H_P^T \quad (20)$$

$$\left[\frac{\partial \mathrm{SSM}}{\partial W_P}\right]^{-}_{f,r_P} = 2F\left[\frac{W_{P_{f-1,r_P}} + W_{P_{f+1,r_P}}}{\sum_{j=1}^{F} W_{P_{j,r_P}}^2}\right] + \frac{2FW_{P_{f,r_P}} \sum_{j=2}^{F}\left(W_{P_{j,r_P}} - W_{P_{j-1,r_P}}\right)^2}{\left(\sum_{j=1}^{F} W_{P_{j,r_P}}^2\right)^2} \quad (21)$$

$$\left[\frac{\partial \mathrm{SSM}}{\partial W_P}\right]^{+}_{f,r_P} = \frac{4FW_{P_{f,r_P}}}{\sum_{j=1}^{F} W_{P_{j,r_P}}^2} \quad (22)$$

$$\left[\frac{\partial d_\beta}{\partial H_P}\right]^{-} = W_P^T\left[(W_P H_P + W_H H_H)^{\beta-2} \odot X_{n_\beta}\right] \quad (23)$$

$$\left[\frac{\partial d_\beta}{\partial H_P}\right]^{+} = W_P^T\left[(W_P H_P + W_H H_H)^{\beta-1}\right] \quad (24)$$

$$\left[\frac{\partial \mathrm{TSP}}{\partial H_P}\right]^{-}_{r_P,t} = \sqrt{T}\frac{H_{P_{r_P,t}} \sum_{i=1}^{T} H_{P_{r_P,i}}}{\left(\sum_{i=1}^{T} H_{P_{r_P,i}}^2\right)^{\frac{3}{2}}} \quad (25)$$

$$\left[\frac{\partial \mathrm{TSP}}{\partial H_P}\right]^{+}_{r_P,t} = \frac{1}{\sqrt{\frac{1}{T}\sum_{i=1}^{T} H_{P_{r_P,i}}^2}} \quad (26)$$

where ^{T} denotes the transpose matrix operator. The terms $\left[\frac{\partial \mathrm{SSM}}{\partial W_P}\right]^{\pm}$ and $\left[\frac{\partial \mathrm{TSP}}{\partial H_P}\right]^{\pm}$ are defined using [22], as adapted to the matrix W_P and H_P.

Detailed terms of the multiplicative update rules of harmonic sounds

Here, each of the terms $\left[\frac{\partial d_\beta}{\partial W_H}\right]^{\pm}$, $\left[\frac{\partial \mathrm{SSP}}{\partial W_H}\right]^{\pm}$, $\left[\frac{\partial d_\beta}{\partial H_H}\right]^{\pm}$ and $\left[\frac{\partial \mathrm{TSM}}{\partial H_H}\right]^{\pm}$ belonging to the harmonic multiplicative update rules are detailed:

$$\left[\frac{\partial d_\beta}{\partial W_H}\right]^{-} = \left[(W_P H_P + W_H H_H)^{\beta-2} \odot X_n\right] H_H^T \quad (27)$$

$$\left[\frac{\partial d_\beta}{\partial W_H}\right]^{+} = \left[(W_P H_P + W_H H_H)^{\beta-1}\right] H_H^T \quad (28)$$

$$\left[\frac{\partial \mathrm{SSP}}{\partial W_H}\right]^{-}_{f,r_h} = \sqrt{F}\frac{W_{H_{f,r_h}} \sum_{j=1}^{F} W_{H_{j,r_h}}}{\left(\sum_{j=1}^{F} W_{H_{j,r_h}}^2\right)^{\frac{3}{2}}} \quad (29)$$

$$\left[\frac{\partial \mathrm{SSP}}{\partial W_H}\right]^{+}_{f,r_h} = \frac{1}{\sqrt{\frac{1}{F}\sum_{j=1}^{F} W_{H_{j,r_h}}^2}} \quad (30)$$

$$\left[\frac{\partial d_\beta}{\partial H_H}\right]^{-} = W_H^T\left[(W_P H_P + W_H H_H)^{\beta-2} \odot X_n\right] \quad (31)$$

$$\left[\frac{\partial d_\beta}{\partial H_H}\right]^{+} = W_H^T\left[(W_P H_P + W_H H_H)^{\beta-1}\right] \quad (32)$$

$$\left[\frac{\partial \mathrm{TSM}}{\partial H_H}\right]^{-}_{r_h,t} = 2T\left[\frac{H_{H_{r_h,t-1}} + H_{H_{r_h,t+1}}}{\sum_{i=1}^{T} H_{H_{r_h,i}}^2}\right] + \frac{2TH_{H_{r_h,t}} \sum_{i=2}^{T}\left(H_{H_{r_h,i}} - H_{H_{r_h,i-1}}\right)^2}{\left(\sum_{i=1}^{T} H_{H_{r_h,i}}^2\right)^2} \quad (33)$$

$$\left[\frac{\partial \mathrm{TSM}}{\partial H_H}\right]^{+}_{r_h,t} = \frac{4TH_{H_{r_h,t}}}{\sum_{i=1}^{T} H_{H_{r_h,i}}^2} \quad (34)$$

where ^{T} denotes the transpose matrix operator. The terms $\left[\frac{\partial \mathrm{SSP}}{\partial W_H}\right]^{\pm}$ and $\left[\frac{\partial \mathrm{TSM}}{\partial H_H}\right]^{\pm}$ are defined using [22], as adapted to the matrix W_H and H_H.

Competing interests

The authors declare that they have no competing interests.

Acknowledgements

This work was supported by the Andalusian Business, Science and Innovation Council under project P2010- TIC-6762 and (FEDER) the Spanish Ministry of Economy and Competitiveness under Project TEC2012-38142-C04-03.

Author details

[1]Telecommunication Engineering Department, University of Jaen, Linares, Jaen, Spain. [2]Music Technology Group, Universitat Pompeu Fabra, Barcelona, Spain.

References

1. N Ono, K Miyamoto, J Le Roux, H Kameoka, S Sagayama, Separation of a monaural audio signal into harmonic/percussive components by complementary diffusion on spectrogram, in *Proceedings of the European Signal Processing Conference* (Lausanne Switzerland, August 2008), pp. 25–29
2. N Ono, K Miyamoto, H Kameoka, S Sagayama, A real-time equalizer of harmonic and percussive components in music signals, in *Proceedings of the Ninth International Conference on Music Information Retrieval (ISMIR)* (Philadelphia, Pennsylvania USA, September 14–18 2008), pp. 139–144
3. L Daudet, Review on techniques for the extraction of transients in musical signals, in *Proceedings of the Third International Conference on Computer Music Modeling and Retrieval* (Pisa, Italy, September 26–28 2005), pp. 219–232

4. M Helen, T Virtanen, Separation of drums from polyphonic music using non-negative matrix factorisation and support vector machine, in *Proceedings of the European Signal Processing Conference* (Anatalya, Turkey, September 4–8 2005)
5. O Gillet, G Richard, Transcription and separation of drum signals from polyphonic music. IEEE Trans. Audio Speech Lang. Process. **3**(16), 529540 (2008)
6. A Ozerov, E Vincent, F Bimbot, A general flexible framework for the handling of prior information in audio source separation. IEEE Trans. Audio Speech Lang. Process. **20**(4), 11181133 (2012)
7. D Fitzgerald, Harmonic/percussive separation using median filtering, in *Proceedings of DAFX* (Graz, Austria, September 6–10 2010)
8. J Yoo, M Kim, K Kang, S Choi, Nonnegative matrix partial co-factorization for drum source separation, in *Proceedings of the IEEE International Conference on Acoustics, Speech, and Signal Processing (ICASSP)* (Dallas, Texas, USA, March 14–19 2010)
9. R Jain, R Kasturi, B Schunck, *Machine Vision*. (McGraw-Hill, New York, 1995)
10. H Tachibana, H Kameoka, S Sagayama, Comparative evaluations of various harmonic/percussive sound separation algorithms based on anisotropic continuity of spectrogram, in *International Conference on Acoustics, Speech and Signal Processing (ICASSP)* (Tokyo, Japan, March 25–30 2012)
11. F Canadas-Quesada, N Ruiz-Reyes, P Vera-Candeas, J Carabias, S Maldonado, A multiple-F0 estimation approach based on Gaussian spectral modelling for polyphonic music transcription. J. New Music Res. **39**(1), 93–107 (2010)
12. Y Ueda, Y Uchiyama, T Nishimoto, N Ono, S Sagayama, *HMM-based approach for automatic chord detection using refined acoustic features*, (Dallas, Texas, USA, March 14–19 2010)
13. D Zhiyao, B Pardo, A state space model for online polyphonic audio-score alignment, in *International Conference on Acoustics, Speech and Signal Processing (ICASSP)* (Prague, Czech Republic, May 22–27 2011)
14. D Lee, S Seung, Learning the parts of objects by nonnegative matrix factorization. Nature. **401**(21), 788–791 (1999)
15. P Hoyer, Non-negative matrix factorization with sparseness constraints. J. Mach. Learn. Res. **5**, 1457–1469 (2004)
16. V Monga, M Mhcak, Robust and secure image Hashing via non-negative matrix factorizations. IEEE Trans. Inf. Forensics Secur. **2**(3), 376–390 (2007)
17. I Kotsia, S Zafeiriou, I Pitas, A novel discriminant non-negative matrix factorization algorithm with applications to facial image characterization problems. IEEE Trans. Inf. Forensics Secur. **2**(3), 588–595 (2007)
18. P Smaragdis, J Brown, Non-negative matrix factorization for polyphonic music transcription, in *IEEE Workshop on Applications of Signal Processing to Audio and Acoustics (WASPAA)* (New Paltz, New York, USA, October 19–22 2003)
19. J Paulus, T Virtanen, Drum transcription with non-negative spectrogram factorisation, in *Proceedings of the European Signal Processing Conference* (Antalya, Turkey, September 4–8 2005)
20. D Lee, H Seung, Algorithms for non-negative matrix factorization, in *Advances in NIPS*, (2000), pp. 556–562
21. C Févotte, N Bertin, JL Durrieu, Nonnegative matrix factorization with the Itakura-Saito divergence. With application to music analysis. Neural Comput. **21**(3), 793830 (2009)
22. T Virtanen, Monaural sound source separation by non-negative matrix factorization with temporal continuity and sparseness criteria. IEEE Trans. Audio Speech Lang. Process. **15**(3), 1066–1074 (2007)
23. J Eggert, E Korner, Sparse coding and NMF, in *Proceedings of the International Joint Conference on Neural Networks (IJCNN4)* (Budapest, Hungary, 25–29 July 2004), pp. 2529–2533
24. J Parras-Moral, F Canadas-Quesada, P Vera-Candeas, N Ruiz-Reyes, Audio restoration of solo guitar excerpts using a excitation-filter instrument model, in *Stockholm Music Acoustics Conference jointly with Sound And Music Computing Conference* (Stockholm, Sweden, 30 July)
25. Activision, Guitar hero World Tour. http://en.wikipedia.org/wiki/Guitar_Hero_World_Tour. Accessed 09/06/2014
26. Activision, Guitar hero 5. http://en.wikipedia.org/wiki/Guitar_hero_5. Accessed 09/06/2014
27. S Araki, A Ozerov, V Gowreesunker, H Sawada, F Theis, G Nolte, D Lutter, N Duong, The 2010 signal separation evaluation campaign (SiSEC2010): audio source separation, in *9th International Conference on Latent Variable Analysis and Signal Separation (LVA/ICA'10)* (Saint-Malo France, September 2010), pp. 114–122
28. E Vincent, Musical source separation using time-frequency source priors. IEEE Trans. Audio Speech Lang. Process. **14**(1), 91–98 (2006)
29. E Vincent, C Févotte, R Gribonval, Performance measurement in blind audio source separation. IEEE Trans. Audio Speech Lang. Process. **14**(4), 1462–1469 (2006)
30. C Févotte, R Gribonval, E Vincent, BSS_EVAL toolbox user guide - Revision, 2.0, Technical Report 1706, IRISA (April 2005)

Permissions

All chapters in this book were first published in EURASIP-JASMP, by Springer; hereby published with permission under the Creative Commons Attribution License or equivalent. Every chapter published in this book has been scrutinized by our experts. Their significance has been extensively debated. The topics covered herein carry significant findings which will fuel the growth of the discipline. They may even be implemented as practical applications or may be referred to as a beginning point for another development.

The contributors of this book come from diverse backgrounds, making this book a truly international effort. This book will bring forth new frontiers with its revolutionizing research information and detailed analysis of the nascent developments around the world.

We would like to thank all the contributing authors for lending their expertise to make the book truly unique. They have played a crucial role in the development of this book. Without their invaluable contributions this book wouldn't have been possible. They have made vital efforts to compile up to date information on the varied aspects of this subject to make this book a valuable addition to the collection of many professionals and students.

This book was conceptualized with the vision of imparting up-to-date information and advanced data in this field. To ensure the same, a matchless editorial board was set up. Every individual on the board went through rigorous rounds of assessment to prove their worth. After which they invested a large part of their time researching and compiling the most relevant data for our readers.

The editorial board has been involved in producing this book since its inception. They have spent rigorous hours researching and exploring the diverse topics which have resulted in the successful publishing of this book. They have passed on their knowledge of decades through this book. To expedite this challenging task, the publisher supported the team at every step. A small team of assistant editors was also appointed to further simplify the editing procedure and attain best results for the readers.

Apart from the editorial board, the designing team has also invested a significant amount of their time in understanding the subject and creating the most relevant covers. They scrutinized every image to scout for the most suitable representation of the subject and create an appropriate cover for the book.

The publishing team has been an ardent support to the editorial, designing and production team. Their endless efforts to recruit the best for this project, has resulted in the accomplishment of this book. They are a veteran in the field of academics and their pool of knowledge is as vast as their experience in printing. Their expertise and guidance has proved useful at every step. Their uncompromising quality standards have made this book an exceptional effort. Their encouragement from time to time has been an inspiration for everyone.

The publisher and the editorial board hope that this book will prove to be a valuable piece of knowledge for researchers, students, practitioners and scholars across the globe.

List of Contributors

Junhong Zhao
State Key Laboratory of Transducer Technology, Institute of Electronics, Chinese Academy of Sciences, No.19 North west Rd., Beijing 100190, China
University of Chinese Academy of Sciences, No. 19A Yuquan Road, Beijing 100190, China

Wei-Qiang Zhang and Hua Yuan
National Laboratory for Information Science and Technology, Department of Electronic Engineering, Tsinghua University, Haidian, Beijing 100084, China

Michael T Johnson
Department of Electrical Engineering, Marquette University, 250 W Wisconsin Ave., Milwaukee, WI 53201, USA

Jia Liu
National Laboratory for Information Science and Technology, Department of Electronic Engineering, Tsinghua University, Haidian, Beijing 100084, China

Shanhong Xia
State Key Laboratory of Transducer Technology, Institute of Electronics, Chinese Academy of Sciences, No.19 North west Rd., Beijing 100190, China

Daniel Pape
Institute of Electronics and Informatics Engineering of Aveiro (IEETA), University of Aveiro, 3810-193 Aveiro, Portugal

Luis M T Jesus
Institute of Electronics and Informatics Engineering of Aveiro (IEETA), University of Aveiro, 3810-193 Aveiro, Portugal
School of Health Sciences (ESSUA), University of Aveiro, 3810-193 Aveiro, Portugal

Upal Mahbub and Shaikh Anowarul Fattah
Department of Electrical and Electronic Engineering, Bangladesh University of Engineering and Technology, Dhaka 1000, Bangladesh

Wei-Ping Zhu and M Omair Ahmad
Department of Electrical and Computer Engineering, Concordia University, Montreal, Quebec H3G 1M8, Canada

Ryo Aihara and Ryoichi Takashima
Graduate School of System Informatics, Kobe University, 1-1 Rokkodai, Nada-ku, Kobe 657-8501, Japan

Tetsuya Takiguchi and Yasuo Ariki
Organization of Advanced Science and Technology, Kobe University, 1-1 Rokkodai, Nada-ku, Kobe 657-8501, Japan

Yi-Ju Lin, Tien-Ming Wang, Ta-Chun Chen, Yin-Lin Chen, Wei-Chen Chang and Alvin WY Su
SCREAM Laboratory, Department of Computer Science Information Engineering, National Cheng-Kung University, Tainan 701, Taiwan

Ning-Yun Ku and Shaw-Hwa Hwang
Department of Electrical Engineering, National Taipei University of Technology, 1, Sec. 3, Chung-hsiao E. Rd., Taipei 10608, Taiwan

Cheng-Yu Yeh
Department of Electrical Engineering, National Chin-Yi University of Technology, 57, Sec. 2, Zhongshan Rd., Taiping Dist, Taichung 41170, Taiwan

Adam Borowicz
Department of Computer Graphics and Digital Media, Faculty of Computer Science, Bialystok University of Technology, Wiejska str. 45A, 15-351 Bialystok, Poland

Toni Heittola Tuomas Virtanen and Dani Korpi
Department of Signal Processing, Tampere University of Technology, P.O. Box 553, Tampere 33101, Finland

Annamaria Mesaros
Department of Signal Processing, Tampere University of Technology, P.O. Box 553, Tampere 33101, Finland
Department of Signal Processing and Acoustics, Aalto University, P.O. Box 13000, Aalto 00076, Finland

Antti Eronen
Nokia Research Center, Visiokatu 3, Tampere 33720, Finland

Zhaofeng Zhang, LongbiaoWang and Masahiro Iwahashi
Nagaoka University of Technology, 1603-1 Kamitomioka-cho, Nagaoka, Niigata, 940-2188, Japan

Atsuhiko Kai and Takanori Yamada
Shizuoka University, 3-5-1 Johoku, Naka-ku, Hamamatsu, Shizuoka 432-8561, Japan

Weifeng Li
Tsinghua University, 1 Qinqhuayuan, Beijing 100084, China

Shi Dong, Ruimin Hu, Xiaochen Wang, Yuhong Yang and Weiping Tu
National Engineering Research Center for Multimedia Software, Computer School, Wuhan University, Wuhan, China

Rupayan Chakraborty, Climent Nadeu and Taras Butko
TALP Research Center, Department of Signal Theory and Communications, Universitat Politècnica de Catalunya (UPC), Barcelona 08034, Spain

Yue Xian Zou, Peng Wang and Yong Qing Wang
ADSPLAB/ELIP, School of Electronic Computer Engineering, Peking University, Shenzhen 518055, China

Christian H Ritz and Jiangtao Xi
School of Electrical, Computer, and Telecommunications Engineering, University of Wollongong, Wollongong 2522, Australia

Yanna Ma and Akinori Nishihara
Department of Communications and Integrated Systems, Tokyo Institute of Technology, Meguro-ku, Tokyo 152-8552, Japan

Bart Ons, Jort F Gemmeke and Hugo Van hamme
Department ESAT-PSI, KU Leuven, Kasteelpark 10, 3001 Leuven, Belgium

Marek Bohac, Michaela Kucharova, JanNouza and Petr Červa
Institute of Information Technology and Electronics, Technical University of Liberec, Studentska 2, 461 17 Liberec, Czech Republic

Zoraida Callejas
Department of Languages and Computer Systems, University of Granada, CITIC-UGR, Periodista Daniel Saucedo Aranda s/n, 18071 Granada, Spain

Jeongook Song and Hong-Goo Kang
School of Electrical and Electronic Engineering, Yonsei University, Seoul 120-749, South Korea

Rafael P Ribeiro, Murilo AP Almeida and Carlos N Silla Jr
Computer Music Technology Laboratory at the Federal University of Technology of Paraná, Av. Alberto Carazzai, 1.640, Cornélio Procópio 86300-000, Brazil

Ali Khodabakhsh, Fatih Yesil, Ekrem Guner and Cenk Demiroglu
Electrical and Computer Engineering Department, Ozyegin University, Orman Street, 34794, Istanbul, Turkey

Francisco Jesus Canadas-Quesada, Pedro Vera-Candeas, Nicolas Ruiz-Reyes and Pablo Cabanas-Molero
Telecommunication Engineering Department, University of Jaen, Linares, Jaen, Spain

Julio Carabias-Orti
Music Technology Group, Universitat Pompeu Fabra, Barcelona, Spain